Climate in Review

Climate in Review

Edited by
Geoffrey McBoyle
University of Waterloo

HOUGHTON MIFFLIN COMPANY
BOSTON
Atlanta
Dallas
Geneva, Illinois
Hopewell, New Jersey
Palo Alto

Cover photograph by W. B. Finch

Printed in the United States of America
Library of Congress Card Number 72-3536
ISBN: 0-395-16007-3

To P.U.D.

Contents

Part 4 **Climatic Classification**

Part 5 **Climatic Change**

Part 6 **Weather Modification —Man-controlled**

Part 7 **Weather Modification –Inadvertent**

Part 8 **Climate and Man**

Preface

Before any book is begun, the prospective author must have some compelling urge in taking up the task. In the case of this set of readings in climatology, two incentives were paramount. The seminal idea was conceived on the strength of a teacher's disillusionment at the deficiency of most standard texts in the fields of recent research, and nurtured by student frustration in library searches for more detailed work on specific topics. So it was inspired by a need for an organized text on something beyond the basic theory of climatology, which is ably treated in a number of basic texts.

Specific research in climatology is documented in a wide variety of publications on an assortment of subjects; often the more obscure of these are not readily obtainable in modest library collections. But not all articles here are of that type, for the major intention of this book is to make available, in a convenient manner, a number of the better articles within their particular field of climatology. The result is a compilation of articles, some recent, some not so recent, some well-known classics, some less renowned treatises, some written at a general level, some of a more detailed and demanding nature. All, it is hoped, exhibit a common characteristic of quality within their particular field.

However, as every editor knows only too well, no one anthology will cater to everyone's desires or satisfy every need. And I freely admit that many of the themes adopted here are intended as inducements to student interest. Hopefully, they will be a stimulus to student enthusiasm in a subject which has considerable relevance and vitality.

The overall approach is a progression from the theoretical background to the practical aspects of each sectional theme. If any minor themes should be apparent, they are the quality of the environment, which particularly pervades those articles dealing with weather modification, and the applied use of conceptual knowledge in climatology, most evident in the final section but implicit elsewhere in the book. I believe that applied climatology has considerable significance in the quality of environments, present and future, and the attendant quality of man's life.

The book is sectionalized along thematic lines, each selected aspect of climatology being treated broadly. To this end the first article in each part is intended, as far as possible, to present a general overview of the subject, followed by a number of articles which either introduce additional conceptual material or develop and expand the original thesis. A knowledge of the basic systematics of climatology is assumed. The part introductions are intended to place each article in a context and to provide some provocative questions for the reader as he begins an article.

This collection of articles spans wide levels of difficulty and is not intended for use at any one level in a university or college. Every article will not be of interest to every reader, at any one point in time, but the breadth of treatment within each section should provide useful material for most levels of learning. In this way it is hoped that the widest choice is made available to cater to various levels of ability and to diverse degrees and topics of interest. Then too the book, which can continue to offer informative material and stimulating discourse as the student gains in knowledge and in maturity of feeling for the subject, is a worthwhile addition to his stock of reading material.

Ending on a personal note, I should like to thank the authors and copyright holders who have generously agreed to the use of their material in this book. I hope that my arrangement of their work in no way detracts from its original quality. Grateful acknowledgments must also go to Houghton Mifflin Company for extending every possible assistance. To all who have offered useful comments and criticism I extend my thanks—their valuable suggestions were never ignored, though, I must confess, sometimes waived. For such vanity the responsibility is all mine.

Geoffrey McBoyle

Contributors

R. G. Barry *Department of Geography and Institute of Arctic and Alpine Research, University of Colorado, Boulder, Colorado.*

Robert P. Beckinsale *Department of Geography, University of Oxford, Oxford, England.*

M. I. Budyko *Main Geophysical Observatory, Leningrad, U.S.S.R.*

F. Kenneth Hare *Department of Geography, University of Toronto, Toronto, Ontario, Canada.*

George C. Holzworth *Division of Meteorology, Air Pollution Control Office, National Oceanic and Atmospheric Administration, Raleigh, North Carolina.*

H. H. Lamb *Meteorological Office, Bracknell, Berkshire, England.*

Philip A. Leighton *Department of Chemistry, Stanford University, Palo Alto, California.*

Heinz Lettau *Department of Meteorology, University of Wisconsin, Madison, Wisconsin.*

Katharina Lettau *Department of Meteorology, University of Wisconsin, Madison, Wisconsin.*

John R. Mather *C. W. Thornthwaite Associates, Elmer, New Jersey.*

W. J. Maunder *New Zealand Meteorological Service, Wellington, New Zealand.*

Geoffrey R. McBoyle *Department of Geography, University of Waterloo, Waterloo, Ontario, Canada.*

Edward A. Morris *Bronson, Bronson, and McKinnon (attorneys), San Francisco, California.*

Jerome Namias *Extended Forecast Division, Weather Bureau, National Oceanic and Atmospheric Administration, Washington, D.C. and Scripps Institution of Oceanography, La Jolla, California.*

M. Neiburger *Department of Meteorology, University of California, Los Angeles, California.*

Hans Neuberger *Department of Geography, University of South Florida, Tampa, Florida.*

James T. Peterson *Assigned to the National Air Pollution Control Administration, Raleigh, North Carolina by the Air Resources Laboratory, National Oceanic and Atmospheric Administration.*

Robert E. Rankin *Department of Psychology, Central Michigan University, Mount Pleasant, Michigan.*

John F. Rooney, Jr. *Department of Geography, Oklahoma State University, Stillwater, Oklahoma.*

John A. Russo, Jr. *Operations Research Department, The Hartford Insurance Group, Hartford, Connecticut.*

J. S. Sawyer *Meteorological Office, Bracknell, Berkshire, England.*

Joanne Simpson *National Hurricane Center, National Oceanic and Atmospheric Administration, Coral Gables, Florida.*

Robert H. Simpson *National Hurricane Center, National Oceanic and Atmospheric Administration, Coral Gables, Florida.*

L. A. Strokina *Main Geophysical Observatory, Leningrad, U.S.S.R.*

Peter W. Summers *Hail Section, Research Council of Alberta, Alberta, Canada.*

Werner H. Terjung *Department of Geography, University of California, Los Angeles, California.*

Peter H. Wyckoff *Atmospheric Sciences Section, National Science Foundation, Washington, D.C.*

N. A. Yefimova *Main Geophysical Observatory, Leningrad, U.S.S.R.*

Gary A. Yoshioka *C. W. Thornthwaite Associates, Elmer, New Jersey.*

L. I. Zubenok *Main Geophysical Observatory, Leningrad, U.S.S.R.*

Part One

Energy Balance

The fuel for the atmospheric engine is its energy balance. The precision of our understanding of this factor is therefore of utmost importance.

Why is the radiation balance lower in arid regions than in humid areas of the same latitude? Why are the maximum values of the heat loss in evaporation from oceans much higher than the corresponding maximum values for land areas? Why do we get higher values of turbulent heat exchange in continental coastal areas coinciding with lower, and even negative, values in offshore ocean areas?

The first article represents a refinement of earlier work in the field of heat balance. It provides an explanation of the spatial distribution of the basic phenomena involved and is illustrated by global maps.

How is insolation affected if the liquid water content of the atmosphere is doubled?—or halved? If we will want to predict the possibilities and implications of climatic modification, then we must first consider the second article which provides an invaluable investigation into the equipoise of forces within the budget-accounting system of short-wave radiation.

For a discussion of the moisture balance aspect of the total energy system, Barry's article in Part 3 should be consulted.

M. I. Budyko, N. A. Yefimova, L. I. Zubenok, and L. A. Strokina

1 The heat balance of the surface of the earth

. . . In the last ten years a large amount of meteorological data became available for computation of heat-balance members, largely as a result of the studies of the International Geophysical Year. The computational methods used for calculating the heat-balance components were also improved during that time. All this made it possible to pose the question of constructing new world maps of the heat balance of the earth's surface that would be much more exact and detailed than the previously compiled maps. The following methods of determining heat-balance components were used during the preparation of these maps.

The radiation balance of the earth's surface was calculated as the difference between the absorbed short-wave radiation and the long-wave effective radiation of the earth's surface (Budyko et al., 1961).

The loss of heat used in evaporation was computed as the product of the latent heat of vaporization by the rate of evaporation. The magnitude of evaporation from the land surface was computed by the water-balance method. Changes in the moisture content of the soil were taken into account in computing the monthly sums of evaporation (see Budyko and Zubenok, 1961).

Evaporation from the surface of seas and oceans was computed by the formula $E = au(e_s - e)$, where E is evaporation, u is the velocity of the wind at the level of shipboard observations, e_s is the relative humidity of saturated air at the temperature of the water surface and e is the relative humidity of the air.

To obtain a more precise magnitude of the coefficient a, we calculated it once again by completing the heat-balance equation for the entire world ocean, i.e., by the method used in our preceding study (Budyko, Beryland, and Zubenok, 1954). By using the new data, Strokina found the value of the coefficient to be equal to $2.5 \cdot 10^{-6}$ g/cm^3. It should be noted that this value is 4% higher than the value obtained in the previous calculation.

The turbulent heat exchange between the surface of oceans and the atmosphere was calculated by the formula $P = ac_pu(\theta_w - \theta)$, where P is the turbulent heat exchange, c_p the heat capacity of the air at constant pressure, and θ_w and θ respectively the temperatures of the water surface and of their air. The turbulent heat exchange between the land surface and the atmosphere for the entire year was taken as the difference between the radiation balance and the loss of heat in evaporation. In calculating the turbulent heat exchange for individual months in areas of the middle and high latitudes, we also subtracted from the radiation balance the values of the heat exchange in the soil and of the heat exchange related to the melting of the snow cover in spring.

The last member of the heat balance for oceans, the gain or loss of heat related to the effect of ocean currents and macro-turbulence for the entire year, was calculated as the difference between the radiation balance and the sum of the heat losses in evaporation and turbulent heat exchange.

The heat-balance components were computed for about 2,000 points of which 300 were at the surface of seas and oceans. As initial data for the computations we used the mean annual values of air temperature, air humidity, cloudiness, precipitation, wind velocity and temperature of the water surface found in various climatological handbooks. For computations of the heat-balance components in ocean areas, we used mainly the data of the *Marine Climatic Atlas of the World* (Washington, 1955–59). . . .

The maps presented show the distribution of the annual values of the main components of the heat balance. Figure 1 shows the distribution of the radiation balance of the earth's surface; Figure 2 shows the distribution of the loss of heat in evaporation; Figure 3 shows the distribution of turbulent heat exchange.

The heat-balance components shown on these maps are expressed in kilocalories per square centimeter per

Reprinted, in edited form, by permission of the American Geographical Society, from *Soviet Geography: Review and Translation*, Vol. 3(5), 1962, pp. 3–16.

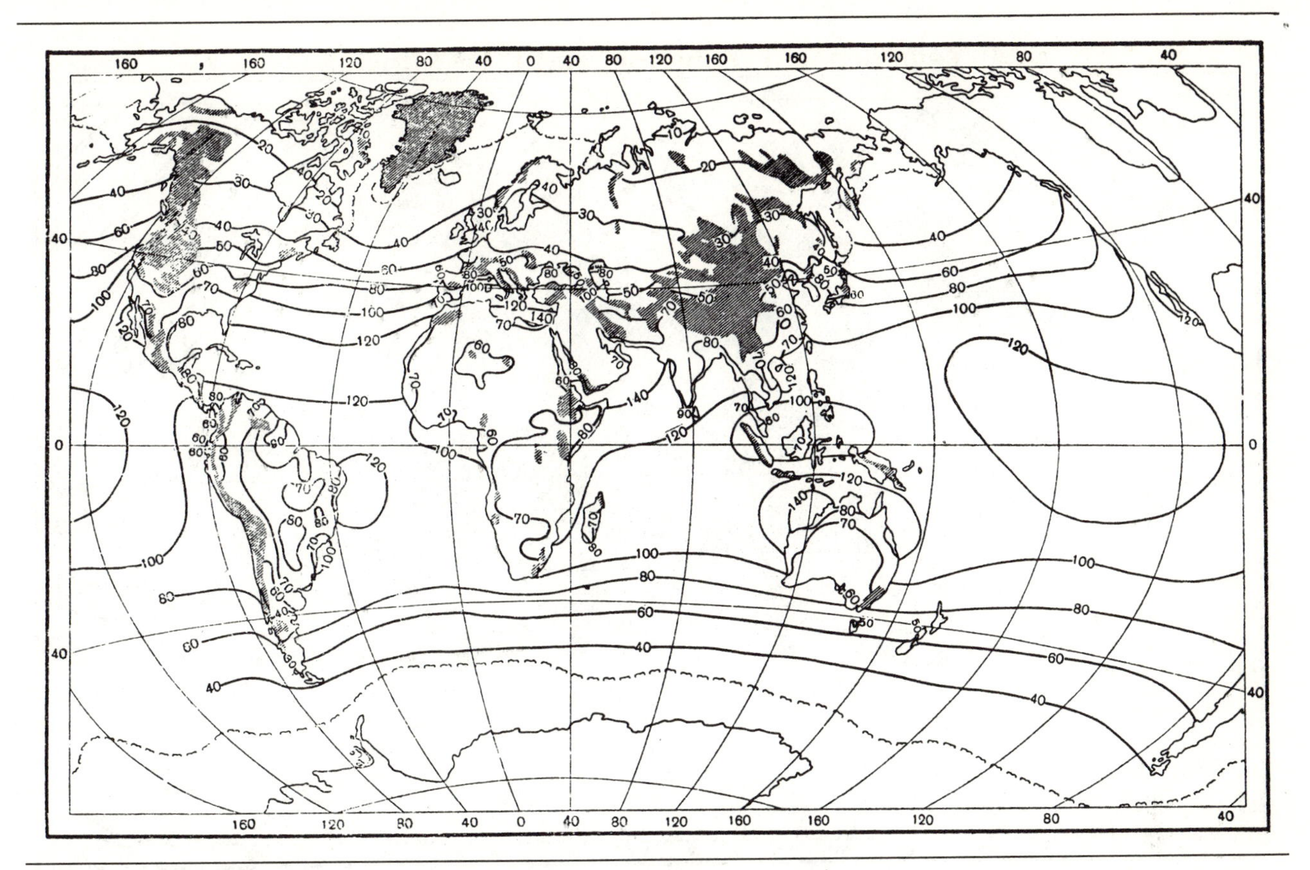

Figure 1 The radiation balance (in kilocalories per square centimeter per year).

year. Mountain areas with elevations exceeding 1,500 meters are hachured on all maps because of the difficulty of mapping the distribution of members of the heat balance in dissected relief on small-scale maps and because of the inadequacy of initial data for such areas.

Figure 1 shows that the radiation balance of oceans is noticeably greater than the radiation balance of land areas at the same latitudes. In extra-tropical latitudes the radiation balance of both oceans and land areas decreases steadily toward the higher latitudes; in the tropics the distribution is rather complex although both on oceans and on land areas the values of the radiation balance vary within rather narrow limits. On the continents the radiation balance is usually somewhat lower in arid areas than in humid areas of the same latitudes. In ocean areas the radiation balance is generally somewhat higher in regions with little cloudiness.

The contrasting effect of cloudiness on the radiation balance of land and ocean areas is explained by the fact that on land areas with an arid climate or little cloudiness, an increase in incoming solar radiation is often more than compensated by an increase in reflected radiation and especially by a rise of effective spaceward radiation. Such compensation is much less evident in ocean areas.

The distribution of the annual sums of the loss of heat in evaporation differs sharply for land and ocean areas (Figure 2). The highest values on oceans are found in high-pressure belts, which have a high income of solar energy and relatively stable wind patterns. The loss of heat in evaporation decreases noticeably near the equator and in extra-tropical latitudes, in the first case, mainly because of lower wind force and a reduction in the humidity deficit, and in the second case, because of a reduction of the income of radiational heat. Ocean currents have a substantial effect on evaporation from the ocean surface by increasing the loss of heat in evaporation in warm-current areas and reducing it in cold-current areas. On the continents the loss of heat in evaporation reaches its highest values in humid and hot regions. Evaporation decreases with an increase in latitude just as it decreases with a drop in precipitation in arid areas.

Of interest is the fact that the maximum values of the heat loss in evaporation from oceans are much higher than the corresponding maximum values in land areas. Several factors are involved: the higher values of the radiation balance on oceans, the possibility that ocean currents provide an additional income of energy for evaporation in some ocean areas, and the high mean wind velocities in some ocean areas.

Like the loss of heat in evaporation, the turbulent heat exchange between the earth's surface and the atmosphere is distributed quite differently on land and

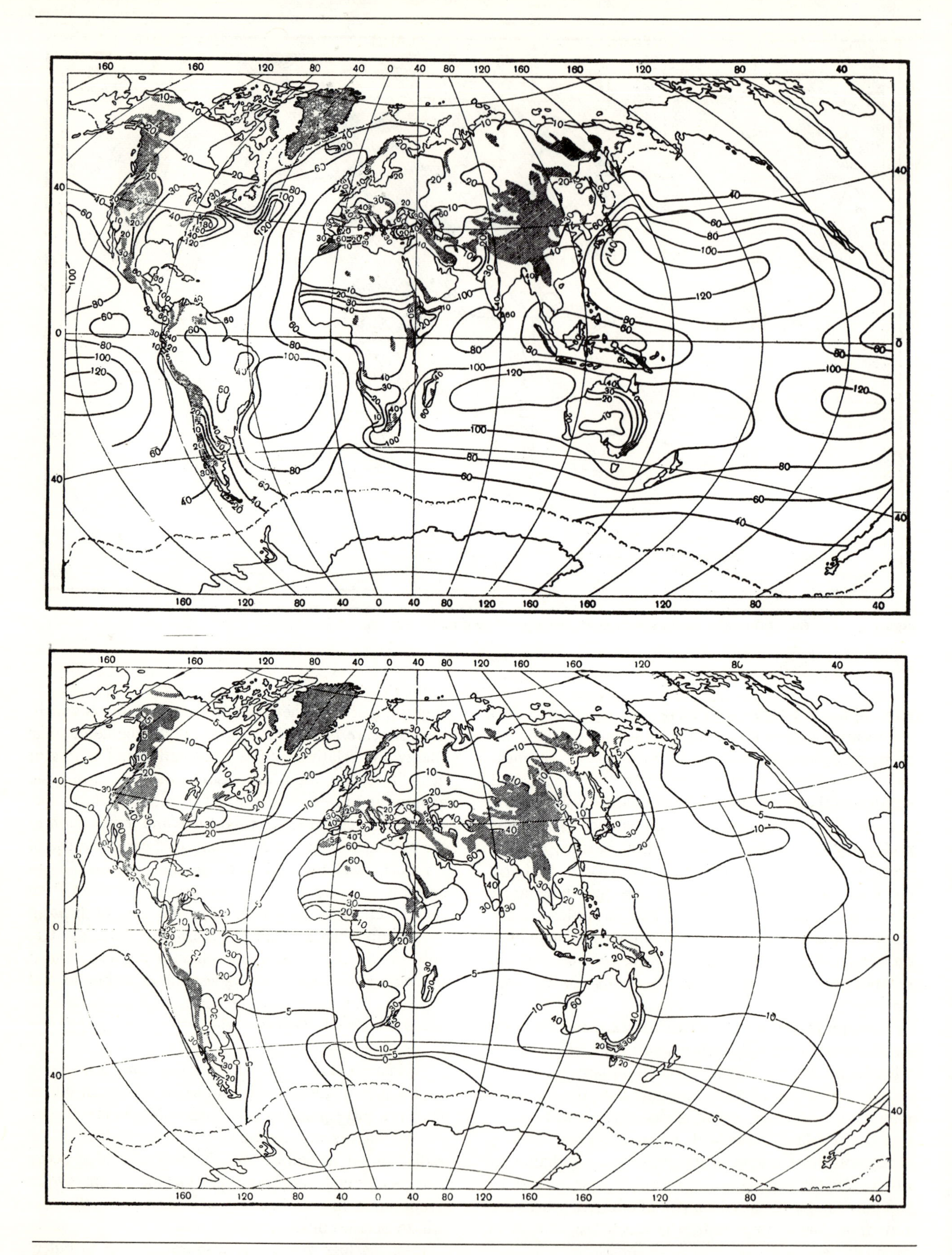

Figure 2 Loss of heat in evaporation (in kilocalories per square centimeter per year).

Figure 3 Turbulent heat exchange (in kilocalories per square centimeter per year).

Latitude	OCEANS				LAND			EARTH			
	R	LE	P	A	R	LE	P	R	LE	P	A
70–60 N	23	33	16	−26	20	14	6	21	20	9	−8
60–50	29	39	16	−26	30	19	11	30	28	13	−11
50–40	51	53	14	−16	45	24	21	48	38	17	−7
40–30	83	86	13	−16	60	23	37	73	59	23	−9
30–20	113	105	9	−1	69	20	49	96	73	24	−1
20–10	119	99	6	14	71	29	42	106	81	15	10
10–0	115	80	4	31	72	48	24	105	72	9	24
0–10 S	115	84	4	27	72	50	22	105	76	8	21
10–20	113	104	5	4	73	41	32	104	90	11	3
20–30	101	100	7	−6	70	28	42	94	83	15	−4
30–40	82	80	8	−6	62	28	34	80	74	11	−5
40–50	57	55	9	−7	41	21	20	56	53	9	−6
50–60	28	31	10	−13	31	20	11	28	31	10	−13
Earth as a whole	82	74	8	0	49	25	24	72	59	13	0

R—Radiation balance
LE—Loss of heat in evaporation
P—Turbulent heat exchange
A—Redistribution of heat by ocean currents

Table 1 Mean latitudinal values of the components of the heat balance of the surface of the earth (in kilocalories per square centimeter per year)

on oceans. The absolute values of turbulent heat exchange are quite small over most of the ocean areas. Only in the areas of strong warm currents (the Gulf Stream and the Japanese Current) does the turbulent heat emission from the water surface into the atmosphere reach higher values. For most of the surface of the world oceans the annual sums of turbulent heat exchange are positive, corresponding to the direction of the heat flow from the water surface to the atmosphere. In the few relatively small areas where the annual sum of the turbulent heat exchange is negative, its absolute values are small.

The largest magnitudes of turbulent heat exchange on the continents are found in tropical deserts. The values decrease noticeably in more humid areas and under conditions of colder climate. It is interesting to note that in some cases higher values of the turbulent heat exchange in coastal areas of continents coincide with lower values and sometimes even negative values in offshore ocean areas. This may be observed, for example, in the northwest part of the Indian Ocean, on the Californian coast, on the west coast of the Sahara and on the southeast coast of South America. One of the reasons for this type of distribution may be found in the laws of transformation of air masses moving from a heated land surface to the cooler ocean or in the reverse direction. Such movements promote stronger heat emission from the land surface and weaken the heat emission from the ocean surface to the atmosphere; in the latter case they may even produce a reversal in direction of the mean heat flow, which would then be from the atmosphere to the surface of the ocean. . . .

On the basis of the spatial distribution of members of the heat balance, we computed the mean latitudinal values of the heat-balance components for the surface of continents, oceans and the earth as a whole. The results appear in Table 1. The table shows that only one member of the heat balance, the radiation balance, has a similar latitudinal distribution on land and on oceans. In both cases the maximum values are found in the tropics; moreover the mean latitudinal values remain virtually constant within the tropical latitudes.

It is interesting to note how the differences between land and ocean values change with latitude. The maximum difference occurs in the tropical zone, where the values of the radiation balance of land areas are only about two-thirds of the values on oceans. The radiation balances of land and oceans come closer to each other with increases in latitude and are virtually identical in the latitudes 50 to 70°. One of the reasons for this correlation lies in the fact that the mean albedo of the water surface increases with latitude.

The mean latitudinal values of the loss of heat in evaporation on land reach their principal maximum on the equator and then decline in the latitudes of the high-pressure belts. A slight increase in evaporation (more noticeable in the northern hemisphere) occurs in the higher latitudes in connection with an increase in precipitation compared with the arid zone of the lower latitudes. Further increases in latitude result in a reduction of the loss of heat in evaporation because of the inadequate supply of heat.

In contrast to continental conditions, the mean latitudinal values of the loss of heat in evaporation on oceans reach a maximum in the belts of high pressure. Of interest is the fact that in the latitudes 50 to 70°, where the radiation balances are virtually equal on land and on oceans, the loss of heat in evaporation is much higher on oceans than on land. This is apparently explained by the fact that a large amount of heat

CONTINENTS AND OCEANS	R	LE	P
Europe	39	24	15
Asia	47	22	25
Africa	68	26	42
North America	40	23	17
South America	70	45	25
Australia	70	22	48
Atlantic	82	72	8
Pacific	86	78	8
Indian	85	77	7

Table 2 The heat balance of continents and oceans (in kilocalories per square centimeter per year)

contributed by ocean currents is used in evaporation from the ocean surface.

The mean latitudinal values of the turbulent heat exchange on oceans increase steadily with latitude. On land these values reach a maximum in the high-pressure belts, decreasing somewhat at the equator and dropping sharply in the higher latitudes. The totally different behavior of these values on land and on oceans reflects the fundamental differences in the mechanism of transformation of air masses on the surface of continents and of oceans.

The distribution of the mean latitudinal values of the gain or loss of heat on oceans, related to the effect of currents, shows ocean currents in general transport heat out of the zone situated between 20° N and 20° S, the maximum heat energy being absorbed by the currents slightly north of the equator. This heat is transported to the higher latitudes and is lost in largest amounts in the latitudes 50 to 70° of the northern hemisphere, where strong cold currents are especially active.

The mean latitudinal distribution of members of the heat balance for the earth as a whole in different latitudes shows patterns typical for continents or oceans depending on whether land or ocean areas predominate in the given latitudinal zone.

The mean values of the heat-balance components for individual continents and oceans are given in Table 2. The table shows that in three continents out of six (Europe, North America, South America) a large part of the heat of the radiation balance is lost in evaporation. The reverse is true in the other three continents (Asia, Africa, Australia) in which climates are typically arid.

The mean conditions of the heat balance of the three oceans differ very little from one another. Of interest is the fact that the sum of the loss of heat in evaporation and the turbulent heat exchange for each ocean is close to the magnitude of the radiation balance. This means that the heat exchange between oceans as a result of the transport of heat by ocean currents does not have any substantial effect on the heat balance of each ocean as a whole. Such a result, which is quite natural from a physical point of view, evidently demonstrates the higher accuracy of this computation of the heat balance of oceans compared with the previous calculation (Budyko, 1956) in which it was impossible to complete the balance for each ocean individually. It may be supposed that the slight excess of the radiation balance over the sum of the loss of heat in evaporation and the turbulent heat exchange in the case of the Atlantic Ocean actually reflects the transport of some heat from the Atlantic to the Arctic Ocean. However, since the difference may lie within the margin of error of the computation, this question requires further study.

Table 2 does not include available data on the heat balance of the Antarctic and the Arctic Ocean because they are somewhat less precise than data for other continents and oceans. It may be noted that the heat balances of the Antarctic and of the Arctic Ocean resemble each other in several respects and differ fundamentally from all other continents and oceans. Both in the Antarctic and in the Arctic Ocean (except for its peripheral areas) all components of the heat balance are quite small in absolute magnitude on the average through the year. Because of the permanent snow or ice cover the albedo reaches such large magnitudes that even with a considerable income of total solar energy the absorbed radiation in the high latitudes is relatively small. This results in small values of the mean annual magnitudes of the radiation balance, which in the polar zone usually fluctuates around zero or reaches small negative values (Gavrilova, 1959; Rusin, 1961, and others).

By using the data presented above, and by taking account of the approximate values of the components of the heat balance for polar regions, we can compute the values of members of the heat balance for all continents, oceans and the earth as a whole. The results appear in the last line of Table 1. From these data it follows that on the oceans about 90% of the heat of the radiation balance is lost in evaporation and only 10% in direct turbulent heating of the atmosphere. On land these two forms of heat losses have almost identical values. For the earth as a whole the loss of heat in evaporation accounts for 82% of the radiation balance and turbulent heat exchange for 18%.

These data may be compared with the findings of previous studies in which the components of the heat balance of the earth as a whole were determined. The

Investigators	Solar radiation absorbed by earth's surface	Effective long-wave radiation	Radiation balance	Loss of heat in evaporation	Turbulent heat exchange
	COMPONENTS OF THE BALANCE				
Dines, 1917	42	14	28	21	7
Alt, 1929	43	27	16	16	0
Baur and Philipps, 1934	43	24	19	23	−4
Houghton, 1954	47	14	33	23	10
Lettau, 1954	51	27	24	20	4
Budyko, Berlyand, and Zubenok, 1954	42	16	26	21	5
Budyko, et al., 1961	43	15	28	23	5

Table 3 The heat balance of the surface of the earth (The components of the balance are expressed in per cent of the amount of solar radiation received at the outer limits of the atmosphere)

results of such comparisons are shown in Table 3. All values in the table are expressed in relative units to exclude the effect of estimates of the solar constant on the results of the calculations. The table shows that the first computation of the heat balance of the earth, made by Dines in 1917 by an extremely approximate method, comes very close to the latest results. Since some of the later results of Alt (1929) and Baur and Philipps (1934, 1935) were far less exact, it may be supposed that the coincidence resulted in part from Dines' remarkable physical intuition and in part from accident.

The findings of the recent studies of Houghton (1954) and Lettau (1954) were close to our latest results for most components of the balance, although some heat flows (Lettau's effective long-wave radiation and Houghton's turbulent heat exchange) differed substantially from the magnitudes we obtained.

In view of the fact that the mean values of the components of the heat balance were computed at the Main Geophysical Observatory by averaging the data of world maps of the components, it may be assumed that they are more firmly grounded than all previous calculations effected either for the earth as a whole or by averaging mean latitudinal magnitudes.

In conclusion we want to offer some data on the water balance of the earth. The mean annual sum of evaporation from continents, in accordance with the material discussed above, is 41 cm. Since the corresponding value for oceans is 125 cm, we find that the average annual evaporation on the earth as a whole is about 100 cm. This value must obviously coincide with the average annual precipitation.

The use of precipitation data available at the Main Geophysical Observatory showed that the average annual amounts of precipitation on continents and on oceans were 72 and 114 cm respectively. These magnitudes yield an average annual amount of 102 cm for the earth as a whole, which agrees closely with the amount of evaporation. In view of the fact that the precipitation data for oceans may be somewhat high because of the methodological problems of determining precipitation in water areas (see Budyko, 1956, and others), it may be assumed that both the annual amounts of precipitation and evaporation are close to 100 cm. In view of this assumption, we took the average annual precipitation on oceans as 112 cm. If we subtract from the average precipitation the average amount of evaporation, we find that the annual amount of water discharged by streams from the land to the ocean corresponds to a mean layer of water of 31 cm for the continents and of 13 cm for oceans. The water balance based on these data appears in Table 4. It should be noted that the amounts of evaporation and precipitation for oceans, derived from these computations, are considerably higher than the results obtained in most previous studies, [as a result of greater accuracy in this computational method]. . . .

Table 4 The water balance of the earth (in centimeters per year)

	Precipitation	Evaporation	Runoff
Continents	72	41	31
Oceans	112	125	13
Earth as a whole	100	100	0

REFERENCES

Alt, E., 1929. The status of the meteorological radiation problem. *Meteorol. Zeitschrift 46(12)*, 504–514.

Baur, F. and Philipps, H., 1934. The heat budget of the atmosphere of the northern hemisphere in January and July at the time of the equinoxes and solstices. *Gerl. Beitr. zur Geophys. 42*, 160–207.

Baur, F. and Philipps, H., 1935. The heat budget of the atmosphere of the northern hemisphere in January and

July at the time of the equinoxes and solstices. *Gerl. Beitr. zur Geophys. 45*, 82–132.

Budyko, M. I., 1956. *Teplovoy balans zemnoy poverkhnosti* (The Heat Balance of the Earth's Surface). Leningrad, Gidrometeoizdat.

Budyko, M. I., Berlyand, T. G. and Zubenok, L. I., 1954. The heat balance of the earth's surface. *Izv. AN SSSR, seriya geogr. 3*, 17–41.

Budyko, M. I., Yefimova, N. A., Mukhenberg, V. V. and Strokina, L. A., 1961. The radiation balance of the northern hemisphere. *Izv. AN SSSR, seriya geogr. 1*, 1–12.

Budyko, M. I. and Zubenok, L. I., 1961. Determination of the evaporation of the earth's surface. *Izv. AN SSSR, seriya geogr. 6*, 3–17.

Dines, W. H., 1917. The heat balance of the atmosphere. *Quart. Journ. Royal Meteorol. Soc. 43*, 151–158.

Gavrilova, M. K., 1959. The radiation balance of the Arctic. *Tr. Gl. geofizich. observatorii* (*Proc. Main Geophys. Observatory*) *92*, Leningrad. Gidrometeoizdat.

Houghton, H. G., 1954. On the annual heat balance of the northern hemisphere. *J. Meteor. 11*, 1–9.

Lettau, H. A., 1954. Study of the mass, momentum and energy budget of the atmosphere. *Archiv für Meteor., Geophys., und Bioklim., A. 7*, 133–157.

Marine Climatic Atlas of the World. 1955, *1*; 1956, *2*; 1957, *3*; 1958, *4*; 1959, *5*. Washington.

Rusin, N. P., 1961. The radiation budget of the snow surface of Antarctica, in *Aktinometriya i atmosfernaya optika* (Actinometry and Atmospheric Optics). (*Trudy 2-go mezhvedomstv. soveshchaniya* (Proc. of 2nd. Inter-Agency Conf.)). Leningrad. Gidrometeoizdat.

H. Lettau and K. Lettau

2 *Shortwave radiation climatonomy*

INTRODUCTION

As far as we know the word "climatonomy"[1] was first used by H. Lettau at the meeting of the American Geophysical Union in Washington, D.C. in 1954. It was suggested in logical expansion of thoughts expressed in a technical study entitled "Synthetische Klimatologie" (1952). In "Review of Climatology, 1951–1955" Landsberg (1957) gave recognition to the term and suggested that "climatonomy" could encompass theoretical, energetic, and circulation climatology. During the last decade a new definition has evolved. Now, "climatonomy" shall be used for the mathematical explanation of the basic elements which determine the physical environment at any planetary surface. This includes the specific climate represented by mean-value and temporal-spatial variations of temperature in the lower atmosphere as well as the upper strata of the underlying lithosphere or hydrosphere.

Mathematical explanation implies the use of numerical models which yield theoretical solutions in the form of a precise "response function," or output, in physical relationship to a precise "forcing function," or input. For any planet the intensity of the forcing function is determined by absolute distance to the central sun, and its time structure by orbital elements and spin or the varying relative positioning of the considered planetary surface section with respect to the sun's energy flux-density. The climatonomic response function, governed by fundamental physical laws of radiation, heat conduction, and universal conservation principles, will be investigated in another paper. The following discussion will be restricted to problems of the forcing function in terrestrial climatonomy which require the parameterization of shortwave radiation flux-density between the top and the bottom of the atmosphere, and its diurnal, seasonal, and multi-annual variation at a given locality.

One of the objectives of climatonomical parameterization is to investigate the possibilities of climate modification and control. It is important to know, for example, how the insolation at a given locality will be affected if one specific climatic parameter were doubled or halved, or changed otherwise in a physically realistic way; such parameter could be cloudiness, amount of precipitable water, aerosol content, surface albedo, etc. This problem can only be solved by budgetary considerations of the complex attenuation processes in the atmospheric column above the region.

Inspection of the international literature shows that the problems of measurement, data interpretation, and climatological analysis of shortwave energy continues to attract attention of investigators in many countries. During the last decade, several world-wide studies and atlases were published, such as the monographs by Budyko (1956), Bernhardt and Philipps (1958), Ångström (1962), and several others. Interest in these problems was stimulated when radiation data (including measurements of "earth-shine" in very satisfactory resolution both geographically and temporally) became available from orbiting satellite stations (Suomi, 1958; Fritz, Rao, and Weinstein, 1964; Hanson, 1967; Vonder Haar, 1968; and others).

Why do we want to add another to the numerous existing and valuable discussions? Because we feel that it is necessary to emphasize the usefulness of a straightforward and reasonably complete yet still elementary budgetary-type method. The quantitative appraisal of attenuation processes in an atmospheric column within

[1] The suffix "nomy," related to the Greek word "nomos," has among other meanings that of "law"; while "logy," derived from "logos," has also several meanings, one of which is "word." Von Ficker argued more than forty years ago that the time has come to proceed from meteorology to "meteoronomy" in order to express increasing reliance on numerical methods at the expense of descriptive methods. More than a decade ago, similar reasoning by S. Chapman resulted in the acceptance of the term "aeronomy" for the mathematical study of physics of the upper atmosphere, restricting "aerology" to a more descriptive approach.

Reprinted with minor editorial modification by permission of the authors and editor, from *Tellus*, Vol. XXI, 1969, pp. 208–222.

the framework set by upper and lower boundary conditions, appears to be most promising for fuller understanding and prediction of how the local radiation-climate will be affected by natural or man-made changes of the environment. Indeed, manipulation of factors which are an integral part of the local shortwave-radiation budget, such as albedo changes by either brightening or darkening of the ground, by shading, smoke-screening, etc., are among the oldest techniques for climate and weather control. In climatonomy, total energies are more important than spectral distributions. Therefore we restrict the discussions to values integrated with respect to wavelength over the band-width of the solar spectrum.

MATHEMATICAL RELATIONS

It is convenient to make the mathematical relationships dimensionless by dividing any energy-flux density by the extra-atmospheric irradiance, $I_0 \cos\theta$, (θ = zenith angle of the sun). To guarantee a reasonable degree of completeness and detail, nine variables must be considered. These can be divided into three groups:

1. Non-dimensional values of shortwave fluxes

A = top albedo, or fraction returned to space,

G^* = global radiation, or fraction received at ground level from sun and sky,

D^* = diffuse radiation, or fraction received at ground level from sky alone,

H^* = solar heating in the air, or fraction absorbed in the atmosphere,

a = energy albedo of the lower boundary; or, $(1 - a)G^*$ = fraction absorbed by the lower boundary.

2. Contributions to the attenuation of non-dimensional beam radiation

σ = part of attenuation due to scatterers in the air,

α = part of attenuation due to absorbers in the air.

3. Coefficients of the scattering process

μ = fraction describing effective outward scattering to space; or $(1 - \mu)$ = fraction of effective downward scattering to the lower boundary,

χ = fraction describing backward scattering of that part of shortwave radiation which has been reflected at least once by the lower boundary.

The physical relationships between the nine dimensionless variables result in five budget equations. The first one deals with total shortwave radiation energy. The fraction not reflected to space must be equal to the sum of absorption by the atmosphere (H^*) and by the ground $(1 - a)G^*$,

$$1 - A = H^* + G^*(1 - a) \qquad (1)$$

The second equation considers beam radiation $(G^* - D^*)$. Its attenuation $(1 - G^* + D^*)$ must equal the depletion by absorbers and scatterers,

$$1 - G^* + D^* = \alpha + \sigma \qquad (2)$$

The third equation deals with radiation reflected back to space, the "earth-shine." The top albedo (A) is the sum of the outward directed part of primary scattering ($\mu\sigma$) plus that part of ground-reflected global radiation which is neither absorbed in the air nor scattered back to the ground,

$$A = \mu\sigma + (1 - \alpha)aG^*(1 - \chi\sigma) \qquad (3)$$

The fourth equation expresses the diffuse radiation arriving at the ground. It consists of the downward directed part of primary scattering $(1 - \mu)\sigma$ and of that part of ground-reflected radiation which is scattered back towards the lower boundary,

$$D^* = (1 - \mu)\sigma + (1 - \alpha)aG^*\chi\sigma \qquad (4)$$

The fifth equation is not independent. The overall balance of the preceding equations requires that the sum of the left-hand sides and the right-hand sides of equations (1) and (4) must be equal, or,

$$H^* = \alpha(1 + aG^*) \qquad (5)$$

Equation (5) gives the heating function (H^*) as the sum of absorption of: beam radiation plus ground-reflected radiation.

With four independent equations and nine unknowns, five variables must be either observed independently, or partly observed and partly prescribed by model assumptions, or externally given in order to calculate the

Set of observed values	Variables to be prescribed	Variables to be calculated	Remarks and examples
G^*, D^*, a	μ, χ	α, σ, A, H^*	Conventional case of ground-based measurements: especially applicable to a uniform lower boundary with uniform albedo; example: desert La Joya, or prairie country O'Neill.
G^*, A, a	μ, χ	D^*, H^*, α, σ	Satellite-based measurements of A, with G^* from network of surface stations, (a) from regional climatological survey flights; example: Hanson's USA data (1967).
G^*, D^*, α, σ	μ	a, H^*, A, χ	G^* and D^* from recordings in complex surroundings; example: Robinson's data for Kew (1963).
$\alpha, \sigma, \mu, \chi$	a	G^*, D^*, A, H^*	Three examples of shortwave radiation-climatonomy for a systematic variation of only one of the prescribed variables at a time.
a, σ, μ, χ	α	G^*, D^*, A, H^*	
a, α, μ, χ	σ	G^*, D^*, A, H^*	
G^*, D^*, a, A, α	—	σ, χ, μ, H^*	Cases of detailed radiation measurements including airborne and satellite data; example: Center of Climatic Research study in Rajasthan desert.
G^*, D^*, a, A, σ	—	α, χ, μ, H^*	

remaining four variables. In the table above, several practically important combinations of variables are listed which can be observed or prescribed.

The fundamental equations (1) to (4) can be transformed by eliminating certain variables. The elimination of D^* in (2) with the aid of (4) produces

$$G^* = (1 - \alpha - \mu\sigma)/[1 - a\chi\sigma(1 - \alpha)] \quad (6a)$$

In the special case of $a = 0$, an ideally black ground surface, (6a) reduces to $G_0^* = (1 - \alpha - \mu\sigma)$. With $d \equiv \chi\sigma(1 - \alpha)$, equation (6a) can be restated as

$$G^* = G_0^*/(1 - ad) \quad (6b)$$

which is Ångström's formula as quoted from Möller (1965) where d represents the "back-scatterance" of the sky. Our equation (6a) shows how backscatterance depends on the efficiency of scatterers and absorbers in the air.

CASE STUDIES

From information on geographical coordinates, time, surface pressure, and solar constant, the optical air mass and the extra-atmospheric irradiance can be derived with the aid of standard tables as included in the IGY manual (1958). These basic numerical data are listed in Table 1 for Kew (England), La Joya (Peru), and O'Neill (Nebraska). The three locations were selected for study because of an interesting contrast in aerosol type. At Kew, a city with presumably high industrial pollution, aerosol absorbs more than it scatters, and at La Joya, where the particulate matter in the desert air consists mostly of mineral dust, the aerosol scatters more than it absorbs. Calculated attenuation coefficients are listed in Table 2 which also shows that optical air mass and total attenuation are similar at the three stations.

The important problem is to separate attenuation by absorbing gases (H_2O, CO_2, O_3, and minor constituents) from attenuation by aerosol. Robinson (1963) has shown that in the presence of appreciable quantities of water vapor the absorption by carbon dioxide and other minor constituents will be negligibly small.

To determine the absorptivity of water vapor we used data summarized and reported by Yamamoto (1962) and derived from them the following semi-empirical formula with the aid of a least-square fit,

$$\alpha_w = 0.102(wM)^{0.276} \quad (7)$$

where w = precipitable water in cm, and M = optical air mass. The actual absorption is obtained as the product of α_w times incident extra-atmospheric irradiance.

The total amount of atmospheric ozone may be estimated from climatic data. Ozone absorption is a complex process which is known to depend on total amount as well as vertical distribution of this constituent gas. Dave and Sekera (1959) have parameterized it with the aid of a simplified model of vertical O_3-distribution. Because ozone absorption is normally small in comparison with water vapor effects, it is suggested here to carry the simplification one convenient

	Kew	La Joya	O'Neill
Location			
Geogr. Latitude	51°30′ N	16°41′ S	42°28′ N
Geogr. Longitude	0°00′ W	71°41′ W	98°32′ W
Elevation (m)	5	1,260	600
Surface Pressure (mb)	x[a]	880	949
Date of Observation			
Month, Year	May 1948	July 1964	Sept. 1953
Day (or Days)	15 thru 19	9 thru 15	7
Extra-Atmospheric Insolation			
Solar Constant (I_{00}, ly/min)	2.00	1.980	1.980
Reduced[b] Value (I_0)	x[a]	1.914	1.950
Sun's Zenith Distance (θ)	36.9°	38.7°	36.5°
Extra-Atmospheric Irradiance ($I \equiv I_0 \cos\theta$, ly/min)	x[a]	1.494	1.568
Observed Shortwave Radiation			
Global Radiation (G, ly/min)	0.728[c]	1.206	1.260
Diffuse Radiation (D, ly/min)	0.110[c]	0.282	0.210
Reflected Radiation (aG)	0.109[c]	0.217	0.277

[a] Not reported by Robinson (1963).
[b] Reduced to actual distance sun–earth.
[c] Reported only as fractions of extra-atmospheric irradiance, i.e. $G^* = G/I$, $D^* = D/I$, and $aG^* = aG/I$.

Table 1 Basic information on location and time of measurements used for case studies of shortwave radiation budgets during near-to-noon conditions under cloudless sky
Data Sources: Kew (Robinson, 1963), La Joya (Stearns, 1966), O'Neill (Lettau and Davidson, 1957)

	Kew	La Joya	O'Neill
Basic Information for Calculation			
Coefficient of Total Attenuation ($\alpha + \sigma$)	0.382	0.381	0.330
Optical Air Mass (M)	1.250	1.112	1.163
Precipitable water (w, cm H_2O)	2.0	0.84[a]	2.10
Total Ozone (equivalent cm of O_3 at NTP)	0.25	0.23	0.26
Aerosol Type	City	Desert	Prairie
Attenuation Coefficients Describing Absorption of Beam Radiation			
Water Vapor Absorptivity (α_w)	0.102	0.100	0.130
Ozone Plus other Gases (α_g)	0.020	0.016	0.018
Aerosol Absorptivity (α_a)	0.103	0.027	0.012
All Absorbers (α)	0.225	0.143	0.160
Attenuation Coefficients Describing Scattering of Beam Radiation			
Neutral Molecules (Rayleigh scattering σ_R)	0.095	0.103	0.107
Aerosol Scattering (σ_a)	0.062	0.135	0.063
All Scatterers (σ)	0.157	0.238	0.170

[a] Calculated as averages of radiosoundings at Lima (Peru) and Antofagasta (Chile).

Table 2 Contributing factors and calculated coefficients of attenuation of beam radiation ($G^* - D^*$); case studies for near-to-noon conditions under cloudless sky at Kew (England), La Joya (Peru), and O'Neill (Nebraska)

step farther by assuming that this absorption is directly related to the total amount of O_3 as well as proportional to the readily calculated intensity of Rayleigh scattering. The tabulations by Dave and Sekera (1959) suggest that, for example, for a total of 0.23 cm, the O_3 absorption will amount to approximately 0.16 of the calculated Rayleigh scattering. Even though this estimate may be wrong by 40%, the effect on the relative error of total attentuation will remain less than 5%.

The atmospheric parameters necessary for the calculation of absorption quantities are listed in Table 2 for the three locations described in Table 1. An important problem is the separation of aerosol attenuation into absorbing and scattering fractions (α_a and σ_a). After determination of α_w and α_g this separation is possible with the aid of diffuse radiation data and albedo values, provided the coefficients of the scattering process μ and χ are prescribed. A tentative value of

$\mu = \frac{1}{3}$ was used while χ was taken to be $\frac{3}{8}$ for Kew, $\frac{3}{3}$ for La Joya, and $\frac{3}{4}$ for O'Neill. Note that the data in Table 2 refer to beam radiation and its depletion under cloudless conditions. It is interesting that the total attenuation $(\alpha + \sigma)$ turns out to be similar for "city" and "desert" conditions while the ratio α/σ is about $\frac{4}{3}$ for the city and only $\frac{2}{3}$ for the desert, and about $\frac{3}{3}$ for prairie land.

For the atmosphere above the continental United States, covering the period from March through May of 1962, Hanson (1967) derived an average top albedo value of 0.40 from TIROS IV low-resolution measurements of shortwave radiation reflected to space by earth, air, and clouds. The Peruvian desert belt normally appears as a rather dark area on satellite pictures. This suggests that local albedo values of the upper boundary should be substantially lower (possibly smaller by more than one half) than in normally cloudier regions. Thus, under cloudless conditions, it will be assumed that the top albedo is smaller than 0.20 above each of the locations considered here. Estimates of actual diffuse radiation fluxes are listed in the shortwave radiation budgets of Table 3, which also specifies seven of the nine variables defined earlier. The remaining two (μ and χ) had been prescribed by model assumptions.

A more secure basis for partitioning aerosol attenuation into absorbing and scattering fractions would be desirable. The numbers presented in Table 2 are the result of trial and error. Obviously, the ratio between aerosol scattering and absorbing is a major factor contributing to climatic differences. Robinson (1963) finds aerosol absorption to be more than twice as large as scatter but reports also that other investigators have suggested various ratios from about equal parts to hardly any absorption.

The question of absolute accuracy of the numerical values reported in Table 3 can be left open if we want to use these data exclusively as background information for the purpose of studying the effects caused by a systematic variation of a parameter, such as surface albedo, or precipitable water, or aerosol content. Controlled parameter variations under cloudless conditions and with clouds present will be discussed later.

Knowing the parameters α, σ, and χ, the "backscatterance" d as defined in equation (6b) can be computed. The result is 0.05 for Kew, 0.11 for O'Neill, and 0.20 for La Joya. This sequence illustrates again differences in aerosol effects between city air and desert air. The range of d-values lies within the extremes of 0.04 and 0.30 for the backscatterance under cloudless sky as quoted by Möller (1965).

SYSTEMATIC VARIATION OF ATTENUATION PARAMETERS UNDER CLOUDLESS CONDITIONS

The surface albedo (a) undergoes naturally a considerable change from summer to winter in the temperate zone. It is also most readily modified by artificial means. Hence, as a first example of shortwave radiation climatonomy, let us consider a stepwise albedo increase from $a = 0$ to $a = 1.00$, with other parameters remaining unchanged. The resulting values of A, H^*, $(1 - a)G^*$, and D^* for Kew, La Joya and O'Neill are listed in Table 4. It must be emphasized that for these calculations the coefficients of scattering (σ, including μ and χ), and absorbing (α) were kept unchanged and equal to the respective local values specified in Table 3. Most strongly affected by surface albedo changes is the fraction absorbed by the ground because of its direct proportionality to $(1 - a)$. However, even in the unlikely extreme case that the earth's surface were perfectly black (with $a = 0$), only 72% to 78% of the extra-atmospheric irradiance would be absorbed; the remainder is partly trapped in the atmosphere by absorbers and partly scattered to space causing a significant minimum "earth-shine" of 5% to 8%. The top albedo increases with the surface albedo although at reduced rate because part of the reflected radiation is re-reflected and absorbed in the air. This causes the intensification of relative heating H^* and diffuse radiation D^* with increasing surface albedo.

It can be concluded from Table 4 that top albedo and diffuse radiation respond strongest to changes in surface albedo if the amount of scatterers in the air is relatively large; this was the case in La Joya. The heating function reacts significantly to the amount of absorbing agents present in the air; this was the case in Kew. Ground absorption shows the strongest response to albedo changes in O'Neill where scatterers and absorbers are almost equally divided. For the maximum

	Kew	La Joya	O'Neill
Beam Radiation (on Horizontal Surface)			
Arriving from Space at Outer Boundary	1.000	1.000	1.000
Transmitted to Lower Boundary ($G^* - D^*$)	0.618	0.619	0.670
Attenuation ($\alpha + \sigma = 1 - G^* + D^*$)	0.382	0.381	0.330
Part of Attenuation Due to Absorbers (α)	0.225	0.143	0.160
Part of Attenuation Due to Scatterers (σ)	0.157	0.238	0.170
Diffuse, Global, and Reflected Radiation			
Diffuse Radiation Arriving at Lower Boundary (D^*)	0.110	0.189	0.134
Global Radiation at Lower Boundary (G^*)	0.728	0.808	0.804
Reflected at Lower Boundary (aG^*)	0.109	0.145	0.177
Energy Albedo of the Lower Boundary (a)	0.150	0.180	0.220
Effective Albedo of the Outer Boundary, Planetary Top-Albedo (A)	0.129	0.175	0.187
Shortwave Radiation Budget			
Outgoing at Outer Boundary (A)	0.129	0.175	0.187
Absorbed in the Atmosphere (H^*)	0.252	0.162	0.186
Absorbed by the Ground, or Transferred into Heat and Other Energy ($(1 - a)G^*$)	0.619	0.663	0.627
Balance, or Incoming at Outer Boundary ($A + H^* + (1 - a)G^*$)	1.000	1.000	1.000

Table 3 Calculated shortwave radiation budgets (expressed as fractions of extra-atmospheric radiation) for near-to-noon conditions under cloudless sky at Kew (England), La Joya (Peru), and O'Neill (Nebraska)

Albedo	Kew ($\alpha = 0.225; \sigma = 0.157$)				La Joya ($\alpha = 0.143; \sigma = 0.238$)				O'Neill ($\alpha = 0.160; \sigma = 0.170$)			
a	A	H^*	$(1-a)G^*$	D^*	A	H^*	$(1-a)G^*$	D^*	A	H^*	$(1-a)G^*$	D^*
0.00	.052	.225	.723	.105	.080	.143	.777	.158	.057	.160	.783	.113
0.04	.073	.232	.695	.106	.100	.148	.752	.164	.080	.165	.755	.116
0.12	.115	.245	.640	.109	.143	.156	.701	.178	.127	.175	.698	.123
0.24	.180	.264	.556	.113	.208	.171	.621	.198	.198	.191	.611	.134
0.36	.245	.285	.470	.117	.277	.188	.535	.216	.272	.206	.522	.145
0.52	.333	.311	.356	.123	.375	.208	.417	.250	.372	.230	.398	.159
0.76	.467	.353	.180	.131	.536	.243	.221	.301	.531	.265	.204	.182
0.88	.536	.373	.091	.136	.625	.261	.114	.329	.613	.283	.104	.194
1.00	.605	.395	.000	.140	.717	.283	.000	.357	.699	.301	.000	.207

Table 4 Example I of shortwave radiation climatonomy
Calculated top albedo (A) and non-dimensional values of: Total absorption in the air (H^*), absorption by the ground ($(1 - a)G^*$), and diffuse radiation (D^*). Assumed for the calculation is that the absorptivity (α) and the coefficient of scattering (σ) remain unchanged (and equal to the values specified in Table 3 for Kew, La Joya, and O'Neill), while the albedo (a) of the lower boundary is varied systematically, as indicated above

possible surface-albedo change (zero to unity), the range of top albedo for the city-type aerosol at Kew is 0.553, that is significantly less than 0.637 for the desert, and 0.641 for the prairie. These findings suggest that measurements of top albedoes from satellites over cloudfree areas of known surface albedoes may permit conclusions on the amount of scatterers in the air.

The data in Table 4 confirm that under otherwise unchanged conditions the global radiation increases with increasing albedo-values of the earth-air interface surrounding a pyranometer station. In nature this can happen as the result of a snowcover build-up. According to Möller (1965) this fact (first demonstrated by Ångström) was confirmed by a comparison of pyranometer data on days with, versus without, snowcover on the ground at the Canadian Station Moosonee. With beam radiation unchanged, the increase of G is caused by a larger D due to backscatter. Equations (6b) and (2) or (4) show that

$$D^* = D_0^* + adG_0^*/(1 - ad) \qquad (8)$$

where the subscripts denote values which would be observed when the albedo of the surroundings were exactly zero. Numerical values of backscatterances d were quoted earlier for the three stations. Obviously, the climatic forcing function, or insolation absorbed by the ground, must decrease with increasing albedo, because $(1 - a)/(1 - a \cdot d)$ is always smaller than unity

	Precipitable water (w, cm)	Absorptivities α_w	α_{total}	A	H^*	$(1 - a)G^*$	D^*
Kew							
$a = 0.15$	0.00	0.000	0.123	0.155	0.139	0.706	0.111
$\sigma = 0.157$	0.42	0.085	0.208	0.135	0.232	0.633	0.110
$\alpha - \alpha_w = 0.123$	0.84	0.103	0.226	0.132	0.250	0.618	0.110
	1.70	0.126	0.249	0.127	0.275	0.598	0.110
	2.55	0.140	0.263	0.125	0.289	0.586	0.109
	3.40	0.152	0.275	0.121	0.304	0.575	0.109
La Joya							
$a = 0.18$	0.00	0.000	0.043	0.200	0.050	0.750	0.196
$\sigma = 0.238$	0.42	0.083	0.126	0.179	0.144	0.677	0.189
$\alpha - \alpha_w = 0.043$	0.84	0.100	0.143	0.175	0.163	0.662	0.188
	1.70	0.122	0.165	0.170	0.188	0.642	0.186
	2.55	0.136	0.179	0.166	0.204	0.630	0.185
	3.40	0.147	0.190	0.164	0.216	0.620	0.184
O'Neill							
$a = 0.22$	0.00	0.000	0.030	0.232	0.036	0.732	0.139
$\sigma = 0.170$	0.42	0.084	0.114	0.202	0.101	0.697	0.134
$\alpha - \alpha_w = 0.030$	0.84	0.101	0.131	0.196	0.122	0.682	0.133
	1.70	0.123	0.153	0.189	0.148	0.663	0.132
	2.55	0.138	0.168	0.184	0.166	0.650	0.131
	3.40	0.149	0.179	0.180	0.179	0.641	0.131

Table 5 Example II of shortwave radiation climatonomy
Calculated top albedo (A) and non-dimensional values of total absorption in the air (H^*), absorption by the ground ($(1 - a)G^*$), and diffuse radiation (D^*). Assumed for the calculations are that surface albedo (a) and the coefficient of atmospheric scattering remain unchanged and equal to the values specified in Table 3, while the coefficient of absorption (α) is varied as indicated above corresponding to a systematic variation of precipitable water (w)

with a and d, both being positive fractions. This can be proven with the aid of equation (6b).

The second example of parameter modification deals with absorption as affected by systematic variation of precipitable water content. Equation (7) was employed to calculate α_w. Although this absorptivity is dependent on optical air mass, the M-values at the three stations are comparable due to near-to-noon conditions (see Table 2). Yamamoto's equation contains only the 0.276th power of M, a relative weak dependency, which was used by Loewe (1963) in his argument that multiple absorption is a factor of minor importance in comparison to multiple scattering. However, it should be emphasized that our budget-type considerations include multiple scattering as well as multiple absorption.

With α_w obtained from equation (7), the total absorptivity was found by adding $(\alpha_g + \alpha_a)$ which remained unchanged at values listed in Table 2. The computational results are summarized in Table 5. Least affected is the diffuse sky radiation, which changes hardly at all, whereas H^* reacts most strongly. In the case of Kew, for example, H^* increases by 0.165 units at the expense of 0.131 units in ground absorption and 0.034 units in top albedo, while D^* decreases only by 0.02 units. Considering the different ratios of α/σ, and the original ratios $\alpha_w/(\alpha_g + \alpha_a)$, the results for the two other stations are self-explanatory.

The third example is one of special importance at a time when more and more particulate matter is released into the atmosphere due to rapid industrialization all over the world. For partial answers to questions of air-pollution effects on climate, the quantitative, budget-type calculations, summarized in Table 6, are basic.

The specific parameter modified in the third example is the attenuation fraction of aerosol scattering σ_a. Keeping the ratio of aerosol absorption to scattering (α_a/σ_a) at the values listed in Table 6, the absorbing fraction α_a was calculated. With attenuation by water vapor and other gases unchanged, $(\sigma - \sigma_a)$ and $(\alpha - \alpha_a)$ were assumed to be local constants; varying total values of σ and α were calculated, and also, with the aid of equations (1) through (5), the values of A, H^*, $(1 - a)G^*$, and D^*. The results are listed in Table 6.

The importance of the ratio α/σ became apparent in the calculations summarized in Table 5. In the third example, the systematic change of aerosol scattering emphasizes the significance of α_a/σ_a. This ratio is affected by the type of particles, their size, structure, and chemical composition. It can be assumed that particulate matter over La Joya consists mostly of mineral dust from the desert soil, whereas at Kew almost all comes probably from stack effluents. At La Joya and O'Neill, the ratios α_a/σ_a are nearly equal and approximately 0.2, and addition of aerosol increases diffuse radiation, top albedo, and atmospheric heating, while ground absorption decreases. This contrasts markedly with results for Kew where α_a/σ_a equals about $\frac{3}{2}$. According to Table 2 observations at Kew and O'Neill yielded the same value

	Attenuation by Aerosol		Total Att.						
	Scatt. (σ_a)	Absorp. (α_a)	σ	α	A	H^*	$(1-a)G^*$	D^*	$D^*/(G^*-D^*)$
Kew									
$a = 0.15$	0.000	0.000	0.095	0.122	0.140	0.137	0.723	0.067	0.01
$\sigma - \sigma_a = 0.095$	0.070	0.116	0.165	0.238	0.131	0.264	0.605	0.115	0.19
$\alpha - \alpha_a = 0.122$	0.135	0.224	0.230	0.346	0.129	0.376	0.495	0.158	0.37
$\alpha_a/\sigma_a = 1.66$	0.200	0.332	0.295	0.454	0.131	0.485	0.384	0.201	0.80
	0.270	0.448	0.365	0.570	0.139	0.597	0.264	0.246	3.78
La Joya									
$a = 0.18$	0.000	0.000	0.103	0.116	0.158	0.134	0.708	0.083	0.01
$\sigma - \sigma_a = 0.103$	0.070	0.014	0.173	0.130	0.166	0.150	0.684	0.137	0.20
$\alpha - \alpha_a = 0.116$	0.135	0.027	0.238	0.143	0.174	0.163	0.663	0.189	0.31
$\alpha_a/\sigma_a = 0.200$	0.200	0.040	0.303	0.156	0.183	0.179	0.638	0.237	0.44
	0.270	0.054	0.373	0.170	0.194	0.193	0.613	0.291	0.64
O'Neill									
$a = 0.22$	0.000	0.000	0.107	0.148	0.179	0.175	0.646	0.083	0.01
$\sigma - \sigma_a = 0.107$	0.070	0.013	0.177	0.161	0.187	0.190	0.623	0.137	0.21
$\alpha - \alpha_a = 0.148$	0.135	0.026	0.242	0.174	0.195	0.204	0.601	0.186	0.32
$\alpha_a/\sigma_a = 0.190$	0.200	0.038	0.307	0.186	0.204	0.217	0.579	0.235	0.46
	0.270	0.051	0.377	0.199	0.216	0.229	0.555	0.287	0.68

Table 6 Example III of shortwave radiation climatonomy
Calculated top albedo (A) and non-dimensional values of total absorption in the air (H^*), absorption by the ground ($(1 - a)G^*$), and diffuse radiation (D^*). Assumed for the calculations is that surface albedo (a) and the coefficients of Rayleigh scattering (σ_R) and absorption by gases (water vapor plus ozone, α_g) remain unchanged and equal to the values specified in Table 3, while coefficients of scattering and absorption by aerosol (σ_a and α_a) are systematically varied as indicated above (keeping the ratio α_a/σ_a constant)

of σ_a about 0.06. However, if σ_a is raised to 0.27 (see Table 6), the beam radiation ($G^* - D^* = 1 - \alpha - \sigma$) decreases over the city from 0.597 to an extremely low value of 0.065, whereas the drop over the open prairie is moderate, from 0.662 to 0.424. Accordingly, the ratio of diffuse to beam radiation, $D^*/(G^* - D^*)$ varies from 0.21 to 0.68 at O'Neill and from 0.19 to 3.78 at Kew. Values larger than unity mean that diffuse radiation exceeds beam radiation. When $D^*/(G^* - D^*)$ passes unity, it will become difficult or even impossible for an observer to see his own shadow. Data in Table 6 suggest that such a state of haziness under "clear" sky may be reached at Kew when the initial level of aerosol scattering (as given in Table 2) would be about tripled. Further increase of turbidity would lead to $G^* = D^*$, which means disappearance of beam radiation; or ($\alpha + \sigma$) approaches unity. Then the haze would cease to be translucent. Attenuation, sufficiently dense to obscure the sun, occurs normally with opaque clouds present. However, the above calculations show that in city air (with strongly absorbing aerosol) a relative modest pollution increase may extinguish the direct sunlight.

The results summarized in Table 6 are calculated for a gradual increase of aerosol scattering (σ_a) by equal increments, with a proportional increase in α_a. In a modified version of this third example, the local value of aerosol attentuaion for both, σ_a and α_a, was first doubled and then increased to the fivefold of the initial level for Kew and La Joya. Figure 1 shows the initial conditions in city versus desert air, Figure 2 and Figure 3 the results computed with the modified values. The graphical scheme selected is the simplest possible which shows for the incoming radiation the relative magnitudes of attenuation (separately for scattered out, scattered down, and absorbed), the resulting beam radiation, and the reflected global radiation (separately for back-scattered and absorbed). The scheme is completed by indicating the top albedo, the diffuse radiation at ground level, the total global radiation, and the absorption by the ground. Budget considerations can be readily verified.

It can be seen that a fivefold increase in aerosol content blocks out beam radiation in both atmospheres. In the desert air, the loss of beam radiation results in a considerable increase in diffuse radiation due to the high scattering quality of desert aerosol, whereas in the city air where considerable more energy is absorbed than scattered, the heating function increases accordingly.

ESTIMATES OF CLOUD EFFECTS ON SHORTWAVE RADIATION FLUXES

If a fraction (c) of the sky is covered by clouds the flux densities of the cloudless area have a weight factor of ($1 - c$). Let σ_c denote the effective coefficient of scattering and α_c the absorptivity of air in the cloud and, in

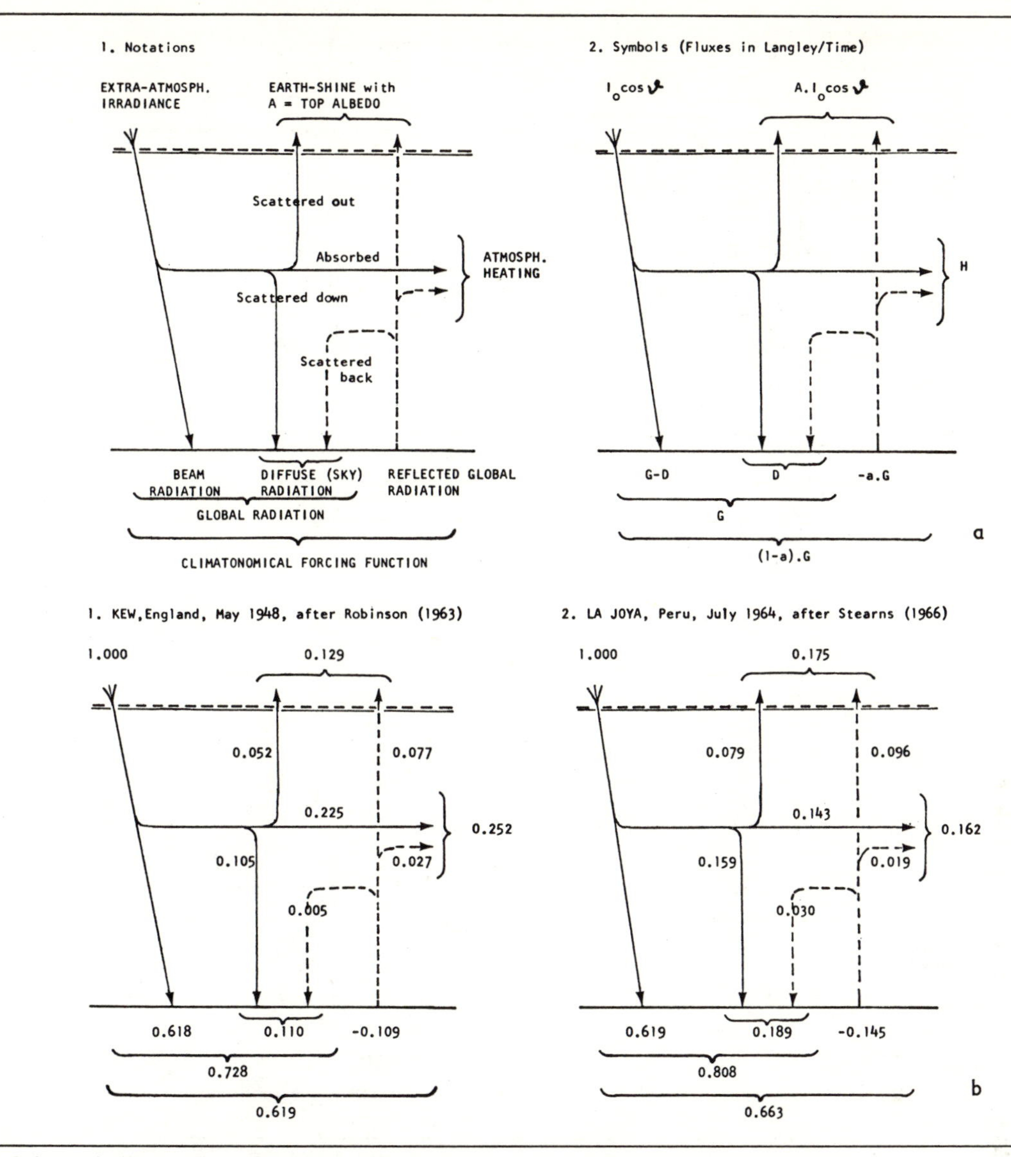

Figure 1 Schematic illustration of the atmospheric short-wave radiation budget. (a) Notation used for the description of radiation flux-densities, (b) example of short-wave radiation budgets for air over a city in comparison with air over a desert. Consult Tables 1 to 3 for information on localities and external conditions.

analogy to μ of an earlier section, let μ' be a fraction so that $\mu'\sigma_c = A'$ will be scattered back to space by the upper cloud surface and $(1 - \mu')\sigma_c$ will reach the ground. The effective albedo of the cloud base will be assumed to equal A'. The employment of "effective" coefficients represents a considerable model simplification of the highly complicated scattering processes which possess pronounced dependencies on particle sizes and their spectral distribution, liquid water content of the cloud, and solar elevation angle. Compressing these characteristics into a few single parameters can be tolerable only for the study of climatonomic gross-effects and overall energy partitions.

In the following energy budget equations, let the subscripts 0 and c indicate values for cloudless sky (with prescribed conditions for clear-air attenuation and ground albedo), and for cloudy sky, respectively. For the calculation of effective coefficients in partly cloudy air, we assume that the contributions by absorbers are additive, while contributions by scatterers are distributive (or must be prorated). Thus, in air with cloudiness (c)

$$\alpha = (1 - c)\alpha_0 + c(\alpha_0 + \alpha_c) = \alpha_0 + c\alpha_c \quad (9)$$

$$\sigma = (1 - c)\sigma_0 + c\sigma_c = \sigma_0 + c(\sigma_c - \sigma_0) \quad (10)$$

From continuity principles developed earlier, it follows that the top albedo is composed of a prorated contribution from the clear area and direct reflection from the cloud surface, plus diffuse radiation reflected from the

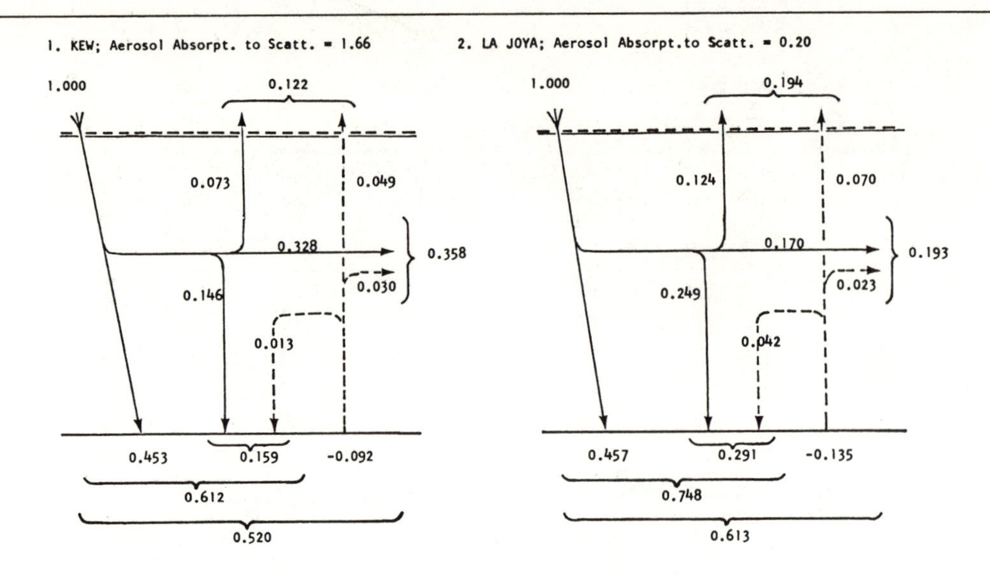

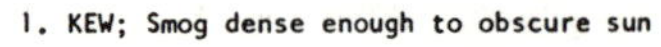

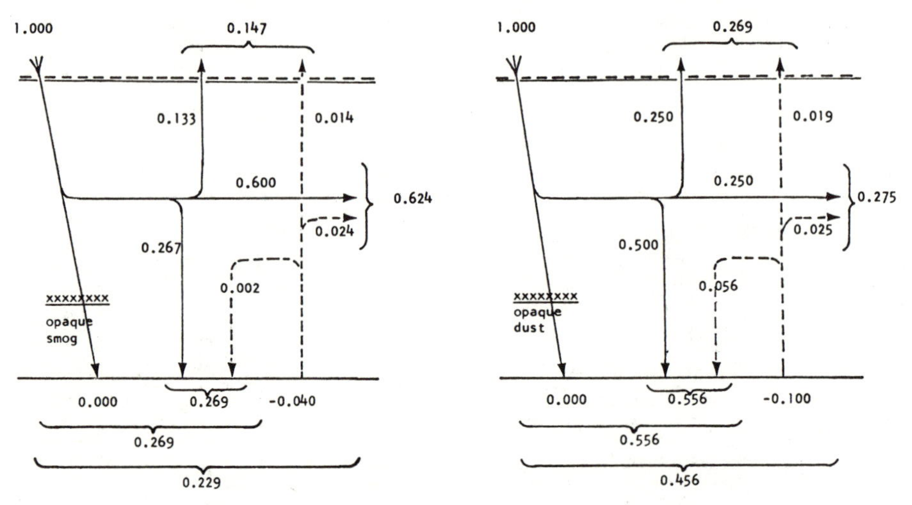

Figure 2 Short-wave radiation budget for air over a city in comparison with air over a desert, calculated under the assumption that aerosol amount is increased by a factor of two while other conditions remain unchanged and as illustrated on the lower half of Figure 1. Unchanged was also the characteristic ratio which expresses scattering efficiency relative to absorbing efficiency of the local aerosol type.

Figure 3 Same as Figure 2 but calculation based on the assumption that the aerosol content at both localities is increased by a factor of five. This happens to be just enough to block out direct beam radiation at both localities. Note the significant differences in diffuse radiation, atmospheric heating, and top radiation, in comparison with the initial conditions shown on the lower half of Figure 1.

ground, and reduced by what is either reflected back from the cloud base or absorbed in the air. Hence,

$$A = (1 - c)A_0 + c[A' + a \cdot G^*(1 - \alpha)(1 - A')] \quad (11)$$

A corresponding development for the heating function H^* yields

$$H^* = (1 - c)H_0^* + c \cdot \alpha(1 + a \cdot G^*) \quad (12)$$

The sum of equations (11) and (12) is used to eliminate the term $(A + H^*)$ in equation (1). The resulting relation can be solved for G^*,

$$G^* = [1 - A_0 - H_0^* + c(A_0 + H_0^* - A' - \alpha)]/[1 - a + c \cdot a(1 - A' + \alpha A')] \quad (13)$$

After G^* has been determined for a given cloudiness, A follows from (11), and H^* either from (5) or (12).

For overcast conditions we must distinguish between opaque and translucent clouds. Equations (9) and (10) apply to both cases and yield $\alpha_1 = \alpha_0 + \alpha_c$, and $\sigma_1 = \sigma_c$. For opaque clouds there is no beam radiation, so that in equation (2) $G^* = D^*$, or $\alpha_1 + \sigma_1 = 1$. Translucent clouds permit the sun to be seen, thus $\alpha_1 + \sigma_1 < 1$.

It follows from equations (9) and (2) that for opaque clouds (with $G_1^* = D_1^*$) D_1^* must be smaller than D_0^*. For translucent clouds D_1^* can be larger than D_0^*. Direct downscatter will be exceptionally large from a partly cloudy sky when the sun shines on lateral boundaries of clouds. Under not unlikely conditions such downscatter may cause $G_c^* > 1$, or to surpass temporarily even the value of the extra-atmospheric irradiance. This may occur when at noon in summer a brilliantly white cumulus tower approaches the pyranometer station from the direction opposite to an unobscured sun. Alternating sunlight and cloud shadows under broken cumulus-cover can cause interesting variations in reflected versus downcoming insolation. Examples of this effect are flight-measurements of ground albedo reported by Bauer and Dutton (1962). For climatonomical purposes, however, these special cases are not as important as statistical relations for global radiation under overcast sky.

Equation (13) is compatible with equation (6a). It is seen more easily from the latter that global radiation responds to albedo changes of the surrounding area under cloudy to overcast sky stronger than under clear sky (other conditions unchanged), because backscatterance from a cloudbase is always larger than from clear sky. Analysing radiation data from Moosonee (Canada), Möller (1965) finds a backscatterance of 0.25 from clear sky, but 0.58 to 0.69 from overcast. Similar values are quoted by Linke (1942).

The problem remains to determine the three parameters μ', α_c, and σ_c. Even if all three would be independent of the degree of cloudiness, equations (10), (11), (12), and (13) show that the change of G_c^*/G_0^* with cloudiness will depend on the turbidity in the cloudfree atmosphere. On page 440 of "Smithsonian Tables" (1963), earlier works by Kimball, Ångström, Mosby and Haurwitz are quoted, all suggesting a simplified universal relationship between insolation and cloudiness. Other authors preferred to relate global radiation linearly to the fraction of actual over possible sunshine-hours per day. These statistical relationships are discussed by Budyko (1956). Gräfe (1956) considers a nonlinear relationship between G_c/G_0 and cloudiness.

Plots of actual G_c/G_0 data versus c (as on page 440 of Smithsonian Tables) usually show a wide scatter of points. At the location of special interest for our study, the pampa de La Joya, clouds cover not more than $\frac{2}{10}$ of the sky on the daily average, and are mostly of the stratiform type. During the period of radiation measurements in July of 1964, no clouds intercepted the beam radiation. However, it became apparent that employment of an average statistical relationship between insolation and cloudcover (as taken from the literature) and disregarding the special nature of the desert aerosol, would not produce reliable results in the computation of G_c for La Joya. A new approach to the general problem of calculating cloudcover effects on global radiation became necessary under due consideration of the shortwave-radiation budget relationships derived earlier. Useful for this purpose are results obtained by Haurwitz (1934) concerning the relationship between global radiation and cloud type during overcast, relative to global radiation under cloudless conditions. Haurwitz included in his statistical analysis the dependency of G_1/G_0 on optical air mass. Our abbreviated summary refers to an average air mass of $M = 1.2$, but the observed variations of G_1/G_0 with M are relatively slight.

Overcast type:	Cirro-Strat.	Alto-Cum.	Alto-Strat.	Strato-Cum.	Stratus	Nimbus
G_1/G_0	0.81	0.51	0.41	0.34	0.25	0.17

One of the few systematic evaluations of attenuation effects of clearly defined cloud masses, or thicknesses, is the study by Neiburger (1949), based on aerological measurements of shortwave radiation fluxes above and below a coastal stratus in California. He finds that absorption is generally small, rarely exceeding 10%, so that transmission is mainly determined by reflection at the upper cloud surface (A') which increases significantly with increasing cloud thickness. For climatonomical calculations, it would be desirable to know the statistical relationship between attenuation parameters and degree of cloudiness at all tropospheric levels, including overcasts of different thicknesses. Neiburger's data refer only to a low stratus. Lacking detailed information, let

	Cirro-Strat.	Alto-Cum.	Alto-Strat.	Strato-Cum.	Stratus	Nimbus
α_c	0.00	0.01	0.03	0.05	0.07	0.10
σ_c	0.68	0.80	0.79	0.77	0.75	0.72
μ'	0.30	0.52	0.60	0.66	0.74	0.82
A'	0.200	0.416	0.474	0.508	0.555	0.590
$G_1^* - D_1^*$	0.140	0.010	0.000	0.000	0.000	0.000
G_1^*	0.620	0.394	0.316	0.262	0.195	0.130
G_1^*/G_0^*	0.81	0.51	0.41	0.34	0.25	0.17

α_c = absorptivity; σ_c = scattering coefficient; μ' = fraction of upward scattering; $\mu'\sigma_c = A'$ = cloud albedo; $G_1^* - D_1^*$ = beam radiation computed as $1 = \alpha_0 - \alpha_c - \sigma_c$ (for $\alpha_0 = 0.180$); G_1^* = global radiation computed as $G_1^* - D_1^* + (1 - \mu')\sigma_c$; ($G_1^*/G_0^*$ was computed assuming that in clear air $G_0^* = 0.766$)

Table 7 Tentative parameterization of shortwave attenuation in cloud overcast of various type

Cloud	Kew					La Joya					O'Neill				
c	G^*	D^*	H^*	$(1-a)G^*$	A	G^*	D^*	H^*	$(1-a)G^*$	A	G^*	D^*	H^*	$(1-a)G^*$	A
Cs 0	0.728	.110	.252	.619	.129	.808	.189	.162	.663	.175	.804	.134	.186	.627	.187
$\frac{1}{3}$	0.677	.234	.251	.575	.184	.760	.289	.162	.623	.215	.750	.250	.186	.585	.229
$\frac{2}{3}$	0.632	.362	.248	.537	.215	.717	.392	.162	.588	.250	.706	.376	.185	.551	.264
1	0.589	.494	.245	.501	.254	.678	.501	.160	.556	.284	.665	.505	.183	.519	.298
St 0	0.728	.110	.252	.619	.129	.808	.189	.162	.663	.175	.804	.134	.186	.627	.187
$\frac{1}{3}$	0.544	.132	.257	.462	.281	.628	.215	.170	.515	.315	.616	.169	.193	.480	.327
$\frac{2}{3}$	0.354	.148	.275	.293	.432	.441	.235	.191	.362	.447	.427	.204	.213	.333	.454
1	0.159	.159	.302	.135	.563	.252	.252	.226	.207	.567	.237	.237	.242	.189	.569

Table 8 Example IV of shortwave radiation climatonomy
Calculated non-dimensional values of global radiation G^*, diffuse radiation D^*, absorption in air H^*, absorption by the ground $(1 - a)G^*$, and top albedo A. Assumed for calculation is that the "clear-air" attenuation parameters remain unchanged (and equal to the values specified in Table 2 for Kew, La Joya, and O'Neill) while cloudiness c varies systematically with cloud attenuation parameters for cirrostratus and stratus (as specified in Table 7)

us assume that separation by cloud types is a substitute for cloud thickness. Table 7 summarizes tentative results of attenuation parameters under overcast ranked according to probable cloud thicknesses. α_c and σ_c were obtained with the aid of equations (9) and (10). The three parameters (albedo, absorption, scattering) were determined so that a monotonic dependency on cloud rank reproduced Haurwitz's empirical data on global radiation ratios. Note that $G_1^* > D_1^*$ indicates overcast of the "translucidus" type, and $G_1^* = D_1^*$ cloud forms of the "opacus" type.

It was assumed that $\alpha_0 = 0.18$ and $G_1^* = 0.766$ (which corresponds approximately to the average clear-air conditions at O'Neill and Kew), would apply also at Blue Hill Observatory near the city of Boston. Not surprisingly, the G_1^*/G_0^*-ratios calculated for the desert atmosphere at La Joya, turn out to be systematically larger than for the two other locations, obviously due to the difference in aerosol scattering. Note that in Table 7 cloud albedo ($A' = \mu'\sigma_c$) has a trend opposite to that of downward scattering, $(\mu' - 1)\sigma_c$. With the exception of cirrostratus, the coefficient of scattering tends to remain nearly constant, between 0.80 and 0.72.

In a tentative and exploratory manner let us discuss now the fourth and last example of shortwave radiation climatonomy. We formally calculated the change in radiation fluxes which occur if cloudiness is increased stepwise from zero to unity ($c = 0, \frac{1}{3}, \frac{2}{3}, 1$) for cirrostratus and stratus at Kew, La Joya, and O'Neill. The results summarized in Table 8 are self-explanatory. Interesting is the difference between translucent and opaque cloud conditions. The top albedo for overcast shows still a dependency on ground albedo, rather significantly so for translucent, and rather weakly for opaque clouds. Atmospheric heating (H^*) decreases slightly with increasing c for translucent clouds, but increases considerably for opaque clouds.

The differences due to clear-air attenuation between the three locations are maintained; global radiation at Kew, for example, is always relatively low. This demonstrates again that one can hardly expect a universal relationship between type of overcast and global radiation.

REFERENCES

Ångström, A., 1962. Atmospheric turbidity, global illumination, and planetary albedo of the earth. *Tellus 14(4)*, 435–450.

Bauer, K. G. and Dutton, J. A., 1962. Albedo variations measured from an airplane over several types of surface. *J. Geophys. Res. 67*, 2367–2376.

Bernhardt, F. and Philipps, H., 1958. Die räumliche und zeitliche Verteilung der Einstrahlung, der Ausstrahlung und der Strahlungsbilanz im Meeresniveau. *Teil I: Die Einstrahlung. Abhandl. d. Meteorol. und Hydrol. Dienstes der DDR, Nr. 45.*

Budyko, M. I., 1958. *The Heat Balance of the Earth's Surface* (1956). Translated by Stepanova, N. A., U.S. Weather Bureau, PB 131692, Washington, D.C.

Dave, J. V. and Sekera, Z., 1959. Effect of ozone on the total sky and global radiation. *J. Meteor. 16*, 211–212.

Fritz, S., Krishna Rao, P., and Weinstein, M., 1964. Satellite measurements of reflected solar energy, and the energy received at the ground. *J. Atmos. Sci. 21*, 141–151.

Graefe, K., 1956. Der strahlungsempfang Vertikaler Ebener flächen; globalstrahlung von Hamburg. *Berichte Deutsch. Wetterdienst 29*, 1–15.

Hanson, K. J., 1967. *The Reflection of Sunlight to Space and Absorption by the Earth and Atmosphere over the United States during Spring, 1962.* M.S. thesis, Univ. Wisconsin, Madison.

Haurwitz, B., 1934. Daytime radiation at Blue Hill Observatory in 1933. *Harvard Meteorol. Studies 1.*

IGY Manual, 1958. *Annals of the Intern. Geophys. Year 5(4).* Pergamon Press.

Landsberg, H. E., 1957. Review of climatology, 1951–1955. *Meteorol. Res. Revs. 3(12).*

Lettau, H. and Davidson, B., 1957. *Exploring the Atmosphere's First Mile.* (2 vols.) Pergamon Press, London.

Lettau, H., 1952. Synthetische Klimatologie. *Berichte Deutsch. Wetterdienst 38*, 127–136.

Linke, F., 1942. *Handbuch der Geophysik 8*, Kap. 7 Die Kurzwellige Himmelsstrahlung, Borntraeger, Berlin. 340–415.

Loewe, F., 1963. On the radiation economy, particularly in ice and snow-covered regions. *Gerl. Beitr. zur Geophys. 72(6)*, 371–376.

Möller, F., 1965. On the backscattering of global radiation by the sky. *Tellus 17(3)*, 350–355.

Neiburger, M., 1949. Reflection, absorption, and transmission of insolation by stratus clouds. *J. Meteor. 6*, 98–104.

Robinson, G. D., 1963. Absorption of solar radiation by atmospheric aerosol as revealed by measurements at the ground. *Archiv. Meteor. Geophys. Bioklim. Series B, 12*, 19–40.

Stearns, C., 1966. *Micrometeorological Studies in the Coastal Desert of Southern Peru.* Ph.D. thesis, Univ. Wisconsin, Madison.

Soumi, V. E., 1958. The radiation balance of the earth from a satellite. *Annals IGY 6*, 330–340.

Vonderhaar, T., 1968. *Variations of the Earth's Radiation Budget.* Ph.D. thesis, Univ. Wisconsin, Madison.

Yamamoto, G., 1962. Direct absorption of solar radiation by atmospheric water vapor, carbon dioxide, and molecular oxygen. *J. Atmos. Sci. 19*, 182–188.

SUGGESTED READING FOR PART 1

Lettau, H., 1969. Evapotranspiration climatonomy—a new approach to numerical prediction of monthly evapotranspiration, runoff, and soil moisture storage. *Mon. Wea. Rev. 97(10)*, 691–699.

Miller, D. H., 1965. The heat and water budget of the Earth's surface. *Adv. in Geophysics 11*, 175–302.

Rauner, Y. L., 1962. The heat balance of forests and its role in the formation of the microclimate of wooded and treeless landscapes of the Moscow region. *Soviet Geography: Review and Translation 3(6)*, 40–47.

Terjung, W. H., 1968. Some maps of isanomalies in energy balance climatology. *Archiv für Meteorologie Geophysik und Bioklimatologie, Serie B, 16*, 279–315.

Part Two

General Atmospheric Circulation

Hare has eloquently spotlighted the lack of a comprehensive theory of world climate. In his article he takes a major step towards rectifying this through a detailed examination of the mid-latitude westerlies in the northern hemisphere. What then is the ultimate significance of the circumpolar vortex?

When is a jet stream not a jet stream? The second article provides a lucid classification of the three main types.

In attempting to find the answers to why there is a tendency for the occurrence of at least one index cycle each year, why these cycles vary in intensity and why the most pronounced index cycle falls in February, Namias has evolved a theory, as yet incomplete, explaining the periodic fluctuations of the westerlies that make up the so-called index cycle.

F. Kenneth Hare

3 The westerlies

It has been customary to treat the climates of the mid-latitudes piecemeal, as though they had little underlying unity. We have known for centuries that the west winds prevail over much of the temperate world, and the weather eye of all mid-latitude nations turns automatically to the west. But we have lacked enough comprehension of the atmospheric circulation to present a unified picture of mid-latitude climates and have organized our conception of them regionally; thus the climates of eastern Asia are always portrayed in terms of their monsoonal characteristics, and little attempt is made to relate them genetically to their analogues elsewhere in the westerly belt. The scientist, however, perpetually seeks unification and consolidation: piecemeal description should always yield in the end to successful synthesis. Rather haltingly, and with the certainty that he will sometimes be proved wrong, the climatologist can soon attempt a synthesis for the westerly belt.

This change in outlook, full of promise for the field, has resulted from an increase in the technical and theoretical resources of the mother science, meteorology. The sudden growth of high-altitude flight has made upperwind forecasting of great economic importance, and large sums are available for the daily exploration of the upper air. In the Northern Hemisphere all the continents are covered by an effective network of radiosonde stations, and with the aid of weather ships the oceans are covered as well, though too thinly for comfort. Daily Northern Hemispheric analysis is now practiced in many countries at the 500-millibar level (roughly 18,000 feet) and can be attempted at much greater heights. Such analysis immediately sweeps away the piecemeal quality of our understanding of the westerly climates, and it becomes apparent that the variants of the mid-latitude regime are but aspects of a single vast, spectacular mechanism, the circumpolar westerly vortex.

Coupled with this daily exploration of the atmosphere there has been a vigorous development of hydrodynamic theory. Such theory is by no means new; many of its theorems were derived in the late nineteenth century, and the Newtonian laws on which they are based are two centuries older. But only in recent years has the meteorologist, aided by the oceanographer, been able to apply these theorems to the complex, unstable, and turbulent westerlies. The culmination of this movement toward theory has been the growth of numerical weather prediction, whereby the electronic computer actually solves the nonlinear equations expressing the dynamics of the flow.

So far, few attempts have been made to use these new ideas in the genetic study of world climate. The standard climatologies are written in the language of twenty-five years ago. As Eady (1957) points out, climatology is a geophysical science in which the main object is comprehension, not description, though the latter is a necessary stage. Genetic climatology requires a knowledge of two great bodies of meteorological process: the general circulation of the atmosphere, which governs the advective element in local or regional climate; and the radiative and turbulent exchanges along the vertical, which dominate the heat and moisture balances. We are here concerned only with the first body, the general circulation. It is clear that the variation of climate within the mid-latitudes can be explained only in the light of the behavior of the general circulation, and, more specifically, of the westerlies. It follows that advances in our understanding of the westerlies should pave the way toward re-examination of the stock explanations of the regional climatic differences. Trewartha (1958) has already attempted this for Monsoon Asia, and many other such studies are in preparation.

In this encouraging situation a survey of the new ideas about the westerly vortex may be of some value.

Reprinted, with minor editorial modification, by permission of the author and editor, from the *Geographical Review*, Vol. 50, 1960, pp. 345–367, copyright by the American Geographical Society of New York.

Such a project is laborious, because it requires a familiarity with the difficult theoretical literature of meteorology. Nevertheless, the writer has attempted it here, in the full knowledge of his inadequacy for the task. His object is not synthesis, but a review of ideas; much work remains to be done before a definitive study of the westerly climates as a whole can be published.

GEOGRAPHICAL EXTENT OF THE WESTERLIES

We shall regard as the westerly belt that irregular circumpolar zone in which the resultant winds of the upper troposphere (that is, 15,000–30,000 feet) are from a westerly point. As so defined, the westerly belt includes some areas where the surface winds are not predominantly from the west—northern Italy, for example—and it is also subject to a seasonal fluctuation in extent and position. Nevertheless, the definition is sound, since cyclonic disturbances (on which precipitation depends) are characteristically similar in direction of motion to the upper tropospheric flow; in other words, they exhibit westerly steering. For reasons of space, the writer has confined himself to the Northern Hemisphere; Lamb (1959) has recently published an excellent empirical study of the southern westerlies. The present account, since the purpose is review, draws freely on numerous sources, among which a recent study of the jet stream prepared under the auspices of the World Meteorological Organization may be given pride of place (Berggren et al., 1958). For stratospheric levels, however, the writer has drawn largely on his own experience.

Figure 1,* redrawn from a map in a recent atlas produced at the University of Wisconsin (Lahey et al., 1958), shows the extent of the westerly belt on the isobaric surface of 500 millibars (roughly in mid-troposphere, near 18,000 feet) in January and July. This surface is now the most important level for synoptic analysis; the patterns are bolder and simpler than at sea level and give a far better picture of the normal extent of the westerly climates. The circulation at 500 millibars is very similar to that at all levels between about 700 millibars (about 10,000 feet) and 100 millibars (about 53,000 feet). It is, however, well below the level of maximum speed, which ordinarily lies near 300 millibars (about 30,000 feet) in temperate latitudes and 200 millibars (about 40,000 feet) in the subtropics.

The diagrams show clearly that the westerlies form a vortex whose center is single in summer and dual in winter, the winter centers being over northeastern Siberia and the Canadian Arctic. The fact that the westerlies flow round a vast polar cyclone was first pointed out by Ferrel (1889), and the vortex is now sometimes named for him. Ferrel expected that this cyclone would exist at sea level, but Von Helmholtz (1893) deduced—and observation subsequently confirmed—that radiative cooling would produce a shallow surface high near the pole. Another striking peculiarity of the diagrams is the seasonal variation in strength and equatorward extent of the westerlies. In winter, they extend at sea level from about 70° N to 35° N, and at 500 millibars from the pole to about 13° N; in summer, by contrast, the surface limits are 65° N and about 40° N, and the limits for 500 millibars the pole and about 25° N. Thus the westerlies shrink in area in summer, largely along their southern margin. The shrinkage is accompanied by a marked weakening; wind speeds are less than half those in winter, and the total kinetic energy below 100 millibars (about 53,000 feet) and between $17\frac{1}{2}$° N and $77\frac{1}{2}$° N has been computed by Bjerknes (1957) to decrease in summer to only 30% of its winter value (see Appendix). The poleward shift and weakening of the westerlies in summer has been known qualitatively for many years; it is the cause of the existence of the Mediterranean climates and the seasonal rainfall inequalities of higher latitudes.

Figure 1 also shows that the westerlies are deformed by standing or "forced" perturbations in both seasons. There is a strong ridge in the longitudes of Alaska and the western cordillera of North America, and another west of Europe, the latter in winter only. Deep troughs occur over eastern North America and east of Japan. These quasi-permanent waves are among the most important climatic controls in the hemisphere. Their origin is still in dispute. The fact that they are stationary suggests strongly that some kind of physiographic

* [Because of the varying nature of the source materials, several systems of units are used in the illustrations. It may therefore be helpful to point out that one meter per second is roughly equal to two knots, and one kilometer equals 3,280 feet.]

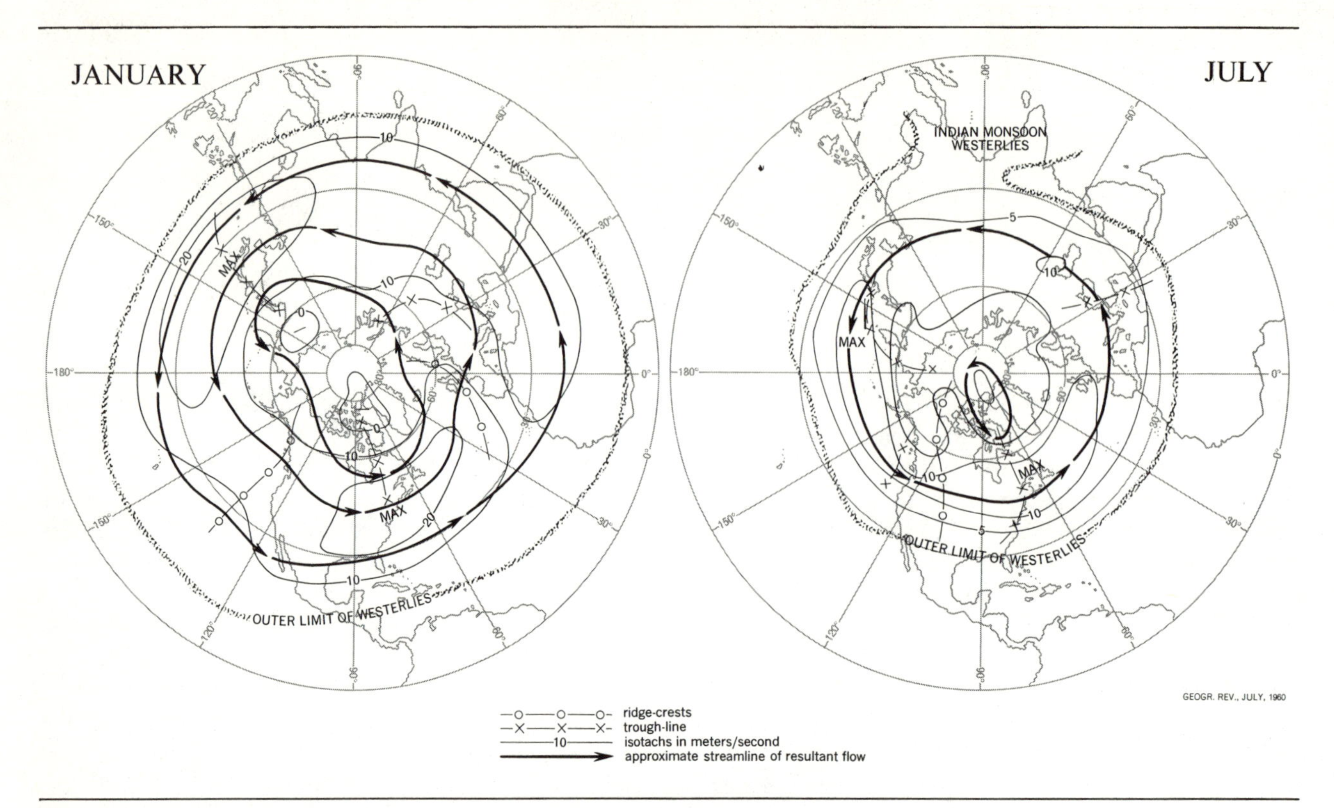

Figure 1 The westerlies in mid-troposphere. Solid lines are isotachs of constant west-component of wind at 500 mb (roughly 18,000 feet, 5.5 km) in January and July. Heavy lines with arrows are selected contours (19,000, 18,000, and 17,000 feet, the last omitted for July) on the same surface; these are very nearly streamlines of the resultant wind, and hence show the prevailing direction of flow. Ridges and troughs in the mean topography are also shown. Note great contraction between winter and summer. From Lahey, Bryson, Wahl, Horn, and Henderson, *Atlas of 500 mb Wind Characteristics for the Northern Hemisphere* (Madison: The University of Wisconsin Press; © 1958 by the Regents of the University of Wisconsin).

control is at work. The most likely view is that they are produced by orographic interference, especially by the barrier effect of the central Asian plateaus and the American western cordillera, reinforced by inequalities in heating and cooling. Several attempts have recently been made to prove these hypotheses by mathematical techniques (Charney and Eliassen, 1949; Smagorinsky, 1953).

The climatological picture tends to show a single westerly belt, with a center of maximum speed somewhere between 25° N and 40° N, in the upper troposphere. Actually, there are often two or three separate maxima within the westerlies. The careless use of such terms as "*the* polar front" and "*the* jet stream" betrays an unfamiliarity with the complexities of the daily weather map. However, this question is better discussed in connection with the vertical structure of the westerlies.

The traveling cyclones and anticyclones of mid-latitudes, and the frontal and air-mass movements associated with them, are largely governed in their behavior by the eastward movement of wave disturbances in the upper troposphere. Petterssen (1950) has carried out a statistical study of the frequency and rate of alternation of cyclones and anticyclones, based on the United States Weather Bureau historical weather-map series. This study, though necessarily imperfect in some areas because of poor analysis on the original maps, enabled him to define an average *perturbation duct* for the westerlies, from which anticyclonic eddies tend to diverge toward the tropics, and cyclonic eddies toward the pole.

VERTICAL STRUCTURE

It is in the analysis of vertical structure that meteorological techniques have made most rapid strides in recent years. The cross section, merely a research tool in the prewar period, has become a vital part of three-dimensional analysis. The writer and his colleagues draw such cross sections daily along the 80° W meridian from Balboa (Panama) to Alert (Ellesmere Island). These sections extend from sea level to 10 millibars (about 100,000 feet). Numerous mean cross sections have been prepared for the Northern Hemisphere in recent years, either by averaging over all longitudes or for specific meridians. All these cross sections, like the daily ones, are based on radiosonde or rawinsonde ascents from the impressive network of stations now available.

From such studies it emerges that the westerlies contain one or more velocity maxima in the upper troposphere, which are called *jet streams* from their likeness to fluid jets. Each jet is associated with a strong poleward temperature gradient in the layers below it. Climatological cross sections, like the familiar world temperature maps of the textbooks, show the temperature contrast between pole and equator to be diffused into a broad gradient extending over 60 to 70° of latitude. In daily analysis, however, one finds that most of this gradient is concentrated in the narrow belts below the westerly jets, the diffuseness in the mean field being due to oscillation in latitude. Beside the jets the tropical and arctic air masses extend with little if any regional temperature gradients. This remarkable and still imperfectly explained tendency was expressed by Bjerknes in the idea of the polar front, the central concept in the pre-World War II Norwegian analytical methods. That concept, however, has been oversimplified, and the term needs re-examination. It is sufficient here to state that (1) sharp concentrations of the poleward gradient are the common rule; (2) these concentrations lead to a rapid increase of westerly flow with height leading to jet streams in the upper troposphere; and (3) the kinetic energy of the entire westerly vortex is derived from the existence of such concentrations, which permit a conversion to kinetic energy of the potential energy due to unequal radiative heating or cooling.

In winter, there are habitually two or three jets. The southernmost, averaging about 30° N, has become known as the *subtropical jet* and is detectable only in the winter half year. It is centered at about 200 millibars (about 40,000 feet) and is associated with moderate northward temperature gradients confined largely to the upper troposphere. The other jet or jets occur in higher latitudes (but usually south of 50° N, except over the semipermanent ridges immediately west of Europe and America), and are associated with temperature gradients extending down to sea level, which constitute the familiar fronts of the surface weather map. In the longitudes of the semipermanent troughs—that is, over eastern North America and east of Japan—the jets are driven into lower latitudes and are often combined into a single broad westerly maximum. The Japanese trough has the most intense jet of all; winds of more than 200 knots are commonplace in the 30,000–40,000-foot layer.

In summer, when the westerlies are constricted into a much narrower latitudinal belt, a single, narrow jet is the commonest structure. Figure 2 gives a cross section along 80° W for 0000 hours Greenwich civil time 31 July 1958. It shows the jet core (the central point of the jet) in latitude 45° N, at 225 millibars (37,000 feet). Below it there is a single zone of strong temperature gradient; on both sides there is very little gradient. The mean cross section for July 1958 (Figure 2), shows the more diffuse structure that results from oscillations in position of the jet system.

The vertical extent of the westerlies in the stratosphere has until recently been little known. It is certain, however, that they decrease in speed with height above the boundary between troposphere and stratosphere, the *tropopause*. Actually there are spatially two tropopauses. North of the jets the arctic tropopause lies at 250 to 300 millibars (30,000 to 35,000 feet), but on the tropical flank there is a much higher tropopause close to 100 millibars (about 53,000 feet) throughout the year. The two tropopauses often fail to meet, the jet core occurring in the gap between them, as in Figure 2. Some analysts identify a third tropopause, in the mid-latitudes intermediate between these systems. Above the arctic tropopause, and immediately north of the jet or jets, is a deep warm layer in the stratosphere. In summer, as Figure 2 shows, this warm layer is obscured because the entire polar stratosphere is warm; there is then a general equatorward gradient. But in winter the warmth north of the jet separates a cold arctic stratosphere from an almost equally cold tropical stratosphere; temperatures between 30,000 and 70,000 feet are highest at about 55° N. Figure 3 shows the typical structure.

The effect of these distributions is to create a north–south temperature gradient in the stratosphere above the Ferrel westerlies. Hence they must decrease with height (for reasons to be discussed later). In summer (Figure 2) they extend to about 70,000 feet, above which winds become easterly. In winter their upward extent is

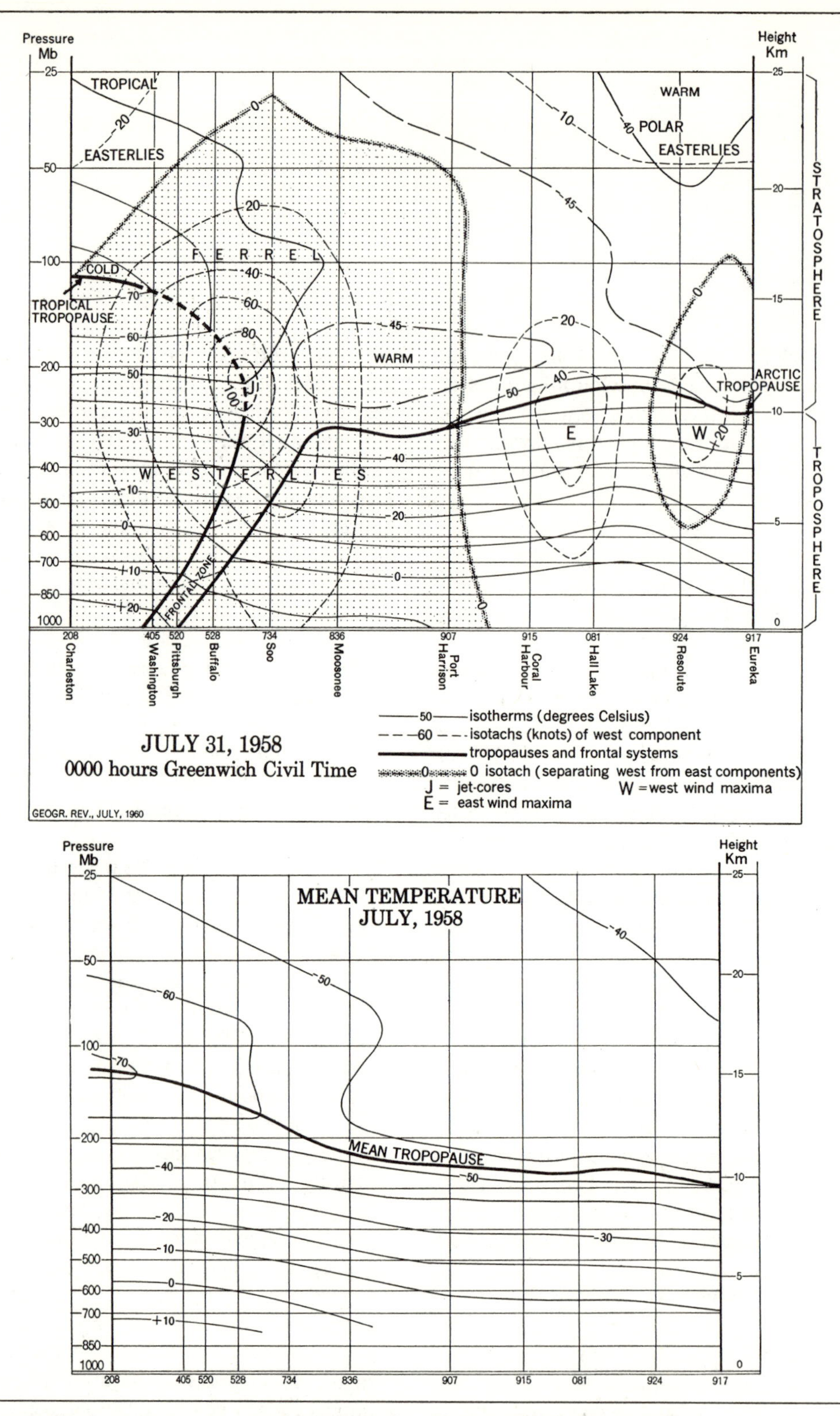

Figure 2 The westerlies in summer: cross sections along the 80° W meridian for 31 July 1958, and for July 1958, as a whole, from Charleston, S.C., to Eureka (80° N). Upper section shows the sharp front separating tropical from arctic air on 31 July, with corresponding jet stream. The vertical and meridional extent of the Ferrel westerlies is shown by shading. Wind speeds are indicated for west-to-east component in knots (negative values indicate east winds). The easterly and westerly currents over the Canadian Arctic are on either side of a sharp ridge centered on about 70° N. Lower section shows the more diffuse thermal gradient in the troposphere typical of a mean cross section because of rapid north-south oscillation of the frontal zone and jet stream.

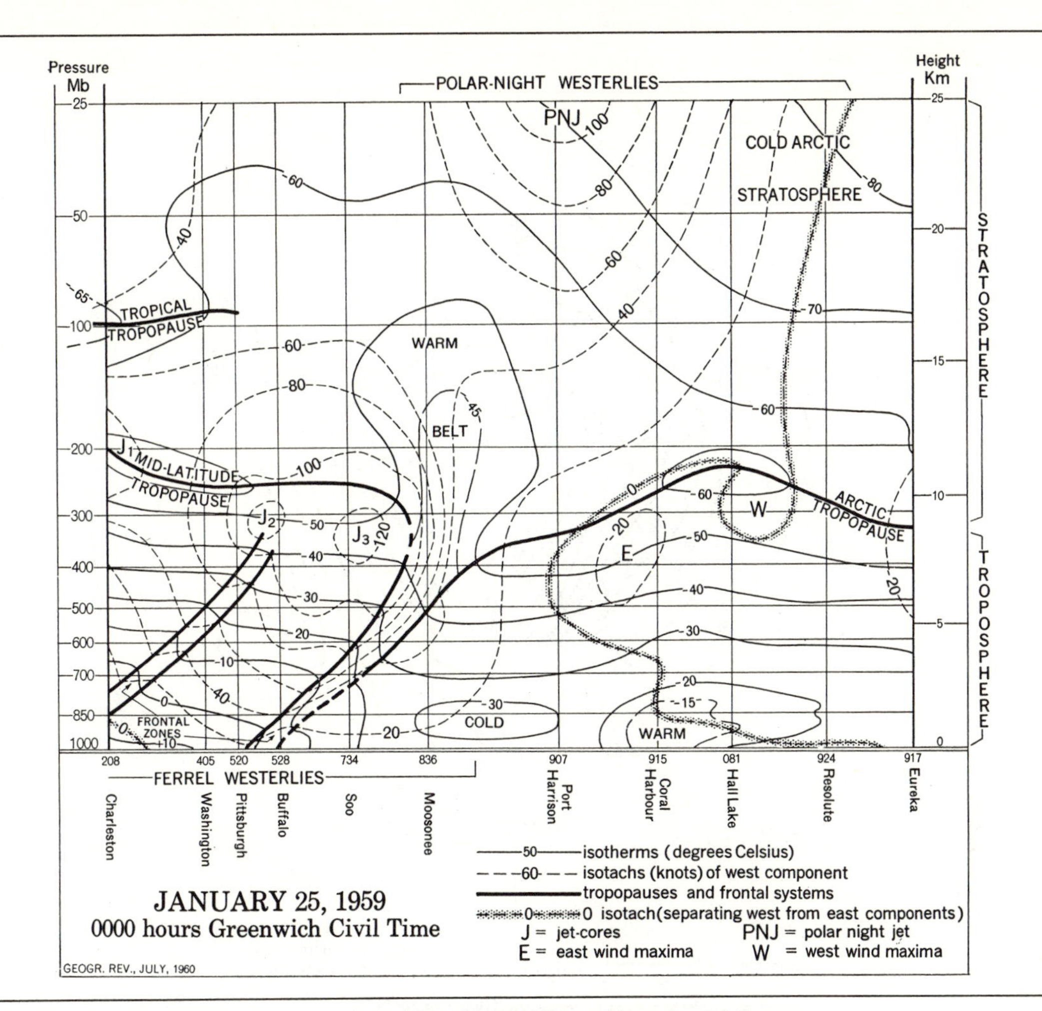

Figure 3 The westerlies in winter: cross section along the 80° W meridian for 25 January 1959. The Ferrel westerlies extend from ground level to above 25 mb, and have three jets on this date. J_1 (south of the section, centered over Key West, Fla.) is the subtropical jet. J_2 and J_3 are mid-latitude jets, each associated with a frontal zone below. The polar-night westerlies of the stratosphere are centered about 60° N, the jet core (PNJ) being above the top of the section. It is clear that the two westerly vortices are distinct, and are separated as usual by a warm belt in about 50° to 55° N in the lower stratosphere.

variable but is usually much deeper, reaching above the limits of direct balloon observation. At both seasons the level of maximum speed is close to the tropopause, where the meridional temperature gradient reverses.

THE POLAR-NIGHT WESTERLIES

A recent discovery of great interest is the presence of another westerly vortex in the arctic stratosphere. This vortex has been called the *polar-night vortex*, to distinguish it from the Ferrel westerlies. During the long polar night the arctic stratosphere (essentially north of 65° N) cools to very low temperatures, often −80° C, in the layers above 70,000 feet. Around the margin of this cold area strong westerlies develop; once again the temperature gradient tends to be concentrated in a narrow belt, and polar-night jet streams occur at levels above 80,000 feet. A good example is shown in Figure 4. It has been customary to assume that the cooling of the arctic stratosphere is due to radiation, but it has recently been shown that the onset of the cold, and with it the polar-night westerlies, comes on different dates in successive years. Similarly, the cold polar-night conditions end almost impulsively, with sudden warmings of more than 30° C in 48 hours. Throughout the winter, in fact, these stratospheric westerlies are subject to large-scale oscillations in temperature, with a characteristic period of about 20 days (Hare, 1959; Godson and Lee, 1958–59). These changes are probably due to vertical motion in association with traveling waves distinct from those which affect the Ferrel westerlies below.

For much of the winter the polar-night westerlies are separated from the Ferrel westerlies by a distinct gap, which shows up on Figure 3; they are then clearly

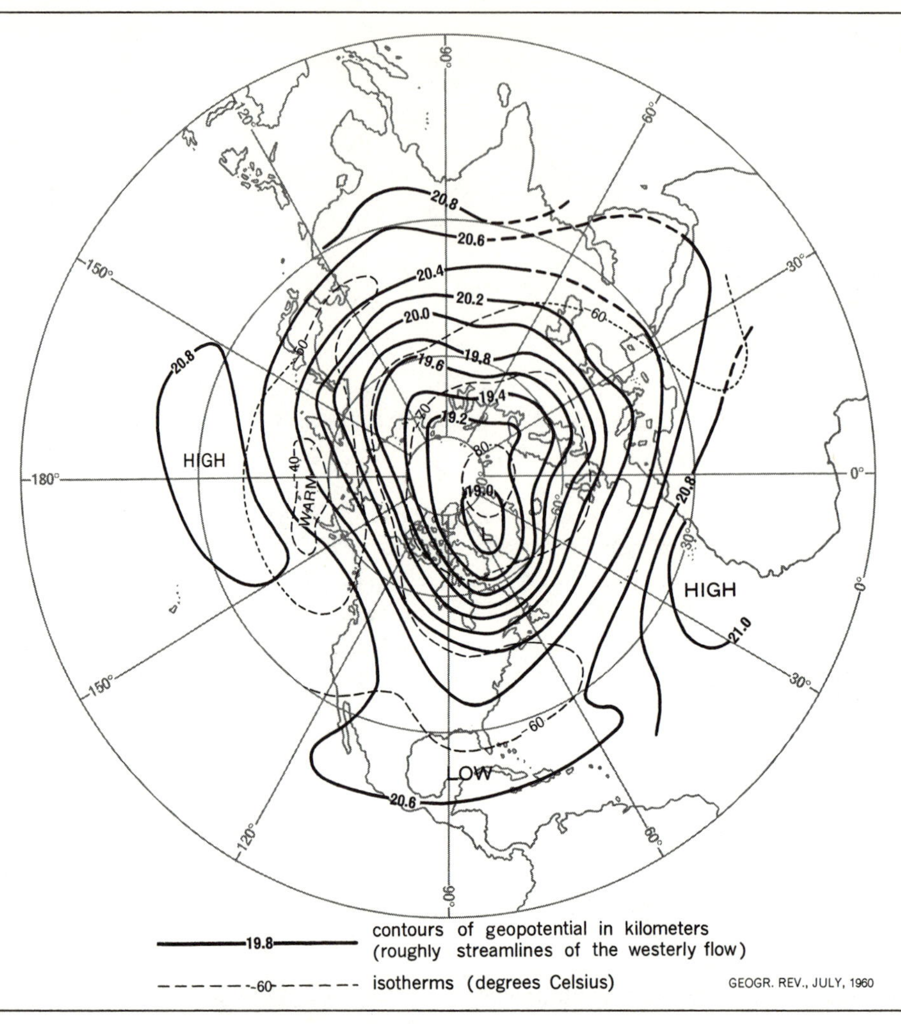

Figure 4 The polar-night westerlies: chart for 50 mb surface, 8 January 1958, 0000 hours, GCT. The chart shows the intense westerly vortex centered near the pole that develops in the middle stratosphere in the polar-night area. The stratosphere over the Aleutians is 40° C higher than that over north Greenland, and the part of the map over Alaska (where wind speëds are about 90 knots) is intensely baroclinic. The vortex is much stronger at greater heights.

independent, and of little interest to the student of surface climates. Late in the winter, however, the two vortices may interact; there is some evidence that the sudden breakdown of the polar-night westerlies in late February or early March may set off the primary index cycle of early spring at lower levels. At present, however, this remains speculative.

SOME DYNAMICAL CONSIDERATIONS

It is impossible adequately to review the theoretical development that has accompanied these exploratory achievements. Certain points, however, require brief mention. First, it is well known that the geostrophic-wind equation, and its extension to greater heights through the thermal-wind equation, are valid for much of the observed flow of the westerlies (see Appendix). The latter equation requires an increase of west-wind component with height whenever there is a horizontal temperature gradient toward the north, and a decrease with a gradient toward the south. Hence an increase of west winds with height to a maximum near the tropopause is a simple consequence of the poleward temperature gradient normally found in the troposphere. It follows that if the poleward temperature gradient is concentrated into a narrow, zonal polar front, as Bjerknes maintained and as observation seems to show, there must also be a narrow belt of strong westerlies above the front in the upper troposphere—the jet stream. In other words, the existence of a polar front requires the existence of a circumpolar jet stream, which was thus deducible from theory before it was observed in practice. Furthermore, the fact that there are often two or three jets crossing the same meridian implies that there is not merely one front at all times, but sometimes two or three (though the front corresponding to the subtropical jet does not normally extend down to ground level).

These facts have become commonplaces among meteorologists. Hydrodynamical theory has provided unifying concepts which enable us to see that things we thought separate are really single entities or systems. The theory itself is old; its relevance, however, has become increasingly clear only in recent years. We understand today that fronts, air masses, jet streams, and unstable disturbances are all aspects of the *baroclinic* structure of the westerlies. A baroclinic fluid is one in which surfaces of constant pressure and constant density intersect. This is virtually equivalent in the present case to saying that air containing a horizontal temperature gradient is baroclinic. From the thermal-wind relationship it follows that the wind must vary with height in the baroclinic state. The complementary state, in which temperature (and hence density) is horizontally uniform, is called *barotropic* and is associated with a constancy of wind with height. Hence from the definition of terms a front is a strongly baroclinic zone, an air mass is barotropic. Nothing is gained, of course, by simply attaching new names to old ideas, but this is not what has happened here. A large body of hydrodynamical theory refers to the baroclinic and barotropic states. Instead of applying rules of thumb to the crude air-mass and frontal concepts of yesterday, we now attempt to apply the baroclinic and barotropic theorems to the entire three-dimensional field of the atmospheric elements. The concept of air masses was a brilliant approximation necessary in an age when upper-air measurement was extremely difficult. Fronts, on the other hand, are unmistakable realities. They are not zero-order discontinuities of density, as they were once regarded. They are zones of very strong temperature gradient, and they are much more complex in structure and properties than many textbook accounts suggest.

Above a shallow surface zone the arctic troposphere is nearly always barotropic over huge distances; so also is the tropical troposphere. In between, the westerlies are necessarily baroclinic. It follows, again from the definition of terms, that the traveling air masses of the daily weather map must originate in either arctic or tropical latitudes, whence they are injected into the westerly stream. Only in a crude and superficial sense are air masses created over "source regions"; arctic air, for example, is remarkably uniform above the surface inversion and shows only minor differentiation over land and ice. This homogeneity, first stressed by Flohn (1952), is probably due to dynamical cooling in the so-called "cold lows" of high latitudes followed by horizontal stirring by drifting highs and lows. Similarly, tropical air is amazingly uniform at levels above 10,000 feet, at least on any one meridian. On the other hand, both arctic and tropical air masses, when injected into the westerlies, lose their barotropic structure rapidly, because of dynamical changes (due to uplift or subsidence) and surface heating or cooling. This accounts for the fact that the air masses of the daily weather map in our own latitudes are nonuniform (that is, baroclinic) and fail to comply with textbook definitions. It is best to think of the westerly belt as made up of weakly baroclinic air streams separated from one another by the strongly baroclinic belts below jet streams, some of which may display the discontinuity properties of fronts (Godson, 1951).

A further point of interest arising from dynamical theory is that vertical motion must occur above, below, and on each flank of jet streams, especially where speed along the streams varies rapidly. This vertical motion affects both the lower stratosphere and the upper troposphere, and it may have important climatic effects. For example, the warm belt in the stratosphere north of the jet-stream belt is probably due in large part to heating through subsidence, though such motion is not continuous. Unfortunately, the nature of the effects cannot be adequately discussed without an extensive foray into theory. A related, though independently derived, concept is that of *dynamic stability*; it has been shown theoretically and empirically that the rate of decrease of the west wind as one travels south from the jet axis—actually along a surface of constant potential temperature rather than horizontally—cannot in the stable condition exceed the local value of the Coriolis parameter (the product of twice the rate of rotation of the earth by the sine of the latitude).

WAVES IN THE WESTERLIES

In 1939, Carl-Gustaf Rossby and his associates pointed out, in a brilliant and history-making paper, that the

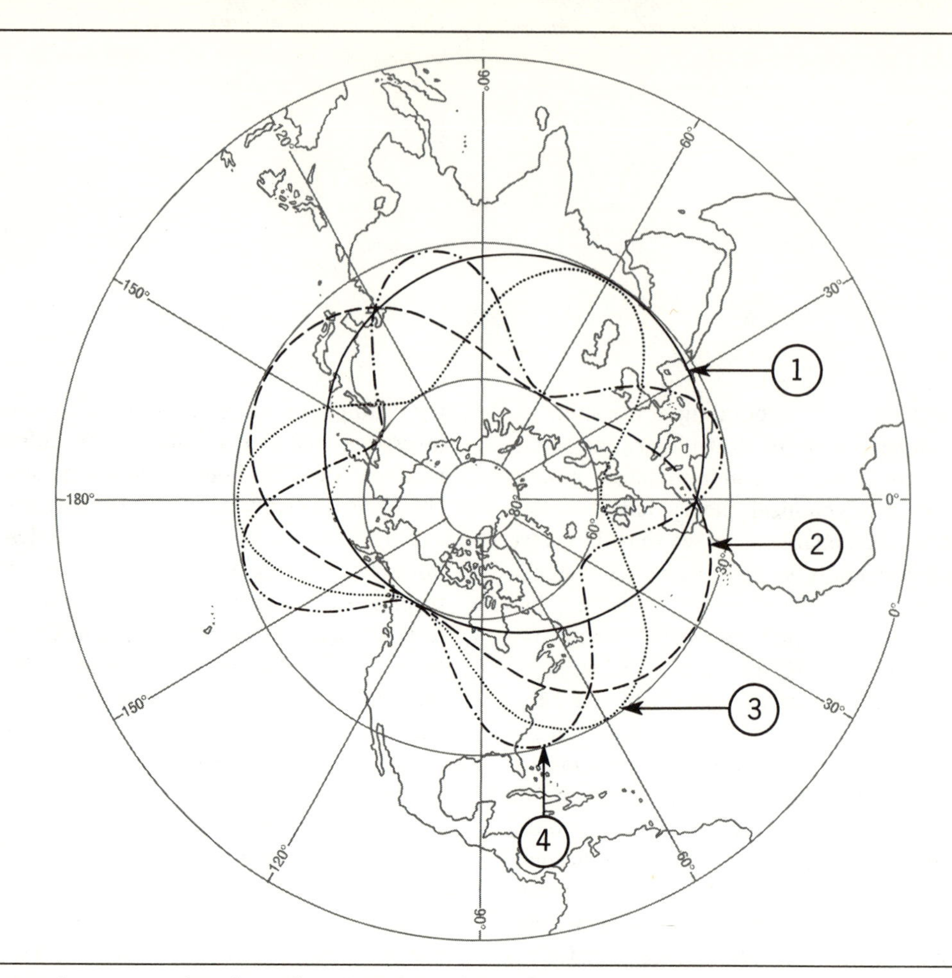

Figure 5 Circumpolar wave trains of numbers 1 to 4 are shown for an amplitude of 15° of latitude, with an origin at 120° W. Number 1 is simply an eccentric circle. Waves of this number are long Rossby waves and move only slowly. Shorter waves (numbers 5 and up) move rapidly east. (After Boville and Kwizak.)

troughs and ridges of the circumpolar westerlies constituted roughly sinusoidal waves, some of which—essentially those of long wavelength—were slow-moving, others fast-moving toward the east. Making use of the concept of vorticity (the kinematic quantity measuring the rate of spin of small fluid elements, in this case about the vertical axis), Rossby developed a means of predicting this motion as a function of wavelength and the speed of the zonal (that is, basic westerly) current. His original treatment was much restricted by simplifying assumptions, but it has since been generalized. In particular, it has been found possible to develop criteria for the stability—or instability—of baroclinic waves. An unstable wave is one whose amplitude grows with time—the familiar phenomenon of "deepening." It has been shown by Charney (1947), Kuo (1953), Fleagle (1955), and others that instability occurs in baroclinic flow whenever certain values of the change of wind with height are exceeded, the values depending also on wavelength and static stability. In other words, above some critical intensity of the poleward temperature gradient or baroclinicity, the westerly current becomes unstable, and small disturbances tend to grow.

These ideas, that the circumpolar westerlies consist of a series of wave trains of various wavelengths superimposed on a basic westerly current and that instability in such waves depends on the rate of increase of the westerlies with height, have become the dominating concepts of modern synoptic practice. Numerical weather prediction, now a concrete reality, depends on a further extension of the theory, and on various simplifying assumptions incorporated into what are called *models*. The eastward travel of low-level weather disturbances, other than small-scale phenomena such as thundershowers, is now thought of as being governed by the motion of the shorter waves of the higher-level westerlies, and numerical prediction aims at predicting this motion; hence it is applied, not at sea level but at 500 millibars, where the flow tends to be nondivergent.

The wave trains are readily identifiable on circumpolar weather maps. Figure 5 shows how the differing wave numbers (number of complete waves along a

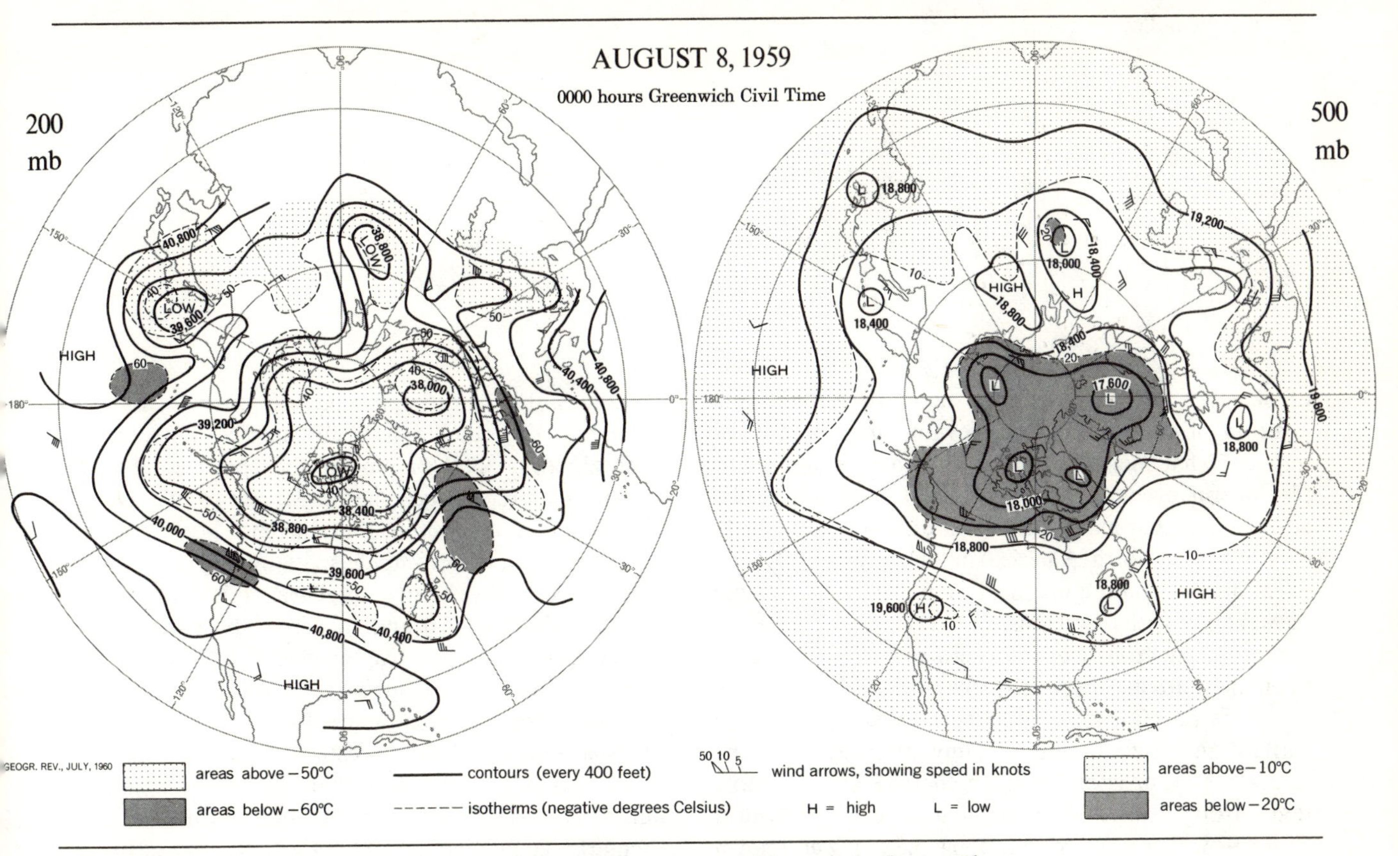

Figure 6 These synoptic charts show a typical summer pattern, with westerly flow weak and confined to latitudes north of about 30° N. The mid-tropospheric map (500 mb, about 18,000 feet) shows a cold arctic cap (areas below −20° C) and a warm tropical area (above −10° C), both highly uniform; in the westerlies, troughs are cold, ridges warm. The 200-mb map (about 40,000 feet) has a warm arctic area (above −50° C), and the coldest parts (areas below −60° C) are on ridges in the westerlies. At 200 mb the arctic area is in the stratosphere. The waves in the westerlies are plainly visible. Over Siberia the westerly stream is split into two, whereas over North America it is a single stream. From the geostrophic-wind law (see Appendix) we may expect winds to flow along the height contours at a speed proportionate to the contour spacing. A few observed winds have been added to convey the sense of the motion.

latitude circle) can be identified visually. Figure 6 shows examples at 500 millibars and 200 millibars; the fact that the two maps are closely similar emphasizes that the more important waves extend without significant tilt through the entire depth of the westerlies, in spite of the changes in strength of the wind. Even the fastest-moving waves move more slowly than the wind, which blows through them. At 500 millibars, in mid-troposphere, the troughs are cold and the ridges warm, as a consequence of vertical motion. At 200 millibars the vertical motion (and hence temperature) behaves in the opposite way, as Figure 6 makes clear.

By means of Fourier analysis, it is possible to achieve a remarkably complete and objective description of these waves, and to show how much of the kinetic energy of the westerlies is associated with each wave number. One such description was recently published by Eliasen (1958) for the period 21 October to 30 November 1950. He found that south of 50° N most of the total energy of the westerlies was contained in the undisturbed zonal flow, whereas at higher latitudes most of the energy was in the waves. About equal amounts of energy were contained in the long quasi-stationary waves (with wave numbers 1, 2, 3, and 4) and in shorter, eastward-moving waves. Wave number 1 is, of course, a measure of the eccentricity of the westerly vortex. Using another technique, La Seur (1954) showed that the "circumpolar" westerlies actually rotate about a centroid (approximately equivalent to a center of gravity) some degrees away from the geographical pole on the 170° W meridian, and that the eccentricity is especially marked at times of strong zonal flow. Thus the circulation pole, like the magnetic pole, departs considerably from the geographical pole.

The relationship of these waves to weather events near the ground is highly complex, and only the briefest of sketches can be attempted here. In general, the position of the standing or slow-moving long waves imposes what might be described as a persistent complexion on events at lower levels. For example, the strong ridge that maintained itself almost continuously over western North America in the summer of 1958 gave nearly continuous heat, and in many places drought, to western regions, and at the same time the compensating trough over eastern regions of the continent permitted a high frequency of polar outbreaks over central Canada, with corresponding cyclonic activity and a cool, cloudy, rainy summer. Thus the ridges and troughs of the mean chart (Figure 1) imply that strong longitudinal differences exist in world climate—which is, of course, confirmed by well-known facts. The position of these "forced" perturbations is influenced by the mountain barriers of western North America and central Asia. Hence any profound diastrophic changes in the earth's crust must alter the world climatic map and may even, by altering the efficiency of horizontal mixing, change the temperature difference between pole and equator, which in turn must influence the strength of the basic westerly current. The study of such relationships has been expedited by the work of the Extended Forecast Section of the United States Weather Bureau under Namias (1951).

The shorter-wave systems, which regularly progress toward the east, are associated in general with the familiar traveling highs and lows of the surface weather map. Thus the eastward-moving troughs tend to accompany cyclones at lower levels, the trough line characteristically lagging behind the cyclone center; similarly, the westerly ridges accompany anticyclones or ridges at lower levels, again with a phase lag. Unfortunately, the theory linking cyclone development, traditionally regarded as an effect of frontal instability (that is, of the stability of disturbances on the frontal surface between differing air masses), to wave development in the overlying westerlies has remained complex, though it has been greatly clarified by the treatments of Sutcliffe (1947) and Petterssen (1956). At present, we depend mainly on rules of thumb derived from these theories rather than on the theories themselves, and the prediction of cyclone deepening and filling is very difficult.

It remains to discuss the so-called *index cycle* of the westerlies, much used in extended-range forecasting (Namias, 1950). Since the Rossby-wave formula and many of the other dynamical theories of recent date contain as a vital parameter the speed of the basic westerly current, it becomes important to find a way of defining and computing that current. The circulation-index method is to compute spatially averaged pressure differences between two parallels of latitude (for example, 35° and 55°) and to derive the corresponding mean flow by the geostrophic law. The Extended Forecast Section of the United States Weather Bureau has computed several such indices for many years. Though they are only generalized estimates of the zonal flow, their use has made clear that the strength and geographical extent of the Ferrel westerlies are subject to remarkable fluctuations, with a wide variety of periods. These fluctuations are distinct from the seasonal variations already discussed. In general, when the zonal current is strong, it tends to be restricted to high latitudes. When it is weaker, it tends to be more extensive, and more complex in form. Riehl and others (1954) have described cyclone paths, intensity of meridional exchange, and other phenomena characteristic of both northward and southward migrations of the main jet or jets. At certain times the wave structure of the westerlies becomes so complex as to permit the cutting off of ridges to the north of the main course, the so-called *blocking highs*, and of closed lows in lower latitudes. The climatic effects of such aberrations have been much discussed, and Willett (1950) and Rex (1951) have constructed plausible hypotheses of European climatology and of Pleistocene climatic change on the basis of the index cycle and the blocking phenomenon.

Unfortunately, there is at present no adequate dynamical theory of the index cycle, and meteorologists are divided about the regularity of development ascribed to it. However, there is no doubt that there are extraordinary fluctuations in the strength and pattern of the Ferrel westerlies, or that these fluctuations are related, as a species of gross turbulence, to the exchanges of momentum, heat, and water vapor between the subtropics and the polar caps.

In fact, the entire disturbed behavior of the westerlies can be looked on simply as a tranfer process by which a net northward flux of heat, water, and momentum is achieved. It is well known that the large tropical excesses of these three physical elements have to be carried pole-

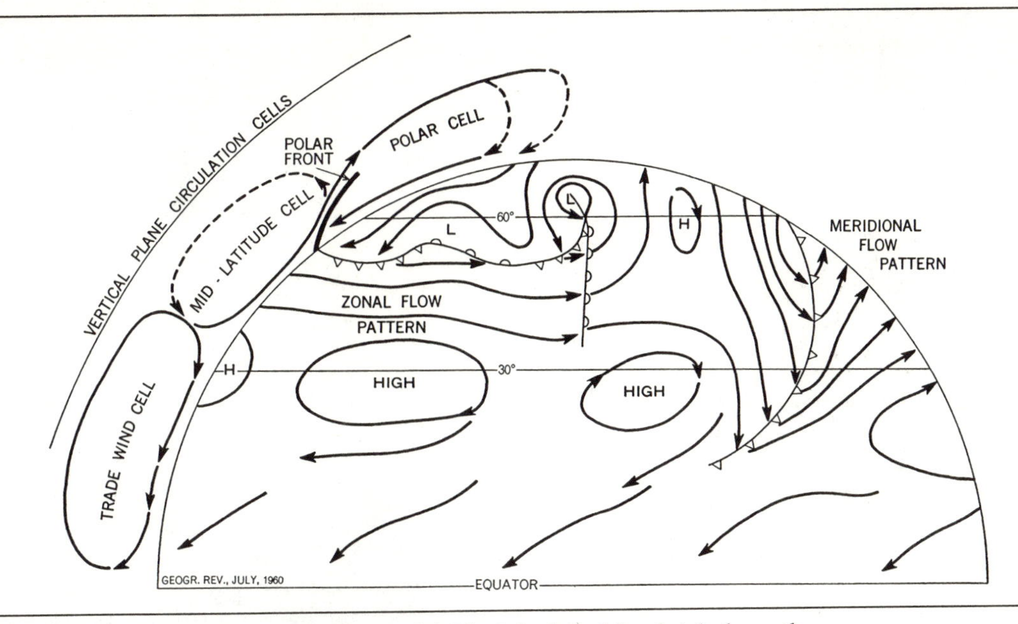

Figure 7 The discredited meridional circulation model. The left edge of the sketch shows the three meridional-plane circulation cells once thought necessary to convey the tropical excess of heat, moisture, and angular momentum toward the polar sink. It is now thought that only the trade wind (or Hadley) cell actually exists. North of the subtropics the traveling westerly waves, cyclones, and anticyclones bring about the required transport by means of predominantly horizontal stirring, especially in areas of strong meridional flow (right). There may be a minor polar front cell just north of the subtropical jet, but this is denied by most authorities.

ward by the atmospheric circulation (with some help from the ocean). Until recently it was assumed that this movement required standing meridional circulation cells in the vertical plane, of the sort still described in school textbooks of geography. Three such cells were envisaged: a trade-wind cell, a mid-latitude cell, and a polar cell (Figure 7). But subsequent observation failed to confirm the existence of the two northern cells, and an alternative solution had to be sought. In the late 1940's three students of the general circulation, Bjerknes at the University of California at Los Angeles, Starr at the Massachusetts Institute of Technology, and Priestley in Australia, revived a brilliant, half-forgotten paper by Jeffreys (1926), who showed that the required transport could be achieved horizontally by means of the stirring by traveling, large-scale eddies—the familiar highs and lows of the daily weather map. It is now generally conceded that this is how the northward flux is maintained in the westerly belt, though the trade-wind cell remains a necessity.

One other development in the climatological study of the westerlies has been the quantitative study of the sources of precipitation within the belt. There is probably no other aspect of climatological writing in which more confusion exists. A recent study by Benton and Estoque (1954) has provided a partial answer for North America. They computed, from aerological soundings at a large number of stations, the flux of water vapor into, out of, and across the continent. Their maps of the flux give a fascinating picture of the relative contributions of Pacific and Gulf air to the moisture budget of the continent. It is still not possible, of course, to assume that the rainfall necessarily falls from the fluxes in equal proportions; in particular, the authors find a large flux over some of the arid southwestern districts. Nevertheless, it is clear that studies of this kind must be multiplied until the source of the vapor utilized in the traveling cyclones of the westerlies is fully understood. Figure 8 is based on one of their characteristic flux maps.

THE WESTERLIES AND WORLD CLIMATE

From the standpoint of the climatologist, the growth in understanding of the westerlies has paid greatest dividends by providing a means of unifying his picture of the mid-latitude climates. The regional climatologies, though many of them have contained praiseworthy exceptions, have tended to put forward purely local and often obscure explanations of the dynamical background of climatic phenomena. Thus the relationship between, for example, the "polar-air depressions" of United Kingdom parlance, the "western disturbances" of India

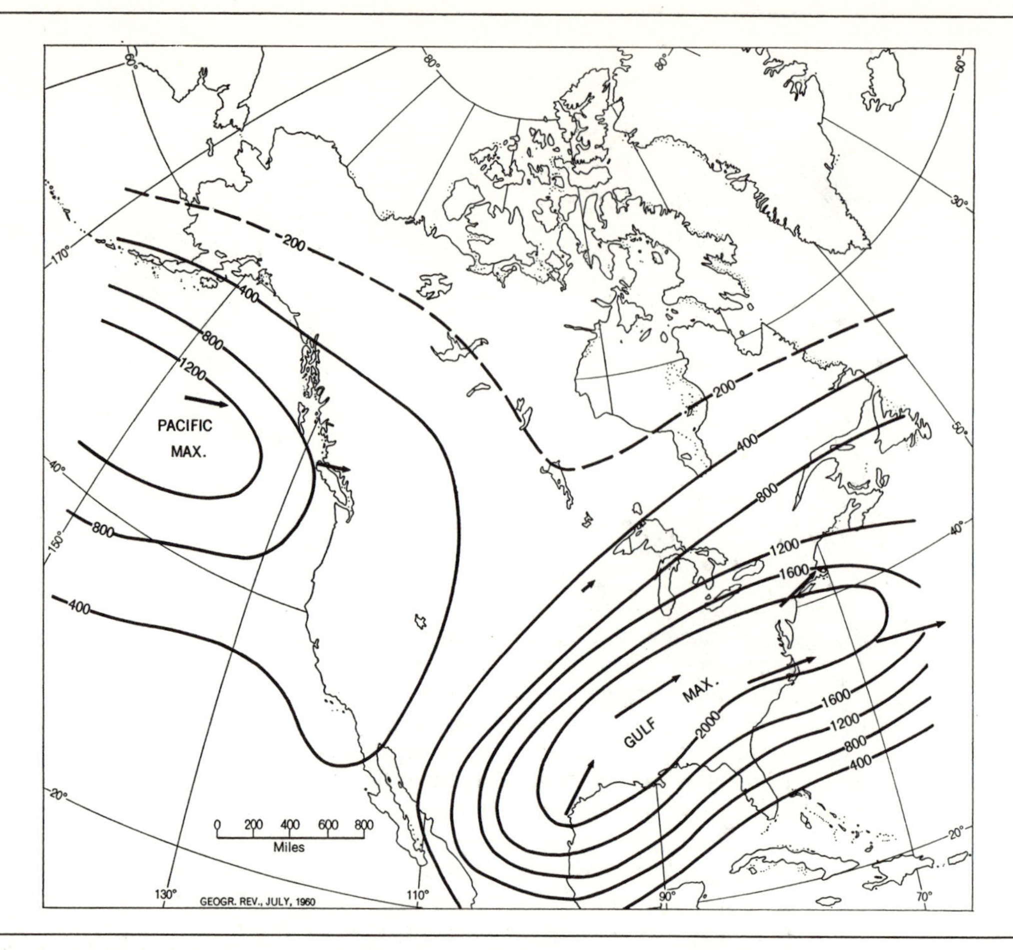

Figure 8 Net integrated water-vapor flux in the lower troposphere, winter months, 1949. The isolines show mean fluxes in grams per centimeter per second over the entire column of air from sea level to the 400-mb surface (about 23,000 feet). There are two centers of rapid flux, one over the Pacific in the main westerly stream, and the other over the southern states, in the form of an arc originating from the Gulf. The arrows show for selected stations the mean direction and magnitude of the vapor flux. The Gulf maximum is, of course, one of the principal points where humidity of tropical origin is injected into the westerlies. (After Benton and Estoque, 1954.)

and Pakistan and the Yangtse depressions of China was obscure thirty years ago. Today the obscurity has been largely removed, and it is possible to construct, at least tentatively, a hemispheric account that connects up these old *ad hoc* ideas. We are still far from a satisfactory picture, but we have advanced.

Nowhere has this been truer than in the Far East, where the prewar picture was obscure in the extreme. The writer had to attempt to penetrate this obscurity during the war, when he prepared, for military purposes, a series of studies of the circulation over Asia (Hare, 1945). This work, based on daily analysis of upper-air charts, convinced him that the so-called "monsoonal" climates were explicable only in terms of the behavior of the overlying westerlies, a view recently confirmed by Trewartha (1958). Since the war, first-class aerological networks have been established by China, Japan, and other Eastern powers. In the writer's stratospheric research the data available from China were found to be plentiful and reliable even for 100 millibars (53,000 feet) and above. Hence it is now possible to see in daily analysis how events over the East are connected with those farther west.

A valuable synthesis of Chinese experience in this field has been prepared by a group of staff members of the Academia Sinica (1957, 1958). In their account stress is laid on the seasonal variation in position and strength of the circumpolar westerlies. They show (as, of course, was less precisely known before) that the westerly stream is typically multiple-cored and is greatly deformed by the Tibetan plateau and the great plexus of ranges diverging from it. The westerlies over northern India in the cooler season appear abruptly south of the Himalayas in October and blow uninterruptedly until May, when they disappear with equal abruptness. The "western disturbances" of India are the shallow surface equivalents of eastward-moving troughs in these westerlies, whose course can be traced from far to the west. North of the upland complex lies the other main course of the westerlies; the eastward motion of short waves in these westerlies accounts for the cold surges of the "northwest monsoon," and for many of the Yangtse depressions. Even the Mai-yü rains of June in China and Japan are satisfactorily related to the baroclinic structure of the westerlies. This masterly paper tears apart many veils in world climatology.

Where, one might ask, do such efforts lead? The attempt to explain climatic patterns does not appeal to all students. There are theoretically minded meteorologists who doubt whether time-averaged systems are truly explicable, and there are geographers who can see no value to their own discipline in such genetic studies. Probably most climatologists would confine themselves to the simple statement that the effort to explain the world climatic pattern stimulates their curiosity and challenges their skill. The possibility that explanation will prove difficult will not discourage the bolder minds.

One thing is clear. There is a great need for a satisfactory *theory* of world climate—of *how* the incoming solar energy and the behavior of the atmosphere govern the observed climatic patterns. Without such a theory, it is difficult to see how one can judge hypotheses concerning past climates, in which so many scientists have a vital interest. This theory must comprehend the general circulation, the heat, moisture, and momentum balances, and the phenomenon of climatic change. Climatic change, in fact, illustrates the unity of climatology with the work and interests of the synoptic meteorlogist. For climatic fluctuations and variations are merely the longest-period oscillations of a continuous spectrum that starts with the fast-moving westerly wave, progresses through the slow movement of the long Rossby waves to the index cycle, and so to climatic variation. Our choice of time scale for climatology is conditioned more by the length of our life span than by logic. In the words attributed to L. F. Richardson, the inventor of numerical weather prediction,

> Big whirls have little whirls
> That feed on their velocity
> And little whirls have smaller whirls
> And so on to viscosity.

His rhyme epitomizes the endless variability in time and space of the westerlies. Which "whirls" one assigns to the synoptic meteorologist and which to the climatologist is a far more arbitrary choice than common usage suggests. In any case, all are of concern to the student of the earth, whether he is geographer or geophysicist.

APPENDIX

The strength of the west component of wind, u, is given approximately for any chosen constant-pressure surface by the geostrophic equation

$$u = -gf^{-1}(\partial z/\partial y) \qquad (1)$$

where g is gravitational acceleration (981 cm sec^{-2}); f is the Coriolis parameter, $2\omega \sin \phi$, where ω is the angular velocity of the earth's rotation and ϕ the latitude; and $(\partial z/\partial y)$ is the gradient of the constant-pressure surface toward north. A similar equation relates the south component to the gradient toward east.

The variation of the west component with height $(\partial u/\partial z)$ in a baroclinic atmosphere is given by the thermal-wind equation

$$(\partial u/\partial z) = -gf^{-1}T^{-1}(\partial T/\partial y) \qquad (2)$$

where $(\partial T/\partial y)$ is the gradient of temperature toward north, and T is the absolute temperature of the layer concerned. Again there is a comparable equation governing the variation of the south component with height $(\partial v/\partial z)$. If $(\partial u/\partial z) = (\partial v/\partial z) = 0$, the atmosphere is barotropic.

The kinetic energy of the west component of wind is $\frac{1}{2}u^2$ per unit mass of air. This energy, which is constantly dissipated by friction, is generated from potential energy due to the existence of horizontal temperature gradients through the agency of baroclinic waves.

The potential temperature of air is the temperature a parcel would attain if it were brought to a pressure of 1,000 millibars *without gain or loss of heat*. Hence in the free air (where radiative temperature changes are usually small) the wind tends to follow surfaces of constant potential temperature (called *isentropic surfaces*).

REFERENCES

Academica Sinica, 1957. On the general circulation over eastern Asia. *Tellus 9*, 432–446.

Academica Sinica, 1958. On the general circulation over eastern Asia. *Tellus 10*, 58–75 and 299–312.

Benton, G. S. and Estoque, M. A., 1954. Water-vapor transfer over the North American continent. *J. Meteor. 11*, 462–477.

Berggren, R., Gibbs, W. J., and Newton, C. P., 1958. Observational characteristics of the jet stream. *Tech. Note 19, WMO-No. 71 TP27*, World Meteorological Organization, Geneva.

Bjerknes, J., 1957. Kinetic energy of the atmosphere in summer. In *Large-Scale Synoptic Processes*, ed. J. Bjerknes. Final Rept., Contract AF 19(604)-1286, University of California at Los Angeles.

Charney, J. G., 1947. The dynamics of long waves in a baroclinic westerly current. *J. Meteor. 4*, 135–162.

Charney, J. G. and Eliassen, A., 1949. A numerical method for predicting the perturbations of the middle latitude westerlies. *Tellus 1*, 38–54.

Eady, E. T., 1957. Climate. In *The Earth and Its Atmosphere*, ed. D. R. Bates. 113–129, New York.

Eliasen, E., 1958. A study of the long atmospheric waves on the basis of zonal harmonic analysis. *Tellus 10*, 206–215.

Ferrel, W., 1889. *A Popular Treatise on the Winds*. New York, 155. (Ferrel's theory was first set forth in 1856. An essay on the winds and the currents of the ocean. *Nashville J. of Medicine and Surgery 11*, 277–301 and 375–389.)

Fleagle, R. G., 1955. Instability criteria and growth of baroclinic disturbances. *Tellus 7*, 168–176.

Flohn, H., 1952. Zur Aerologie der Polargebiete. *Meteorol. Rundschau 5*, 81–87 and 121–128.

Godson, W. L., 1951. Synoptic properties of frontal surfaces. *Quart. Journ. Royal Meteorol. Soc. 71*, 633–653.

Godson, W. L. and Lee, R., 1958–1959. High-level fields of wind and temperature over the Canadian Arctic. *Beiträge zur Physik der Atmosphäre 31*, 40–68.

Hare, F. K., 1945. Aviation meteorological report on South China. *M.O.M. 365/29*. Air Ministry, London.

Hare, F. K., 1959. The disturbed circulation of the arctic stratosphere. *J. Meteor. 17*, 36–51.

von Helmholtz, H., 1893. Ueber atmosphärische Bewegungen. Translated by Cleveland Abbe. *Smithsonian Misc. Colls. 34(843)*, 78–93.

Jeffreys, H., 1926. On the dynamics of geostrophic winds. *Quart. Journ. Royal Meteorol. Soc. 52*, 85–104.

Kuo, H. L., 1953. The stability properties and structure of disturbances in a baroclinic atmosphere. *J. Meteor. 10*, 235–243.

Lahey, J. F., Bryson, R. A., Wahl, E. W., Horn, L. H., and Henderson, V. D., 1958. *Atlas of 500 mb Wind Characteristics for the Northern Hemisphere*. Madison.

Lamb, H. H., 1959. The southern westerlies: a preliminary survey; main characteristics and apparent associations. *Quart. Journ. Royal Meteorol. Soc. 85*, 1–23.

La Seur, N. E., 1954. On the asymmetry of the middle-latitude circumpolar current. *J. Meteor. 11*, 43–57.

Namias, J., 1950. The index cycle and its role in the general circulation. *J. Meteor. 7*, 130–139.

Namias, J., 1951. General aspects of extended-range forecasting, in *Compendium of Meteorology*. 802–813, American Meteorological Society, Boston.

Petterssen, S., 1950. Some aspects of the general circulation of the atmosphere. In *Centenary Proceedings of the Royal Meteorological Society* 120–155, London.

Petterssen, S., 1956. *Weather Analysis and Forecasting*. New York (2nd ed., 2 vols.), *1*, 320–370.

Rex, D. F., 1951. The effect of Atlantic blocking action upon European climate. *Tellus 3*, 100–112.

Riehl, H., Alaka, M. A., Jordan, C. L., and Renard, R. J., The jet stream. *Amer. Meteorol. Soc. Meteorol. Monographs 2(7)*, 48–54.

Rossby, C. G. and others, 1939. Relation between variations in the intensity of the zonal circulation of the atmosphere and the displacements of the semi-permanent centers of action. *J. Marine Res. 2*, 38–55.

Smagorinsky, J., 1953. The dynamical influence of large-scale heat sources and sinks on the quasi-stationary mean motions of the atmosphere. *Quart. Journ. Royal Meteorol. Soc. 79*, 342–366.

Sutcliffe, R. C., 1947. A contribution to the problem of development. *Quart. Journ. Royal Meteorol. Soc. 73*, 370–383.

Trewartha, G., 1958. Climate as related to the jet stream in the Orient. *Erdkunde. 12*, 205–214.

Willett, H. C., 1950. The general circulation at the last (Würm) glacial maximum. *Geografiska Annaler 32*, 179–187.

J. S. Sawyer

4 Jet stream features of the Earth's atmosphere

At the University of Chicago in 1946 regular series of upper-air charts were drawn for the first time to cover the whole of the northern hemisphere. It was found that the upper westerlies were concentrated into a narrow meandering belt which encircled the earth in temperate latitudes and the term "jet stream" was introduced (Staff members of the Department of Meteorology at the University of Chicago, 1947). It is not clear from the original paper whether the term "jet stream" was intended to describe the whole system of meandering upper westerlies or merely local sections of them where the wind was particularly strong and was concentrated into a narrow belt with a well defined structure.

Since the original investigations at Chicago there have been many studies of "jet streams" but when we come to relate their findings and to interpret their theories for the existence of "jet streams", it is important to realize that all of the authors are not necessarily describing the same thing, for each may interpret the term "jet stream" in his own way. It does, indeed, seem likely that at least two, and possibly three, distinct phenomena are being described under the title "jet streams" and that each is set up by its own characteristic process.

Before attempting to discuss the causes of the jet stream, it will be useful to illustrate the various types of phenomena by means of charts relating to a particular occasion, namely 19 December 1953, for which appropriate data happen to be available.

It is perhaps worth noting here that the World Meteorological Organization has recently adopted a conventional definition of the "jet stream". This restricts the term arbitrarily to wind streams exceeding 30 m/sec (approx. 60 knots) but otherwise includes all narrow wind-currents in the upper troposphere (*W.M.O. Bulletin*, 1956).

Reprinted, with minor editorial modification, by permission of the author and editor, from *Weather*, Royal Meteorological Society, Vol. 12, 1957, pp. 333–344.

THE JET STREAMS OF 19 DECEMBER 1953

Figure 1 is a chart of the 500-millibar contours over most of the northern hemisphere. On this chart the shaded areas indicate where the wind speed exceeded 50 knots, either as observed or as estimated on the basis of the geostrophic wind equation. The areas with wind exceeding 100 knots are also shown.

The rather narrow circumpolar westerly current is clearly shown with two main southward excursions, one over the east coast of America and the other over east Asia. (For clarity in reproduction geographical features have been omitted from Figures 1, 3, and 4.) Apart from a small break over the Rockies, the wind exceeds 50 knots throughout, and there are small areas where it just exceeds 100 knots. To distinguish this large-scale aspect of the jet stream current, this meandering upper westerly current may be called the "circumpolar jet stream".

A second important area of strong winds is found extending across Africa and Asia in lower latitudes. This may be regarded as a subsidiary branch of the main "circumpolar jet stream" and could easily be overlooked in view of the scanty observations from the areas concerned. This current is shown more clearly on charts for higher levels.

A third feature which is of interest is the small area of strong wind over the British Isles. It is difficult to regard this as part of the main circumpolar jet stream, but it has most of the characteristics usually attributed to jet streams. In this paper, this is called a "local jet stream". There is also an isolated area of strong winds over Arizona.

Figure 2 is a cross-section across the circumpolar jet stream over America along the line AA′ in Figure 1, and it shows a typical wind- and temperature-distribution. Inspection of the isotherms shows that the jet stream lies above a region where temperature decreases rapidly northward along a horizontal surface and there

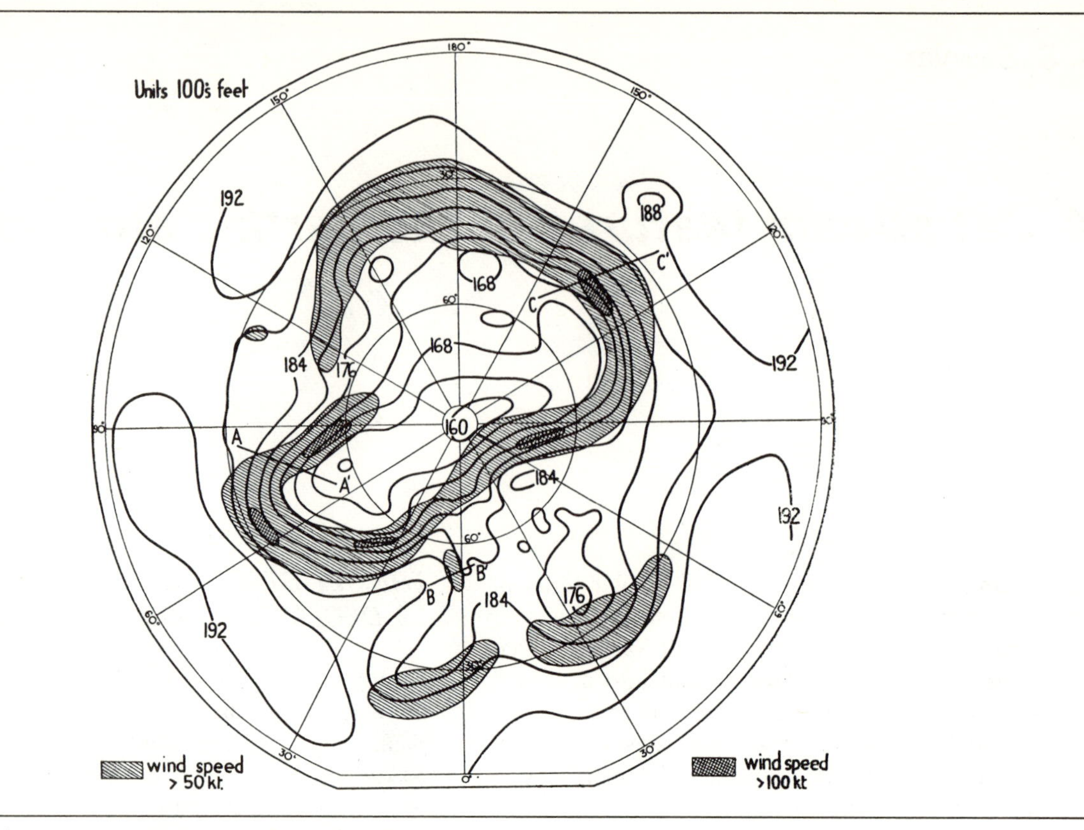

Figure 1 500 mb contours 0300 GMT 19 December 1953.

is a sloping frontal zone beneath the jet maximum in which the temperature change is particularly rapid. Such a horizontal temperature gradient must accompany the vertical wind shear in the jet stream if the flow is to be approximately geostrophic—this is a direct consequence of the dynamical equations. Only part of the horizontal temperature-gradient occurs in the sloping frontal zone which lies beneath the jet stream—there are also considerable horizontal changes of temperature in both the warm and cold air masses—i.e., in technical language these air masses are strongly baroclinic.

The maximum of the jet stream occurs a little below the tropopause. In this case the maximum of about 140 knots occurs at about 350 millibars. The tropopause level changes abruptly in the region of the jet stream from high levels around 220 millibars on the warm side to low levels around 300 or 400 millibars on the cold side—a change in level of around 12,000 feet in the present case.

Figure 3 shows the contours of the 300-millibar surface on the same occasion as that of Figure 1. This is near the level at which the "circumpolar jet stream" reaches its greatest speed. The general flow pattern is similar to that at 500 millibars, but the wind on the axis of the jet stream exceeds 100 knots over long stretches. The "local jet stream" over the British Isles appears as a less important feature. The strong flow over North Africa is more striking than at the 500 millibar level and exceeds 100 knots over two bands which probably exceed 1,000 miles in length; there is also a weaker continuation of this stream across Northern India. The winds over Arizona, which appeared rather anomalous at 500 millibars, now form part of a stream linking the Pacific with the Caribbean at 300 millibars.

Going higher again to the 200-millibar level (Figure 4), the same pattern is repeated; but clearly the emphasis has been transferred to the strong belt of wind which encircles the earth in subtropical latitudes. 50 knots is exceeded everywhere in this belt and there are extensive areas with winds over 100 knots. The 200-millibar level is above the maximum winds in the "circumpolar jet stream", and winds over 100 knots are mainly in the lower latitudes. The sub-tropical belt of strong winds clearly deserves consideration independently of the "circumpolar jet stream" and the title which is now often applied to it of the "sub-tropical jet stream" seems quite appropriate.

In the stratosphere temperature generally increases from equator to pole, the reverse of conditions in the troposphere so that the westerly winds tend to decrease with height. There is an important exception in the arctic regions in winter when the arctic stratosphere becomes colder than the stratosphere in middle latitudes. The westerlies thus tend to increase again with height.

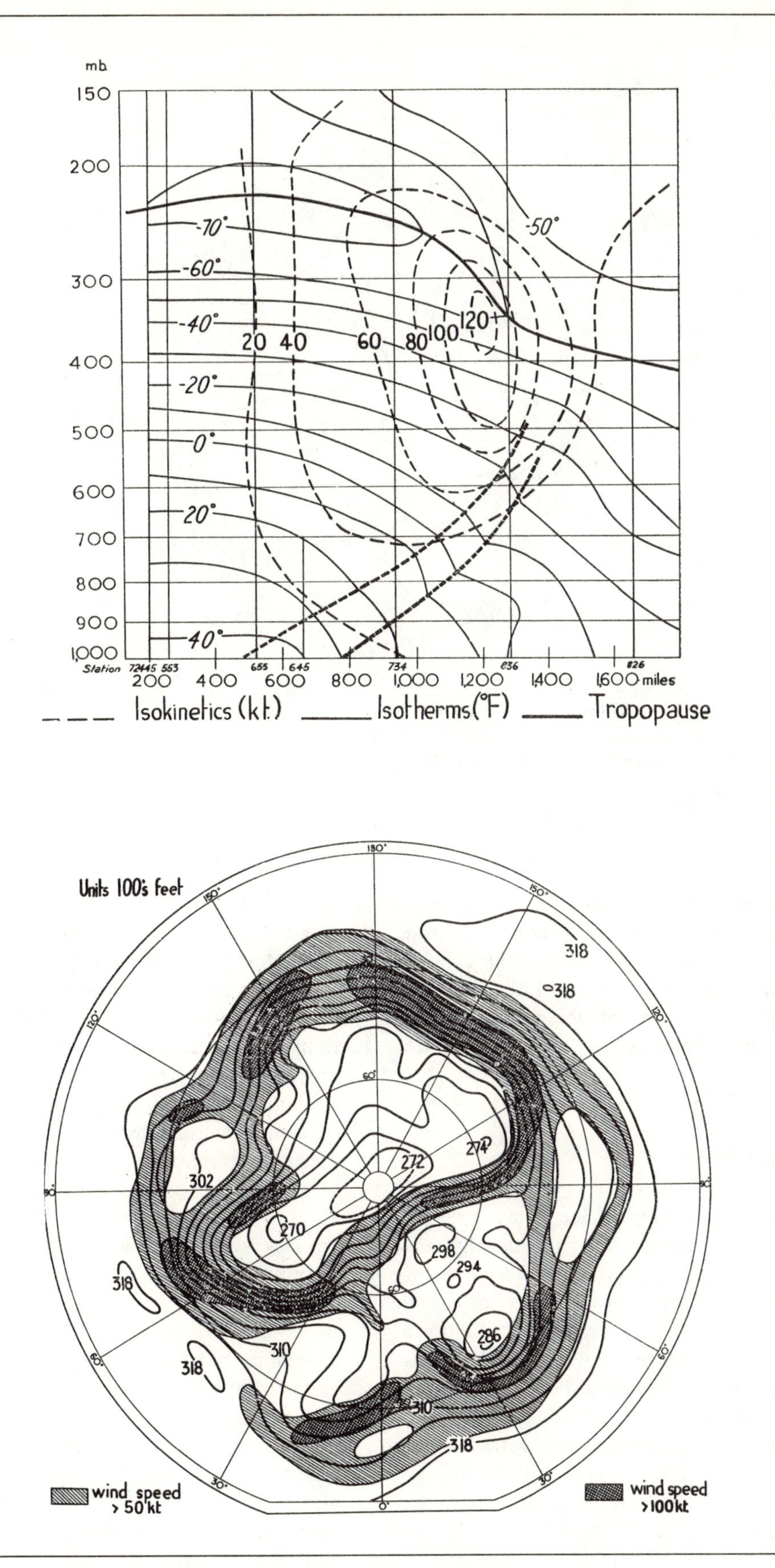

Figure 2 SW-NE cross section of jet stream over North America 0300 GMT 19 December 1953.

Figure 3 300 mb contours 0300 GMT 19 December 1953.

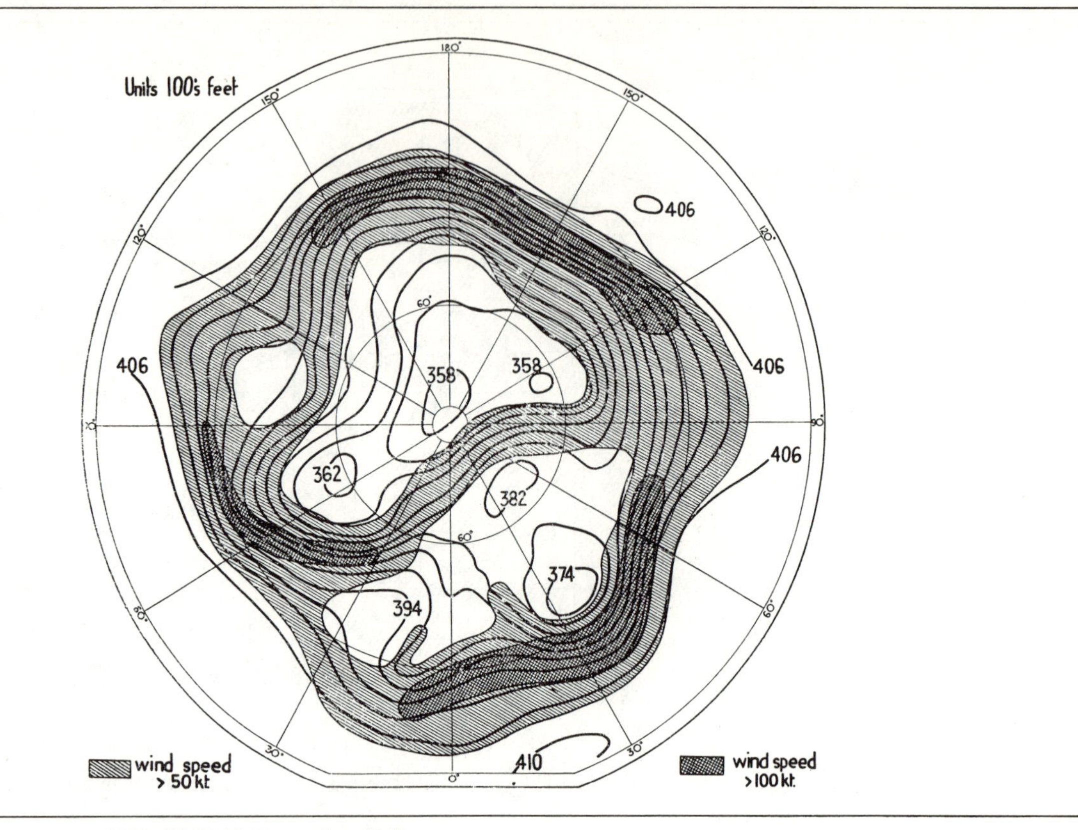

Figure 4 200 mb contours 0300 GMT 19 December 1953.

At 50 millibars there is probably often a belt of strong upper westerlies of 50 to 100 knots around latitude 60° N with irregular excursions north and south similar to those of the circumpolar jet stream. Little is known about the detailed structure of these stratospheric westerlies which have been referred to as a "stratospheric jet stream".

LOCAL JET STREAM

Cross-sections drawn through "local jet streams" detached from the main "circumpolar jet" usually show much the same structure as illustrated in Figure 2 as typical of the circumpolar jet stream. A section drawn along the line BB′ of Figure 1 is shown in Figure 5. On this occasion some of the features of the cross-section are not altogether typical, although the general form of velocity distribution is preserved. The horizontal temperature changes are very largely concentrated in an intense sloping frontal zone resulting in a velocity maximum at 420 millibars, a rather lower level than usual. The tropopause distortion is also less than usual.

At all levels throughout the troposphere, not only near jet streams, the isotherms are usually in much the same direction and the "thermal wind" (a fictitious wind in the direction of the vertical wind shear) blows along them. Thus the component of wind across the isotherms (that is the component which is responsible for their advective movement) is much the same at all levels in the troposphere. It is therefore convenient to think of the changes in the temperature distribution arising from horizontal advection as though they were arising solely from the wind at one level (just above the friction layer for example).

If then we have a horizontal temperature field represented by the dashed isotherms in Figure 6 and an advective wind field shown by the continuous stream lines, it is clear that a concentration of isotherms will be built up along the line AB. Conditions of geostrophic balance then require either that a strong vertical wind shear and a narrow belt of strong winds should be formed in the upper troposphere or that a belt of strong wind from the opposite direction must occur near the ground. Strong surface winds are indeed sometimes observed under these circumstances, but more commonly the atmosphere already has a general motion along the isotherms in the direction of the thermal winds upon which the new wind shear is superimposed and the strongest winds occur aloft. Since the horizontal temperature gradient in the stratosphere is usually opposed to that in the troposphere, a strong thermal wind is built up in the lower stratosphere opposed to

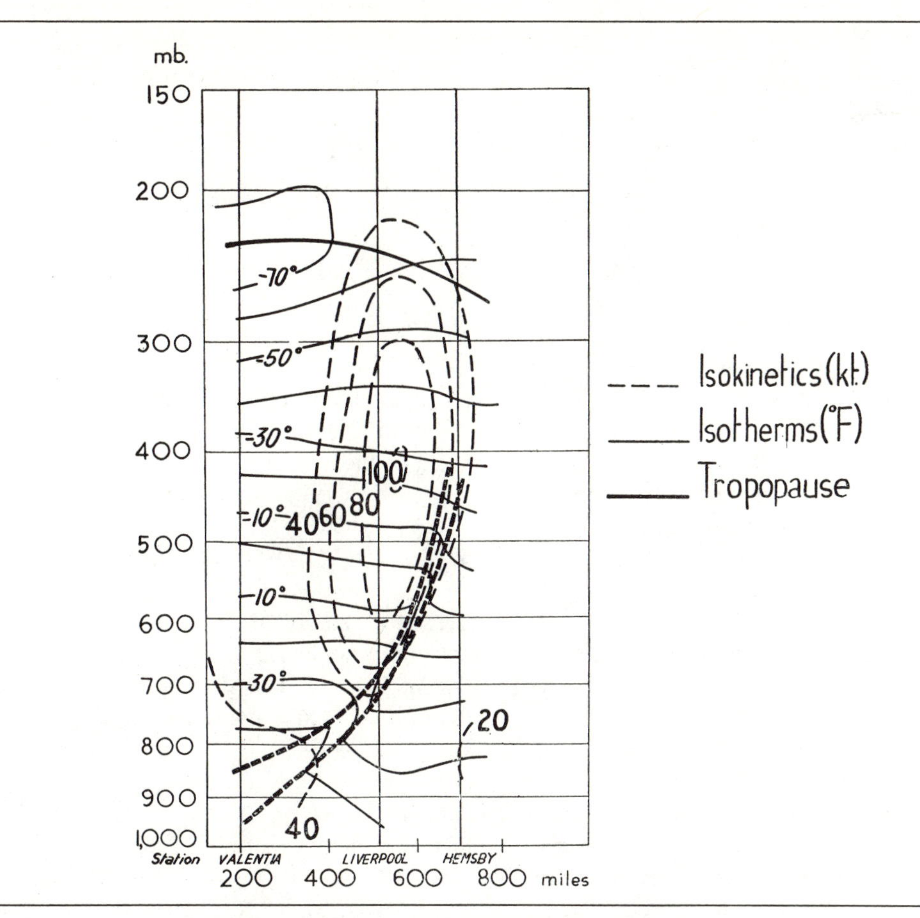

Figure 5 Cross section, Valentia-Liverpool-Hemsby 0300 GMT 19 December 1953.

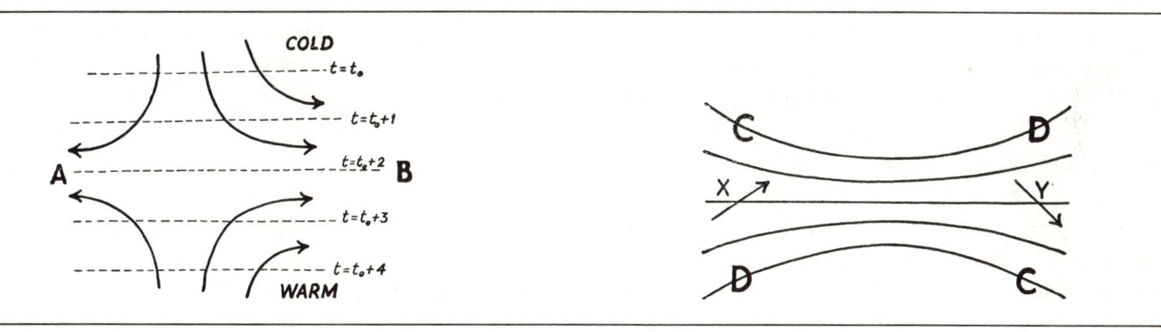

Figure 6 The concentration of isotherms by the wind field.

Figure 7 Cross-isobaric motion in jet streams.

that in the troposphere. As a consequence the jet stream winds fall off rapidly above the tropopause.

This is the so-called confluence theory of "jet streams" usually attributed to Namias and Clapp (1949). However, it is not easy to trace all local jet streams to an origin in a pattern of confluent flow, because the process clearly operates best when some concentration of temperature differences is already present and nearly all "local jet streams" seem in some way to be connected with the circumpolar jet stream, either as off-shoots from it, or as reintensifications of sections which have become divorced from the main flow.

On the other hand, the confluent process can easily be seen at work in intensifying and maintaining existing jet streams, and it has some interesting consequences in respect of the vertical air movement in jet streams. Individual maxima of wind velocity usually progress more slowly than the wind speed. Thus the wind may be considered as flowing through a quasi-stationary pattern of isobaric contours as shown in Figure 7. As a

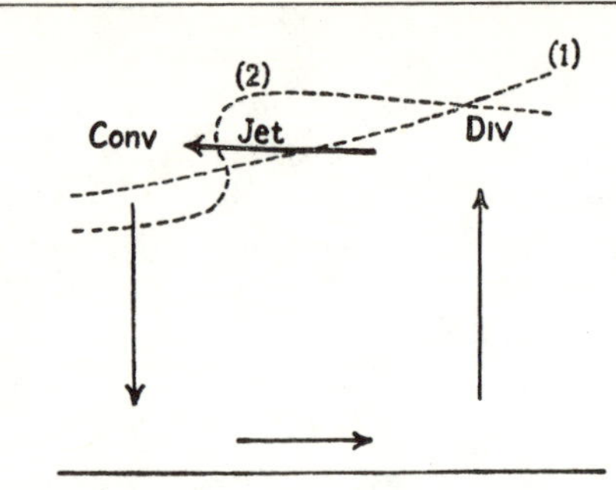

Figure 8 Vertical circulation at the jet entrance.

consequence of its acceleration in the confluent region X the air has a component of motion across the isobars to the left and similarly in the diffluent region Y, it has a component towards the higher pressure—to the right. The existence of this cross-isobaric motion has been demonstrated by Murray and Daniels (1953) directly from the wind observations. It averages about 10 knots.

The transverse component of the flow leads to convergence in the left entrance and divergence from the right entrance to the jet. This appears to form part of a vertical circulation with descent below the left entrance and ascent below the right entrance, as shown in Figure 8 which is sketched looking downstream. The circulation below the exit region is believed to be the reverse of that illustrated. One effect is probably to distort an initially slightly inclined tropopause (dotted line (1)) to a tropopause with an abrupt step (dotted line (2)). Indirect evidence for these vertical circulations is found in the dryness of the air and the absence of high cloud to the left of the jet stream axis (Murray, 1955) and in the distribution of rainfall in the neighborhood of jet streams (Johnson and Daniels, 1954).

The horizontal profile of wind in the jet stream deserves comment. There is now plenty of evidence that the horizontal wind shears are of the order of 20 to 60 knots in 100 miles—sometimes more (Murray and Johnson, 1952; Johnson, 1953; Hurst, 1952). Theory suggests that there is an upper limit to the wind shear on the warm side of the jet stream. This occurs when the rotation of a small portion of the air stream in the region of shear becomes as rapid as the component of the earth's rotation about the vertical. With greater wind-shears the individual elements of the air would be rotating in the opposite direction to the earth and theory indicates that the flow would break up into eddies. The limit to the shear is expressed numerically by the Coriolis parameter. Observational evidence suggests that the limiting shear (about 40 knots in 100 miles in latitude 50°) is frequently approached on the right hand (anticyclonic) side of the jet stream although the shear is not always so great. No such limit exists to the cyclonic shear on the cold (left hand) side of the jet stream and this is found usually to be greater than the anticyclonic shear on the warm side—the cyclonic shear averages 150% of the anticyclonic.

THE CIRCUMPOLAR JET

Cross-sections of the circumpolar jet stream which have been published usually show a structure similar to that in Figure 2. However authors have usually chosen to examine intense and fairly straight sections of the jet, and it seems likely that a more complicated structure exists in the weaker or strongly curved parts of the circumpolar jet. Some studies have shown double maxima of velocity (Murray, 1953).

At first sight it seems difficult to explain a band of strong wind which encircles the hemisphere on the basis of local confluence as discussed in relation to the "local jet streams" because it is difficult to envisage confluence operating simultaneously round the whole of the hemisphere. This led Rossby (Staff members of the department of Meteorology of the University of Chicago, 1947) to seek an explanation in the transport of momentum and vorticity by lateral mixing in which the depressions and anticyclones formed the turbulent elements. An essential factor in Rossby's explanation was the variation of the Coriolis parameter with latitude arising from the spherical shape of the earth, and yet experiments with water in *flat* rotating dishpans have reproduced features very similar to both the circumpolar jet stream and the local jet streams discussed earlier (Fultz, 1953).

The main factors which are common to the dishpan experiments and the atmosphere are, rotation, baroclinity and bottom friction, and there is some evidence for the belief that these are the factors which give rise to the jet stream to be found in the numerical experiment conducted by Phillips (1956). Phillips set up a system of equations to represent atmospheric motion and carried out a numerical integration on an electronic computer to determine the behavior of the atmosphere over several weeks. Some simplification was inevitable and Phillips used a two-parameter representation of the atmosphere which effectively assumes similar horizontal temperature-gradients at all levels and quasi-geostrophic motion. Phillips included friction with the ground and internal friction in the air. He also allowed for the difference in heating by the sun in different latitudes. Introducing only an arbitrary small disturbance on an initial zonal motion, Phillips showed that large depressions and anticyclones should form similar to those observed in the atmosphere and that in the process the meridional temperature differences should become concentrated in a sinuous band as observed in the atmosphere. As a result of surface friction the velocities near the ground remain moderate, and the sinuous band of strong temperature gradient is associated with a sinuous jet stream similar to that observed in the atmosphere.

In Phillips' treatment the jet stream becomes more intense as the cyclonic and anticyclonic systems develop, and it seems that the characteristics of the type of disturbances which form in the atmosphere play an essential part in building up the circumpolar jet stream. Any system of eddies will set up regions of confluence and diffluence and thereby form regions of intensified temperature differences. These regions are particularly favorable for development of new eddies which may tend further to concentrate and extend them. Friction with the ground ensures that strong easterly winds are not built up in the lower layers, and that the strong horizontal temperature gradient of the circumpolar jet is associated with strong upper westerlies rather than with low-level easterlies.

The circumpolar jet stream undergoes many changes in form. Characteristic distortions occur during "blocking" when a closed cyclonic vortex is cut-off south of the main jet stream current as illustrated in Figure 9. Occasions also occur when jet streams exist in two distinct latitude bands around most of the hemisphere.

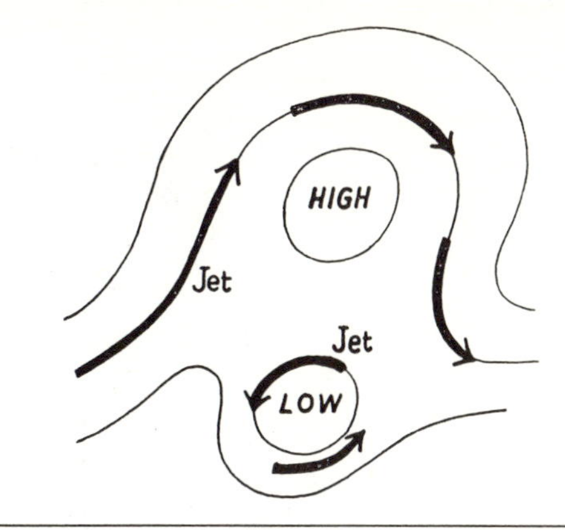

Figure 9 Typical distortion of the jet stream during "blocking".

THE SUBTROPICAL JET STREAM

Figure 4 shows that a strong westerly current encircles the earth in latitudes around 30° N at a pressure level of about 200 millibars. This current exceeds 50 knots almost all round the globe and exceeds 100 knots over substantial areas. It differs from the circumpolar jet stream in that its position is much less variable, it does not undergo such large distortions and disruptions and it is not usually associated with active disturbances such as the depressions and mobile anticyclones of higher latitudes. It seems difficult to regard the subtropical jet stream as set up by confluent wind fields associated with large scale eddy motion.

It seems much more likely that the subtropical jet stream is a direct consequence of a slow northward drift of air in the middle and upper troposphere from near the equator to latitudes of 20 or 30°. At the equator the air shares the rotation of the earth and if it retains its angular momentum it must move faster as it reaches the latitude circles of 20° or 30° which are smaller than the equator. The slow northward drift is part of a circulation first remarked by Hadley, in which air approaches the equator in the lower atmosphere in the trade-wind currents, ascends there and returns aloft. Such circulations are observed in dishpan experiments at lower rotation rates than those which give rise to a "meandering jet stream," and it therefore appears reasonable that a simple Hadley-type cell should appear in lower latitudes on the earth where the Coriolis parameter is small. A ring of air displaced northward from the equator to latitude 30° and conserving its angular momentum would attain a velocity relative to the earth exceeding 200 knots so that the observed wind speeds are not unreasonable as arising from this cause. Rogers (1954) gives a theoretical treatment of forced flow in a thin layer of viscous fluid on a rotating sphere in which the dynamics of a meridional Hadley-type cell are discussed. Despite numerous artificialities and dissimilarity from the atmospheric regime, it is interesting to note that Rogers calculates a concentration of westerly velocity in the upper layers of the fluid in quite low latitudes—namely around 12°—considerably nearer to the equator than the subtropical jet stream—not as might have been expected farther north.

A cross-section (Figure 10) drawn along line CC′ in Figure 1 is particularly interesting as it shows both the circumpolar and the subtropical jet streams. This

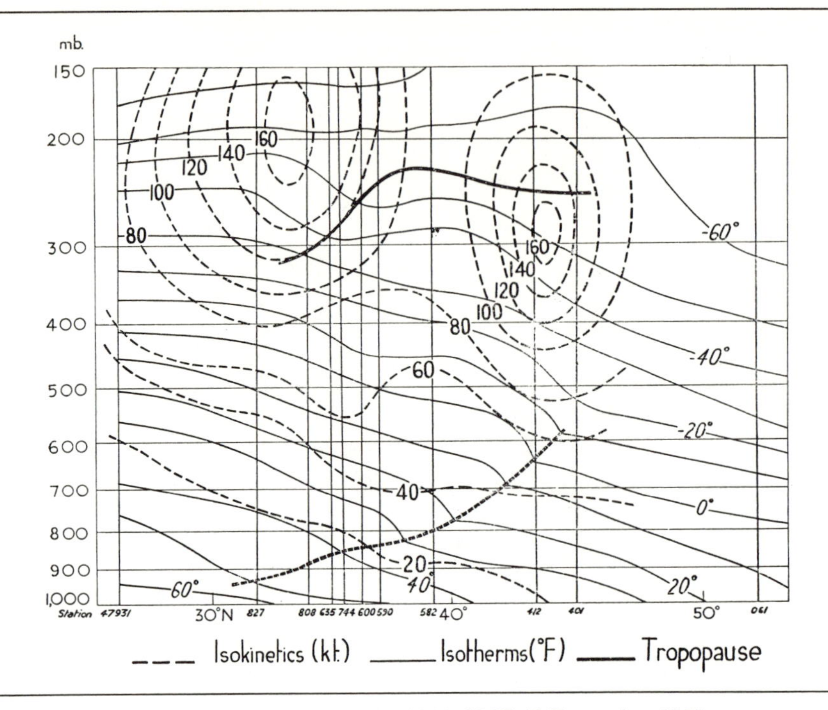

Figure 10 NNE-SSW cross section of jet streams near Japan 0300 GMT 19 December 1953.

suggests that the two streams retain their independent character even in longitudes where the circumpolar jet stream is displaced southward as over Japan.

Lack of observations precludes the construction of such detailed cross-sections through the subtropical jet stream in other longitudes as are desirable—although this may be remedied by the observations made for the Geophysical Year. However a number of studies of the subtropical jet stream have been made. Such studies (Gilchrist, 1955) show that the subtropical jet stream is associated with a discontinuity in the tropopause—the tropopause being around 100 millibars on the equatorward side but around 200 millibars or lower to the north. On the other hand, unlike the circumpolar jet stream it has no close association with any frontal region in the lower troposphere.

Accurate observations of wind by aircraft flying through the subtropical jet stream in the Middle East (Harding, 1955) suggest that the belt of strong wind has a broader maximum than is usually found in the circumpolar jet stream—the maximum wind is found to be maintained over a belt 200 miles or more in width. However the cyclonic wind shear to the north is almost as strong as is observed at the jet streams in higher latitudes and to the south the magnitude of the anticyclonic wind shear seems to be from a half to three-quarters of the Coriolis parameter.

The preceding discussion has related entirely to observations from the northern hemisphere but the more limited evidence from the southern hemisphere reveals a similar structure there. Undoubtedly the continents, and possibly also the presence of the Himalayas and Rocky Mountains, have an effect on the jet stream winds. The extraordinarily high wind speeds observed near Japan in winter must be partly due to the strong temperature contrast between Asia and the Pacific. However it seems likely that were the topography of the earth much more uniform there would still be a subtropical jet, a circumpolar jet and probably also local jets disconnected from the main circumpolar stream.

REFERENCES

Fultz, D., 1953. *Proc. 1st Symp. on Geophys. Models.* Baltimore, John Hopkins University.

Gilchrist, A., 1955. Winds between 300 and 100 mb in the tropics and subtropics. *Meteorol. Rept. 16*, Meteorol. Office, London.

Harding, J., 1955. The profile of jet streams in the Middle East. *Meteorol. Res. Pap. 932*, London.

Hurst, G. W., 1952. The profile of a jet stream observed 18 January 1952. *Quart. Journ. Royal Meteorol. Soc. 78*, 613–615.

Johnson, D. H., 1953. A further study of the upper westerlies; the structure of the wind field in the eastern North Atlantic and western Europe in January 1950. *Quart. Journ. Royal Meteorol. Soc. 79*, 402–407.

Johnson, D. H. and Daniels, S., 1954. Rainfall in relation to the jet stream. *Quart. Journ. Royal Meteorol. Soc. 80*, 212–217.

Murray, R., 1953. Jet streams over the British Isles during June 14–18, 1951. *Meteorol. Mag. 82*, 129–140.

Murray, R., 1956. Some factors of jet streams as shown by aircraft observations. *Geophys. Mem. 97*, Meteorol. Office, London.

Murray, R. and Daniels, S., 1953. Transverse flow at entrance and exit to jet streams. *Quart. Journ. Royal Meteorol. Soc. 79*, 236–241.

Murray, R. and Johnson, D. H., 1952. Structure of the upper westerlies; a study of the wind field in the eastern Atlantic and western Europe in September 1950. *Quart. Journ. Royal Meteorol. Soc. 78*, 186–199.

Namias, T. and Clapp, P. F., 1949. Confluence theory of the high tropospheric jet stream. *J. Meteor. 6*, 330–336.

Phillips, N. A., 1956. The general circulation of the atmosphere: a numerical experiment. *Quart. Journ. Royal Meteorol. Soc. 82*, 123–164.

Rogers, M. H., 1954. The forced flow of a thin layer of viscous fluid on a rotating sphere. *Proc. Royal Soc., A. 224*, 192–208.

Staff Members, Dept. of Meteor. Univ. Chicago, 1947. On the general circulation of the atmosphere in middle latitudes. *Bull. Amer. Meteorol. Soc. 28*, 255–280.

World Meteorological Organization, 1956. *W.M.O. Bull. 5(3)*, 103. Geneva.

Jerome Namias

5 The index cycle and its role in the general circulation

INTRODUCTION

The concept of the zonal index was introduced into meteorology by Rossby (1939). This quantity, expressing numerically (at sea level or aloft) the strength of the temperate-latitude (35° N to 55° N) westerly winds over a hemisphere, was found to be related to the form of the general circulation, particularly in the longitudinal positions and extent of the great centers of action. The terms "high" and "low" index patterns have emerged to describe the state of the sea-level and mid-troposphere isobaric patterns associated respectively with strong and weak mid-latitude westerlies.

More recently, as emphasized by Willett (1948), variations in the zonal index have been found to be associated with pronounced *latitudinal* as well as longitudinal shifts in the characteristic branches of the general circulation. For example, the low-index pattern referred to above is not only associated with certain longitudinal characteristics of the centers of action but also with a displacement southward of the midtropospheric band of maximum westerly winds. Thus, strictly speaking, low index of the temperate latitudes, for the most part, is identified with strong westerly flow in subtropical latitudes. In view of the fact that at least in the middle and high troposphere the zonal westerlies often reach their greatest strength at the time of "low index" as defined above, it would appear that the terms "high" and "low" index are misnomers.

While the index concept has been of considerable help in classifying and studying general circulation patterns, it must be confessed that its usefulness in the practice of extended forecasting has been somewhat disappointing. Inasmuch as general circulation patterns were found to be related to the zonal index, a vast research effort has been directed towards developing statistical methods of predicting this quantity. Aside from a small degree of persistence combined with a tendency to return to a normal seasonal value, attempts to find statistical time-lag relationships have failed (Willett, 1948). In 1947 the writer stated (Namias, 1947a): "In spite of the fact that index forecasts have been made for about seven years, it is noteworthy that up until recently no physical method for forecasting this quantity (the zonal index) has been developed."

In the report in which the above statement was made, some preliminary suggestions were given as to the physical causes of index fluctuations, and this thesis was expanded upon in subsequent papers (Namias, 1947b; Namias and Clapp, 1949). Although not previously stated in such broad terms, in essence this thesis is: *The total hemispherical zonal index is not a primary parameter (independent variable) whose variations can be statistically accounted for on the basis of its own past behavior; rather, it is a derived quantity (dependent variable) which represents the degree of latitudinal organization of certain large-scale energy producing mechanisms in the middle and upper troposphere.*

It is the purpose of this paper to describe and explain this organization in terms of the observed transformations of the component parts of the wintertime general circulation as the zonal index at some period during winter proceeds from high to low values and then recovers.

THE INDEX CYCLE

From week to week each winter there are large variations of the zonal index about its seasonal normal. Such variations at the 700-millibar level for means of 5 days computed twice weekly are shown for the period November through March 1944–49.[1] From these graphs it

Reprinted, with minor editorial modification, by permission of the author and editor, from *Journal of Meteorology*, American Meteorological Society, Vol. 7, 1950, pp. 130–139.

[1] Only half of the hemisphere from 0° westward to 180° had sufficiently adequate upper-air coverage upon which to draw synoptic charts on a current basis over the six-year period 1944–49.

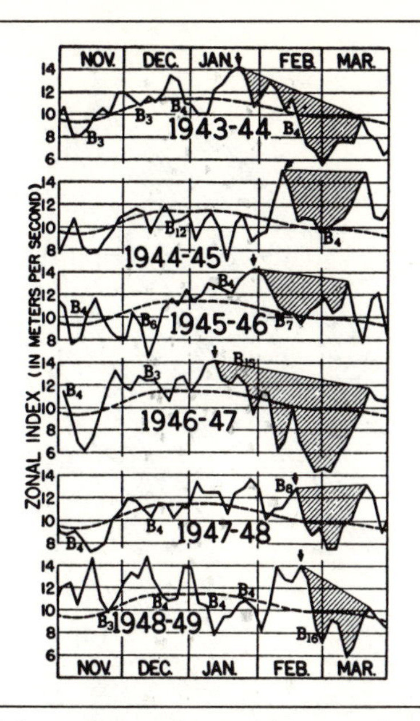

Figure 1 5-day mean and monthly normal zonal indices (strength of the zonal westerlies in m sec^{-1} between 35° N and 55° N) at 700 mb over the northern hemisphere from 0° westward to 180° for colder months of the years 1944 through 1949. Values are plotted against the last day of the 5-day period. Heavy arrows point to beginning of primary "index cycles," and the shaded area gives a measure of their duration-intensity. B's refer to periods of blocking as defined in text.

appears that in each year there is at least one period lasting several weeks in which there takes place a more or less gradual decline of index from high to low values followed by a similar rise. The beginnings of the most pronounced period of this kind are indicated in Figure 1 by arrows, and some measure of their intensity-duration is indicated by the shaded area. Other cycles occurring earlier in the winter would apparently be less pronounced in terms of a similarly constructed shaded area. For want of a better term, such an extended period of falls and rises has been called an "index cycle." Figure 1 poses a number of problems vital to an understanding of the general circulation:

1. Why during each year is there a tendency for the occurrence of at least one index cycle lasting several weeks?

2. Why do these cycles vary in intensity (i.e., length of duration and departure from normal) from one year to the next?

3. Why does there appear to be one particular time of year (late February) when the most pronounced index cycle occurs?

Before an attempt can be made to answer any of these questions, it is necessary to look into the flow patterns associated with high and low index. While considerable work of this character was done in the early period of the "index" decade, this work was mostly connected with the characteristic features of sea-level hemispherical maps. In the extended forecasting work at the U.S. Weather Bureau, it has been possible during the years 1944 through 1949 to construct, with little extrapolation, hemispheric or nearly hemispheric maps for mid-troposphere levels. Similar maps, obtained in part through use of extrapolation techniques over large areas, have also been prepared by Willett (1947) for the colder half of the years from 1932–39 as part of the Weather Bureau-Massachusetts Institute of Technology extended forecasting research project. A study of these latter maps, along with sea-level data, led Rossby and Willett (1948) to the following description of the index cycle:

> . . . four principal stages of the index cycle are recognized, each of which can be briefly characterized essentially as follows:
>
> (1) Initial high index (strong sea-level zonal westerlies), characterized by (*a*) sea-level westerlies strong and north of their normal position, long wavelength pattern aloft; (*b*) pressure systems oriented east–west, with strong cyclonic activity only in high latitudes; (*c*) maximum latitudinal temperature gradient in the higher middle latitudes, little air mass exchange; and (*d*) the circumpolar vortex and jet stream expanding and increasing in strength, but still north of the normal seasonal latitude.
>
> (2) Initial lowering of sea-level high-index pattern, characterized by (*a*) diminishing sea-level westerlies moving to lower latitudes, shortening wavelength pattern aloft; (*b*) appearance of cold continental polar anticyclones in high latitudes, strong and frequent cyclonic activity in middle latitudes; (*c*) maximum latitudinal temperature gradient becoming concentrated in the lower middle latitudes, strong air mass exchange in the lower troposphere in middle latitudes; and (*d*) maximum strength of the circumpolar vortex and jet stream reached near or south of the normal seasonal latitude.
>
> (3) Lowest sea-level index pattern, characterized by (*a*) complete breakup of the sea-level zonal westerlies in the low latitudes into closed cellular centers, with corresponding breakdown of the wave pattern aloft; (*b*) maximum dynamic anti-

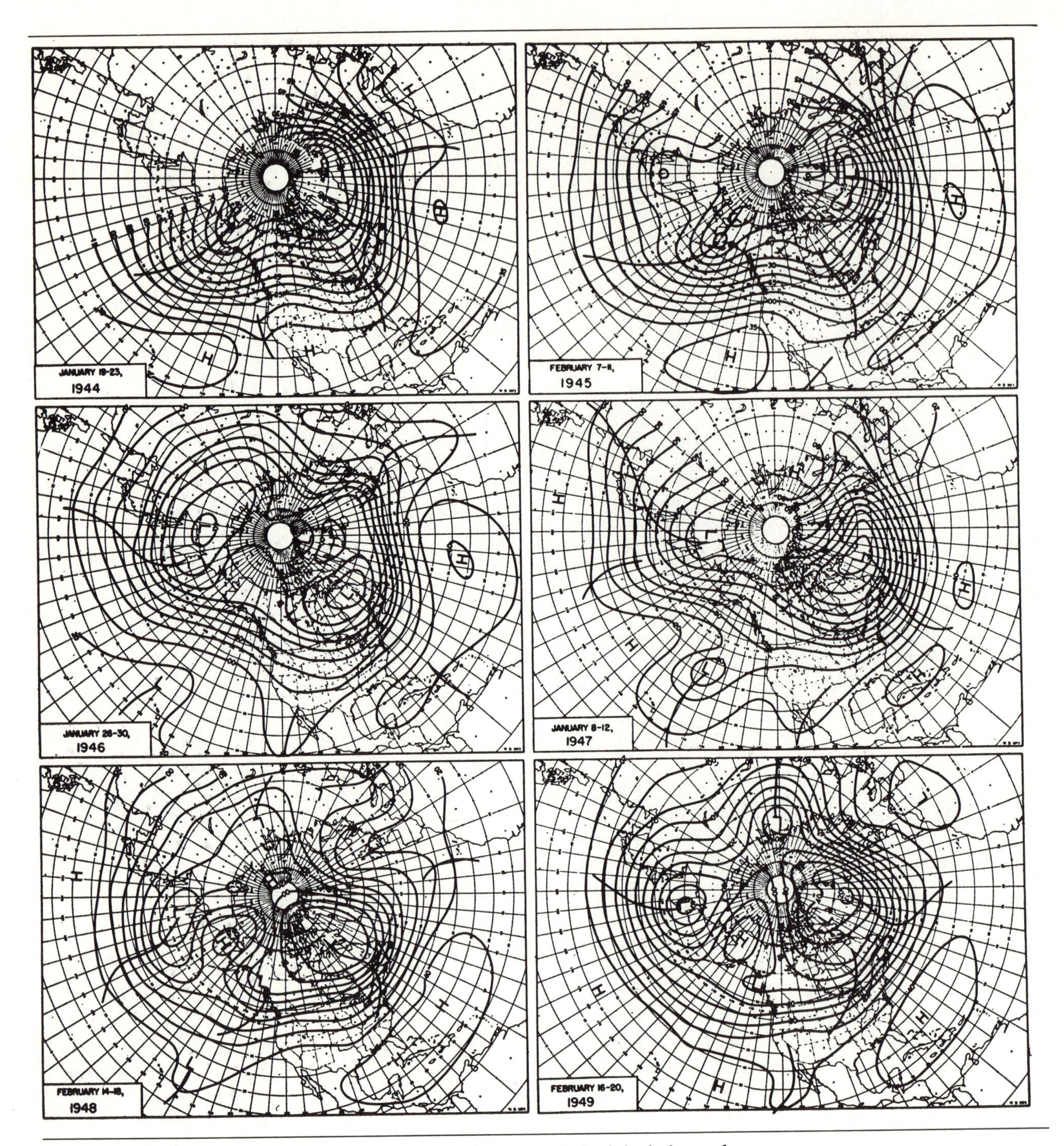

Figure 2 700-mb 5-day mean charts at the initial stage (high index) of the index cycle.

cyclogenesis of polar anticyclones and deep occlusion of stationary cyclones in middle latitudes, and north–south orientation of pressure cells and frontal systems; (*c*) maximum east–west rather than north–south air mass and temperature contrasts; and (*d*) development of strong troughs and ridges in the circumpolar vortex and jet stream, with cutting off of warm highs in the higher latitudes and cold cyclones in the lower latitudes.

(4) Initial increase of sea-level index pattern, characterized by (*a*) a gradual increase of the sea-level zonal westerlies with an open wave pattern aloft in the higher latitudes; (*b*) a gradual dissipation of the low-latitude cyclones, and merging of the higher-latitude anticyclones into the subtropical high-pressure belt; (*c*) a gradual cooling in the polar regions and heating of the cold air masses at low latitudes to re-establish a normal poleward temperature gradient in the higher latitudes; and (*d*) dissipation of the high-level cyclonic

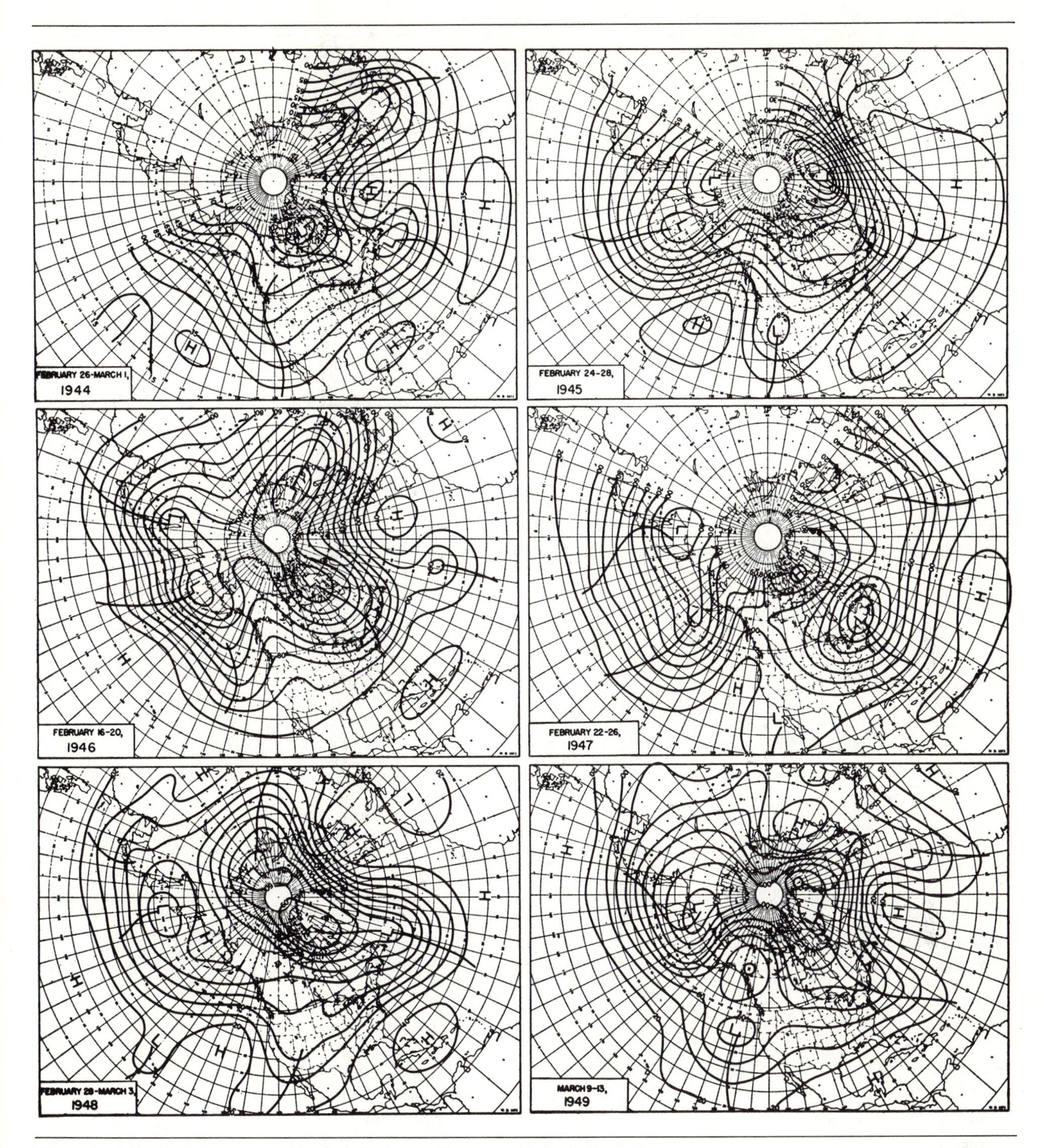

Figure 3 700-mb 5-day mean charts at the low index stage of the index cycle.

and anticyclonic cells with a gradual re-establishment of the circumpolar vortex jet stream in the higher latitudes.

In addition to the above description, certain other facts concerning the index cycle seem to be of importance for any treatment of its mechanics. These facts are brought to light in a study of high- and low-index patterns as reflected in 700-millibar 5-day mean maps during 1944–49. Some of these maps, representing the beginning high point of the index cycle (corresponding to the arrows of Figure 1) and the subsequent low point, are shown in Figures 2 and 3.

It appears that high index requires mid-latitude bands of confluence (Namias, 1947a) where cold, polar-continental air masses are drawn beside warm, tropical (or polar-maritime) air masses to form high speed "jet streams" which organize into the strong westerlies. From Figures 2 and 3 and other data, there appear to be preferred geographical sites for such confluence—just off the northwest coast of the US (confluence

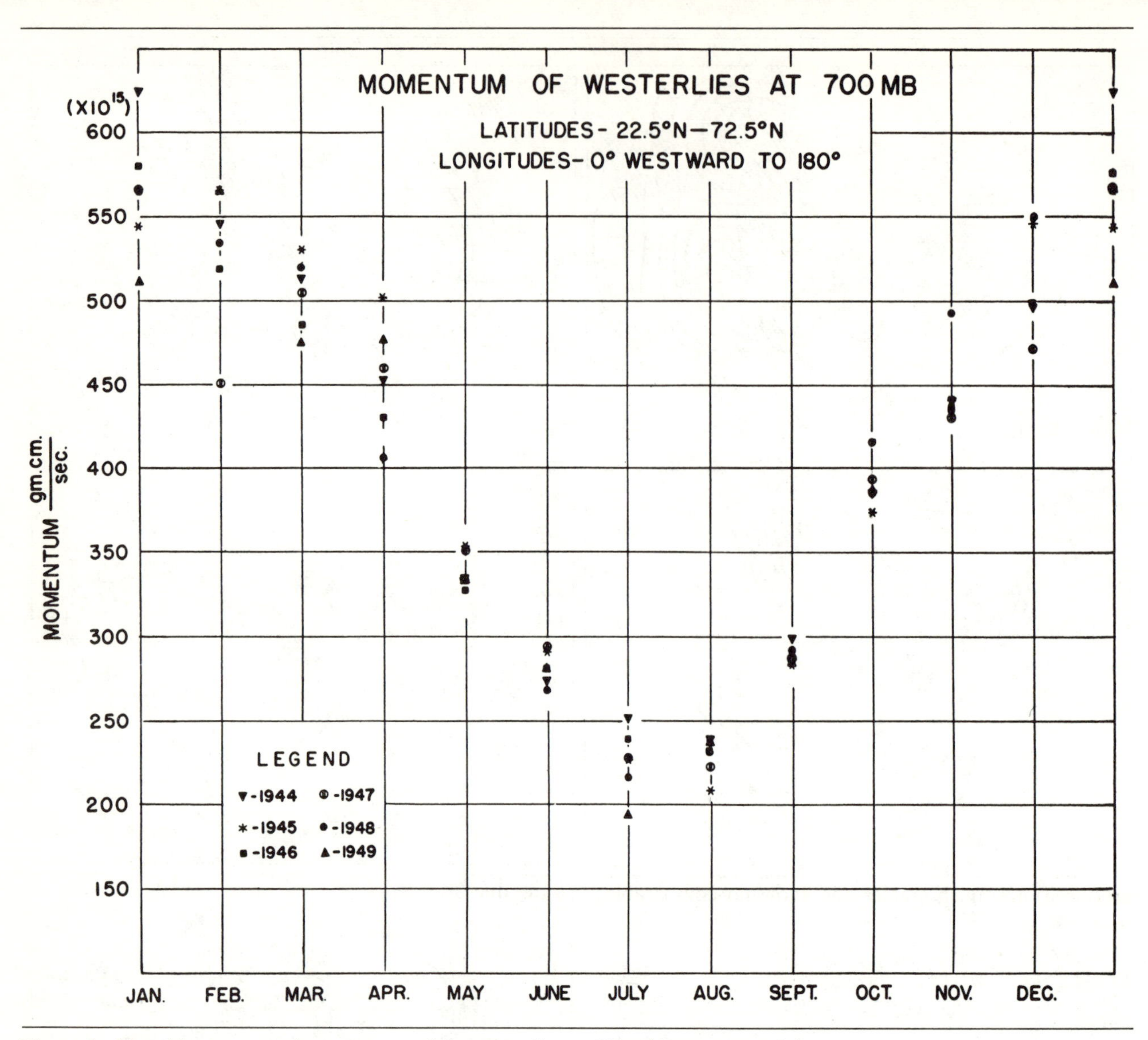

Figure 4 Monthly mean total momentum of the westerlies at 700 mb as computed from zonal windspeed profiles for the zone from 20° N to 75° N and from 0° to 180° W for the years 1944 through 1949. Momentum is expressed in units of g cm sec^{-1} × 10^{-15}.

associated with the Gulf of Alaska surface low surmounting the eastern cell of the Pacific High) and in eastern North America (where an Arctic air flow is directed beside polar-Pacific and often tropical-Gulf air to form a strong jet which breaks into the Atlantic).

On the other hand, low-index patterns (Figure 3) are freer from mid-latitude confluence, and those confluence bands which do exist seem to be developed chiefly at low latitudes. Since confluence requires out-of-phase wave systems in different latitude bands, it appears from geometrical considerations that during high index the amplitude of wave systems increases with distance poleward and equatorward from the jet stream.

Other important facts concerning index variations are brought to light through an analysis of 700-millibar zonal-windspeed profiles. In working with these profiles in the 5-day forecast routine, one is impressed by the tendency for the total momentum of the westerlies aloft (neglecting density), as represented by the area under one of these profiles, to remain nearly constant from week to week, or in the case of monthly mean profiles, for the total area to agree with the area under the normal zonal-wind profile for that month. To illustrate this concept quantitatively, computations of the total momentum of the 700-millibar flow from 22½° N to 72½° N between 0° and 180° W have been made from monthly mean zonal-wind profiles with the help of a planimeter. The mean density has been computed from corresponding 700-millibar temperature profiles. Variations in density due to moisture have been considered negligible. The results of these computations are shown in Figure 4, where the individual monthly mean values

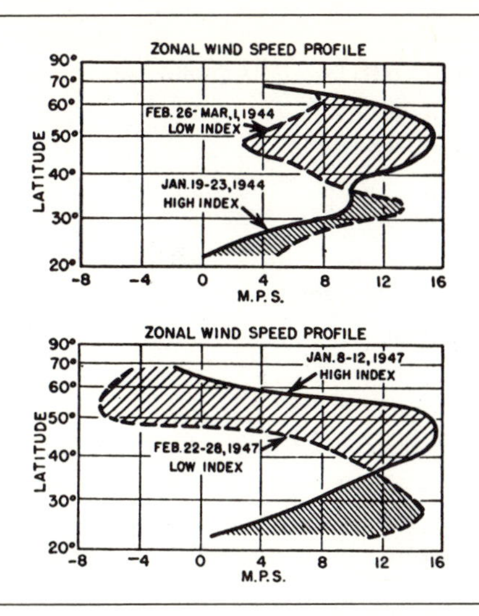

Figure 5 5-day mean windspeed profiles for high- and low-index stages of the index cycles of 1944 (upper) and 1947 (lower).

for 1944–49 are indicated. From this figure it appears that, at least for the transition seasons of the year, the total momentum for a given month has a strong tendency to fall into one general range of values peculiar to that month, and these values do not overlap with those of adjacent months. On the other hand, the variations in monthly mean zonal index at 700 millibars in the same month of different years (figure not reproduced) are much greater and show no comparable tendency to cluster. Indeed, a statistical analysis indicates a much greater variability of zonal index than of total momentum even when account is taken of the difference in area over which computations are made.

Another illustration of the conservative nature of the total momentum of the mid-troposphere westerlies is furnished by a comparison of individual windspeed profiles for the high- and low-index cases of the index cycle. Two such cases, for 1944 and 1947, are reproduced in Figure 5. They suggest that there is a net loss of momentum at intermediate latitudes which is made up by an increase in low latitudes. In fact, these and other comparisons of monthly windspeed profiles indicate that the compensation to some extent occurs at latitudes even farther south than our data permit computations (20° N). In this connection, it is to be noted that the area represented in the computations constitutes less than one-third of the entire hemisphere. In all probability, the compensation would be more exact if one considered the entire hemisphere.

The conclusion suggested by the above data is that, in spite of large departures from normal of the zonal index between fixed latitudes in the same month of different years, the total momentum of mid-troposphere westerlies over the hemisphere in any month does not vary much from year to year; or, from another point of view, it appears probable that the *total momentum of the mid-troposphere westerlies around the hemisphere reaches a certain value at any given period each year, and it is only the distribution of this momentum with latitude that varies.*

In this concept we find evidence for the feature of low index stressed by Willett—namely, that low, temperate-latitude westerlies are generally associated with an expanded circumpolar vortex (i.e., westerlies which are displaced southward). It is noteworthy that of the index cycles shown in Figure 1, only one (1944–45) showed a contraction of the circumpolar vortex (i.e., the polar westerlies aloft, rather than those of the subtropics, increased).

If it is only the form of the westerly circulation which varies appreciably and not its absolute, overall, integrated strength (considering the entire hemisphere), it would appear that variations in the zonal index (or for that matter any zonal wind flow between fixed latitudes) are due to the degree of organization (or lack of organization) of the large-scale mid-tropospheric wave systems into confluence or diffluence patterns, for it is the confluence patterns which usually concentrate the energy of the westerlies in narrow zones (jet streams) where peak speeds are reached and the diffluence patterns which represent areas of strong "blocking."

Granting the above line of argument, the problem of index variations is reduced one step further to the fundamental problem of what factors lead to the peculiar long-wave structures favorable to sustained confluence, preferably in more than one area (for high index) and what factors are associated with the breakdown of these wave structures into diffluence (blocking) patterns with their resulting low index.

INITIAL STAGE OF THE INDEX DECLINE

By examining the 700-millibar 5-day mean maps subsequent to the periods of high index at the start of the index cycle (arrows of Figure 1), it became clear that each case of index decline was attended by a wave of "blocking" in the Atlantic, i.e., by a strong localized diminution in zonal index over the Atlantic. Since the Atlantic constitutes a portion of the index band (0° to 180° W) with which we are concerned, this conclusion may at first glance appear to be a statistical necessity.

Period	Atlantic	North American	Eastern Pacific
19–23 January to 26–30 January 1944	−2.0	−2.7	−4.6
7–11 February to 14–18 February 1945	−4.4	−3.0	−4.6
26–30 January to 2–6 February 1946	−1.2	−1.3	+0.1
8–12 January to 15–19 January 1947	−8.3	+0.6	−0.6
14–18 February to 21–25 February 1948	−8.8	−5.1	+4.1
16–20 February to 23–27 February 1949	−10.0	−4.4	+2.4
Mean	−5.8	−2.6	−0.5

Table 1 Changes in regional sea-level zonal indices in m sec^{-1} during the first week of the index cycle

Period	Atlantic	North American	Eastern Pacific
19–23 January to 26–30 January 1944	−4.5	−2.9	−2.9
7–11 February to 14–18 February 1945	−5.9	−2.4	+3.6
26–30 January to 2–6 February 1946	−1.9	−5.3	+0.4
8–12 January to 15–19 January 1947	−8.3	+1.8	+2.4
14–18 February to 21–25 February 1948	−9.3	−8.2	+4.1
16–20 February to 23–27 February 1949	−8.8	−7.0	−1.3
Mean	−6.4	−4.0	+0.2

Table 2 Changes in regional 700-mb zonal indices in m sec^{-1} during the first week of the index cycle

The question may be raised as to what extent the initial index decline begins in the Atlantic rather than in North America or the eastern Pacific—areas which also compose the total index used. In Tables 1 and 2 are given the local changes in index for the three equal areas, the Atlantic (5° W to 55° W), North America (65° W to 125° W), and the eastern Pacific (135° W to 175° E), during the first week of the index cycle (one-week change following the week indicated by the arrows in Figure 1). Both at sea level and aloft, the bulk of evidence suggests that the Atlantic usually contributes most heavily to the initial index decline. Notable exceptions occur in 1944 at sea level and 1946 at 700 millibars. But we are expecting too much if we ask the atmosphere to release its secrets completely to a "straight-jacketing method" such as this, in which we have set up restricted zones and time intervals. Indeed, it is surprising that the results are as uniform as they are in Tables 1 and 2.

Thus there is a suggestion that blocking is a necessary condition for the inception of an index cycle. But the long-range weather forecaster is well aware of the fact that blocking is not a *sufficient* condition. There arise during winter many more blocking waves in the Atlantic than there are hemisphere index cycles. For example, indicated by a "B" in Figure 1 are the beginnings of those cases of blocking taken from an independent study (Aubert, 1949). Here a period of Atlantic blocking was defined as one in which the 5-day mean sea-level pressure anomaly in the area 50° N to 70° N and 20° E to 50° W has a positive value of 15 millibars or greater and persists for at least one and one-half weeks. The "B" is placed at the beginning of such a blocking wave (i.e., the first period which meets the criterion stated). It is noteworthy that one of these so-defined blocking periods occurs during each one of the index cycles of Figure 1, but there are other such blocking waves that do not materialize into an extended index cycle. An inspection of the maps (not reproduced) associated with all the blocking cases indicated by the letter B fails to disclose a fundamental difference in the nature of initial Atlantic blocking waves which are part of index cycles and those which are not. It thus appears that some external factors (external in the sense of being outside the Atlantic area) must be of importance. These external factors may be expressed in the more complete state of the general circulation, involving areas far distant from the blocking, and/or possibly the effect of differing quality and quantity of radiation from the sun. Notwithstanding the growing popularity of solar hypotheses, it appears to the author that there is sufficient cause for variations in the earth-atmosphere system to explain the effectiveness of Atlantic blocking in producing an index cycle. Such a theory follows.

The strength of the mid- and upper-troposphere westerlies is to a considerable extent a measure of the degree of imprisonment or containment of cold air masses composing the polar cap. A special case of such containment (during the winter 1946–47) was discussed in some detail in an earlier report (Namias, 1947b). During these periods, strong westerlies forbid any extensive meridional transport, and the radiational cooling permits the air overlying northern land areas to grow into a vast cold-air reservoir. To some extent we may draw an analogy to a condenser where the containment mechanism of the strong westerlies is the most efficient method of operation to store cold air. Since an exchange of air between pole and equator is an obvious necessity for the atmospheric heat balance, it would appear that those *blocking waves which occur simultaneously with a great reservoir of cold polar air are the ones which materialize into an index cycle.*

Some substantiation of this idea may be found in the fact that the year's primary index cycle usually occurs considerably later than the December solstice, i.e., after

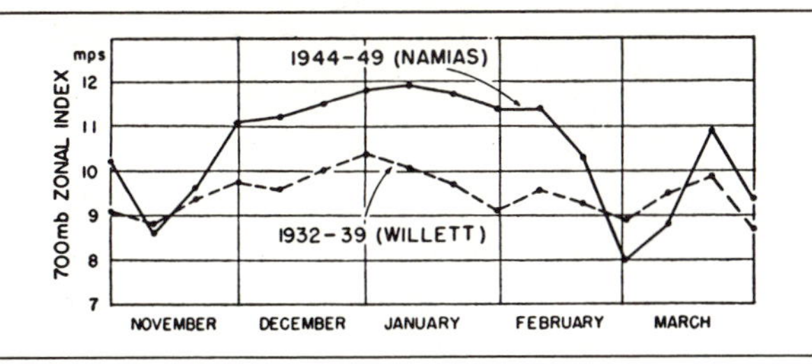

Figure 6 Average zonal indices taken at 10-day intervals from the graphs of Figure 1 (solid curve) and from similar data for 1932–1939 given by Willett (dashed curve).

the polar atmosphere has had time to develop into an extensive cold pool. Not only during this six-year period from 1944 through 1949 does the index cycle prefer February, but it shows up even in the period 1932–39 (see Figure 6). Here are plotted means of the index curves of Figure 1 for successive 10-day intervals for the entire six years 1944–49. It might be argued that such a phenomenon might be peculiar to this 6-year period. While no strictly comparable upper-air data for other years are available, an attempt to obtain upper-air maps for the period 1932–39 was made during the war by the U.S. Weather Bureau in collaboration with the University of California at Los Angeles and the Massachusetts Institute of Technology. Willett (1947) has made considerable use of this material in his exhaustive statistical studies of this period and lists 7-year average 3-km zonal index values for successive twice-weekly 5-day periods for the years 1932–39. Values taken directly from his data (with some small interpolation when necessary) are given in Figure 6. While it is apparent that the same major index cycle is present in the 1932–39 data as in the 1944–49 data, the amplitude of variation in earlier years is considerably less. The principal explanation of this suppression is to be found in the fundamental assumption under which the maps comprising Willett's data were prepared—namely, that the lapse rate in data-sparse areas (Atlantic and Pacific) was the moist adiabatic. Hence, under low-index conditions of winter, when warm highs appear in northern latitudes, these assumptions are erroneous in the direction of giving fictitiously high zonal indices. Similarly, with high index, subtropical anticyclones are well developed and have lapse rates more stable than the moist adiabatic, and computed indices are thus too low. Operating in the same direction (i.e., to reduce the variability) is the fact that Willett's data embrace about one-third again more data in longitudinal expanse than do ours. Both factors minimize index variations. It appears that the scale of the curve for Willett's data should be multiplied by some factor and this would bring it into a better agreement with the 1944–49 curve. Thus, there is a distinct suggestion of a global "singularity" in the general circulation.

As further evidence for support of the condenser theory, one might examine the observed temperatures of air comprising the polar cap preceding or at the onset of the index cycles. Precisely what thermal indices should be used for this test is difficult to decide, for certainly large areas and long periods of time must be considered. The extent of available data during this period makes it impossible to obtain reliable upper-air temperatures from the polar cap north of 70° N and also for a large portion of the northern hemisphere between 0° eastward to 180°. Thus our data are restricted to the polar cap south of 70° N and lying between 0° and 180° W. This area considers one important large source of manufacture of cold air, northern North America, but neglects the perhaps more important cold air factory of Siberia. Considering, therefore, the restricted area between 55° N and 70° N and between 0° and 180° W as representative of the entire polar cap, the average thicknesses and corresponding mean virtual temperatures of the air between 1,000 and 700 millibars over this area for the six Januarys from 1944 through 1949 are given in Table 3. (January is the month which immediately precedes most of the cycles, and such data were readily available.)

From these data it is apparent that 1947, the year of the largest and most intense cycle which led to worldwide record-breaking weather in many areas (Namias, 1947b), was characterized by the coldest January[2] while 1945, the year of a very weak cycle (particularly in departure from normal of the low point of the index) was preceded by one of the warmest Januarys in the sub-polar region. However, the year 1944, with its fairly intense and long-lasting cycle but warm January polar reservoir, and also 1946, with a weak cycle preceded by a moderately cold reservoir, appear out of line.

A better and more quantitative measure of the intensity of an index cycle can be determined from the

[2] Temperatures in the latter half of January and early February 1947 in the Canadian Yukon reached an unofficial minimum of −84° F at Snag Airport and during the period 29 January–5 February a mean of −62° F was recorded. The manner of production of this cold air was described by H. Wexler (1948), who traced originally very cold air from Siberia into the Canadian Yukon via the polar cap and computed from aerological soundings minimum surface temperatures in fair agreement with those observed.

Period	Thickness (ft)	Mean virtual temperature (° C)
January		
1944	8,920	−12.8
1945	8,940	−12.1
1946	8,870	−14.2
1947	8,820	−15.7
1948	8,890	−13.7
1949	8,870	−14.2

Table 3 Monthly mean thickness and corresponding mean virtual temperature between 1,000 and 700 mb for the Polar reservoir between 0° and 180° W and between 55° N and 70° N

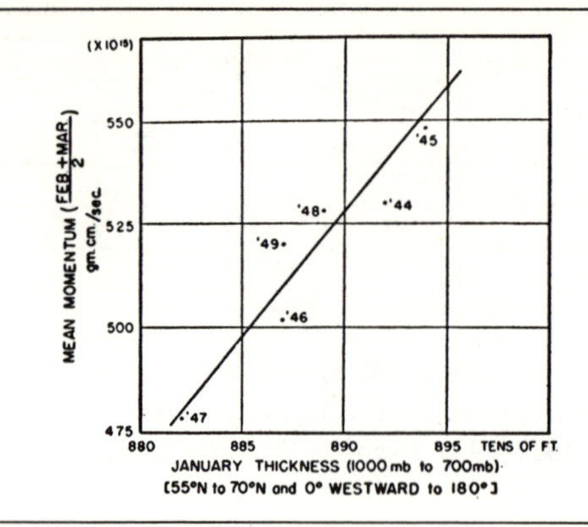

Figure 7 Mean monthly momentum of the flow at 700 mb from mid-February to mid-March (ordinate) as a function of mean thickness of the layer 1,000–700 mb over sub-polar regions (55° N to 70° N and 0° to 180° W) during the preceding January.

wind profiles during the period encompassing a good part of the low index portion of the cycle. Data of this sort have already been presented in Figure 4 for monthly mean wind profiles during the 6-year period. Returning to Figure 4, it is apparent that the month of February has the greatest year-to-year variability in momentum between latitudes 22½° to 72½° N and between 0° and 180° W—apparently a reflection of the marked year-to-year differences in the intensity of the cycle. From the wind profiles (not reproduced) upon which Figure 4 was based, it is clear that, particularly at times of low index, when the jet stream is displaced far to the south, westerly momentum is apparently lost by the atmosphere north of 20° N and is gained (reappears) south of here. Examples appear in Figure 5. The momentum computed and indicated in Figure 4 may therefore in a rough sense be a measure of the intensity of the index cycle—low values representing an intense cycle in which the upper-level westerlies of the tropics are strengthened, and high values a weak cycle. Now, the index minima during these six years (1944–49) were generally reached around the end of February or the first part of March. Interpolating in Figure 4 between February and March, we obtain the following values:

	Momentum from mid-February to mid-March (gram cm/sec)
1944	530×10^{15}
1945	548
1946	502
1947	478
1948	528
1949	520

These values are plotted against the original January mean temperatures of the sub-polar cap (from Table 3) in Figure 7. Thus it appears that *the intensity of long index cycles is largely determined by the reservoir of cold air preceding their onset.*

Since the reservoir of cold air is assumed to be largely a function of the period of high-index imprisonment, here is a clue to the possible effectiveness of a blocking wave in producing an extended index cycle. Thus, in the blocking waves marked B in Figure 1, it is evident that many of those which failed to produce a major cycle were too close to a recent period in which the atmospheric condenser had discharged in the form of an index minimum well below normal. It would thus appear that a certain recovery time is needed before the condenser can restore its reservoir of cold air and produce another discharge in the form of a great index cycle. This conclusion, arrived at independently, is reminiscent of the one by Rossby and Willett (1948):

> Rossby visualizes this cycle of index change as having a natural period which depends upon the relative effectiveness of the radiational cooling processes in the higher latitudes. This period appears to be shorter at the beginning of the winter and because the rate of cooling is most rapid at that time. Each index cycle tends somewhat to expand the effective polar cap and to diffuse the thermal contrasts equatorward.

The flow patterns in the Atlantic during the initial decline of the index tend to favor cellular structures with warm pools (highs) at high latitudes and cold pools (lows) at low latitudes. The well-known stability of such great vorticities is especially favored during late winter when the surface stability, because of increased Arctic ice and cold surface water, can reach a maximum in northern latitudes under the warm highs and thus, as Wexler (1943) points out, permit little air to escape from the friction layer of the anticyclones. At the same time, the cold lows at low latitudes (which are prevailingly found over the ocean between the Azores and Spain), because of strong vertical temperature gradients, can probably be maintained against the filling inflow in the friction layer by vigorous convective showers with their vast supplies of latent heat. As pointed out by Namias and Clapp (1944), the decline of circulation first appearing in the Atlantic progresses slowly westward at about 60° per week. But it is important to note that this

Period	Atlantic	North American	Eastern Pacific
26 February–4 March to 4–8 March 1944	8.4	3.4	−3.4
24–28 February to 3–7 March 1945	−6.5	4.5	5.2
16–20 February to 23–27 February 1946	6.0	0.6	−2.1
1–5 March to 8–12 March 1947	5.0	1.0	1.2
28 February–3 March to 6–10 March 1948	4.0	2.0	0.0
9–13 March to 16–20 March 1949	0.9	6.0	3.9
Mean	3.0	2.9	0.8

Table 4 Changes in regional zonal index at 700 mb in m sec^{-1} during the week immediately following the minimum of the index cycle

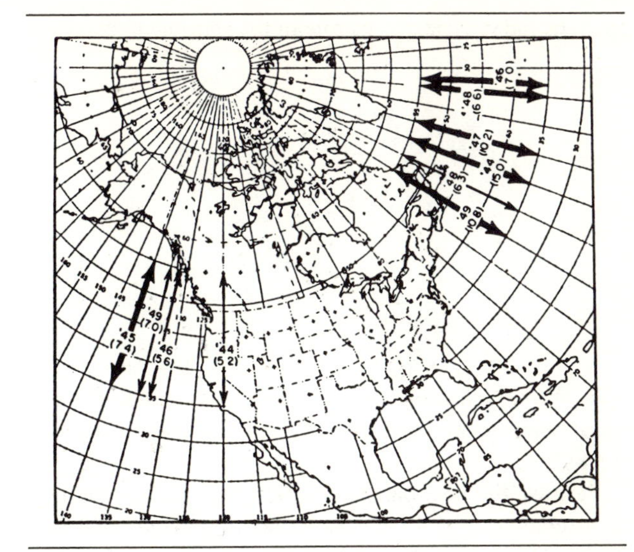

Figure 8 Longitudes of primary and secondary increase of zonal index at sea level for the week following the index minimum (upper figure gives year, lower figure the rise in m sec^{-1}).

westward local diminution in index during an index cycle proceeds without disrupting the initial cellular blocking pattern set up in the Atlantic. This contention is strengthened by the large number of half-week periods the Atlantic blocking persisted within index cycles as compared with the smaller number elsewhere. These numbers are indicated by subscripts attached to the B in Figure 1. Here again only 1945, the unique cycle year, is out of line. Thus we may consider the initial Atlantic wave of blocking of the index cycle as an infection of the westerlies which spreads upstream as a malignant growth, progessively decaying the zonal circulation.

INITIAL INCREASE OF ZONAL INDEX FOLLOWING THE MINIMUM

The recovery of zonal index from its minimum of the cycle appears to take place in late February or early March. Some light is shed on the manner in which this increase is effected by regional changes in index for the week subsequent to the minimum. The changes are shown in Table 4 in a similar fashion to those for the initial decline shown in Tables 1 and 2. Apparently the initial increase is favored in the Atlantic (4 out of 6 cases and the highest mean change), to a lesser degree in North America, and seldom (only 1945) in the eastern Pacific. Remembering that 1945 was a contracted circumpolar-vortex cycle, it appears that the Atlantic is normally favored for the initial index increase, just as it was for the initial decrease. A further study of sea-level pressure-change maps suggests that an even more localized area of major increase can be found in the western Atlantic. The longitudes of greatest increase and secondary increase in local index are shown in Figure 8. It appears that the greatest initial increases in index occur in the western Atlantic with a secondary area of increase just off the Pacific coast. These two areas are precisely those in and to the east of common confluence zones—in the western Atlantic between cold, Labrador and tropical-Gulf (and Atlantic) air and off the Pacific coast between cold, Arctic air from Alaska and warmer, sub-tropical air from the eastern cell of the Pacific High.

The above paragraph is merely a *description* of how the index begins its upward climb from the minimum of the cycle. It does not explain *why* the flow patterns evolve in a manner favorable to the initiation of these zones of confluence. Indeed, this problem is a most difficult one to solve, and at present we can only bring up for consideration some apparently relevant points. One of these is that around the time of the index minima (late February and early March), total incoming solar radiation begins to increase rapidly and thereby tends to destroy the very elements responsible for low index—the cold pools of low latitudes and the warm pools of high latitudes. Perhaps this factor becomes dominant at this time of year. It is quite probable that the increasing insolation may produce different effects in different areas of the upper-level flow patterns and that these effects favor the initiation of the flow patterns favorable to the confluence cited above.

SUMMARY

To sum up, we may now return to the three questions posed earlier in this study.

1. Why during each year is there a tendency for the occurrence of at least one index cycle lasting several weeks?

Because the heat balance of the atmosphere demands an exchange of air between polar and equatorial latitudes. This exchange could operate in short week-to-week oscillations of the index, were it not for slow-moving blocking waves with their strong, deep, meridional components of flow gradually operating to deplete vast reservoirs of cold air which are imprisoned at high latitudes during extended periods by strong mid-latitude westerlies.

2. Why do these cycles vary in intensity from one year to the next?

Because, for reasons unknown, the quasi-permanent anchor troughs and ridges of the mid-troposphere are fixed in different regions in different years and thus in some years the flow patterns become more highly organized into confluence (high index) bands than others. In this manner, the cold reservoir becomes stored or depleted at different times of the year. Since the intensity of the cycle appears to be dependent upon the amount of stored cold air, which in turn depends both on the net outgoing radiation and upon the storage time, it is clear that there must be considerable variation in the character of index cycles through the years. What we do not yet know is why the quasi-permanent anchor troughs and ridges of mid-troposphere are fixed in different regions in different years; it is quite possible that lag effects of ocean currents, snow cover, etc., are dominant factors.

3. Why does there appear to be one particular period (late February) when the most pronounced cycle occurs?

Because the two primary factors responsible for the cycle, an extensive cold air reservoir and the initial, stable, cellular-type, Atlantic blocking-mechanism are both favored at this time of year. The cold reservoir is favored by the long polar night over a pre-established, extensive, snow cover. The Atlantic warm anticyclone to the north and the cold low to its south, concomitant features of blocking, are both favored by considerations of vertical stability in the friction layer as deep pools of polar air move into southerly latitudes and deep pools of tropical air move into high latitudes. The stability under the warm highs results in less "leakage" from below, while the instability at low latitudes, extending through a deep layer of air over water surfaces, more than compensates the frictional filling by producing copious showers releasing much latent heat.

REFERENCES

Aubert, E. J., 1949. *Solar-weather relationships.* Unpublished manuscript in the Extended Forecast Section, U.S. Weather Bureau, Washington, D.C.

Namias, J., 1947a. Physical nature of some fluctuations in the speed of the zonal circulation. *J. Meteor. 4*, 125–133.

Namias, J., 1947b. Characteristics of the general circulation over the northern hemisphere during the abnormal winter 1946–47. *Mon. Wea. Rev. 75*, 145–152.

Namias, J. and Clapp, P. F., 1944. Studies of the motion and development of long waves in the westerlies. *J. Meteor. 1*, 57–77.

Namias, J. and Clapp, P. F., 1949. Confluence theory of the high tropospheric jet stream. *J. Meteor. 6*, 330–336.

Rossby, C.-G., 1939. Relations between variations in the intensity of the zonal circulation and the displacements of the semipermanent centers of action. *J. Marine Res. 2*, 38–55.

Rossby, C.-G. and Willett, H. C., 1948. The circulation of the upper troposphere and lower stratosphere. *Science 108*, 643–652.

Wexler, H., 1943. Some aspects of dynamic anticyclogenesis. *Dept. Meteor. Univ. Chicago, Misc. Rept. 8*, 28 pp.

Wexler, H., 1948. A note on the record low temperature in the Yukon Territory January–February 1947. *Bull. Amer. Meteorol. Soc. 29*, 547–550.

Willett, H. C., 1947. *Final report of the Weather Bureau–MIT extended forecasting project for the fiscal year July 1, 1946–July 1, 1947.* MIT. Cambridge. 110 pp.

Willett, H. C., 1948. Patterns of world weather changes. *Trans. Amer. Geophys. Union 29*, 803–809.

SUGGESTED READING FOR PART 2

Defant, F. and Taba, H., 1958. The strong index change period from January 1 to January 7, 1956. *Tellus 10(2)*, 225–242.

Dzerdzeevskii, B., 1962. Fluctuations of climate and of general circulation of the atmosphere in extra-tropical latitudes of the northern hemisphere and some problems of dynamic climatology. *Tellus 24*, 328–336.

Lamb, H. H., 1961. Fundamentals of climate. In *Descriptive Paleoclimatology*, ed. by A. E. M. Nairn, Interscience Publishers, New York, 8–44.

Lorenz, E. N., 1966. The circulation of the atmosphere. *Amer. Scientist 54(4)*, 402–420.

Namias, J., 1964. Seasonal persistence and recurrence of European blocking during 1958–1960. *Tellus 16(3)*, 394–407.

Reiter, E. R., 1967. *Jet Streams.* Doubleday and Company, New York, 189 pp.

Part Three

Evapotranspiration and Plant Growth

Part 3 begins with Barry's concise and lucid review of the method and instrumentation involved in measuring evaporation and transpiration.

It continues with Mather and Yoshioka's study of the relationships between climate and vegetation in the conterminous United States. They find temperature and precipitation to be by themselves poor descriptors of the relationship of climate to the growth of vegetation. Those climatic indices active in growth of vegetation, that is, potential evapotranspiration and moisture index, are more relevant monitors and may provide an accurate basis for differentiating mid-latitude vegetation groupings. Why then is it difficult to explain climatically the distribution of Redwood and Douglas fir?

In reviewing the literature on agroclimatological relationships, Maunder is forced to conclude that many relationships are indeed complex and that the expression of weather as a resource is far from simple. As a step toward identifying the potential economic importance of significant departures from the climatic norm, he develops a statistical model which can help predict, for instance, the effect of increased or decreased rainfall on crop yields.

R. G. Barry

6 Evaporation and transpiration

Investigation of the transfer of moisture from the surface of the earth to the atmosphere concerns workers from a number of disciplines. On the practical side there are agriculturalists, foresters, and hydrologists, and on the theoretical side meteorological physicists and plant physiologists. The physical controls on evaporation have been recognized since 1802, when John Dalton first stated the basic principles, but it is only during the last twenty or so years that the active exchange of practical and theoretical findings between research workers has begun to provide a coherent body of knowledge. Inevitably there are innumerable specialized papers on evaporation reflecting these different approaches, and the treatment here is necessarily restricted to a statement of the basic concepts and an outline of some applications and results. More detailed accounts of the theory are provided by King (1961), Sellers (1965), and Thornthwaite and Hare (1965).

Basic Mechanisms of Evaporation

Net transfer of water molecules into the air occurs only if there is a vapor-pressure gradient between the evaporating surface and the air, i.e., evaporation is nil when the relative humidity of the air is 100%. Evaporation from a moist surface involves a change of state from liquid to vapor, and therefore necessitates a source of latent heat. To evaporate 1 g of water requires 540 cal of heat at 100° C and 600 cal at 0° C. An external heat source must therefore be available. This may be solar radiation, sensible heat from the atmosphere, or from the ground. Alternatively, it may be drawn from the kinetic energy of the water molecules, thus cooling the water until equilibrium with the atmosphere is established and evaporation ceases. In general, solar radiation is the principal energy source for evaporation.

In addition to the two primary controls, the evaporation rate is affected by wind speed, since air movement carries fresh unsaturated air to the evaporating surface. Within approximately 1 mm of the surface the upward movement of vapor is by individual molecules ("molecular diffusion"), but above this surface boundary layer turbulent air motion ("eddy diffusion") is responsible. The temperature of the evaporating surface also affects evaporation. At higher temperatures more water molecules can leave the surface due to their greater kinetic energy. Salinity depresses the evaporation rate in proportion to the solution concentration. For seawater the rate is about 2–3% lower than for fresh water.

Plant Factors

Water loss from plants—*transpiration*—takes place when the vapor pressure in the air is less than that in the leaf cells. About 95% of the diurnal water loss occurs during the daytime, because water vapor is transpired through small pores, or *stomata*, in the leaves, which open in response to stimulation by light. Transfer of water vapor to the atmosphere is the initiating process in the movement of water from the soil via the plant. It is a passive process so far as the plant is concerned, but it performs a vital function in effecting the internal transport of nutrients and in cooling leaf surfaces. Transpiration considerably exceeds the direct water needs of the plant. Nevertheless, the transfer of water to the air is unavoidable. In the absence of a plant cover evaporation would still occur from the soil.

Interaction between soil-moisture content and root development is a complicating factor. If soil water is not replenished over a period of weeks vegetation with deeper roots, especially trees, will transpire more than shallow-rooted plants, other things being equal. Some support for this idea is provided by catchment studies. Run-off from catchments under grass generally exceeds that from catchments under woodland. However, this problem remains a subject of considerable controversy.

Reprinted, with minor editorial modification, by permission of the author and publisher, from *Water, Earth, and Man*, R J. Chorley (editor), Methuen and Co., 1969, pp. 169–184.

Resistances to water movement, both in the soil and in plant tissues, must be considered. These include soil-water tension, the resistance of cell walls in the roots and leaves to water transport, and the resistance of stomata to vapor transfer. The *internal* (stomatal) *resistance* of a single leaf to diffusion is an important control on transpiration, and it is dependent on the size and distribution of the stomata. For a crop or vegetation cover with several leaf layers the effective stomatal resistance (r_s) is reduced to approximately 30% of that of an individual leaf, owing to the decreased ventilation within the cover. Seasonal variations associated with changes in the leaf area affect r_s, as do diurnal variations. The latter result partly from the opening and closing of the stomata with light intensity and partly from the effects of transpiration stress on the stomata when water uptake lags behind transpiration. A separate *external resistance* (r_a), which is the time for a unit volume of air to exchange water vapor with a unit area of surface, depends on both wind speed and surface geometry. A decrease in r_a may be due to higher wind speeds or greater "roughness" of the vegetation surface, which causes increased turbulence in the air flow. Generally the stomatal resistance r_s is larger than r_a, although the interaction of r_s and r_a is an important determinant of evaporation rates.

A further effect of a vegetation cover is that it intercepts precipitation before it can reach the surface. A forest canopy may retain up to 30% of total precipitation (more for conifers than deciduous species), and the proportion is larger for light, showery precipitation.

The amount reaching the ground via stem flow varies according to tree-type, but the bulk is evaporated without entering into the soil–plant part of the cycle. This might be regarded as an excessive loss of moisture compared with a grass cover, especially in the winter. However, the radiant energy used in evaporating intercepted water is unavailable for other evapotranspiration, and hence interception is not as serious a problem as it might appear to be.

Potential Evapotranspiration

Moisture transfer from a vegetated surface is often referred to as evapotranspiration,[1] and when the moisture supply in the soil is unlimited the term potential evapotranspiration (PE) is used. It has been suggested that PE can be defined more specifically as the evaporation equivalent of the available net radiation, i.e., $PE = R_N/L$, where L is the latent heat of vaporization (59 cal cm^{-2} $\approx$ 1 mm evaporation). In some cases this equivalence may be invalid. For example, if an irrigated area is surrounded by dry fields evaporation rates can exceed R_N/L by 25–30%. Air heated by passing over the dry areas upwind maintains the high rates through the downward transfer of sensible heat to the irrigated section—the so-called "oasis effect". Horizontal transport (advection) of sensible heat *through* the vegetation cover (Figure 1) can also cause anomalous evaporation rates—termed the "clothesline effect". This occurs when a study plot is not surrounded by a zone with identical vegetation cover and environmental conditions. The "buffer zone" necessary to eliminate these effects varies in size, but may exceed 300 m radius. Nevertheless, for all short crops of approximately the same color and completely covering the ground the *PE* rate is essentially determined by the total available energy as long as there is unlimited soil water. Plant physiology is important in the case of specialized crops, such as rice and sugar cane (high water use rates) and pineapple (low usage).

Actual Evaporation

It is known that when the moisture supply in the soil is limited plants have difficulty in extracting water, and the evaporation rate (E) falls short of its maximum value (PE). The precise nature of this relationship is controversial. One view is that the potential rate is maintained until soil-moisture content drops below some critical value, after which there is a sharp decrease in evaporation, while another is that the rate decreases progressively with diminishing soil moisture. At field capacity (maximum soil moisture content under free drainage) $E/PE = 1$, i.e., evaporation proceeds at the maximum potential rate. Veihmeyer and Hendrickson consider that no change takes place in this ratio until the plant is

[1] In agricultural studies the term "consumptive use" (CU) of water by crops is commonly used. However, in irrigation engineering, where the term is applied to irrigated crops, $CU = PE$. At certain times of year CU is less than PE. Consequently, there is a risk of confusion and misinterpretation.

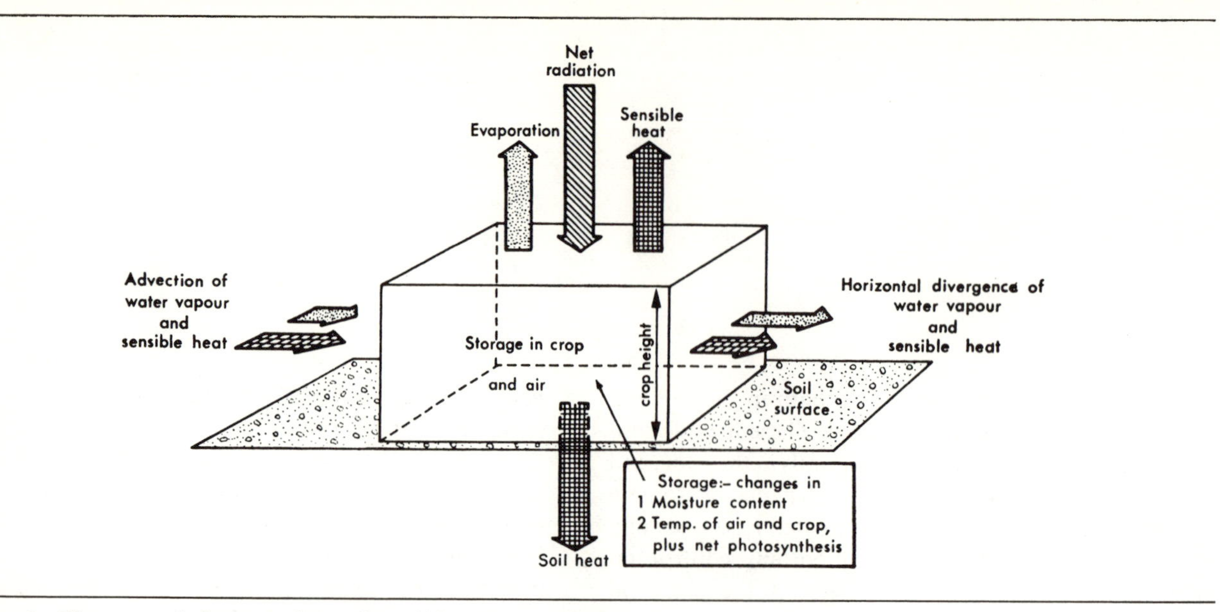

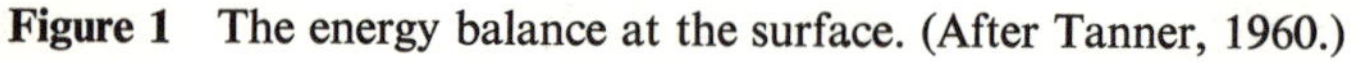

Figure 1 The energy balance at the surface. (After Tanner, 1960.)

near wilting point (Figure 2). Thornthwaite and Mather assume the decrease below field capacity to be a logarithmic function of soil suction, but recent work suggests that $E/PE \approx 1$ as long as the moisture content is at least 75% of field capacity. Undoubtedly the soil type and climatic conditions are important; field capacity ranges from 25 mm in a shallow sandy soil to 550 mm in deep clay-loams. Chang (1965) indicates that Veihmeyer and Hendrickson's results may apply to a heavy soil with vegetation cover in a humid, cloudy region, whereas in sandy soils with a vegetation cover under arid conditions a rapid decline in E/PE is likely. Experimental work by Holmes (1961) supports this view; see lines (1) and (2) on Figure 2.

Figure 2 The relationship between the ratio of actual to potential evapotranspiration (E/PE) and soil moisture. (After Holmes, 1961, and Chang, 1965.)
V and H = Veihmeyer and Hendrickson (1955).
Th and M = Thornthwaite and Mather (1951).
1 and 2 = Schematic curves for a vegetation-covered clay-loam under low evaporation stress and a vegetation-covered sandy soil under high evaporation stress, respectively.
1 Bar = 1,000 millibars (10^6 dynes/cm^2).

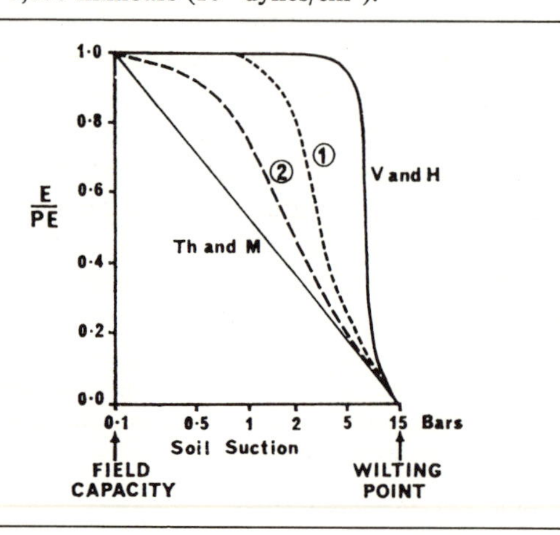

Meteorological Formulae

There are two principal lines of approach to estimating evaporation through physical relationships; one is the aerodynamic (or mass transfer) method, the other is the energy budget method.

Aerodynamic method

This method considers factors controlling the removal of vapor from the evaporating surface. These are the vertical gradient of humidity and the turbulence of the air flow. The mathematical expression relates evaporation from (large) water bodies to the mean wind speed at height $z(u_z)$, and the mean vapor pressure difference between the water surface and the air at level $z(e_w - e_z)$,

$$E = Ku_z(e_w - e_z) \qquad (1)$$

where K is an empirical constant. e_w is calculated for mean water surface temperature. The method has been applied to ocean areas in particular, but only for monthly averages, since it assumes that the temperature lapse rate is adiabatic, and this does not apply on many individual occasions. More elaborate forms of equation (1) incorporating complex wind functions have been developed for land surfaces and other lapse-rate conditions, but their value is limited mainly to the provision of independent estimates of evaporation for research purposes.

Energy budget method
From fundamental principles of the conservation of energy it follows that the net total of long- and short-wave radiation received at the surface (R_N) is available for three processes (Figure 1): the transfer of sensible heat (H) and of latent heat (LE) to the atmosphere and of sensible heat into the ground (G). That is,

$$R_N = H + LE + G \qquad (2)$$

The fraction of R_N used in plant photosynthesis is generally negligible. Accordingly, evaporation can be determined by measurement of the other terms

$$E = \frac{R_N - H - G}{L} \qquad (3)$$

R_N can be measured by the use of a net radiometer, and G is calculated from data on the soil-temperature profile or by direct measurement of soil heat flux, but H cannot readily be estimated. An indirect method is to employ Bowen's ratio $\beta = H/LE$. This is calculated from the ratio of the vertical gradients of temperature and vapor pressure. However, the determination is unreliable when the surface is dry and H is large. On substitution of β in (3)

$$E = \frac{R_N - G}{L(1 + \beta)} \qquad (4)$$

The use of Bowen's ratio assumes that the vertical transfer of heat and water vapor by turbulence takes place with equal efficiency. Recent work in Australia (Dyer, 1967) shows that this assumption is universally valid. Given the requisite observational data, the energy budget approach is a practicable one for determining evaporation over periods as short as an hour.

Combination methods
A number of methods have been developed to combine the aerodynamic and energy budget approaches, thereby eliminating certain measurement difficulties which each presents. The most widely used combination method was derived by Penman (1963). He expresses PE as a function of available radiant energy (R_N) and a term (E_a) combining saturation deficit and wind speed.

$$R_N = 0.75S - L_N \qquad (5)$$

where
$0.75S$ = solar radiation absorbed by a grass surface;
L_N = net long-wave (terrestrial) radiation from the surface.

$$E_a = f(u)(e_s - e) \qquad (6)$$

where
$f(u) = 0.35(1 + 0.01u)$ for short grass;
u = wind speed at 2 meters (miles/day);
e_s = saturation vapor pressure (mm mercury) at mean air temperature;
e = actual vapor pressure at mean air temperature and humidity.

The expression for PE from short grass[2] is

$$PE\,(\text{mm/day}) = \frac{\left(\frac{\Delta}{\gamma}\frac{R_N}{L} + E_a\right)}{\frac{\Delta}{\gamma} + 1} \qquad (7)$$

where
$\frac{\Delta}{\gamma}$ = a weighting factor for the relative effects of energy supply and ventilation;[3]
γ = 0.27 (mm mercury/° F), the psychrometric constant;
$\Delta = \frac{de_s}{dt}$, the change of saturation vapor pressure with mean air temperature (mm mercury/° F).

It is worth while examining certain of these terms further. In equation (5) the 0.75 weighting of the incoming solar radiation is due to the 25% albedo (reflection coefficient) of short grass. Most green crops have a similar reflectivity, but values for coniferous forest and heath are approximately 15%, while an average value for water is 5%. This factor should augment evaporation from a water surface compared with grass or crops, but at least a partial compensation is provided by the greater aerodynamic roughness of these surfaces. Rutter (1967),

[2] In Penman's original formulation evaporation was determined first for an open-water surface (PE_0), then a weighting factor (f) of 0.6–0.8 was used according to season and type of surface. $PE = f \,.\, PE_0$.

[3] Equivalent to $1/\beta$ when evaporation is independent of wind speed and surface roughness.

for example, finds an annual evaporation of 679 mm from Scots Pine (*Pinus sylvestris*) in Berkshire for 1957–63, compared with a calculated open-water evaporation of 597 mm. Observational evidence is by no means unanimous, however, on this point. The meaning of the saturation deficit term $(e_s - e)$ in equation (6) is a common source of misunderstanding. It represents the "drying power" of the air, but this need not be directly related to evaporation. In fact, $(e_s - e)$ is likely to be greatest when the surface is very dry and no moisture is available for evaporation. Certain aspects of this approach require further comment. First, the basis of the formulation involves the assumptions of Bowen's ratio, discussed in the previous section. Second, the transfer of sensible heat into the ground is neglected. Third, the use of mean temperatures and humidities makes it unsuitable for short-period estimates (< 24 hours) of evaporation rates.

Budyko (1956) independently derived a similar expression for *PE*, and Figure 3 illustrates the accuracy of the Penman–Budyko approach compared with lysimetric observations in Australia.

A much simplified approach incorporating saturation deficit and radiation has been developed by Olivier (1961). The equation for *PE* (mm/day) is

$$PE = (T - T_w)\frac{\bar{L}}{L^2} \qquad (8)$$

where

$L = \dfrac{S}{S_v}$, $\bar{L} = \dfrac{\bar{S}}{\bar{S}_v}$;

S = total solar radiation under clear skies for the latitude of the station for a particular month;
S_v = vertical component of S;
$\bar{S}$ = average of the 12 monthly values of S;
$\bar{S}_v$ = average of the 12 monthly values of S_v;
T = mean monthly temperature (° C);
T_w = mean monthly wet-bulb temperature.

Figure 4 shows the estimate of *PE* obtained in three different regimes using Penman's and Olivier's formulae.

Temperature formulae

One of the best-known methods of estimating *PE* was developed by Thornthwaite (1948). He related observations of consumptive use of water in irrigated areas in the western United States to air temperatures, with adjustments for daylength.

$$PE\,(\text{mm/month}) = 16\left(\frac{10T}{I}\right)^a \qquad (9)$$

where

T = mean monthly temperature (° C);
a = an empirical function of I;
$I = \sum_{1}^{12}\left(\frac{T}{5}\right)^{1.514}$

The values can be readily calculated from published tables or nomograms. The method has been widely applied, although in some climatic regimes it gives unreliable results, as Figure 3 indicates for south-east Australia.

A more soundly based relationship has been illustrated by Budyko. He shows that if heat storage and sensible heat transfer are effectively zero annual *PE* is given approximately by

$$PE\,(\text{mm/year}) \approx \frac{R_{NO}}{L} \approx 0.18 \sum T \qquad (10)$$

where

R_{NO} = the net radiation budget of a wet ground surface;
$\sum T$ = the sum of daily mean temperature which exceeds 10° C.

Another temperature formula has been developed by Turc for *actual* evapotranspiration. Annual amounts (mm) for catchments are given by

$$E = \frac{P}{\sqrt{\left\{0.9 + \left(\frac{P}{I}\right)^2\right\}}} \qquad (11)$$

where

P = annual precipitation (mm);
$I = 300 + 25T + 0.05T^3$;
T = mean air temperature (° C).

This method has not been tested sufficiently to make any sound assessment possible.

Evaporation Measurement

There are four main types of measuring device, although each has its limitations and disadvantages.

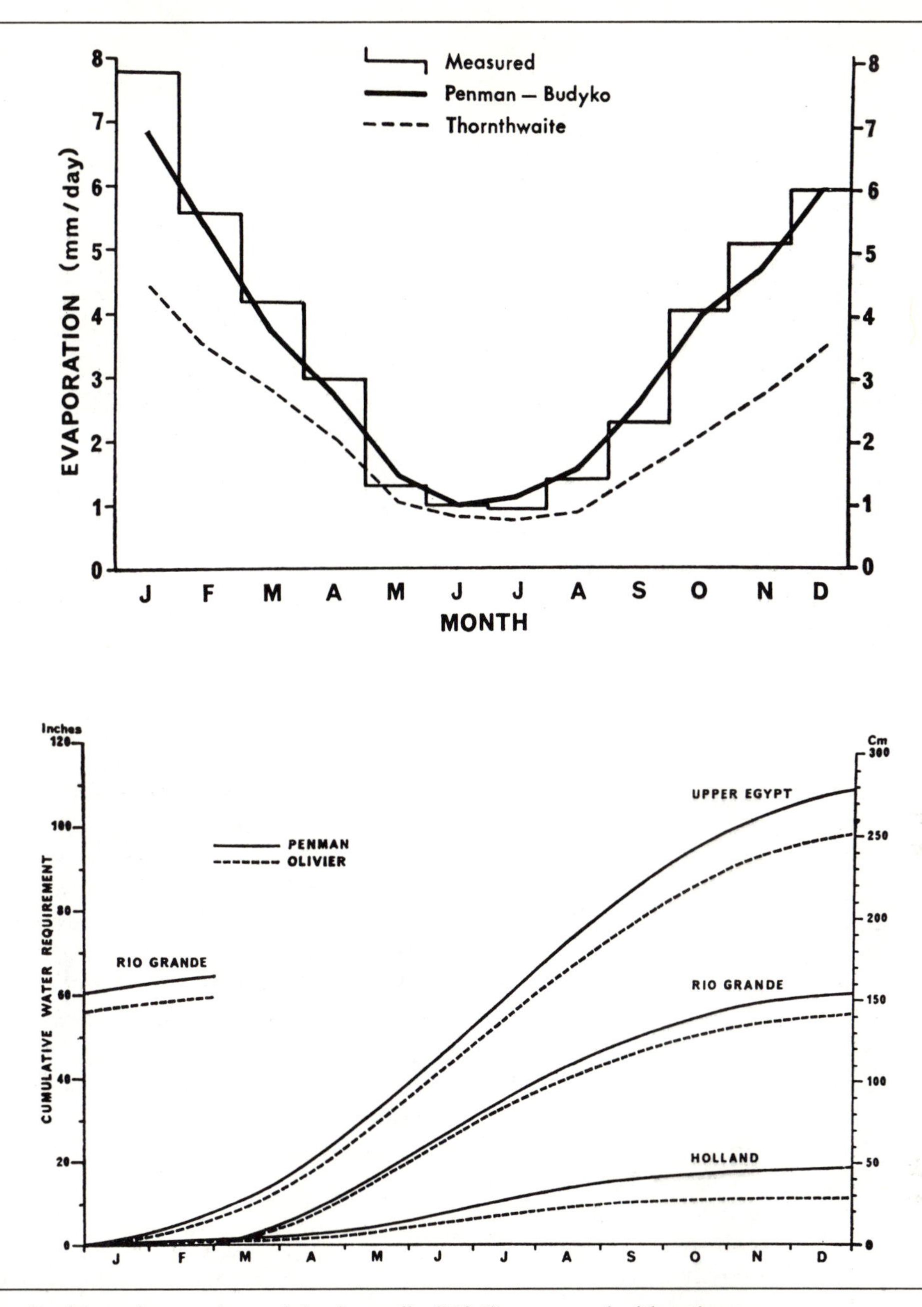

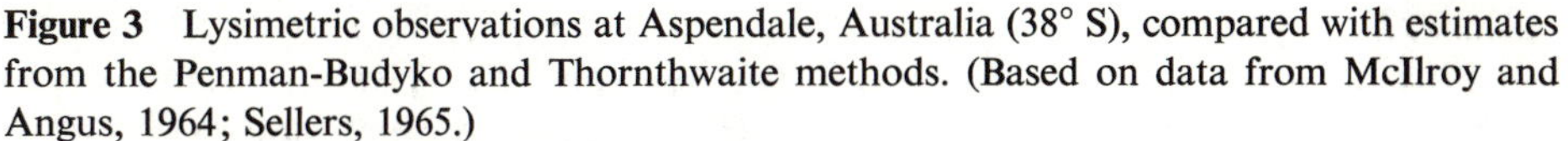

Figure 3 Lysimetric observations at Aspendale, Australia (38° S), compared with estimates from the Penman-Budyko and Thornthwaite methods. (Based on data from McIlroy and Angus, 1964; Sellers, 1965.)

Figure 4 Estimates of potential evapotranspiration in three different climatic regimes using Penman's and Olivier's methods. (After Olivier, 1961.)

Atmometers

These are water-filled glass tubes having an open end through which water evaporates from a filter-paper (Piche type) or porous plate (Bellani type). The tube supplying water is graduated to read evaporation in mm, but the evaporation (termed "latent evaporation") can only be compared with readings from another such instrument in a similar exposure, and may bear little or no relation to evaporation from land or water surfaces, since it only reflects the saturation deficit of the air. The instrument is apparently more responsive to wind speed than radiant energy.

Evaporation pans

The "Class A" pan of the United States Weather Bureau, which is approved by the World Meteorological

Organization, is 122 cm (48 in) in diameter and 25 cm (10 in) deep. Problems can arise through splashing, heating of the pan walls, and interference by birds or animals, while the installation position (sunken or mounted on or above the surface) is particularly critical. Unfortunately, pan evaporation (E_p) is not related to lake evaporation (E_L) in any simple or constant manner. For the United States E_L/E_p is generally within the range 0.6–0.8.

In general, evaporation decreases as the size of the water body increases. In part, this arises from the "oasis" effect. Air travelling over a large water surface picks up sufficient moisture to reduce the evaporation rate towards the leeward shore. Water depth is another cause of pan–lake differences. Much energy in spring goes into heating a deep lake, thereby suppressing evaporation rates. Morton (1967) calculates the average annual evaporation from Lake Ontario as 813 mm (32.0 in), whereas that from Lake Superior is only 546 mm (21.5 in). In the case of pan–lake comparisons these seasonal effects are even more serious, and it is only safe to use pan data for annual estimates of lake evaporation.

Even allowing for the regional and other factors which affect the ratio E_L/E_p (the "pan coefficient"), the measurements provide no indication of evaporation from a land surface. The other two approaches now to be described are directed to this end.

Lysimeters
A lysimeter is an enclosed block of soil with a vegetation cover (usually short grass) similar to that of its surroundings. Figure 5 illustrates the installation at Hancock, Wisconsin. At regular intervals the weight change (ΔS), precipitation (P), and percolation (r) are measured. By means of the moisture-balance equation, evapotranspiration is determined as a residual.

$$E = P + \Delta S - r \quad (12)$$

The use of a large block allows an accuracy of 0.01 in of water-depth.

A simpler version for PE determination is the "Thornthwaite" type of evaporimeter. Here the moisture supply is maintained by "irrigating" the block when necessary, so that the soil-moisture storage can be regarded as constant. The percolation (r), precipitation (P), and added water (W) are measured.

$$PE = P + W - r \quad (13)$$

With both types of device the presence of the tank's base may interfere with the soil moisture profile compared with that in a natural soil unless special precautions are taken.

The "evapotron"
Attempts have recently been made to measure the vertical transfer of moisture directly. Instruments developed by C.S.I.R.O. in Australia measure the magnitude and direction of vertical eddies which transfer water vapor upwards. There are many difficulties with this approach. In particular, there is a need for instruments that measure instantaneous changes of both the vapor content and vertical velocity of the air. The subsequent determination of average evaporation rates requires a computer to integrate the results. Moreover, effects of advection and storage below the measuring level (see Figure 1) may present serious difficulties unless measurements are made near the surface. This technique seems likely to be limited to research applications.

Budget Estimates

The moisture-budget equation already referred to can be used in two very different ways to estimate evapotranspiration from large areas over a time period of the order of a month.

Catchment estimates
If suitable allowance can be made for storage in the catchment system, or if it is assumed constant, over a sufficiently long time interval,

$$E = r - P \quad (14)$$

where r is the runoff measured by river gauging. Checks of this kind showed that estimates of evaporation from the Thornthwaite formula in northern Finland and northern Labrador–Ungava were 80% too high. This may reflect low snowfall estimates as well as low water use by moss and lichen surfaces.

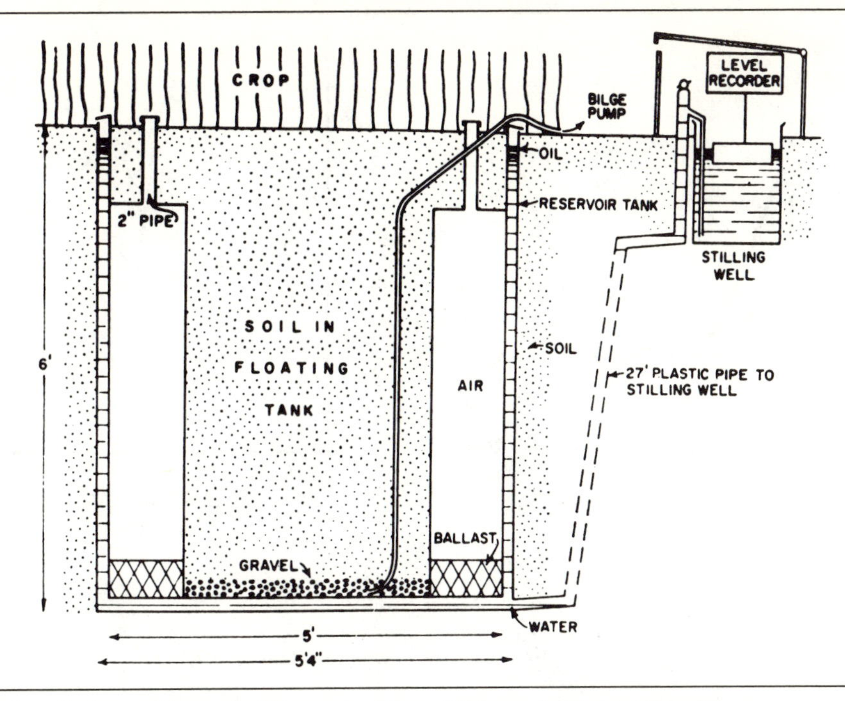

Figure 5 Lysimeter installation at Hancock, Wisconsin. (From King *et al.*, *Transactions 37*: p. 739, 1956.) Here the soil block floats in a tank of water. Changes of water level are recorded instead of weighing the block.

Aerological estimates
In analogous manner evaporation can be estimated from data on atmospheric moisture.

$$E = \Delta D - P \pm \Delta S \qquad (15)$$

where ΔS is the storage change in the overlying air column and ΔD is the net divergence (or convergence) of water vapor out of (or into) the column. This method requires very complete aerological records.

Evaporation Rates in Different Macroclimates

The global pattern of annual evaporation has been outlined elsewhere and we can now look in more detail at variations in seasonal regime.

For the ocean areas of the world an average of 90% of annual R_N is used for evaporation, whereas for the continents the figure is only just over 50% and the remainder represents sensible heat transferred to the soil and the atmosphere. The inverse correlation of the seasonal march of R_N and LE over the oceans is therefore at first sight unexpected (Figure 6). This is a result of the complex interaction of heat storage and heat transfer by ocean currents. Much of the required energy for ocean evaporation is derived from the water itself, and the rate is mainly determined by wind speed and the vapor-pressure gradient. Over the Indian Ocean the summer maximum is caused by the higher wind speeds, while cloudiness diminishes the radiation receipt. The secondary winter maximum is due to the advection of dry trade-wind air. Off the eastern shores of Asia and North America there are large evaporation losses in winter as cold, dry continental polar air flows across the warm Gulf Stream and Kuro Shio Currents. In summer, however, reduced wind speeds and low air–sea vapor-pressure differences suppress evaporation rates. It may be recalled that this is the season when large horizontal moisture fluxes are directed from the continent.

Over land, the seasonal regime generally reflects the occurrence of maximum net radiation receipts and maximum surface-air vapor pressure difference. Where precipitation occurs mainly in summer, or has an even distribution throughout the year, there is a simple summer maximum and winter minimum of evaporation. Figure 7 illustrates typical profiles for West Palm Beach, Florida, and Paris. In areas of summer drought and winter rains there is a spring evaporation maximum, such as at Lisbon, while in districts with rains in autumn and winter there may be a double maximum in spring and autumn as at Yuma, Arizona.

Regional and local differences in evaporation rate arise not only from variations in the meteorological controls but also from largely independent soil and vegetation factors. For example, observations in the Canadian Subarctic indicate that $LE/R_N \approx \frac{1}{3}$ over lichen (*Cladonia* spp.) surfaces. The low evaporation rate is apparently due to the negligible extraction of moisture from the soil by nonvascular vegetation. Local

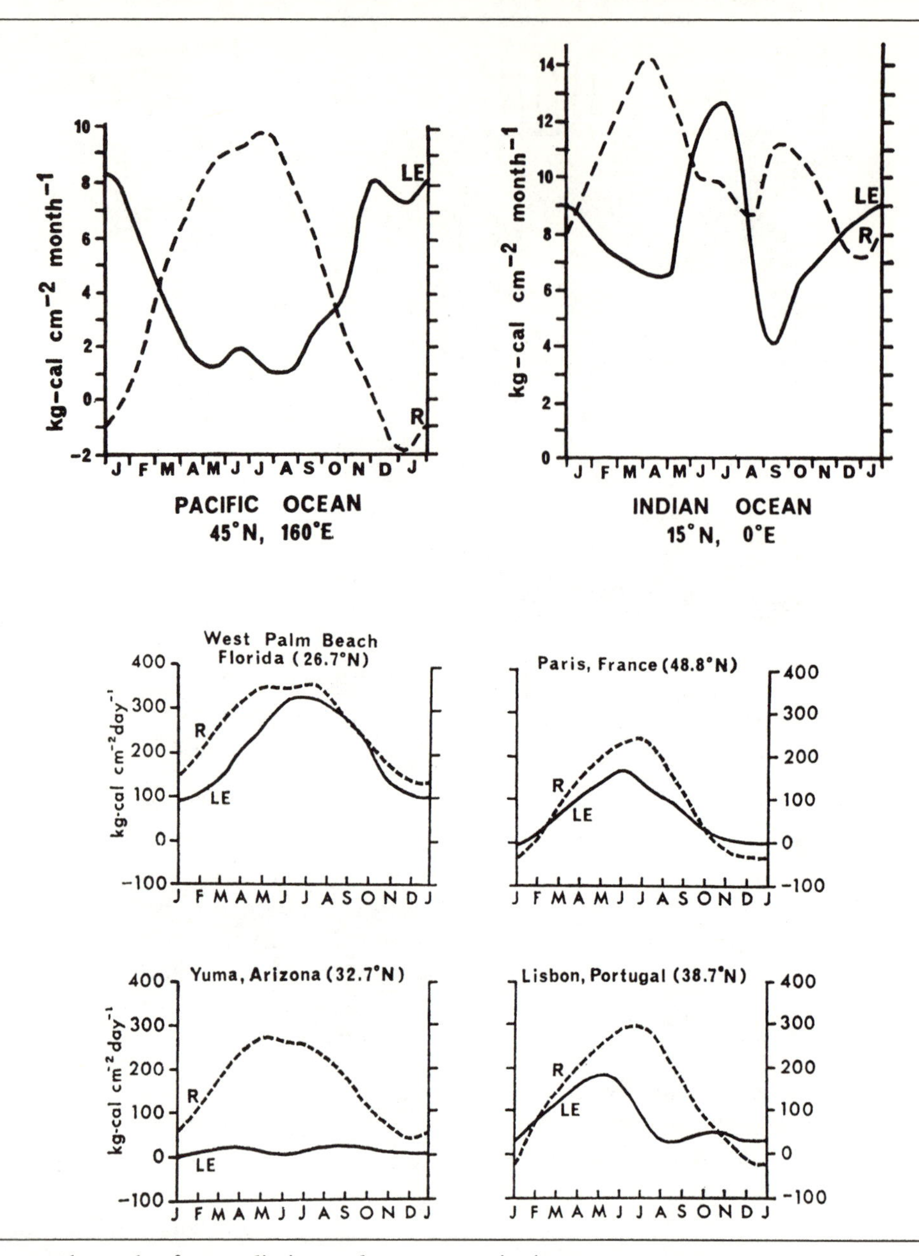

Figure 6 The seasonal march of net radiation and evapotranspiration over ocean areas. (After Budyko, 1956.)

Figure 7 The seasonal march of net radiation and evapotranspiration in different climatic regimes. (After Budyko, 1956; Sellers, *Physical Climatology*, University of Chicago Press, 1965.)
West Palm Beach, 27° N, 80° W; Paris, 49° N, 2° E; Yuma, 33° N, 115° W; Lisbon, 39° N, 9° W.

differences may also reflect the varying external (r_a) and internal (r_s) resistances of vegetation surfaces to vapor diffusion. Theoretical computations by Monteith (1965) indicate a potential transpiration in the Thames valley, England, of 47 cm/year for short grass, compared with 58 cm/year for a tall farm crop with smaller r_a due to greater surface roughness. The loss from a pine forest (48 cm/year) is almost the same as from grass because increased stomatal resistance offsets the influence of lower albedo (15% compared with 25% for grass and green crops) and greater surface roughness of the forest which otherwise tend to promote transpiration.

The Moisture Balance and Some Applications

The principal difficulty encountered in computing moisture budgets for individual localities is the problem of assessing soil-moisture storage and actual evapora-

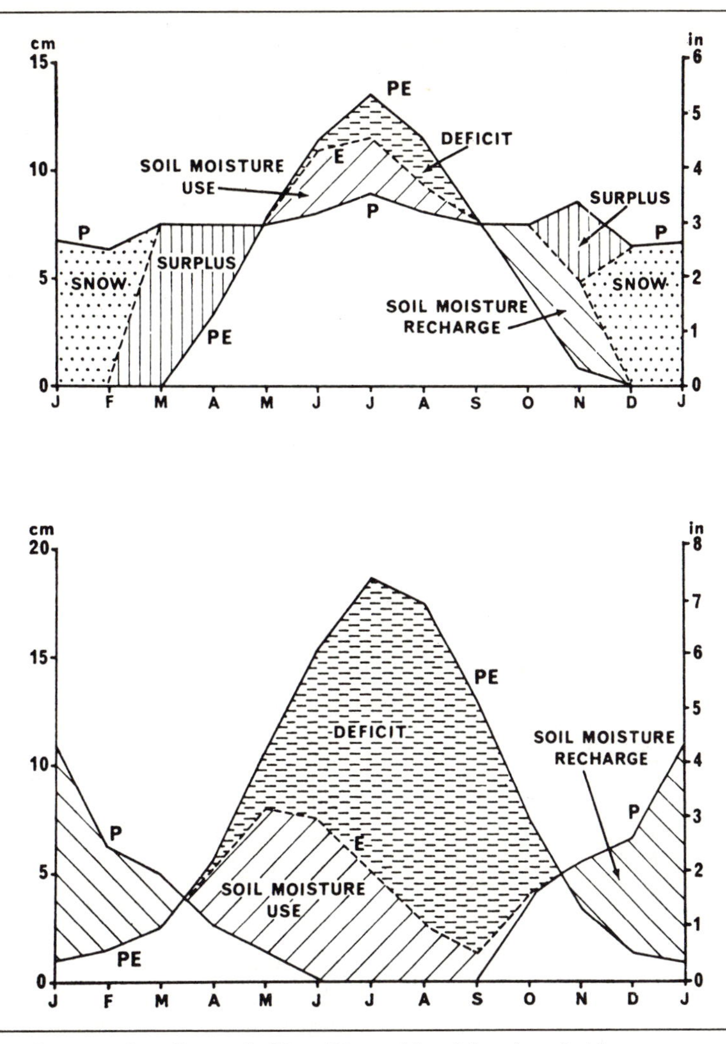

Figure 8 Moisture budget diagrams for Concord, New Hampshire (*above*) and Aleppo, Syria (*below*). (Based on data in Thornthwaite and Hare, 1965, and Mather, 1963.) The method assumes that 50% of the soil water surplus runs off in the first month, 50% of the remainder in the next, and so on, unless additional surplus forms.

tion. Only the simplest of the models has been extensively applied in practice—namely that of Thornthwaite and Mather (Mather, 1963). Details of Budyko's more complex method are summarized by Sellers (1965). Figure 8 illustrates typical moisture budget diagrams for mid-latitude stations in humid (Concord, New Hampshire) and semi-arid (Aleppo, Syria) regimes. The relative amounts of annual moisture surplus (S) and deficit (D) provide one of the bases for Thornthwaite's 1948 classification of climates. In the revised 1955 version of this classification the moisture index (Im) is

$$Im = \frac{100(S - D)}{PE} \qquad (16)$$

In North America forest predominates in regions where the humidity index ($100S/PE$) > 35 and the aridity index ($100D/PE$) < 10, so that there is ample soil moisture in nearly all months. Where S and D are small and approximately equal, the vegetation is typically tall grass prairie, but where the aridity index is of the order of 30 this gives way to short grass. Sagebrush and other desert vegetation occurs with aridity indices > 40. For regions where data are inadequate to calculate a complete budget, estimates of ($P - PE$) may provide a useful climate parameter (Wallén, 1966; Davies, 1966). Davies, for example, discusses the relationships between vegetation and ($P - PE$) isopleths in Nigeria. In the Near East, however, Perrin de Brichambaut and Wallén (1963) find that the limit of dry-land farming is determined more by the amount and reliability of rainfall and soil-moisture storage than by potential evapotranspiration, which does not vary much over short distances.

REFERENCES

Budyko, M. I., 1956. *Teplovoy balans zemnoy poverkhnosti* (The Heat Balance of the Earth's Surface. Translated by Stepanova, N. I., 1958. U.S. Weather Bureau, Washington.) Leningrad, Gidrometeoizdat.

Chang, J. H., 1965. On the study of evapotranspiration and the water balance. *Erdkunde 19*, 141–150.

Davies, J. A., 1966. The assessment of evapotranspiration for Nigeria. *Geografiska Annaler, Ser. A 48*, 139–156.

Dyer, A. J., 1967. The turbulent transport of heat and water vapour in the unstable atmosphere. *Quart. Journ. Royal Meteorol. Soc. 93*, 501–508.

Holmes, R. M., 1961. Estimation of soil moisture content using evaporation data. In *Proc. Hydrology Symposium No. 2 Evaporation.* Dept. Northern Affairs and National Resources, Ottawa. 184–196.

King, K. M., 1961. Evaporation from land surfaces. In *Proc. Hydrology Symposium No. 2 Evaporation.* Dept. Northern Affairs and National Resources, Ottawa, 55–80.

King, K. M., Tanner, C. B., and Suomi, V. E., 1956. A floating lysimeter and its evaporation recorder. *Trans. Am. Geophys. Union 37(6)*, 738–742.

Mather, J. R., 1963. Average climatic water balance data of the continents, No. 2 Asia (excluding USSR). *Publications in Climatology XVI(2)*, 262 pp.

McIlroy, I. C. and Angus, D. E., 1964. Grass, water and soil evaporation at Aspendale. *Agr. Meteor. 1(3)*, 201–224. Amsterdam.

Monteith, J. L., 1965. Evaporation and environment. In *The State and Movement of Water in Living Organisms*, Society for Experimental Biology, 19th Symposium (Cambridge), 205–234.

Morton, F. I., 1967. Evaporation from large deep lakes. *Water Resources Res. 3*, 181–200.

Olivier, H., 1961. *Irrigation and Climate.* London, 250 pp.

Penman, H. L., 1963. *Vegetation and Hydrology.* Commonwealth Bur. Soils (Farnham Royal) Tech. Com. 53, 124 pp.

Perrin De Brichambaut, G. and Wallén, C. C., 1963. *A study of Agroclimatology in Semi-Arid Zones of the Near East.* W.M.O. Tech. Note 56, 64 pp. Geneva.

Rutter, A. J., 1967. Evaporation in forests. *Endeavour 26*, 39–63.

Sellers, W. D., 1965. *Physical Climatology.* Chicago. 272 pp.

Tanner, C. B., 1960. Energy balance approach to evapotranspiration from crops. *Soil Sci. Soc. Amer. Proc. 24*, 1–9.

Thornthwaite, C. W., 1948. An approach towards a rational classification of climate. *Geographical Review 38*, 55–94.

Thornthwaite, C. W. and Hare, F. K., 1965. The loss of water to the air. In Waggoner, P. E. (ed.), *Agricultural Meteorology, Meteorological Monographs 6(28)*, 163–180.

Thornthwaite, C. W. and Mather, J. R., 1951. The role of evaporation in climate. *Archiv für Meteor., Geophys., und Bioklim., Serie B, 3*, 16–39.

Veihmeyer, F. J. and Hendrickson, A. H., 1955. Does transpiration decrease as the soil moisture decreases? *Trans. Amer. Geophys. Union 36*, 425–448.

Wallén, C. C., 1966. Global solar radiation and potential evapotranspiration in Sweden. *Tellus 18*, 786–800.

John R. Mather and Gary A. Yoshioka

7 The role of climate in the distribution of vegetation

Climate, soil, fire, drainage, man, animals, and many other factors all influence the distribution of vegetation. The present study undertakes a detailed examination of the relationship between natural vegetation and climate over the conterminous United States in an effort to identify more clearly the nature of the relationship that exists between these two factors under different environmental situations. The phrase natural vegetation in the present context refers to the predominantly existing vegetation under the normal pattern of climate, environmental (including natural fires), and edaphic conditions generally uninfluenced by the directed or willful action of man. It does not necessarily refer to a so-called climatic climax vegetation association.

The present study will not utilize directly the statistics of temperature and precipitation or relate annual averages of these statistics to vegetation distribution. Rather, derived factors of climate, factors concerned (1) with the plant water need as it is related to the supply of moisture and precipitation and soil moisture storage, and (2) with the energy for plant development, will be correlated with the distributions of vegetation. It is hoped that the careful selection of climatic indices that are active in vegetation growth and development can result in a new understanding of the role of climate in vegetation development, and possibly place in better perspective the roles of other environmental factors and man as influencers of vegetation distribution.

ACTIVE CLIMATE INDICES

The 1948 climatic classification developed by Thornthwaite made use of average monthly values of temperature and precipitation, as do most such classifications. Thornthwaite, however, emphasized the moisture factor in climate by introducing the concept of the climatic need or demand for water (which he called potential evapotranspiration) and using it in the development of a moisture index. He compared the climatic need for water with the climatic supply of water (the precipitation) in a monthly or daily climatic water balance in order to determine whether a climate was moist or arid.

Determinations of potential evapotranspiration were so limited both in time and space that Thornthwaite concluded "the only alternative is to discover a relation between potential evapotranspiration and other climatic factors for which there are abundant data." Thornthwaite found a close relation between mean monthly air temperature and potential evapotranspiration which he employed to provide world-wide information on the climatic demands for water. He realized the limitations of such a correlation approach and continued to search for more adequate ways of determining potential evapotranspiration. Since a fully rational predictor of potential evapotranspiration is not yet available, Thornthwaite's original relation must still be used. The general equation for potential evapotranspiration is

$$e = 1.6(10t/I)^a$$

where e is unadjusted monthly potential evapotranspiration in centimeters, t is mean monthly temperature in °C, I is the annual heat index obtained from the summation of the twelve monthly values $i = (t/5)^{1.514}$, and $a = 675 \times 10^{-9} I^3 - 771 \times 10^{-7} I^2 + 1792 \times 10^{-5} I + 49239 \times 10^{-5}$. To obtain adjusted potential evapotranspiration, the value of e must be corrected for the actual number of days in the month and the number of hours of daylight in each day. As Thornthwaite acknowledged, the "equation is completely lacking in mathematical elegance. It is very complicated and without nomograms and tables as computing aids would be quite unworkable." Tables to evaluate potential evapotranspiration as well as detailed instructions to work out all the steps of the climatic water balance are

Reprinted, with minor editorial modification, by permission of the authors and the Association of American Geographers, from *Annals* of the Association of American Geographers, Vol. 58, 1968, pp. 29–41.

to be found elsewhere (Thornthwaite and Mather, 1957).

The climatic water balance is merely a monthly comparison of the precipitation with the potential need for water or potential evapotranspiration. It provides quantitative information on the magnitude of the periods of water surplus and water deficit during the year. An actual example of the bookkeeping procedure used in determining the climatic water balance is included in Table 1. Thornthwaite combined the derived factors of annual surplus, deficit, and potential water need into a moisture index,

$$I_m = \frac{100(S - D)}{PE}$$

where S is the annual water surplus, D is the annual water deficit and PE is the potential evapotranspiration.

The climatic moisture index serves as the basis for Thornthwaite's 1948 classification of climate. Thornthwaite acknowledged that the climatic types derived from the moisture index had the same meaning as those utilized in his 1931 classification, but whereas the limits of the earlier classification had been determined in part by reference to vegetation boundaries, those resulting from the 1948 study "are rational and are established solely in terms of the relation between potential evapotranspiration and precipitation."

In a further discussion of the differences between his 1948 classification and Koeppen's classifications or his own 1931 classification, Thornthwaite wrote

> The difference may be illustrated by the change in point of view respecting vegetation. The earlier study adopted Köppen's position that the plant is a meteorological instrument which integrates the various factors of climate and which, with experience, can be "read" like a thermometer or a rain gauge. In the present study, vegetation is regarded as a physical mechanism by means of which water is transported from the soil to the atmosphere; it is the machinery of evaporation as the cloud is the machinery of precipitation.
>
> Climatic boundaries are determined rationally by comparing precipitation and evapotranspiration. The subdivisions of the older classification were justly criticized as being vegetation regions climatically determined. The present climatic regions are not open to this criticism, since they come from a study of the climatic data themselves and not from a study of vegetation.
>
> . . .
>
> There is an encouraging prospect that this climatic classification which is developed independently of other geographical factors such as vegetation, soils, and land use, may provide the key to their geographical distribution.

Since the publication of Thornthwaite's 1948 classification, a large number of geographic studies relating vegetation distributions to the climatic indices available from the water balance bookkeeping procedure have appeared. Some of these have shown a close relation between climate and vegetation whereas others have used the vegetation distribution to question the validity of the classification itself (Fukui, 1960; Rao and Subrahmanyam, 1962; Daubenmire, 1956; Kollmeyer, 1958; Hare, 1950; Vernet, 1965).

Few investigators would really question that climate has an influence on the distribution of vegetation, although the exact nature of this influence under different environmental conditions may be quite variable. Eyre (1963) felt that chemical composition and physical characteristics of the soil and parent material could be as important as climatic factors. He cited several examples of different vegetation covers on different soil areas under exactly the same climatic conditions in verification of this point. He did not eliminate the basic control of vegetation by climate, however. Thornthwaite, however, writing as a climatologist, had a more deterministic view of the influence of climate. He felt that climatic factors almost by themselves could be used to distinguish vegetation associations. In writing about grassland associations in 1952 he said:

> The evidence is unmistakable that there is a definite grassland climate. It is marked by a specific balance between moisture deficit and surplus which results in few periods during the year of either very dry or moist soil conditions. As a result of this environment grasses flourish. It is possible to explain the presence of grasslands on the basis of climate alone, without bringing in other

(All values except T in mm)

	J	F	M	A	M	J	J	A	S	O	N	D	Y
$T°$ F	46.8	48.4	55.2	63.5	72.0	78.8	80.8	79.9	75.2	64.6	54.4	47.2	63.9
PE	13	15	37	65	115	158	172	157	114	64	29	13	952
Precip.	76	95	91	80	78	99	141	139	98	61	73	77	1,108
$P - PE$	63	80	54	15	−37	−59	−31	−18	−16	−3	44	64	
Storage	300	300	300	300	265	217	196	184	175	173	217	281	
St. change	+19	0	0	0	−35	−48	−21	−12	−9	−2	+44	+64	
Act. evapo.	13	15	37	65	113	147	162	151	107	63	29	13	915
Deficit	0	0	0	0	2	11	10	6	7	1	0	0	37
Surplus	44	80	54	15	0	0	0	0	0	0	0	0	193

Explanation of Terms in the Climatic Water Balance

$T°$ F—Average monthly temperature in °F.
PE—Potential evapotranspiration, in mm, adjusted for length of day and number of days in month.
Precip.—Average monthly precipitation in mm.
$P - PE$—Precipitation minus potential evapotranspiration—negative when precipitation fails to meet needs of PE.
Storage—The amount of water held in the soil, obtained by simple bookkeeping procedure and tables to account for decreasing withdrawals as soil dries. A 300 mm storage capacity is assumed here.
St. change—Change in storage from month to month.
Act. evapo.—Actual evapotranspiration, equal to the potential evapotranspiration when $P - PE$ is positive; equal to the sum of precipitation and storage change, otherwise.
Deficit—Water deficit or $PE - AE$.
Surplus—Water surplus or positive $P - PE$ when storage capacity is full.

Table 1 Climatic water balance, Columbia, South Carolina. Source: Calculated by authors.

agencies such as man or fire. Grasslands are a natural phenomenon, in equilibrium with their climatic environment. Thus, through climatic information it is possible to differentiate the grasslands in general and even the small scale variations in different grasslands within the broad group. This must lead to a more rational interpretation of the vegetation associations of the world.

The views expressed by Eyre and Thornthwaite are not necessarily irreconcilable. Eyre considered the climatic factors of temperature and precipitation and found that these were not always sufficiently definitive in terms of vegetation distribution to permit an unqualified statement to be made about the influence of climate on vegetation. Thornthwaite, however, considered soil moisture distribution, plant water need, and the seasonal course of moisture surplus and deficit. He felt that these were climatic factors that might truly influence vegetation distribution. Soil moisture content is related to the precipitation and evapotranspiration as well as to edaphic factors such as the depth of rooting of the vegetation and the type of soil. Thus soil moisture can combine both the climatic and edaphic factors. Since soil moisture storage is a climatic factor, according to Thornthwaite, his reference to a definite grassland climate becomes more understandable.

Hare (1954) and later Thornthwaite and Hare (1955) attempted to correlate potential evapotranspiration and the moisture index (I_m) with the distribution of vegetation. They found that in a temperate, humid climate, I_m can roughly distinguish different forest types, whereas the sub-arctic, boreal forest subdivisions are better correlated with PE. Major (1963) in an alternative approach has considered a third factor, actual evapotranspiration (AE), and has suggested that vascular plant activity and growth might be related to the actual water loss of the vegetation since this factor expresses the real water activity of the plant rather than some unrealized potential value.

VEGETATION DISTRIBUTION AND CLIMATE IN CONTERMINOUS UNITED STATES

Most previous studies have been concerned only with the direct correspondence between vegetation associations and broad climatic zones defined in terms of both temperature and moisture factors. Less effort has been expended on trying to relate subtypes of vegetation with individual factors of the water balance at particular places because of the complexities of the problem. Thornthwaite (1952) attempted to relate actual vegetation to the aridity and humidity indices which combine to form the moisture index for selected stations across the United States at latitude 40° N but the effort still resulted in only a broad-scale agreement between climate and vegetation.

Relation Between Vegetation and the Thornthwaite Climatic Indices

The present study has been concerned with an extensive investigation of the relation among natural vegetation, the climatic moisture index, and potential evapotranspiration in all major vegetation regions across conterminous United States. The data have been plotted on Figure 1 where the abscissa shows the moisture index and the ordinate gives the potential evapotranspiration. Information concerning the characteristic natural vegetation at each station has been indicated by means of a symbol. Data from tundra and tropical rainforest areas around the world as well as coniferous and hardwood areas in Canada have also been included in Figure 1 for comparison purposes although these vegetation associations do not, of course, occur in conterminous United States.

Vegetation distributions were obtained from a map by Shantz and Zon in the Atlas of American Agriculture (1924). A more recent map by Küchler (1964) became available during the course of this study. The original study was first completed using the Shantz map. The program was then enlarged to consider the results available from the more recent Küchler study.

Using the Shantz map, sixteen principal vegetation areas were identified. Climatic stations were located in each of these sixteen vegetation areas (in the case of discontinuous regions some stations were sought in all segments of the regions) and the data of I_m and PE were obtained for each station (Mather, 1964).

The moisture index, I_m, equals

$$100\left(\frac{S - D}{PE}\right)$$

Since S equals $P - AE$ if there is no change in storage and D equals $PE - AE$, the moisture index can also be written as

$$100\left(\frac{P}{PE} - 1\right)$$

Consequently, PE occurs in both the abscissa and ordinate scales of Figure 1. But since the precipitation, P, is entirely independent of PE, the ordinate variable is independent of the abscissa variable. Different climatic stations within the same vegetation association are found to show similar active thermal (PE) and moisture (I_m) factors. Thus these two indices demonstrate an ability to distinguish among different vegetation associations.

The range of annual PE for stations within conterminous United States is from about 400 mm to 1,300 mm. Below 400 mm, the vegetation is, in general, tundra. Above 1,300 mm tropical rainforests and semiarid or arid vegetation associations are found. The moisture index across conterminous United States ranges from well over $+100$ in the mountainous areas of both the northeastern and northwestern parts of the country to values close to -100 in the arid areas of the southwest. The range of stations used in Figure 1 is sufficient to include all but one portion of these distributions of PE and I_m. Within the United States there appear to be no stations with PE values over 800 mm and with I_m values over 60. These values represent a fairly warm, moist climate and should be found near latitude 30° to 35° N, close to the coast where sufficient moisture is present to give the high values of I_m. However, such climates do not generally exist in the world since these latitude belts fall in the so-called subtropical high pressure regions where precipitation is low to moderate. I_m values above 50 seldom occur in these areas except possibly where orographic influences prevail.

Stations in the lower left-hand corner of Figure 1 are found in the southern desert shrub region. The most important plants in this desert are creosote bush and mesquite. The region extends from southern California to the valley of the Rio Grande, and the selected stations range from Barstow, California, in the Mojave Desert, to Mission, Texas, in the extreme southern part of the state. Just as creosote bush areas are surrounded by desert savanna, bunch grass, desert grassland, and sagebrush areas in Figure 1, these latter four types of vegetation all border creosote bush on the vegetation map.

The desert savanna region is identified by mesquite-mesquite grass and thorn bush-mesquite grass associations. It lies in a wide belt from the Gulf of Mexico near Corpus Christi, Texas, northwest to southeastern New Mexico, and north to the Red River. To the north and east, short grass, tall grass, and grassland-forest transition vegetation are found. The thermal and moisture indices for stations in this zone indicate that this savanna requires more water than desert shrubs, but less

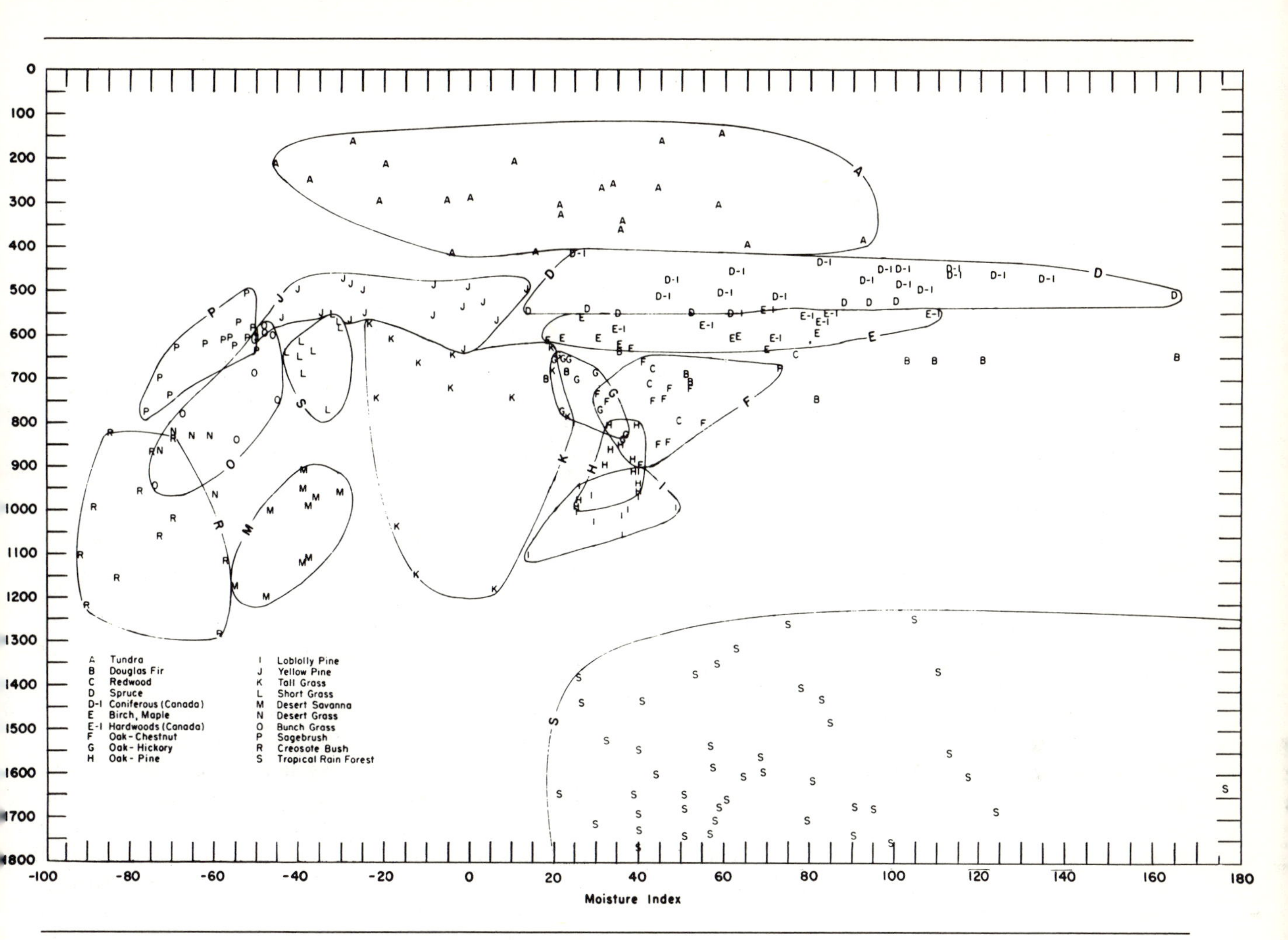

Figure 1 Relation between natural vegetation and the climatic factors of potential evapotranspiration and the moisture index at selected stations in the United States, Canada, and the Tropics.

than tall grass, and that it can endure a higher *PE* than most short grasses.

The dominant northern desert shrub is sagebrush. Just as a *PE* of 800 mm seems to separate sagebrush from creosote bush in Figure 1, geographically the two types of vegetation lie roughly on either side of latitude 37°. Sagebrush occupies most of the basin land between the Cascade-Sierra ranges and the Continental Divide.

The bunch grass and wheat grass of Washington, Oregon, and Idaho are found at elevations above the sagebrush but below the conifer zone. In addition, they mark a transition between desert shrub and plains grassland (short grass). Both these facts are confirmed by the climatic data. Bunch grass, with stipa and poa, is also believed to be the original vegetation of the central and coastal valleys of California. This Pacific grassland, together with the desert grassland (black grama, crowfoot grama, and curly mesquite) vegetation found in southern Arizona and New Mexico, seem to be transitional with respect to both moisture and thermal conditions.

The short grass of the plains is represented in Figure 1 by stations as remote as Dickinson, North Dakota, and Amarillo, Texas. The great variety of grasses found in this area between the Rockies and the 100th meridian are all characterized by low growth and shallow roots. In many areas plains grassland begins at elevations immediately below yellow pine, a position occupied by bunch grass in the northwestern states.

The western pine forest, dominated by yellow pine, Douglas fir, and lodgepole pine, is found at elevations in the Rockies, Sierras, and Cascades. Stations selected to reflect the climate of this forest were from nine different states. The plot of climatic indices for the stations shows that the pine is able to withstand quite arid conditions, though in general it receives more moisture than desert shrubs. It is *PE* which seems to distinguish this forest from most grasses. The thermal index is in the range 450–550 mm, which is apparently favored by boreal conifers.

Tall grass, according to Küchler, is the natural vegetation in a large part of the Mississippi Valley. The climatic data from this vegetation region cover a wide geographic area—from Devils Lake, North Dakota, to La Salle, Illinois, to Fort Worth, Texas—with Okeechobee, Florida, representing an isolated pocket. The variation in climatic factors is shown by the scatter of the plotted points. The change from tall grass to forest vegetation, particularly the oak-hickory forest, is not an abrupt one. Large portions of Texas, Oklahoma, Missouri, and Illinois lie within the transition zone between trees and grasses.

The southern hardwood forest (subdivided into the chestnut-oak-poplar, oak-hickory, and oak-pine forests) covers a large region from the Mississippi River to the Atlantic Ocean and from Michigan to Georgia. Geographically, stations range from Connecticut to Alabama for oak-chestnut, from Ohio to Arkansas for oak-hickory, and from Virginia to Texas for oak-pine. That there is overlap among the subdivisions is not really surprising. Some generalizations may still be drawn. Oak-chestnut associations seem to dominate in a wetter environment than do oak-hickory forests. Oak-pine appears to prefer a higher *PE* than either of the others. The longleaf, loblolly, and slash pines of the southeastern pine forest have approximately the same range of I_m as the southern hardwoods, but in general the *PE* is lower for the hardwoods.

The northeastern hardwoods (birch, beech, maple, and hemlock) are found in New York, New England, and the northern portions of Pennsylvania, Michigan, and Wisconsin. The points associated with this forest in Figure 1 represent stations in each of these areas. The climate of the region can be described as having a narrow range of annual *PE* and a wide range of moisture conditions.

In colder areas, such as northern Maine, the Adirondack Mountains, and the upper peninsula of Michigan, species representing the northern coniferous forest are found. Spruce and fir are the dominant trees of this forest. Like those of the northeastern hardwood forest, stations in the conifer zone seem to have a narrow *PE* range. Note that in Figure 1 the western pine forest could almost be considered a drier extension of the northeastern forest.

Redwood and Pacific Douglas fir are two important members of the northwestern coniferous forest. The redwood dominates along the northern California coast and the Douglas fir is found along the coastal ranges of Oregon and Washington. The forest as a whole (including the white pine and larch areas of western Montana and northern Idaho) is considered to be a cedar-hemlock association. Although there may be some evidence that these latter species are the natural vegetation in these areas, it is still difficult to explain climatically the presence of redwood and Douglas fir points (or cedar-hemlock points) in Figure 1 in the region occupied by southern hardwood stations.

The pattern of distribution shown on Figure 1 is quite interesting. It is possible to locate discrete and, in many cases, non-overlapping vegetation areas that can be defined by the annual potential evapotranspiration and the climatic moisture index. If active factors of climate can be selected it appears possible to obtain a good relation between certain climatic and botanic distributions in mid-latitude. From the distributions shown in Figure 1, one might conclude that it is possible to speak not only of a forest climate, as opposed to a grassland or a desert climate, but that it may be possible to identify a birch-maple as opposed to a spruce or oak-chestnut forest climate in a region such as the United States. Where the areas on Figure 1 are fairly distinct as in the case of the tundra, spruce, birch-maple, oak-chestnut, oak-hickory, tall grass, yellow pine, short grass, desert savanna, creosote bush, sagebrush, bunch grass-desert grass, and tropical rainforest, one even feels more confident in emphasizing the dominant role that climate must play in vegetation zonation. It does not, of course, eliminate the possibility of edaphically influenced vegetation associations within large homogeneous climatic regions or the possibility of more than one vegetation association being able to survive and compete successfully within a given climatic range. Thus an overlap of Douglas fir and redwood points or an overlap of bunch grass and desert grass station points is found in Figure 1. In part, this overlap may be caused by poor definition of the vegetation association itself or by the choice of less descriptive climatic parameters to delimit the particular vegetation association. Edaphic and other environmental factors must also contribute to these areas of overlap, but the general lack of overlap bears witness to the responsiveness of the botanic environment to significant climatic stimuli.

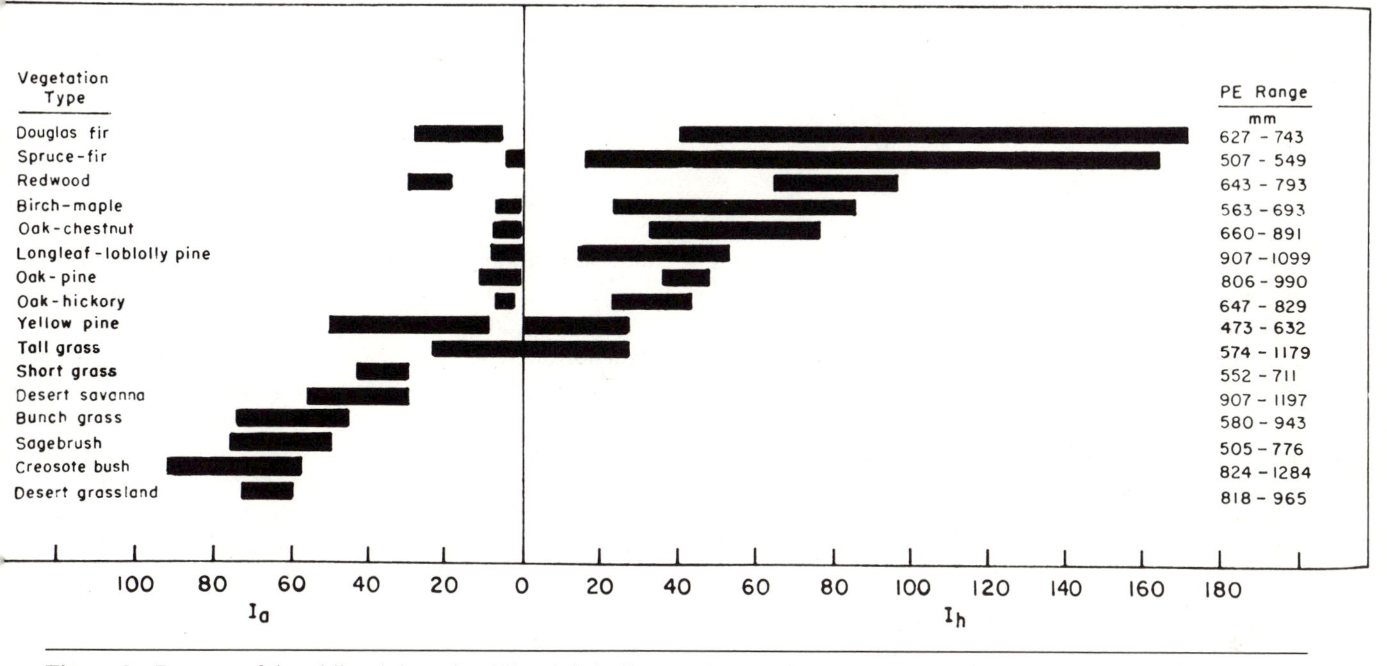

Figure 2 Ranges of humidity (I_h) and aridity (I_a) indices and annual potential evapotranspiration for representative vegetation types in the United States.

Relation Between Vegetation and the Humidity and Aridity Indices

Both Thornthwaite (1952) and Mather (1954) have shown that the vegetation distribution can be related to the two factors that make up the moisture index—namely the aridity index (I_a) and humidity index (I_h).[1] The average values of the aridity index and humidity index have been obtained for the United States stations on Figure 1 and the results are given in Figure 2. The vegetation classes have been ranked by decreasing humidity index and increasing aridity index as much as possible. Figure 2 clearly shows the unique position of the redwoods and Douglas fir among the forest association. They can exist in a climatic region that results in as much aridity as grasslands experience but at the same time they are also found in areas of very high rainfall with some of the highest humidity indices in the United States. All of the other forest types except the yellow pine and the pinon-juniper (not included) associations exist in areas with significant humidity indices and insignificant aridity indices.

As expected, the tall grass vegetation areas occur in those climatic regions with moisture indices very close to 0. Because of the lack of accord between the distribution of precipitation and evapotranspiration, periods of both moisture surplus and deficit exist but since no long periods of deficit occur the vegetation is luxurious and the soil is rich and productive.

[1] $I_a = D/PE$ while $I_h = S/PE$ where D is annual water deficit, S is annual water surplus, and PE is annual potential evapotranspiration.

The short grass, bunch grass, and other semiarid and arid vegetation associations occur in areas with markedly negative moisture indices. Essentially no moisture surplus occurs at stations where these vegetation associations are found while increasing aridity indices accompany a shift in the vegetation from short grass, to bunch grass, to sagebrush, and creosote bush.

Figure 2 indicates clearly that it is possible to speak about a forest climate as opposed to a grassland climate and a desert shrub climate in mid-latitudes. The occurrence of certain forest associations such as the yellow pines between grassland associations on Figure 2 points up the need for caution in making all-inclusive statements but it does not vitiate the conclusion that one can speak of forest or grassland climates. Within the land biochores there would appear to be distinct breaks between certain types of forest associations and other forest groups, as well as between certain semiarid vegetation associations and others which are slightly more moist or dry. The distribution shown on Figure 2, of course, cannot really distinguish the loblolly-longleaf pine from the oak-hickory hardwoods in southeastern United States. Figure 2 would indicate that possibly either oak-hickory or pine would be well adapted to the region and either group might exist in the region as a result of the climatic conditions alone. The breaks between the spruce-fir and birch-maple associations and between the sagebrush and desert grassland associations, however, seem very real and significant. Especially when viewed in conjunction with the values of annual potential evapotranspiration it becomes clear that many vegetation associations can be identified in large measure from knowledge of the annual and seasonal ranges of just a few significant climatic parameters.

Relation Between Küchler's Vegetation Distributions and the Thornthwaite Climatic Indices

The results in Figures 1 and 2 can be questioned because of the use of the early Shantz and Zon vegetation map and the possibility of some bias in the selection of the climatic stations that fall within the sixteen vegetation regions. Therefore, the study was extended to make use of the more recent Küchler vegetation map and to compare the vegetation and climatic indices at a completely random selection of stations across the conterminous United States. To obtain a random selection of stations, the list of stations included in "Average Climatic Water Balance Data of the Continent, Part VII" which is given alphabetically by states was used. Every fifth station was selected. Climatic or vegetation bias in the selection of stations was, thus, removed.

In all, 237 stations were available for analysis. The moisture index, the annual potential evapotranspiration, and the vegetation association from the Küchler map were obtained for each of these stations. The results, grouped by broad vegetation categories and by I_m and *PE* are shown in Figure 3, along with the number of stations included in each category. Five categories of I_m are included (>75, very moist; 26–75, humid; 1–25, moist subhumid; −49–0, dry subhumid to semiarid; and −50 and below, semiarid to arid). The humid, semiarid, and arid groups have been further divided on the basis of potential evapotranspiration; the semiarid and arid categories on the basis of an annual potential evapotranspiration above or below 800 mm, and the humid category on the basis of < 600, 601–800, 801–1,000, and > 1,000 mm of potential evapotranspiration. It was shown in Figure 2 that annual potential evapotranspiration could help distinguish certain vegetation associations within the forest and grassland biochores.

For easier identification, seven broad vegetation classes were used to divide the data although in the final analysis, the actual description of the vegetation association used by Küchler was included on Figure 3.

As would be expected, the results are not as clear-cut or definite as in the case of the study using the Shantz and Zon map. Fewer vegetation associations were utilized in the Shantz and Zon study and the stations were selected to occur within recognized vegetation associations. In the present case, the station selection was random and a much more detailed vegetation mapping was utilized. Still, the results show a certain orderliness about them that clearly indicates that within certain limitations, the type of natural vegetation in a mid-latitude area can be determined on the basis of annual values of potential evapotranspiration and the moisture index.

The greatest lack of correlation was found with the needleleaf forest. Whereas spruce-fir associations were found only in the more moist and cooler areas, the pine-Douglar fir and cedar-hemlock associations occurred under very moist, moist, and semiarid climatic conditions. Climate is possibly not as definitive in the distribution of these conifer species as is soil, relief, cultural activities, fire, or other factors.

The mixed broadleaf and needleleaf forests have a wide climatic distribution also. The maple-birch-beech-hemlock association is found under fairly cool conditions (low potential evapotranspiration) although the moisture index can range anywhere from 1 to above 75. Oak-hickory-pine associations are confined to the more moist (I_m between 1–75) climates with moderate temperatures (*PE* 600–1,000 mm). As the temperature and water demands increase, these associations are replaced by a tupelo-oak-cypress or beech-gum-magnolia-oak associations. Oak-juniper associations are found under semiarid conditions.

Broadleaf forests are not so widely distributed from a climatic point of view. All broadleaf associations require I_m above 0 (hence moist subhumid or humid) with the exception of the cottonwood-elm-willow group. These, however, are found in flood plain areas with

I_m	above 75	26 to 75				1 to 25	-49 to 0		-50 and below	
PE		600 and below	601 – 800	801 – 1000	above 1000		800 and below	above 800	800 and below	above 800
DLELEAF ORESTS	spruce-fir (4) pine-Douglas fir (3) cedar-hemlock-Douglas fir (3)	spruce-fir (1) pine (2) Douglas fir (1)				pine (2)	juniper-pinon (5) pine-Douglas fir (8) cedar-hemlock-Douglas fir (1)		juniper-pinon (1) pine-Douglas fir (1)	
OADLEAF AND DLELEAF ORESTS	maple-birch-beech-hemlock (3)	maple-birch-beech-hemlock (3)	maple-birch-beech-hemlock (3) cedar-hemlock-Douglas fir-oak (2) oak-hickory-pine (4)	oak-hickory-pine (10) tupelo-oak-cypress (5)	beech-gum-pine-magnolia-oak (6) tupelo-oak-cypress (5)	maple-birch-beech-hemlock (1) beech-gum-pine-magnolia-oak (2) oak-hickory-pine (3) tupelo-oak-cypress (1)		oak-juniper (1)	oak-juniper-mahogany-oak scrub (1)	
OADLEAF ORESTS			beech-maple (11) oak-hickory (12)	oak-hickory (6)		maple-basswood (2) oak-hickory (7) cottonwood-elm-willow (2)	cottonwood-elm-willow (7)			
ASSLAND AND ORESTS			oak-hickory-bluestem (2)			oak-hickory-bluestem (4) sawgrass-bay (2)	oak-hickory-bluestem (2)	oak-bluestem (3) juniper-oak (2) mesquite-buffalo grass (2)		
ASSLAND						bluestem (5) bluestem-sacahuista (2)	bluestem (3) wheatgrass-needlegrass-fescue-grama-buffalo grass (30)	bluestem-grama-needlegrass (7) tule (1)	grama-fescue-wheatgrass-needlegrass-buffalo grass (5)	needlegrass (2)
ASSLAND AND HRUB							sandsage-bluestem (1) wheatgrass-sagebrush (4)	bluestem-oak (1) mesquite-acacia (1)	wheatgrass-sagebrush (6)	grama-tobosa-creosote (2) tarbush-creosote (2) wheatgrass-sagebrush (1)
SHRUB							mountain mahogany-oak scrub (2) sagebrush (2)		saltbush-greasewood (4) sagebrush (5)	creosote-bur sage-tarbush (3) saltbush-greasewood (1) sagebrush (1)

Figure 3 Relation between natural vegetation (Küchler) and the climatic factors of potential evapotranspiration and the moisture index at random stations in conterminous United States.

adequate water supplies for the root systems. Thus, while seven such forested stations with I_m below 0 are included on Figure 3, the existence of the unique soil moisture conditions of the flood plain should be taken into account in evaluating their climatic requirements.

Grassland-forest or just grassland associations require moisture supplies in balance with the water demands. Oak-hickory-bluestem vegetation can occur under humid, or subhumid conditions with moderate temperatures. The mesquite-buffalograss association requires definitely drier conditions. Bluestem, sawgrass, and sacahuista prairie are the principal grasses that will be found with I_m greater than 0. As the moisture index drops into the dry subhumid, semiarid, and arid range the grasses predominate, but they change from bluestem, wheatgrass, grama, and fescue to needlegrass and buffalograss. Grassland and shrub associations are entirely confined to the drier moisture regions.

Figures 2 and 3 clearly show the broad relation between climatic conditions and vegetation distribution that has been generally recognized for many years, but they also provide certain climatic limits or ranges that begin to make possible a more quantitative evaluation of the vegetational response to climate. Figure 3 has much less definition than does Figure 2 because of its method of preparation but both show that even relatively simple climatic parameters, in mid-latitudes, can be utilized to identify vegetation associations. If stations with limiting drainage or topographic controls had been eliminated, more conclusive results would, of course, have been achieved. At the same time if frequencies of dry or moist conditions, or if annual ranges of I_m or PE values had been utilized rather than just average annual values, more definitive results might have been achieved. A study of climatic frequencies is, however, beyond the scope of the present paper.

LIMITATIONS IN THE TROPICS AND SUBTROPICS

Figure 1 has already shown how tundra, boreal forest vegetation, and tropical rainforests from various parts of the world fit generally into the climate-vegetation relationship found in conterminous United States. Such a useful relationship is not found everywhere. Within the tropics and subtropics and especially when the problem of tropical savanna vegetation associations are considered, the role of climate becomes of less import. Correlations found in mid-latitudes between climate and vegetation do not necessarily exist there.

Parsons (1955) has described in some detail the Miskito pine savanna of the Nicaragua and Honduras coast. Not only is this pine savanna significant in that it marks the most tropical penetration of pines in the Western Hemisphere but it

> . . . is probably the rainiest area of its size in the New World with a savanna-type vegetation. For so extensive a tropical grassland, either with or without trees, to occur under an average rainfall of 100 to 150 inches with so abbreviated a dry season clearly contradicts once more the traditional concept of the "savanna climate."

Parsons felt that the limits of the pine savanna are well marked by the termination of the gravel shelf on which this vegetation association is found:

> The interior boundary or "bush line" between pine savanna and high evergreen forest . . . is almost everywhere sharp. Silicious sand and gravel soils, often containing a mottled, impervious subsoil with iron concretions, characteristically support a savanna vegetation; the high forest occurs on the crumb-structured humic clays of superior water-holding capacities which are found towards the interior.

Parsons also considered fire, hurricanes, and impeded subsurface drainage as possible contributing factors. There was a suggestion that hardwoods might be able to take over the area if fire and other cultural destructive agents were eliminated.

The Miskito pine savanna represents a challenge to geographers, climatologists, botanists, and others who seek neat causal relationships between distributions of active factors whether they be climatic, edaphic, or cultural. Puerto Cabezas is located on the coast of Nicaragua near the middle of the area of pine savanna. It experiences 3,293 mm of precipitation on the average through the year, ranging from 458 mm in June to 50 mm in March. The moisture index for Puerto Cabezas is +106 with a humidity index of 109 and an aridity index of 3. With a potential evapotranspiration of 1,545 mm this area by all rights should experience a tropical rainforest type of vegetation. Climate, in this case, may be a permissive factor but it certainly is not a determinant one since savanna rather than rainforest exists there.

Denevan (1961) in a study of the upland pine vegetation of Nicaragua emphasized the role of fire in the development of the pine savanna. He considered the frequency of burning as the determining factor:

> If fires are very frequent, only grass survives, and if they are very infrequent, hardwoods begin to take over a pine stand. Thus both the survival and the density of the pine forests, is related to the frequency, severity, time and extent of burning.

The existence of a pine savanna in this region poses two separate problems. First the problem of conifers in tropical and subtropical climatic regions and second, the problem of the savanna vegetation and its relation, if any, to climate. Denevan felt that pines are not able to compete successfully against broadleaf vegetation in tropical and subtropical areas without the aid of fires and man. Other vegetation associations in tropical areas should be restudied with this possibility in mind.

CONCLUSIONS

It is clear that vegetation develops in response to many different stimuli among which we might list climatic, edaphic, and cultural conditions as possibly the most important. The response of vegetation to climate is both direct and indirect—direct through the role that the factors of temperature, radiation, moisture, and wind play on the growth and development of the vegetation and indirect through the influence that the climatic factors have on soil conditions, disease organisms, competing botanic associations, and cultural practices.

In addition, the reciprocal influence of vegetation on the microclimate of the particular area and on the other factors of the microenvironment creates another level of influence that must be considered in evaluating the factors that contribute to the distribution of vegetation.

The degree to which the distribution of vegetation can be explained on the basis of climatic conditions depends in part on the proper selection of climatic factors. Temperature and precipitation by themselves are poor descriptors of climate. Precipitation does not indicate whether a climate is moist or dry unless one is able to compare it with the water need of a place. Temperature does not really reveal the energy that is useful for plant development unless one knows also the moisture condition of the soil at the time. Thus, the active factors of climate from a vegetation viewpoint are water surplus or water deficit (water supply in relation to water need), actual evapotranspiration (actual plant water use), and soil moisture storage. Knowledge of these factors is available from computations of the climatic water balance and combinations of these factors are expressed in shorthand form in the moisture index or the humidity and aridity indices of the Thornthwaite climatic water balance.

The foregoing analyses of vegetation distributions have shown that the moisture index along with the evapotranspiration is useful in differentiating among the three land biochores, grassland, forest, and desert. Even within each of these biochores, these active climatic parameters are able to reveal differences which seem to be related to the different formation classes. For example, careful analysis of the vegetation and climatic records reveal the oak-hickory vegetation association as developing under a different climate from birch-maple, or bunch grass in a different climatic environment from the sagebrush.

Use of more expressive climatic parameters or finally the use of the frequencies of some of these elements to describe distributions in environmental biology should yield increased appreciation of the role of climate. The ability to identify some individual formation classes within the biochores provides a new and powerful tool for analysis and classification of the vast body of existing environmental data from all parts of the world. It introduces another aspect of rationality and simplification into the complex world of environmental influences and distributions.

The foregoing should not depreciate the role of edaphic, cultural, and other factors in influencing vegetation distribution. There are many examples of nonclimatic controls on vegetation such as the savanna and the low latitude pine associations. It is, of course, only through a complete interpretation of all contributing influences that the most understandable picture of the geographic distribution of vegetation can be obtained.

As more basic studies of climatic influences on vegetation growth and development are completed, a better understanding of the exact relation between the moisture and energy balance and the vegetation will be obtained. Such a development may well lead to more uniform world-wide classifications of vegetation, eliminating many of the problems that now exist as a result of loose terminology. It will also provide us with more powerful tools for evaluating the botanic potential of newly developing areas and for increasing productivity.

REFERENCES

Daubenmire, R., 1956. Climate as a determinant of vegetation distribution in eastern Washington and northern Idaho. *Ecological Monographs 26*, 131–154.

Denevan, W. M., 1961. The upland pine forests of Nicaragua. *Univ. California, Publications in Geography 12*, 251–320.

Eyre, S. R., 1963. *Vegetation and Soils, A World Picture.* Aldine, Chicago. 324 pp.

Fukui, E., 1960. Climate and vegetation of the Pescadores Islands. *Geog. Res. Bull. 4*, Tokyo Kyoiku Univ. 41–56.

Hare, F. K., 1950. Climate and zonal divisions of the boreal forest formation in eastern Canada. *Geographical Review 40*, 615–635.

Hare, F. K., 1954. The boreal conifer zone. *Geographical Studies 1(1)*, 4–18.

Kollmeyer, K. O., 1958. Climatic classification and the distribution of vegetation in Ceylon. *Ceylon Forester 3(2,3,4)*, 144–288.

Küchler, A. W., 1964. *Potential Natural Vegetation of the Conterminous United States.* American Geographical Society, New York.

Major J., 1963. A climatic index to vascular plant activity. *Ecology 44*, 485–498.

Mather, J. R., 1959. The moisture balance in grassland climatology. In Sprague, H. B. (ed.), *Grasslands*. American Association for the Advancement of Science, Washington, D.C.

Mather, J. R. (ed.), 1964. Average climatic water balance data of the continents, Part VII, United States. *Publications in Climatology 17(3)*, 415–615.

Parsons, J. J., 1955. The Miskito pine savanna of Nicaragua and Honduras. *Annals*, Association of American Geographers *45*, 36–63.

Shantz, H. L. and Zon, R., 1924. Natural vegetation. In *Atlas of American Agriculture*. U.S. Dept. of Agric., Washington, D.C.

Subba Rao, B. and Subrahmanyam, V. P., 1962. A climatic study of arid zones in the central Deccan. *Proc. Natural Institute Science India 28*, 568–572.

Thornthwaite, C. W., 1931. The climates of North America according to a new classification. *Geographical Review 21*, 633–655.

Thornthwaite, C. W., 1948. An approach toward a rational classification of climate. *Geographical Review 38*, 55–94.

Thornthwaite, C. W., 1952. Grassland climates. *Publications in Climatology 5(6)*, 14 pp.

Thornthwaite, C. W. and Hare, F. K., 1955. Climatic classification in forestry. *Unasylva 9(2)*, 50–59.

Thornthwaite, C. W. and Mather, J. R., 1957. Instructions and tables for computing potential evapotranspiration and the water balance. *Publications in Climatology 10(3)*, 185–311.

Vernet, A., 1965. *Climats et Vegetation*. UNESCO.

W. J. Maunder

8 Agroclimatological relationships: a review and a New Zealand contribution

Mankind is dependent to a very large extent on the products of the land. These products are, however, closely dependent on the prevailing weather and climatic conditions, and it is clear that climatic variations are an important if not the most important cause of the variations which occur from the over-all trend in per unit production.

The rapid increase in productivity in the United States during the past 35 years has, however, raised the question as to the relative contributions of technology and weather. Bean (1967), in discussing this question asks: "How much of this lift in per acre yields is due to the doings of man,—with what he does with his land, his seeds, his fertilizer, his pest and disease control, his row spacings, and how much may be the result of a fundamental change in weather?"

Commenting further on this question, Thompson (1964) says that there has been a growing tendency to believe that technology has reduced the influence of weather on grain production so that we no longer need to fear shortages due to unfavorable weather. Thompson reminds us, however, that "there is increasing evidence . . . that a period of favorable weather interacted with technology to produce our recent high yields, and that perhaps half of the increase in yield per acre since 1950 has been due to a change to more favorable weather for grain crops."

Irrespective of the importance of the overall trends in the seasonal weather, however, there remains the day to day, week to week, and season to season weather variations. And, it is these variations—the possibility of more accurate long-range weather forecasting and the actual modification of the weather notwithstanding—that are of fundamental importance.

The importance of agroclimatological studies is therefore not lessened by advances in technology. Indeed, increasing pressure of populations on world food supplies suggests that such studies will be of even more value in the future than they are today.

THE SETTING

The general problem of agroclimatological relationships is complex both in its biology and in its physics. Penman (1962) states

> Physically, at least five divisions are possible. After a zero order crop-weather interaction, to be defined when it is discussed later, there is the obvious first-order effect in which we know that crop growth depends on weather, and that seasonal and secular changes in yield are caused by weather changes. To be able to demonstrate the dependence may require the absence, avoidance or neutralization of second-, third-, and fourth-order effects, any of which might be dominant. Second-order effects are those arising from pests and diseases, nearly all of which have strong associations with weather in their incidence, in their development, or in their spread. . . . Third-order effects, too, have a biological origin—man. Soil management and crop husbandry frequently have to go on whatever the weather, and though soils and crops will tolerate a lot of ill-treatment, their resilience is not unlimited. . . . The fourth-order effects may be peculiar to a species or even to a variety of plant. They are the effects that involve "trigger action" or the existence of threshold values; to which for completeness, might be added the effects of meteorological abnormality, sometimes manifest in frost, rain, hail or wind damage.

Agroclimatological relationships can, therefore, be extremely complicated, and as Hogg (1964) has said "the answers are not easily come by. . . ." However, Frisby (1951) says that Klages (1942) voices generally

Reprinted, with minor editorial modification, by permission of the author and the Canadian Association of Geographers, from *The Canadian Geographer*, Vol. XII, 1968, pp. 73–84.

accepted agricultural opinion when he deals unhesitatingly with climatic before physiographic, edaphic, or biotic factors; and Frisby further comments that if farming practices, seed, and soil for a given locality are constant, "climate must be the variable resulting in year-to-year fluctuations in crop yield."

Yet, according to Frisby, despite his recognition of the importance of climate, the agriculturalist concerns himself principally with soils, seeds, pests, farming practices, and the intricacies of physiological development of the plant, and shows less willingness even than the climatologist to become involved in questions of climate-agriculture relationships. Frisby continues: "It would seem, therefore, that here there is scope for the imagination of the geographer. His role must be to appreciate and bridge the gaps between climatologist, agriculturist, and statistician."

Similar thoughts are echoed by Reeds (1964) who suggested that ". . . clearly . . . a more accurate assessment of world agricultural resources is urgently needed in order that reasonable potentials may be predicted more accurately." These resources include, of course, those of the weather and the climate and its variations. The evaluation of this resource may be made in several ways, some of these ways being discussed in this paper. These evaluations are, however, generally based on the more conventional climatological data and it is appropriate at this point to mention the growing importance of heat and moisture balance studies, as ably reviewed by Tweedie (1967), and their undoubted significance in agroclimatological studies particularly as more and better measurements of this aspect of the weather and climate become available.

The importance of climatic variations on agricultural production is stressed by many observers and Sanderson (1954), who made a detailed study of the methods of crop forecasting, comments that among the various branches of economic activity, agriculture is the only one which seems to be destined forever to be subject to wide and irregular fluctuations of output, the year-to-year variations of agricultural output being largely determined by physical factors. Sanderson further comments:

> We need concern ourselves only with those environmental factors which are variable from year to year. Among these, weather factors are by far the most important. Diseases and insect pests are next in importance, but their year-to-year variations in turn are frequently dependent on meteorological conditions. The same is true of competition by weeds. Economic factors must also be mentioned. Year-to-year variations in the acreage planted to a given crop in response to fluctuations in price, are, as a rule, confined to land which is marginal from the point of view of that crop, and thereby affect the average yield. Economic conditions such as the price received in the previous year may also be responsible for variations in the use of input factors, such as fertilizers and sprays, for instance.

Further comments on agroclimatological relationships are made by Smith (1961), who in discussing climatic variations says that the "only constant feature of any climate is its tendency to change." Smith continues: "The weather has an infinite variety of items in its repertoire and although the range may be restricted, the variations within the range are rarely repeated, so that the climate, which is the digest of weather, is always in a state of change."

The *significance* of climatic variations is however difficult to determine. Smith, for example, in looking at the general question of climatic change says that if we are asked the question "Are climatic changes significant?" the answer depends on how we define the word "significant" and as Smith so aptly expresses it: ". . . if . . . we examine recent climatic variations through the eyes of a statistician, we may be forced to conclude that none of them are 'significant'. . . . This may be acceptable to the pure scientist, but it could tend to infuriate the man who has experienced what *appears to him* [italics mine] to be a radical change in the circumstances in which he has to live and work." Similar comments could also be made with regard to the significance of climatological variations, and their association with agricultural production. Some relationships may not be statistically significant, but they may nevertheless be very significant to the plant and the farmer.

Although most of the references in this paper are to American or British Commonwealth authors and experience, it must be emphasized that many other important studies have been made. In this regard it is useful to note (especially if compared with the comments of Tweedie discussed above) that Sinelshikov (1965), in discussing agroclimatological trends in the USSR indicates that: "The great and economically important agroclimatic investigations carried out by Soviet agroclimatologists require a more comprehensive and profound study of the theory of the origin of agroclimatic conditions, on the basis of the radiation, heat, and water balance of agricultural fields."

APPROACHES

There are, broadly speaking, three ways of approaching the issues and establishing whether climate-agriculture relationships are in fact significant. The first is the study of the fundamentals of plant-climate relationships, namely, the radiation and moisture balance for various crops in various climatic environments. Monteith (1965) indicates, in fact, that the analysis of the radiation climate is a central problem of agricultural meteorology since the rates of photosynthesis depend on the receipt of visible light and the rates of transpiration depend on the net exchange of radiation by a crop canopy. These views are further explored by Monteith (1966), who discusses the importance of the radiant energy absorbed by a leaf and stored chemically in photosynthesis, the estimation of the potential or maximum photosynthesis of a crop canopy, and the potential rate of transpiration.

The second method of determining climate-agriculture relationships is by studying agricultural data and climatic data for a number of places within a given area, for as long a period as consistent records of both agriculture and climate allow, and deducing agroclimatological relationships from analyses of the data. Despite the many useful agroclimatological relationships obtained using such methods, there are, as pointed out in the following section, many associated problems. In particular, it is very difficult to isolate the contribution made by any one of the various components of the climate such as temperature, day length, light intensity, and precipitation. However, controlled environments, the third method, do permit in many cases an analysis of the importance of the various climatic components, because the individual climatic factors can be varied one at a time. For this method, expensive climate control chambers are necessary, and in many countries these control chambers have only recently been established. Control chambers theoretically can give the optimal and marginal conditions for the growth stages of a plant, nevertheless McWilliam (1966) sounds a note of warning about the extrapolation of results derived in the controlled environment to field situations. He states: "It is quite obvious that it is impossible to reproduce natural environments under controlled conditions. For this reason it is not possible to predict whole plant performance."

STATISTICAL METHODS

In the past, agroclimatological studies were greatly handicapped by the lack of theoretical insight into the processes of plant growth and the responses of plants to weather influences, the inadequacy of the available data, the lack of efficient statistical methods of analysis, and the lack of coordination of the efforts made by experts working on the various phases of the problem. However, with the recognition of the need for close cooperation between agronomists, plant ecologists, plant physiologists, meteorologists, climatologists, economists, statisticians, and geographers, recent progress in this field has been encouraging. Further, the advent of electronic computers has made possible the assessment of the importance of a large number of climatic variables in their effect on agricultural production, a task which before was impracticably laborious.

One of the first to apply correlation methods to the problem of the effects of the meteorological factors current during growth was Hooker (1922), who systematically correlated rainfall and accumulated temperature for overlapping eight-weekly periods throughout the year with the yields of a number of farm crops in eastern England. Fisher (1925) modified this method, and computed partial regressions of the climate/wheat yield for each period of the year. Many others have also applied statistical methods to agroclimatological relationships.

The application of correlation analysis to geographical areas, rather than to individual experimental stations, however, was not attempted until fairly recently, and one of the first papers in this field was by Rose (1936) who assessed climate/corn yield correlations for several counties in the mid-west of the United States. A further paper along similar lines was given by Zacks (1945) who discussed the effect of climatic factors on oat production in Southern Ontario, while Frisby (1951) assessed the weather crop realtionships for wheat in the northern Great Plains of the United States.

In recent years, there have been many more papers published on agroclimatological analyses applied to geographical areas, particularly in the main crop areas of the United States. In particular, multiple regression techniques have been used extensively and Thompson (1963a), commenting with regard to analyses of the evaluation of weather factors in the production of grain sorghums in Texas, Nebraska, Kansas, Oklahoma, and Missouri, states that "results from five states analysed simultaneously provide a degree of replication which is desirable and practical in this instance."

PROBLEMS

There are many difficulties facing the research worker in agroclimatological studies, particularly when statistical techniques are used. This has long been appreciated, Kincer and Mattice (1928) stating:

> It is . . . well known that there are critical periods of growth during which certain weather influences are more marked than during other times. These critical periods, in some crops at least, are of comparatively short duration, and, consequently, it is necessary for best results to use weather variants based on similar short intervals of time, so that their greater importance may be reflected in the final result. In most weather and crop correlations the month is used as the basic unit of time, principally by reason of the fact that weather data are usually compiled and published in this way. It is preferable, however, that shorter intervals of time be used in most cases.

The limitations of statistical correlations in studying the influence of weather on crops arise to a considerable extent, because of the large number of weather factors which apparently influence the yield. In addition, the varying importance of different periods of growth necessitates the use of comparatively short time intervals, and Sanderson (1954) comments: "Plants pay little attention to the calendar; they germinate, blossom, ripen their seeds according to the seasons not according to the calendar. The progress of the seasons is different from year to year and from locality to locality, so that seeds germinate, flowers bloom and fruits ripen on different dates in different places and in different years at the same place."

The answer to this problem is, according to Sanderson, to select as time intervals stages of equal development of the plant, such as (for wheat): (a) formation of seed, (b) from germination up to formation of first pair of leaves, (c) from formation of first pair of leaves to heading, (d) from heading to flowering, (e) from end of flowering to milky stage, and (f) from milky stage to maturity. However, in many areas the comparative lack of phenological records, together with the paucity of soil moisture and evaporation data, represents a most serious obstacle to further progress in weather-crop research.

This problem is partly overcome by Odell (1959) who used two approaches in an investigation of corn and soybean yields in Illinois. The first was to study the relationships within fixed calendar periods each year (this being the usual method and the one followed, by necessity, in many analyses); the second was to study a period which consisted of a specific number of days before and after corn tasseling so that the dates varied from year to year. Odell comments that by using a variable calendar period, tied to the date of tasseling each year, we are able to pinpoint more exactly the effects of weather conditions. In addition, Sanderson and others comment that it would also be desirable that the data be grouped by climatically and agriculturally homogeneous regions, rather than by political units.

It can therefore be seen that time units smaller than the month, and regions smaller (or at least more homogeneous) than political or administrative units such as counties, are highly desirable for agroclimatological studies. However, both these ideals may not be met by the available range of data; and Malcolm (1947), in his ecological study of the barley crop in New Zealand, noted that the county is the smallest unit for which reliable and complete crop statistics are available. This, he says, is a serious limitation, especially in large counties where climatic and soil variations may cause yield differences which cannot be determined from district to district within the county.

Equally serious is the small amount of available published climatic data relating to smaller units of time than the month. In most countries climatic data are available on a monthly basis, and although daily data are available in the original records or occasionally in data processing banks, compilation of, say weekly data would in many cases be difficult. Consequently, many agroclimatological analyses are more or less restricted to taking the month as the smallest unit of time. An even more serious problem is that monthly climatic data may not necessarily give a "true" indication of the climate for that particular month. For example, in two consecutive months average rainfalls may be recorded, but one month may be completely rainless for (say) the last twenty-five days of the month, while the next month may be completely rainless for the first twenty-five days. There would, therefore, be a period of fifty days without rain, even though the two months had average rainfalls. Such inadequacies are of course well recognized but they are not always taken into account by enquirers, especially when looking at past climatic records.

COMPLEXITIES

The complexities of the agroclimatological relationship are further emphasized by some recent studies. The advent of electronic computers, for example, enabled Buck (1961) to re-examine the results of Fisher's Broadbalk experiment. He calculated multiple regression of yield on seventeen variates—five representing the time trend of yield, together with six rainfall and six temperature variates—for six plots in sixty-seven years, and also repeated Tippett's analysis of sunshine effects (1926). The only significant partial correlation, however, was on total annual rainfall, Watson (1963) concluding accordingly that "after eighty years of intermittent but intensive study of a set of data that appears uniquely suited for the purpose, all that has been established with statistical certainty about the dependence of the wheat yield of Broadbalk on weather is that it decreases with increase in annual rainfall above the average." Watson also says that similar studies on long-continued experiments on other crops at Rothamsted have been no more successful. However, he indicates that some studies such as that by Cornish (1950) have been more formative, and indeed one could cite many similarly formative agroclimatological analyses (Doepel, 1959; Johnson, 1959; Millington, 1961; Thompson, 1962). Nevertheless, it is true that many correlation and regression analyses have proved to be unrewarding, some of the reasons for this being given by Watson:

> Past experience . . . does not encourage us to expect that knowledge of how yield depends on weather can come from measurement of yields in naturally varying environments. The fundamental defect of this approach is that the dependence of yield on climatic factors is usually far too complex to be described adequately by linear regressions on a few gross measures of climatic variation, except perhaps when one factor, most likely to be rainfall or lack of it, dominates over all others. At best, any correlations so established are empirical, difficult to interpret reliably in terms of known effects of climatic factors on plant growth, and do not necessarily describe a direct influence of weather on the plants.

A similar view is expressed by Chang (1965) who says that the customary procedure of correlating yield with temperature and rainfall is so crude that it contributes little in actual operation and long-range regional planning; and he comments that only through the application of the water balance and other sophisticated concepts can the effect of climate on yield be expressed in exact, quantitative terms.

PREDICTION OF AGRICULTURAL PRODUCTION

Agroclimatological investigations are usually designed to show the association between several climatic elements and various agricultural factors. In particular, emphasis can be placed on the effects of variations from the average climate on variations from the expected (i.e., time trend deviations) agricultural unit production. In such cases, the actual climate or the actual agricultural production is of secondary importance.

Theoretically, however, the prediction (either within the period of record, or beyond the period of record used for assessing the regression coefficients) of agricultural production based on the various regression analyses, is possible by substituting in the various regression equations the specific climatic and time values for a particular season. However, it is doubtful if such prediction through the use of multiple linear regression analyses is justified. Thompson (1962) for example, states that although multiple linear regression coefficients are useful in gaining a perspective of the influence of weather variables on crop yields, they should not be used for predicting yields. In this context, Thompson is referring specifically to "prediction" within the period of record—that is, a comparison between the actual yields and the expected (or predicted) yields of the regression equation—and forecasting beyond the period of the regression is not considered. Prediction at least within the period of record is, however, possible with some success, by using multiple curvilinear regression coefficients, as has been demonstrated by Thompson (1963b) and others. Watson suggests, however, that only when more detailed knowledge has been acquired of how and at what growth stages climatic factors influence yield, will it be possible to derive complex variates that give appropriate weight to the different factors for correlation with yield in naturally varying climates, and eventually use them to predict yields from meteorological records.

A NEW ZEALAND EXAMPLE

In a recent study (Maunder, 1966a) an "agroclimatological model" was formulated for different agricultural factors and for different areas in New Zealand, based on variations in agricultural production and climate, mainly since the 1930's. Variations from this model were then studied in order that probability assessments could be made of certain variations in agricultural production.

Specifically, three aspects of agroclimatology in New Zealand were examined: (1) an assessment of the variations in climate, and in agricultural production, (2) the agroclimatological associations, and (3) an assessment of the probability of occurrence of significant climatic variations, and the effect upon agricultural incomes of the consequent variations in agricultural production.

In order to "select" the appropriate climatic variables six "trial" regression analyses were made. Analysis of the results obtained showed in general that under New Zealand conditions (1) monthly data were more significant than seasonal data, (2) the substitution of linear aspects of the climatic variables with their second-degree terms, or second-degree terms of other climatic variables, did not increase to any great extent the significance of the results, and (3) the use of mean maximum temperatures rather than mean temperatures did not necessarily increase the level of significance.

These findings therefore suggested that the multiple regression model should include monthly climatic data instead of seasonal data, linear rather than second-degree variables (except time), and mean temperatures instead of mean maximum temperatures.

The following multiple regression model (Maunder, 1966b) was therefore used: $y = a_0 + a_1x_1 + a_1'x_1^2 + a_2x_2 + a_3x_3 + \cdots + a_{15}x_{15}$ where y = agricultural factors (wheat yield per acre, butterfat yield per cow, etc.).

x_1 = season (1933/34 = 1, 1934/35 = 2, etc. in most cases)

$x_2 \ldots x_6$ = rainfall (Oct., Nov., Dec., Jan., Feb.)

$x_7 \ldots x_{11}$ = mean temperature (Oct., Nov., Dec., Jan., Feb.)

$x_{11} \ldots x_{15}$ = sunshine (Nov., Dec., Jan., Feb.)

The model therefore assessed the association between various agricultural variables and 14 aspects of the climate: linear aspects of rainfall, and mean temperature, for each of the months October to February, and linear aspects of sunshine for the months November to March. In addition, two time factors were included.

Although a multiple linear relationship is finally used to ascertain the correlation between agricultural factors and climatic elements, it must be realized that linear relationships are generally the exception rather than the rule, and that given a sufficient climatic range, optimum values will generally become apparent. Such threshold, optimum, and limiting values are well established in the American mid-west for corn (Thompson, 1963b); and it is reasonably apparent that such values apply to other crops in other areas. In New Zealand it is probable, however, that the climatic range is not sufficient to affect seriously the findings of a linear relationship.

The detailed methods and results of the investigation are given elsewhere (Maunder, 1966a) but, in brief, the coefficients in the multiple regression equations already discussed were used to estimate the effect of specific climatic variations on agricultural production, a specific climatic variation being taken as a positive or negative departure from the average of one standard deviation.

For example, in the case of October rainfall at East Gore climatological station and the associated oat production per acre in the Southland county, the coefficient for October rainfall in the multiple regression model was −2.50. Accordingly, it follows that a rainfall one standard deviation (1.10 inches) *above* the average, would, if all other factors remained constant, "decrease" the oat yield by 2.8 bushels (2.50 × 1.10) from what it would have been if the rainfall had been equal to the average of 2.74 inches. Similarly, a rainfall one standard deviation *below* the average, would "increase" the oat yield by 2.8 bushels from what it would have been with a rainfall equal to the average.

The utilization of such a method enables one to suggest that once in, say, five, six, or seven years the effect on production of a specific climatic variation such as a "dry"* October would be in the order of x bushels. Further, the value or economic significance in terms of agricultural income of such a variation in production per acre, can be assessed by multiplying the per acre or per animal variation by the particular total acreage or total livestock population.

One striking example of the application of such a concept is the "economic value" of "dry" and "wet" conditions in the major dairying area of New Zealand—the Waikato. In this case, the analysis showed that a "wet" January was associated with an "increase" in butterfat production of about 9 ± 2 pounds per cow, and this increase, at the 1964 farm-to-factory price of NZ$0.28 per pound butterfat, was valued at about NZ$2.6 ± 0.6 per cow. Now, the dairy cow population in the Waikato county and surrounding areas has in recent years totalled over 750,000. Accordingly, if the "values per cow" in the Waikato county are taken as an index of the variations in butterfat income per cow for the South Auckland area, it can be seen that a significant climatic variation such as a "wet" January is "worth" about NZ$2 million to the area (750,000 cows × $2.6 per cow). It must be remembered, however, that this is simply a very approximate value of a significant variation in one aspect of the climate for only one month.

Based on the period 1936/37–1959/60 at the Ruakura climatological station the possibility of a "wet" January (the significant month discussed above) is one in six, compared with a probability of one in eight for a "dry" January. It can therefore be suggested, that if Ruakura is taken to be representative of the climate of South Auckland, then once in about six seasons a "wet" January will be associated with an "increase" of about NZ$2 million in the income of dairy farmers of South Auckland. Conversely, it migh tbe expected that a comparable "fall" in income would occur once in about eight seasons as a "result" of a "dry" January. These substantial variations in income thus provide a measure of the potential economic importance of significant climatic departures from the mean, as they affect butterfat production in New Zealand's major dairying region.

* [Defined as a month with a rainfall one standard deviation (or more) below average.]

THE FUTURE

In the past there has been neglect of studies of the dependence of yield on climate and weather, apart from the technical difficulties. One reason for this has been the apparent impossibility of applying the results to improve the productivity of field crops, since farmers cannot control climatic conditions. However, this is surely not a valid reason for it is possible to alter crops or methods of husbandry to suit the prevailing weather and climatic conditions. It is also possible to alter the weather, and although there are claims and counterclaims as to what can and cannot be done, there now seems general agreement that in certain areas, under certain circumstances, and at certain times, weather modification in the form of an increased precipitation of the order of 10% can be achieved (Sewell, 1966).

Recent investigations have re-emphasized the importance of agroclimatological relationships. A comprehensive study by the Center for Agricultural and

Economic Development (1964), for example, assessed the effect of recent climatic trends on grain production in the United States. Wadleigh, in an introduction to the study, states that the evaluation of weather in relation to food production is more than a need, it is a must. Further, Wadleigh says that we need to be far more knowledgeable on (1) the phenology of crop plants, (2) weather probabilities, (3) making alternative decisions in management practices in relation to sequential weather events, and (4) weather prediction.

The forecasting of agricultural yields on the basis of meteorological conditions is a problem which has fascinated agricultural climatologists for a long time, although as Bourke (1963) has said: "In this type of investigation, we are groping on the very outside frontiers of our knowledge, and even beyond." However, according to Bourke, the importance of the subject to agriculture is so great that "much of our efforts in the next ten years must necessarily be devoted to finding some reliable basis for extrapolating beyond the short-term weather forecasts adequate for agricultural tactics, to estimates of long-term weather trends which will be applicable to the even more important field of agricultural strategy."

The world is dependent to a very large extent on agricultural production, agricultural production in turn being closely related to climate. The normal climate is however rarely experienced, and what occurs most of the time are variations from the normal. It is reasonably clear that these climatic variations while not random, are not rhythmic enough to produce a prediction equation; nevertheless, significant departures of one or more consecutive seasons from the normal climate are especially important in the world's major agricultural areas; and Thompson, in his 1966 study on weather variability and the need for adequate food reserves in the United States, warns that it is important for policy-makers to recognize the contribution of favorable weather particularly during the recent 1961–65 period to United States corn and soybean yields. This, he says, is particularly true now that many other countries are becoming dependent on the United States for their grain supply.

That the weather plays an important part in the growth of a crop is well recognized. Further, several recent studies have assessed the more basic physical and biological relationships as, for example, in an analysis of meteorology and grassland farming in Britain by Smith (1967).

The estimation of the precise economic value of climate and weather in their influence on crop and livestock production has not, however, received as much support from researchers as has that of the physical and biological relationships. It is, of course, a particularly difficult problem, and as Curry (1958) comments: "All over the world, and as a matter of course, farmers assess weather as a resource but the formal expression of this evaluation is far from easy." Despite these difficulties, however, solutions to this problem can be found even if only very approximately (Maunder, 1968), and it is imperative that more research along these lines be done, for as Watson states: "Climate determines what crops the farmer can grow; weather influences the annual yield, and hence the farmers' profit, and more important, especially in underdeveloped countries, how much food there is to eat."

REFERENCES

Bean, L. H., 1967. Crops, weather and the agricultural revolution. *Amer. Statist. 21(3)*, 10–14.

Bourke, P. A. M., 1963. Agricultural biometeorology. *Intern. Journ. Biometeor. 7*, 121–125.

Buck, S. F., 1961. The use of rainfall, temperature, and actual transpiration in some crop weather investigations. *J. Agr. Sci. 57*, 355–365.

Chang, J.-H., 1965. On the study of evaporation and water balance. *Erdkunde 19*, 141–150.

Cornish, E. A., 1950. The influence of rainfall on the yield of wheat in South Australia. *Aust. Journ. Scient. Res. B3*, 178–218.

Curry, L., 1958. *Climate and Livestock in New Zealand—A Functional Geography*. Unpub. Ph.D. thesis, Univ. New Zealand.

Doepel, G. C., 1959. Regression analysis of fleece weight on rainfall. *Quart. Rev. Agr. Econ. 12*, 57–60.

Fisher, R. A., 1925. The influence of rainfall on the yield of wheat at Rothamsted. *Phil. Trans. Royal Soc. London B 213*, 89–142.

Frisby, E. M., 1951. Weather–crop relationships: forecasting spring wheat yield in the northern Great Plains of the U.S. *Trans. Inst. British Geogrs. 17*, 77–96.

Hogg, W. H., 1964. Meteorology and agriculture. *Weather 19*, 34–43.

Hooker, R. H., 1922. The weather and crops in eastern England: 1885–1921. *Quart. Journ. Royal Meteorol. Soc. 48*, 115–138.

Johnson, W. C., 1959. A mathematical procedure for evaluating relationships between climate and wheat yields. *Agronomy Journ. 51*, 635–639.

Kincer, J. B. and Mattice, W. A., 1928. Statistical correlations of weather influence on crop yields. *Mon. Wea. Rev. 56*, 53–57.

Klages, K. H. W., 1942. *Ecological Crop Geography*. New York.

Malcolm, J. P., 1947. *An Ecological Study of the Barley Crop in New Zealand.* Unpub. M. Agric. Sci. thesis. Univ. New Zealand.

Maunder, W. J., 1966a. Climatic variations and agricultural production in New Zealand. *N.Z. Geogr. 22*, 55–69.

Maunder, W. J., 1966b. An agroclimatological model. *Sci. Record 16*, 78–80.

Maunder, W. J., 1968. The effect of significant climatic factors on agricultural production and incomes. *Mon. Wea. Rev. 96*, 39–46.

McWilliam, J. R., 1966. The role of controlled environment in plant improvement. *Aust. Journ. Sci. 28*, 403–407.

Millington, R. J., 1961. Relations between the yield of wheat, soil factors and rainfall. *Aust. Journ. Agr. Res. 12*, 397–458.

Monteith, J. L., 1965. Radiation and crops. *Exp. Agr. Rev. 1*, 241–251.

Monteith, J. L., 1966. The photosynthesis and transpiration of crops. *Exp. Agr. Rev. 2*, 1–4.

Odell, R. T., 1959. Effects of weather on corn and soybean yields. *Illinois Res.*, Univ. Illinois Agr. Exp. Station 3–4.

Penman, H. L., 1962. Weather and crops. *Quart. Journ. Royal Meteorol. Soc. 88*, 209–219.

Reeds, L. G., 1964. Agricultural geography: progress and prospects. *Can. Geogr. 7*, 51–63.

Rose, J. K., 1936. Corn yield and climate in the corn belt. *Geographical Review 26*, 88–102.

Sanderson, E. H., 1954. *Methods of Crop Forecasting.* Cambridge, Massachusetts.

Sewell, W. R. D., ed., 1966. *Human dimensions of weather modification.* Univ. Chicago, Dept. Geography Res. Series, *105*.

Sinelshikov, V. V., 1965. Agroclimatological trends in the USSR, problems, methods and scientific and practical results. *Agr. Meteor. 2*, 73–77.

Smith, L. P., 1961. Measuring the effects of climatic changes. *New Scient. 12*, 608–611.

Smith, L. P., 1967. Meteorology and the pattern of British grassland farming. *Agr. Meteor. 4*, 321–338.

Thompson, L. M., 1962. Evaluation of weather factors in the production of wheat. *J. Soil and Water Conservation 17*, 149–156.

Thompson, L. M., 1963a. Evaluation of weather factors in the production of grain sorghums. *Agronomy Journ. 55*, 182–185.

Thompson, L. M., 1963b. Weather and technology in the production of corn and soybeans. *Center for Agr. and Econ. Development, Rept. 17.*

Thompson, L. M., 1964. Weather and our food supply. *Center for Agr. and Econ. Development, Rept. 20.*

Thompson, L. M., 1966. Weather variability and the need for a food reserve. *Center for Agr. and Econ. Development, Rept. 26.*

Tippett, L. H. C., 1926. On the effect of sunshine on wheat yield at Rothamsted. *J. Agr. Sci. 16*, 159–165.

Tweedie, A. D., 1967. Challenges in climatology. *Aust. Journ. Sci. 29*, 273–277.

Watson, D. J., 1963. Climate, weather, and plant yield. In Evans, L. T. (ed.). *Environmental Controls of Plant Growth*, 337–349. New York.

Zacks, M. B., 1945. Oats and climate in southern Ontario. *Can. Journ. Res. C23*, 45–75.

SUGGESTED READING FOR PART 3

Barry, R. G., 1969. The world hydrological cycle. In *Water, Earth and Man*, ed. by R. J. Chorley, Methuen & Company, London. 11–29.

Biel, E. R., 1961. Microclimate, bioclimatology, and notes on comparative dynamic climatology. *Amer. Scientist 49*, 326–357.

Chang, J.-H., 1968. *Climate and Agriculture*. Aldine Publishing Company, Chicago. 304 pp.

Chang, J.-H. and Okimoto, G., 1970. Global water balance according to Penman approach. *Geogr. Analysis 2*, 55–67.

Thornthwaite, C. W., 1956. Modification of rural microclimates. In *Man's Role in Changing the Face of the Earth*, ed. by W. L. Thomas jr., The University of Chicago Press. 567–583.

Part Four

Climatic Classification

What have been the objectives behind the major climatic classifications? How successful are they? The answers can be found in Hare's critique of the subject.

Is it possible to devise a technique of classification that is wholly objective and relies purely on climatic data? McBoyle's classification of the Australian climate achieves this through the use of factor and grouping analyses. But does it work for both the macro- and the micro-scale?

How is man affected by climate? Is there in climate a basis for stress? Terjung's article explores this fascinating topic and reflects on the possibility of using physioclimatic regimes to determine recreation and retirement regions.

F. Kenneth Hare

9 Climatic classification

Introduction

This essay contains a critical summary of existing methods of climatic classification, a subject hitherto almost ignored in British geographical literature. Systems of classification of any natural complex make little appeal to the British temperament, which is rarely inclined to like pigeon-holes or card-indexes. This conservative pragmatism sometimes leads us, however, to ignore real necessities imposed by progress in research and by the elaboration of established knowledge. We are then compelled to adopt with an ill grace the systems of classification produced by foreign workers, especially the Germans and the Americans, who excel in such matters. It has been this way with physiography, where rational systems of nomenclature came from the geologists who were exploring the western interior of the U.S.A., culminating in the genius of William Morris Davis, and from the German and Austrian school, typified by Walther Penck. And it is equally true of climatology, in which the names of Wladimir Köppen and C. Warren Thornthwaite stand out. It is noteworthy that the most widely read British works on climatology, W. G. Kendrew's *Climate* and his *Climates of the Continents*, contain no reference of any length to this important topic.

Climatic classification is an essentially geographic technique. It allows the simplification and generalization of the great weight of statistics built up by the climatologists. These figures do not mean much to the geographer unless a way can be found of reducing them into an assimilable form. The real purpose of classification is hence to define climatic types in statistical terms, in which climate as a geographic factor is to be regarded as having definite and uniform characteristics: only by such a classification can rational climatic regions be defined.

Reprinted, with minor editorial modification, by permission of the author and publisher, from *London Essays in Geography*, L. D. Stamp and S. W. Wooldridge (editors), The London School of Economics and Political Science, 1951, pp. 111–134.

Since climates are faithfully reflected by soil and climax vegetation, all modern systems of classification attempt to give the climatic limits of characteristic soil or vegetation regions. This has the effect of rendering the classifications suitable for use in a discussion of the economic consequences of climate, particularly as regards the agricultural potential of a region. They are less useful in a study of climate as a factor influencing human distributions directly, through physiological effects. The earliest workers in this field were plant-geographers, and even today, interest in the subject is mainly found among ecologists, soil scientists and hydrologists. During the war years, however, much progress has been made in the study of medical climatology, and it is possible that further progress in classification will have to take account of these developments.

As has already been said, two names dominate all others in this field. Wladimir Köppen, a St. Petersburg-trained biologist, was the pioneer of comprehensive classification, publishing his earliest results in 1900. His scheme has been repeatedly modified, both by himself and his well- or ill-wishers. C. Warren Thornthwaite, of the U.S. Soil Conservation Service, did not publish his first classification until 1931, and his second was deferred until 1948. Although there have been other attempts at classification (such as the recent work of Gorczynski) few are statistical, and they cannot in consequence be regarded as adequate for geographical purposes. The three major classifications are discussed in detail below. It is hoped that the discussion will be useful by making the subject more easily accessible and by showing some of the limitations of the existing schemes.

Köppen's Classification of Climate

It is not proposed to make here a detailed review of early attempts at climatic classification; a good sketch of their history has been given by Thornthwaite (1943) in a recent study of Köppen's work. The whole idea of a systematic analysis of world climate is in any case very recent, as the necessary statistical basis has only been available for the past seventy or eighty years. Even today analysis is handicapped by the paucity of observations over much of the earth's surface, especially over the oceans and the continental interiors. Until the nineteenth century, therefore, there could be no effective substitute for the classical concept of the thermal zones. The first world-wide pictures of temperature and rainfall distributions began to appear early in the latter half of the nineteenth century, and it was from these pictures that the classification of climates was first attempted. The work was undertaken by the group of naturalists who were concerned with the world distribution of species, especially of plants. Among these we can select the work of de Candolle (1875), Linnser (1905), and Drude (1887), whose contributions have been admirably summarized by Thornthwaite (1943).

Köppen, whose classification of climate has dominated the subject for many years, was one of these biologists, and the earliest motive underlying his work was to reduce the complex facts of climate to a simple, numerically-based classification that could be used in the study of plant geography. This may be regarded as Köppen's major contribution to modern climatology, *that schemes of classification must rest on an objective, numerical basis*, so that different climatologists can refer the same set of climatic figures to the same climatic class. His other important contribution to the ideal basis of the science is the view that the climatic classification should be drawn up with respect to the known facts of phytogeography. Though by no means the originator of the idea, Köppen was the first systematically to investigate the climatic limits of the great regional divisions of the plant cover. The idea that climax vegetation and soil groups are reflections of the climatic régime is now commonplace, and permeates most modern discussions of physical geography. In Köppen's youth, however, this idea had nothing but probability and deduction to support it. By showing that definite climatic values could be assigned to the boundaries between natural vegetation regions, he proved the general validity of the idea.

Köppen's earliest attempt at classification dates from 1900. He has subsequently much changed the scheme, and has greatly added to its complexity and scope. In its latest form, it appears in his masterpiece, the *Handbüch der Klimatologie* (1936), and it is discussed below primarily as it appears in this most recent form. There are several good summaries in the English language, notably those of Leighly (1926), and Haurwitz and Austin (1944). It is advisable to exercise care in using English re-statements of Köppen's views, as they often refer to different stages in the development of the scheme, and are in many cases made so that the classification can be criticized: a friendly approach has not been too common. The above writers, however, have been eminently fair. The reader is referred to the Leighly and Haurwitz-Austin versions for nomograms and other aids in deciding the class of a particular climate.

A sketch of Köppen's classification is given in diagrammatic form in Figure 1, which gives the critical limiting values for each climatic type.

As will be seen from Figure 1, Köppen envisaged five major climatic types:

Letter	Type
A	Tropical rainy climates (Megathermal)
B	Dry climates (Xerophilous)
C	Warm temperate rainy climates (Mesothermal)
D	Cold snow-forest climates (Microthermal)
E	Snow climates (Ekistothermal)

which broadly correspond to the five major vegetation regions of the earth. For a definition of these latter regions, Köppen referred to the work of de Candolle, whose vegetation-groups are indicated in brackets behind Köppen's corresponding climatic type in the above table. De Candolle was a plant physiologist, and his regions were based on the climatic requirements of each group of plants for their internal functions. This approach is essentially distinct from that of the ecologists, who dealt in the field of plant associations. Of these, the greatest was Schimper (1903), whose *Plant Geography* remains a standard work. As Thornthwaite (1943) pointed out, it is a pity that Köppen based his scheme on de Candolle's work rather than Schimper's: the latter has dominated ecological thought and method, and Köppen's life work has therefore been constantly out of harmony with that of the ecologists.

Köppen's method can be summed up as the effort to find the climatic correspondences of the boundaries between de Candolle's regions. From the start, therefore,

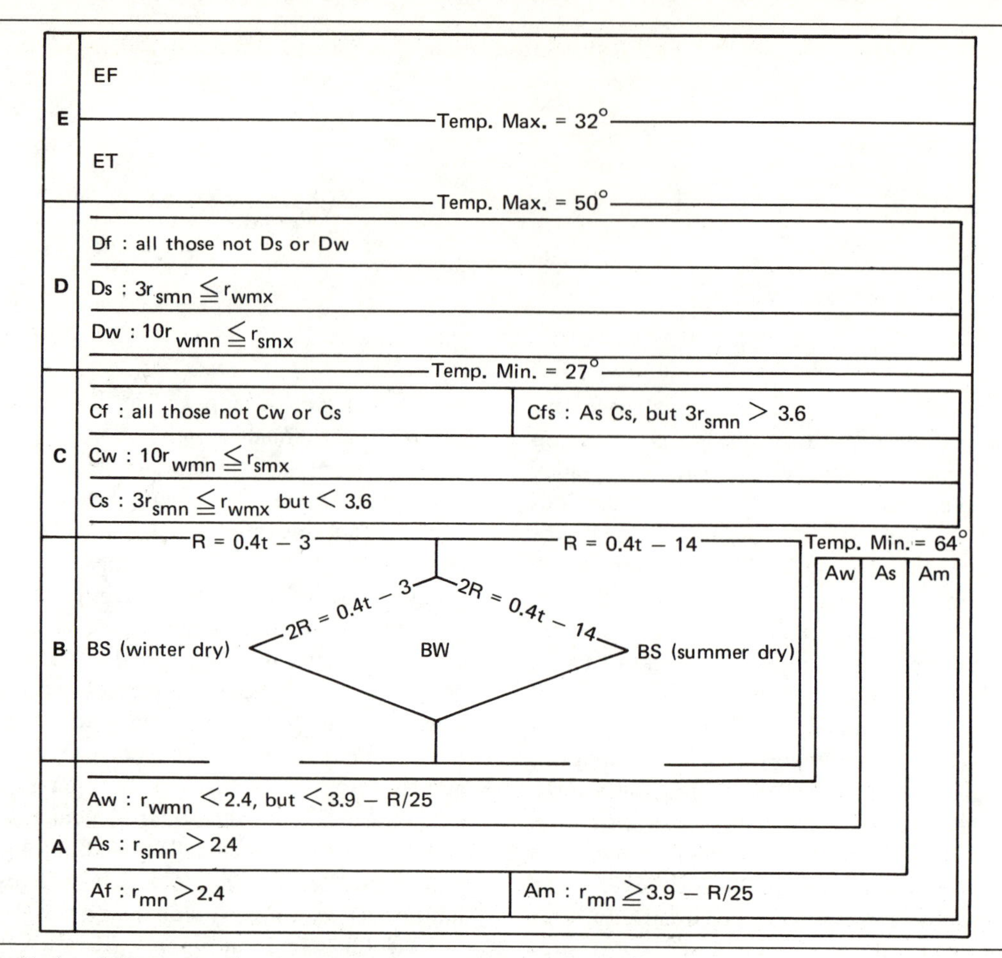

Figure 1 The limits of Köppen's climatic types.
Some idea of the complexity of Köppen's scheme is given by this diagram, which goes only as far as the first order sub-division in most cases. The diagram shows the limiting values corresponding to each of the possible boundaries between types. An exception is that the BS climates can also come into contact with the D climates; the boundary between them is the same as that between the B and C climates. The first-order sub-divisions of the forest climates (A, C, and D) are shown as enclosed areas within the space allotted to the major types corresponding. The complexity of the classification is also brought home by the formidable number of climatic elements listed in the key below.

KEY

Temperature Values (° F)

Temp. max = avg. temperature of warmest month
Temp. min = avg. temperature of coldest month
t = avg. annual temperature

Rainfall Values (inches)

r_{mn} = avg. rainfall, driest month
r_{mx} = avg. rainfall, wettest month
r_{smn}, r_{wmn} = avg. rainfall of driest summer or winter month
r_{smx}, r_{wmx} = avg. rainfall of wettest summer or winter month
R = avg. annual rainfall

he was dedicated not to synthesis of climatic controls but to the finding of arbitrary climatic limits. The result has been that Köppen has used a wide variety of different elements in defining his boundaries, and has often changed his mind about which element to use in defining a particular boundary.

Köppen's major types are sub-divided several times, and it is a great help in remembering the details to refer these sub-divisions to a hierarchy of orders. The first order sub-divisions usually refer to special features of the rainfall régime, and the second order to thermal characteristics. The third and remaining orders refer to various characteristics of a more variable and local kind, and they are not discussed here. The first and second order sub-divisions are summarized in Table 1.

The major types in the Köppen scheme are always indicated by the capital letters A–E and H. The first and second order sub-divisions are usually indicated by

Major Type	First Order Sub-divisions	Second Order Sub-divisions
A. (Tropical, rainy)	Af (wet at all seasons) Aw (dry winters) As (dry summers—rare) Am (monsoon type)	Axi (isothermal)
B. (Dry)	BS (steppe) BW (desert)	Bxh (hot) Bxk (cool)
C. (Warm, temperate, rainy)	Cf (wet at all seasons) Cs (dry summers) Cw (dry winters) Csf (rather dry summer, very wet winter)	Cxa (summers long and warm) Cxb (summers long and cool) Cxc (summers short and cool)
D. (Cold snow-forest)	Df (wet at all seasons) Ds (dry summers) Dw (dry winters)	Dxa (summers long and warm) Dxb (summers long and cool) Dxc (summers short and cool) Dxd (ditto, but great winter cold)
E. (Snow climate)	ET (tundra) ES (permanent snowfields)	—
H. (Mountain climates)	—	—

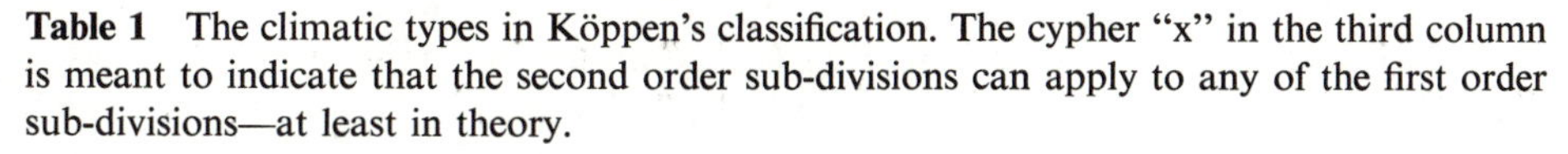

Table 1 The climatic types in Köppen's classification. The cypher "x" in the third column is meant to indicate that the second order sub-divisions can apply to any of the first order sub-divisions—at least in theory.

small letters (except in the Dry and Snow climates where the German substantival capital usage requires a departure).

The boundaries between the first order sub-divisions as well as the major types themselves are summarized in Table 1. The major types can be sub-divided into the forest (A, C, D), dry (B) and cold (E, H) climates, which are discussed in turn below.

(1) *The Forest Climates.* The forest climates are distinguished from one another (Figure 1) on the basis of temperature. The A climates are those in which the coldest month has a mean temperature of over 64° F, and the D climates are those whose coldest month has a mean temperature below 27° F. The C climates lie between. These climates then represent the selva, warm temperate rain forest and cold snow-forest régimes (including taiga), though the selected temperatures are justified on the basis of the optimum for human comfort (64° F) or the possibility of prolonged winter snow-cover, for which 27° F is regarded as the upper limit.

As shown in Figure 1, the forest climates are further divided on the basis of rainfall régime. Those whose rainfall is abundant at all seasons are given the first order sub-division f (Af, Cf, Df), though the meaning of the cypher differs slightly from group to group. The w sub-divisions are winter dry, and the s sub-divisions, summer dry. Here again the quantitative limits indicated in Figure 1 differ from group to group. Among the C and D climates, summer is considered dry if its driest month gets a third of the rain of winter's wettest month, whereas the ratio is a tenth in the reverse case of winter drought. This is to allow for the greater evaporation and lower rainfall effectiveness in summer.

Special first order sub-divisions are provided in the A and C climates for those climates which have a wet and dry season, but which are humid throughout because of an excessive winter rainfall.

(2) *The Dry Climates.* Köppen's dry climates (B) are defined by an empirical aridity criterion which he derived from a study of the climates of the steppe-forest boundary. The criterion adopted has been changed several times, and has had to meet considerable criticism, especially in the United States, where it fits the steppe-desert boundary rather poorly. Although the B climates are mainly within the warm or hot latitudes, they are not specifically confined to these latitudes by the aridity criterion, which is of the form

$$R \leq 0.44\,T - K$$

where R = the mean annual rainfall (inches) and T the mean annual temperature in degrees Fahrenheit. K is an empirical constant depending on the seasonal rainfall distribution. If the rainfall maximum is in summer (and is therefore less effective due to evaporation) K = 3, but if rain falls mainly in winter K = 14. This relationship is based on the observed fact that in warm climates a greater rainfall is necessary to maintain abundant vegetation than farther north in cooler climates.

Within the dry climates a distinction is made between BW (desert) and BS (steppe) climates, the criterion adopted being that

$$2R \leq 0.44\,T - K$$

where R, T, and K have the same meanings as before.

(3) *The Cold Climates* (*E*). Köppen observed that the taiga of the D climates passed into tundra very close

Major Type	Second Order Sub-division	Characteristics and Limits
A x	i	Mean monthly temperatures all within a 9° F range (isothermal)
B x	h	Hot desert or steppe (mean annual temperature > 64° F)
	k	Cool desert or steppe (mean annual temperature < 64° F)
	a	Warmest mo. > 72° F : > 4 mo. > 50° F
C x	b	Warmest mo. > 50° F : > 4 mo. > 50° F
D x	c	Warmest mo. > 50° F : < 4 mo. > 50° F
	d*	As c, but coldest month < −36° F

* refers only to D climates.
x indicates any first order sub-division.

Table 2 The second order (thermal) sub-divisions of the Köppen climates.

to the 50° F isotherm for the warmest month, and this boundary has not been seriously challenged. He drew a further distinction between the tundra climates (ET) and the frost climates (EF): the latter, characterized by permanent snow-fields, lay north of the 32° F isotherm for the warmest month. This again is a fairly obvious boundary, though Jones (1932) has pointed out that permanent snow-fields may occur well south of the 32° F isotherm in areas of heavy snowfall. A special class of cold climates (H) occurs on high ground in middle and low latitudes: Köppen used different computing formulae for the H climates, which cannot be described in detail here.

The second order sub-divisions. The further sub-division of the climates is a matter chiefly of interest to pure climatologists. Most geographers appear to regard the scheme as adequate as it stands after the first sub-division has been made. Thus Trewartha (1943) has prepared a map of world climates on a simplified first order classification. But Köppen himself attached considerable importance to the thermal, second order sub-divisions, which undoubtedly add greatly to the precision of the classification. The third cypher in a Köppen label (As*i*, Cw*b*), etc., refers to the thermal characteristics of the climate within its major division. Table 2 gives a summary of the principal second order sub-divisions, each of which can in theory (though not in practice) be applied to any first order derivative of the group.

Further sub-divisions have been suggested by Köppen, but it is extremely doubtful whether the further complication of the scheme can be justified. They will not be discussed here.

The Geographic Validity of Köppen's Classification

For at least two decades the Köppen classification has been widely accepted by geographers, most of whom have used it as the standard system. The great majority of simplified classifications, like Blair's (1942) and Trewartha's, are based on Köppen's system. So widespread has it become that some American writers have begun to regard it as an international standard, to depart from which is scientific heresy. There has none the less been very extensive criticism of the classification, some of it constructive, in the U.S.A. America is a land of climatic contrasts, and it has been clear for years that Köppen's boundaries do not correspond to the real natural limits of the major climatic zones as they are expressed by the soil and vegetation. To remedy this (or to try to) many American workers have suggested moving a boundary one way or the other by slightly changing its numerical basis. Others have attacked the scheme on the grounds of its empiricism, and have in some cases attempted to replace its methods by others more logical.

Among the first group, who can be dubbed the line-shifters, we may list Van Royen (1927), Russell (1926, 1931–32, 1934, 1945), and Ackerman (1941). Van Royen constructed a map of the climatic regions of the eastern U.S.A., and showed that there is no sort of fit between the boundaries and those of the natural vegetation. Russell applied the classification to California (1926), and later to the dry western half of the U.S.A. (1931–32). Like Van Royen he found a bad fit, especially along the B/C and BW/BS boundaries. He attempted to improve the fit by using an older expression for the aridity criterion, but it cannot be pretended that this modification has improved matters very much. Russell (1934) also introduced the concept of "climatic years." He had been concerned at the substitution on climatic maps of lines for what in the landscape are the broad transition zones between climates. He tried to remedy this by showing how the boundary migrates from year to year, and defining the zone over which it oscillates. He also drew attention to the fact that it is the occasional drought, flood or other climatic extreme which is most significant in limiting development of the plant cover and of land-form: hence Köppen's extensive use of annual and monthly means may not give a true picture of the climatic boundaries. Ackerman (1941) prepared a map of the U.S.A. showing the Köppen divisions, but used the 32° F isotherm for the coldest month as marking the C/D boundary, following Russell. All these writers, however, accepted the classification as the best possible at present, while admitting its obvious defects.

Group	Decimal Type	Description	Limiting Factors
I (Warm group)	1	Wet tropical type	Mean temperature > 68° F in three coolest months
	2	Savannah type	
II (Arid group)	3	Desert type	Aridity criterion (see above) high throughout year
	4	Steppe type	Ditto, for several months only
III (Maritime group)	5	Dry winter type	Mean temperature of coldest month between 59° F and 23° F; or less than four months below 32° F
	6	Dry summer type	
	7	Wet temperate type	
IV (Continental group)	8	Heavy winter snow type	Mean temperature of coldest month below 23° F, of warmest above 50° F
	9	Light winter snow type	
V (Sub-polar or mountain group)	10	—	Mean temperature of warmest month below 50° F

Table 3 Gorczynski's decimal classification of climates. This table is not intended to define the types precisely, but merely to indicate the general character of the classifications.

More radical objections have been voiced by Gorczynski (1935, 1945), who especially quarrels with the aridity criterion. He himself employs a criterion which depends on temperature range, rainfall variability and latitude. It is of the form

$$A \propto (T_{max} - T_{min}) \frac{(R_{max} - R_{min})}{(R)} \operatorname{cosec} \phi$$

where

T_{max} = highest monthly mean temperature
T_{min} = lowest monthly mean temperature
R_{max} = highest annual rainfall so far recorded
R_{min} = lowest annual rainfall so far recorded
ϕ = latitude
R = annual rainfall average.

The proportionality constant is chosen so that A = 100 under mid-Saharan conditions, and the criterion then becomes a percentage fraction of maximal aridity. Gorczynski (1942) introduces this concept into a new classification of climates that he calls the decimal system. His arguments are hard to follow, and the scheme seems even more empirical than Köppen's. His main preoccupation is to allow a differentiation between continental and maritime types; otherwise his scheme resembles Köppen's closely. The ten types are summarized in Table 3 (Gorczynski, 1935).

Far more damaging, far more explicit, is the criticism of Thornthwaite (1943), who bases his attack on the arbitrary nature of Köppen's limits, on their empiricism and on what he regards as their unnecessary complexity. He objects to the use of temperature to define the major types, pointing out that changes in temperature régime do not in fact lead to abrupt boundaries of soil or vegetation. Thornthwaite himself has in the last twenty years developed alternative systems of classification that attempt to remedy the empiricism of Köppen's methods. The whole question of climatic classification has become closely identified with Thornthwaite's name; his work must be considered the most advanced and most complete as yet available.

The Thornthwaite Classifications

The fundamental idea behind Thornthwaite's methods (1931, 1933, 1948) is that of climatic efficiency, i.e., the capacity of the climate to support the growth of plant communities. The factors which govern this growth, soil and ecological factors being ignored, are the available moisture, the annual march of temperature, and the degree of correlation in time of the optimum periods in both temperature and rainfall régimes. The reduction of these facts to a single, numerical standard is achieved in the earlier classification by the use of precipitation and temperature efficiency indices; a third factor expresses their time-correlation. In the second of his classifications Thornthwaite (1948) radically changes his position, while still seeking a measure of efficiency.

(*i*) *The first classification: precipitation efficiency.* The moisture available for plant-growth, in Thornthwaite's view, is conditioned by (*a*) the precipitation and (*b*) the loss due to evaporation. The efficiency of an annual rainfall of 20 in, for example, is obviously low if the corresponding evaporation is 30 in and high if the latter is only 10 in. The P–E index is essentially of the form, precipitation divided by evaporation, and is unity when the two are equal. In detail, Thornthwaite proposed (*a*) a P–E ratio, which is the ratio of the two elements in a month and (*b*) a P–E index, which is the sum of the twelve monthly ratios. This idea was by no means new; Transeau (1905) had used similar ratios early in the century, and Thornthwaite also acknowledged the work of Lang (1920), de Martonne (1927), Meyer (1926), and Szymkiewicz (1925). All these workers had been handicapped by the lack of accurate and numerous evaporation data. Thornthwaite proposed to overcome this difficulty by expressing evaporation as a function of temperature. He found that the

P–E ratios were related in a simple way to the mean monthly temperature. In his final formula for the P–E index, evaporation does not appear; it is replaced by the temperature-function which he considers expresses it adequately. The P–E index is of the form

$$(\text{P–E index}) = \sum_{n=1}^{12} 115\left(\frac{P}{T - 10}\right)_n^{10/9}$$

where

$\sum$ = summation of twelve monthly ratios
P = monthly mean precipitation (inches)
T = monthly mean temperature (degrees F)
n = the particular month.

The P–E index was then computed for areas along the boundaries of recognized vegetation regions, and the following humidity provinces determined:

Humidity Province	Vegetation	P–E Index
A	Rain forest	> 127
B	Forest	64–127
C	Grasslands	32–63
D	Steppe	16–31
E	Desert	< 16

These cyphers A–E are the first members of the three-cypher group used to indicate the climatic type (except in cold climates); thus the major divisions in Thornthwaite's scheme depend on the humidity rather than on the thermal characteristics of the climate. Although Thornthwaite is well ahead of Köppen in allowing for the evaporation, there is no denying that this P–E index is as empirical as anything in Köppen, especially as regards the functional expression for evaporation. The fit of the provinces with the vegetation zones is also achieved by purely empirical methods. A major arbitrary element is that the P–E index is regarded as being equivalent to the value for 28.4° F if the temperature is below that figure, on the grounds that snow is of no value to plants until it melts; how the value of 28.4° F was obtained, Thornthwaite does not say.

The summer concentration of precipitation-effectiveness (rightly regarded as a significant element) is indicated by small letters placed third (after temperature-effectiveness) in the climatic labels. They are "r" for all-year-round rainfall, "s" for summer drought, "w" for winter drought, and "d" for all-year-round drought. Figure 2 (after Thornthwaite, 1933), gives the precise limits of these classes.

(*ii*) *The first classification: thermal efficiency.* The thermal efficiency index (also called the temperature-effectiveness) is a totally empirical quantity, computed by analogy with the P–E index. Temperature is highly significant in its influence on plant-growth, operating through many different physiological processes. Thornthwaite claimed that its influence is closely analogous to that of available moisture; he therefore sought for an arbitrary scale which would give numerical values similar to those of the P–E index. The scale he eventually adopted gives zero effectiveness along the poleward limit of the tundra and a value of 128 along the outer limit of the selva. The index is computed by the summation of twelve T–E ratios, i.e.

$$(\text{T–E index}) = \sum_{n=1}^{12}\left(\frac{T - 32}{4}\right)_n$$

where

T = monthly mean temperature (degrees F)

and

n = the particular month
$\sum$ as before indicates summation over the twelve months.

The use of this T–E index leads to the definition of thermal provinces as follows:

Temperature (Thermal) Province	T–E Index
A′ (Tropical)	> 127
B′ (Mesothermal)	64–127
C′ (Microthermal)	32–63
D′ (Taiga)	16–31
E′ (Tundra)	1–15
F′ (Frost)	0

Note that these closely resemble Köppen's major types. The dash employed in the symbol is to distinguish the index from the P–E index. The seasonal distribution of temperature is expressed by small letters, a, b, c, d, or e, which stand for varying degrees of summer concentration of temperature-effectiveness:

Sub-province	% Concentration in 3 Summer Months
a	25–34
b	35–49
c	50–69
d	70–99
e	100

Thornthwaite, however, makes little use of this subdivision.

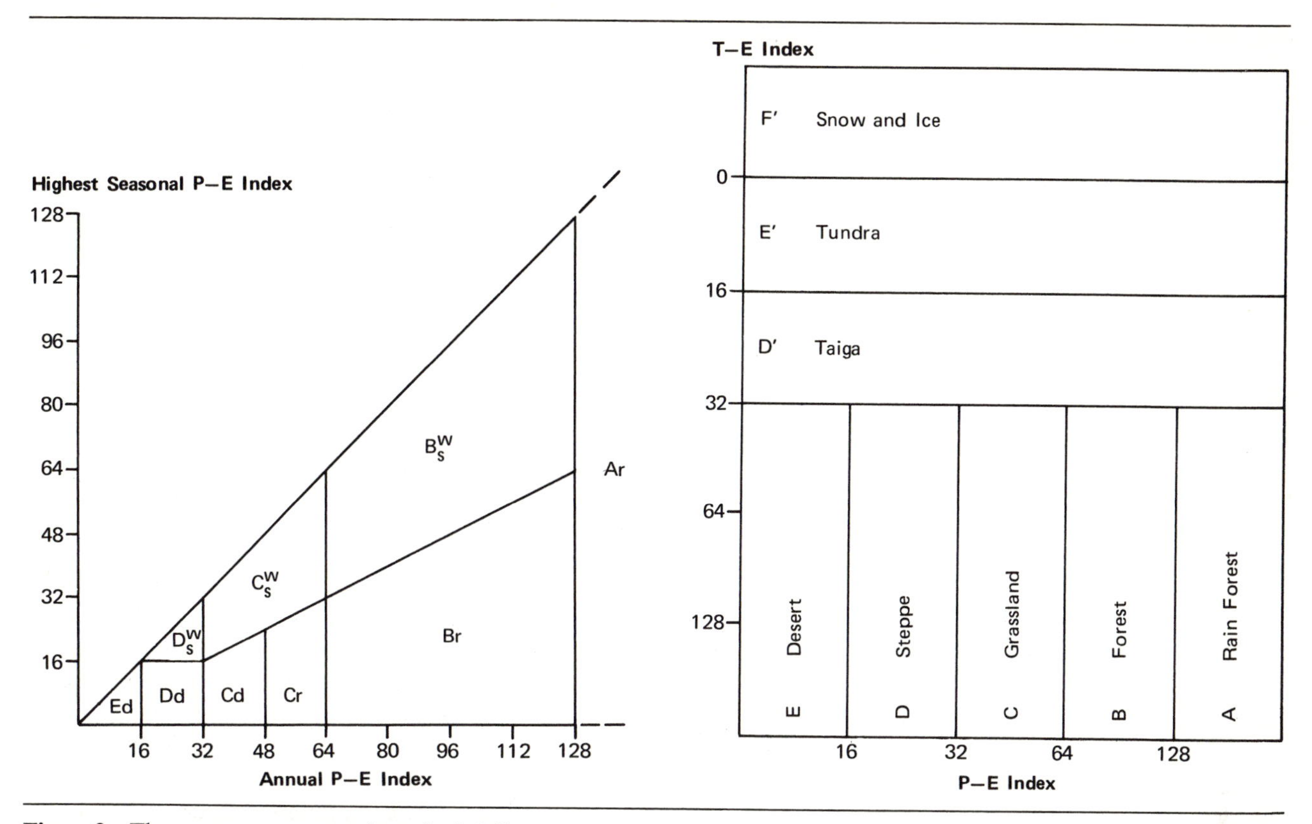

Figure 2 The summer concentration of rainfall.
This diagram shows how the various sub-divisions on the basis of summer rainfall-concentration are determined. The scale on the vertical is the sum of the P-E ratios for the three summer months, and that on the horizontal, the P-E index.

Figure 3 The primary types in Thornthwaite's classification.
When the T-E index is below 32, the major divisions disregard the P-E index, and appear on the horizontal F^1, E^1, and D^1 zones. With the T-E index above 32, however, the primary types depend on P-E index first; the thermal divisions become secondary. Hence the primary types appear as the vertical columns A-E.

(iii) The Overall Form of the First Thornthwaite Classification. The indices described above are combined to give the overall description of the climates. The policy adopted by Thornthwaite is as follows:

(*a*) With T–E indices of 32 or over, the climates are divided on a primary basis of the humidity province, viz., AB′, CC′, etc., the P–E cypher coming first, the T–E cypher (dashed) second. The seasonal distribution of the P–E index is indicated by the third (small) cypher. The seasonal distribution of the T–E index is shown as the final cypher. Thus AB′ra means rain forest, mesothermal, even distribution of precipitation-effectiveness, little summer concentration of temperature-effectiveness.

(*b*) With T–E ratios below 32, Thornthwaite regards P–E as being of little significance; very little moisture is needed to maintain the limited plant growth possible at these temperatures. Thus the main climatic groups in this classification become:

AA′ BA′ CA′ DA′ EA′
AB′ BB′ CB′ DB′ EB′
AC′ BC′ CC′ DC′ EC′

Adequate temperature-effectiveness

D′ E′ and F′

Inadequate temperature-effectiveness

The equivalence of these to the main vegetation groups is indicated in Figure 3, from Thornthwaite's original paper.

The Merits of the Earlier Thornthwaite Classification

There can be no question that Thornthwaite's earlier classification was the most far-reaching and theoretically adequate scheme available until recently. In many ways it resembles the Köppen system, but it is a more refined tool for use in geographical analysis. Both classifications rest on the assumption that the world distribution of plant associations is the real clue to the distributions of climates. In both cases the central method is to determine the quantitative climatic limits of the major plant regions. There, however, the parallelism ends. Köppen's limits are essentially arbitrary; he uses a wide variety of climatic elements to define his boundaries, and there can be no denying that the result is an illogical, unnecessarily complex method. Moreover, arbitrary limits cannot be used in extrapolation, and it has often been found that limits determined in one area do not work when applied to another. Thornthwaite, on the other hand, attempted (i) to define the climatic elements which were active in limiting plant growth, (ii) to find logical parameters for the main controls, viz., precipitation- and temperature-effectiveness, and (iii) to determine empirically the numerical scale which completes the fit of these parameters to the natural plant regions. The empiricism of Thornthwaite's methods is due to our imperfect knowledge of soil-moisture relationships, of plant physiology, and of the climatic requirements of specific plant associations. Since we cannot hope to learn these requirements in the near future, it is no criticism of his methods to declare them empirical in this particular. Criticism is better directed at the effectiveness parameters, which rest on premises which appear to be in part fallacious.

The most serious criticism refers to the assumptions made about evaporation, chiefly as regards two points: (i) the means of obtaining the evaporation and (ii) the view that evaporation from soil is simply a loss to plant growth. These points will be considered in turn.

(i) The evaporation is expressed by an empirical formula based on the precipitation and temperature. It is of the form

$$\frac{P}{E} = 11.5\left(\frac{P}{T-10}\right)^{10/9}; \quad \text{or } E = \frac{P}{11.5}\left(\frac{P}{T-10}\right)^{-10/9}$$

which makes the evaporation a function of the temperature and precipitation. Although the exact functional character of evaporation is not yet known, it is true that it depends on the temperature; but it also depends on the windspeed and the moisture content of the air, neither of which is allowed for. The presence of precipitation in the formula is quite unjustifiable, for in this sense evaporation cannot be regarded as a function of precipitation. The "evaporation" data on which this formula is based are for losses from open pans run by the U.S. Weather Bureau; the formula can refer only to such figures, and it is of little value to say that the presence of P can be justified on the grounds that if there is no precipitation there can be no evaporation. This would be true of soil surfaces, but it is not of pans in which water supply is unlimited. There are now many expressions by which evaporation from open water surfaces can be predicted; all are to some extent empirical, but they at least make some attempt at physical reality. One such expression is due to Thornthwaite himself in collaboration with Holzman (1937). Penman (1948) has recently described encouraging results from England.

(ii) It may also be doubted whether evaporation from pans is in any way representative of the evaporation from plant-covered soils. Furthermore, evaporation takes place primarily through the transpiration of the plants themselves, so that it is not really a loss to be deducted from the rainfall, the balance being available for plant growth. In fact the opposite is more nearly true; the real evaporation from a soil represents almost entirely water that *has* been used in growth. For these reasons it is highly improbable that a simple precipitation-pan evaporation ratio can represent the humidity régime of the soil.

The evaporation from a soil surface is in practice a highly complex process. An upper limit to the loss of moisture to the soil through evaporation is afforded by the capacity of the air to transport the moisture away. This transport is effected almost solely by eddy diffusion. Near the ground there is a very shallow laminar layer in which evaporation is by pure molecular diffusion, the rate being determined by the temperature of the soil surface. Above this laminar layer, however, eddy diffusion vastly outweighs molecular diffusion, the rate of upward transfer depending on the rate of windspeed and the hydrolapse (i.e., the decrease in moisture content of the air with height). The effectiveness of the eddy diffusion hence limits the evaporating power of the air, which in turn limits the capacity of the soil to evaporate its water. The evaporation off an open water surface is sometimes regarded as being a measure of the evaporating power; even if this proposition is accepted, however, it can be regarded as sound only for unit area of an infinite water surface. The evaporation off small, limited surfaces like that of an evaporation pan is affected by many special boundary conditions, and also by the turbulence set up by the flow of air across the rims of the pan. The evaporation-pan measurements tell us something about evaporating power, but we are not yet sure what it is.

The capacity of a soil surface to take up this limited evaporating power depends on the nature of the soil and on its plant cover. Evaporation from saturated fallow soils is rapid, equalling or exceeding that of an open water surface, but as the soil dries out the rate diminishes; capillary rise may maintain the supply for a short time, but the rate is very slow indeed after the initial rush. The fallow soil, then, creates for itself an insulating dry cap, retaining the moisture below it almost intact. Penman (1940), for example, showed that loss from the loamy soil of Rothamsted diminished almost to zero after about 0.5 in had been lost. It is this property of fallow soils that is made use of in dry farming operations.

From plant-covered surfaces, direct evaporation from the soil is much reduced, being largely replaced by transpiration from the plant leaves. The plants are able to withdraw moisture through their roots from a considerable depth, and can hence maintain full rate evaporation for a longer period than can a fallow soil. In the absence of rain, however, they eventually abstract all the moisture not held too firmly by colloidal adsorption, and wilting begins. If rain does not come, the plant cover dies or becomes dormant, and evaporation is once again negligible. Both on fallow and plant-covered soils, then, there is a limit to the amount of moisture that can be removed by evaporation, and the amount of pan evaporation is a poor clue to this limit. The maintenance of full-rate evaporation from either depends on constant replenishment by rainfall.

The idea that evaporation is an inescapable charge that has to be paid before plant growth can begin is also plainly unsound. If the soil stays fallow, evaporation is negligible after a brief initial flush; and if it becomes plant covered, the plants themselves bring about the evaporation. In most mid-latitude areas there are definite soil seasons. In spring the soil is normally saturated, either because of snow-melt, or because (as in Britain) the winter rains have far exceeded the winter evaporation loss. As temperatures rise, the growing season begins; rich vegetation springs up on the moisture-loaded soil, and rapidly evaporates off much of the moisture. In the mid-summer months, the evaporating power of the air exceeds the rainfall in all but the wettest areas, and the plant cover is hence able to reduce the soil well below its field capacity. Any rains that fall are greedily absorbed by the drying soil. Vegetation may wilt if the drought is severe. In the autumn the fall of the evaporating power allows the soil moisture content to rise again; in warm climates (like those of Mediterranean countries) there may even be a second growing season. But finally the cold of winter supervenes and plant growth becomes restricted.

The Second Thornthwaite Classification

It is apparent that Thornthwaite's P–E ratios and indices can give little hint of the complexity of soil moisture relationships, and they cannot be regarded as adequate measures of precipitation efficiency. In an effort to provide a rational basis for classification he has recently proposed a radical change of viewpoint. Recognizing that the task of determining the climatology of individual plant associations is almost hopeless, he has formed his new classification on certain rational assumptions of the greatest interest. As before, he classifies climates on the basis of precipitation and thermal efficiency, but he has entirely altered his way of evaluating these indices. Since the new classification has only recently been published, it is as yet far too early to pass judgment upon it. It has already been extended to Canada by Sanderson (1948), but there has not as yet been much published debate as to its value. The objections raised above to the earlier classification have very largely been met, and there can be no doubt that the second scheme is by far the most refined analytical tool yet devised. The treatment given below is necessarily brief, and for a more extended review the student is referred to Thornthwaite's 1948 paper.

Evaporating Power

A cardinal feature of the earlier Thornthwaite classification was, as we have seen, the use of computed "evaporation" in the calculation of precipitation efficiency. In the foregoing paragraphs exception has been taken to this step on the grounds that (i) the computation is based on insecure physical grounds and (ii) the actual evaporation off a plant-covered soil is rarely the same as the evaporating power of the air. In his second scheme, Thornthwaite proposes as a fundamental quantity the evaporating power of the vegetation over a moist soil surface, i.e., one in which the supply of moisture is unlimited, and evaporation is limited only by the capacity of the vegetation for vigorous transpiration. For this quantity he suggests the term "potential evapo-transpiration," which is similar in objectives to Penman's "evaporating power" (1940) except that it includes a term allowing for the effect of daylight on transpiration.

It is necessary to be very clear as to the meaning of this new concept: Thornthwaite in effect rejects the attempt to estimate real evaporation, and bases his classification on its upper limiting value. This limit is almost wholly dependent on plant agencies, though it depends on the capacity of the atmosphere to carry the moisture away, and on the length of day. It is hence an unfamiliar measure climatically. The main difficulty it raises is that the measurement or computation of evaporating power is an obscure problem. Thornthwaite makes no attempt at a theoretical treatment of this difficulty, and computes the potential evapo-transpiration as a function of temperature alone:

$$e = 1.6(10t/\mathrm{I})^a$$

where

e = 30-day evapo-transpiration (for 12-hour days) in centimeters,
t = mean monthly temperature in degrees C,
a = an arbitrary constant varying from place to place,
$\mathrm{I} = \sum_0^{12}(t/5)^{1.514}$, the sigmoid indicating summation over the twelve months.

This equation is empirical and rather unwieldy. The arbitrary constant a is determined by a manipulation based upon the so-called heat index I, mentioned above. Its place to place variation is very natural, for evaporating power is a function of windspeed, the hydrolapse and the eddy diffusivity of the atmosphere, none of which appears in the equation. Insofar as the latter works, it does so as a regression equation expressing the degree of correlation between e and t; it cannot be regarded as expressing a complete functional relationship. It must be noted also that e has still to be corrected for length of day and month.

The Moisture Index

Having thus defined the *capacity* of the atmosphere to absorb evaporation, Thornthwaite goes on to define a moisture index* relating the *water need* of the climate (i.e., that amount of rain necessary to meet the demands of potential evapo-transpiration) to the available water (viz., rainfall together with approximately 10 cm of soil-water). The steps he uses to get his index are briefly these:

(i) At certain times of year the rainfall r exceeds the water need n; there is hence a surplus of available water

$$s = r - n$$

and the climate is humid. A humidity index can be defined as follows:

$$\mathrm{I}_h = \frac{100s}{n}$$

which expresses the surplus as a percentage fraction of the water need.

(ii) In other months the rainfall may very well be less than the water need, and there is a deficit

$$d = n - r$$

and the climate is arid. An aridity index comparable with the humidity index defined above is

$$\mathrm{I}_a = \frac{100d}{n}$$

Both I_h and I_a can be computed for the year by adding together the twelve months values. In normal North American conditions the summer months have moisture deficits, and the winter months moisture surpluses. Both I_h and I_a can therefore be computed for a typical station. At some stations in the humid south, however,

* [The moisture index was modified in 1955 (see Mather, J. R., ed., Average Climatic Water Balance Data of the Continents, Part VII, United States. *Publications in Climatology, 17(3)*, pp. 419–428.]

and at many in Highland Britain, rainfall exceeds water need throughout the year, and I_a is zero.

(iii) The moisture index I_m is the difference between I_h and I_a. Since, however, deep-rooted perennial plants may be able at times of drought to draw upon subsoil moisture, a surplus at one season may be able to compensate for a somewhat larger deficit in others. Thornthwaite claims that a surplus of 6 cm in one season may compensate for a deficit of 10 cm in another, but no evidence is adduced in support of the claim. The index of aridity is hence weighted by a factor 0.6, and the moisture index is defined as

$$I_m = I_h - 0.6\,I_a = \frac{100s - 60d}{n}$$

This index is the primary basis of classification in the new scheme, replacing the P–E index in the earlier classification. The two are related by the simple formula

$$(\text{P–E index}) = 0.8\,I_m - 48$$

provided that E is computed in the same manner in both schemes. A similar set of climatic types can be defined on the basis of the new classification:

Climatic Type	Moisture Index
A Perhumid	> 100
B_4 Humid	80–100
B_3 Humid	60–80
B_2 Humid	40–60
B_1 Humid	20–40
C_2 Moist sub-humid	0–20
C_1 Dry sub-humid	−20 to 0
D Semi-arid	−40 to −20
E Arid	−60 to −40

The line dividing dry from moist climates ($I_m = 0$) plainly occurs where $I_h = 0.6\,I_a$, not as one might expect where $I_h = I_a$.

A sub-division of these primary types can be devised on the basis of the summer or winter concentration of water surplus or deficit. These limits are rendered quantitative by reference to the I_h or I_a values:

MOIST CLIMATES

Symbol		Value of I_a
r	Little or no deficit	0–16.7
s	Moderate summer deficit	16.7–33.3
w	Moderate winter deficit	16.7–33.3
s_2	Large summer deficit	> 33.3
w_2	Large winter deficit	> 33.3

DRY CLIMATES

Symbol		Value of I_m
d	Little or no surplus	0–10
s	Moderate summer surplus	10–20
w	Moderate winter surplus	10–20
s_2	Large summer surplus	> 20
w_2	Large winter surplus	> 20

Thermal Efficiency

One of the most surprising features of Thornthwaite's second scheme is his treatment of thermal efficiency. Since potential evapo-transpiration is expressed as a function of temperature and duration of daylight, Thornthwaite claims that it will also serve as a measure of thermal efficiency; the thermal divisions are hence based on the values of *e* computed from the relationship discussed above. The T–E index of the second classification is expressed in inches or centimeters of potential evapo-transpiration. For a boundary between the megathermal and mesothermal climates Thornthwaite chooses the 23° C (73.4° F) annual isotherm, which is equivalent on the equator to a potential evapo-transpiration of 114 cm (45 in approximately). The remaining divisions (listed below) are made in a descending arithmetic progression:

Annual Potential Evapo-transpiration (T–E Index)		Climatic Type	
cm	in	Symbol	Title
		E′	Frost
14.2	5.6		
		D′	Tundra
28.5	11.2		
		C_1'	Microthermal
42.7	16.8		
		C_2'	Microthermal
57.0	22.4		
		B_1'	Mesothermal
71.2	28.1		
		B_2'	Mesothermal
85.5	33.7		
		B_3'	Mesothermal
99.7	39.3		
		B_4'	Mesothermal
114.0	44.9		
		A′	Megathermal

Summer concentration of thermal efficiency is also regarded as a prime factor. The fourth digit in the label of a Thornthwaite class is a small letter indicating the degree of summer concentration of the T–E index:

Type	Summer Concentration of T–E Index %
a′	< 48.0
b'_4	48.0–51.9
b'_3	51.9–56.3
b'_2	56.3–61.6
b'_1	61.6–68.0
c'_2	68.0–76.3
c'_1	76.3–88.0
d′	> 88.0

The label describing a climate therefore has four digits, as in the first classification. The label

$$B_3\ B'_2\ s\ b'_4$$

means that the climate is third-order humid, second-order mesothermal, with a moderate summer moisture deficit and a low summer concentration of thermal efficiency.

Conclusion

We have briefly traced the progress of overall climatic classification from the pioneer work of Köppen through to the work of Thornthwaite in the U.S.A. Little or no reference has been made to the work of the many biologists who have sought on a more limited scale to find climatic equivalents for the major structural or formational boundaries in the plant cover. The work described in this essay stands alone in its attempt to define a comprehensive classification sufficient to divide the world into unique climatic regions. Only Köppen and his successor Thornthwaite have maintained this completely geographic outlook.

The geographer may well wonder whether the basis used by these climatologists for classification is the most adequate for geographic purposes. Köppen, following de Candolle, began with the assumption that the living plant—the living individual plant species—is, as Thornthwaite puts it, "a meteorological instrument which integrates the various factors of climate"; Thornthwaite himself, working in a day when plant ecologists stress the structure of the great vegetation formations rather than the relations of the individual plant, first of all sought empirically to find climatic equivalents for observed formational boundaries, and later to find a rational classification independently of observed biotic phenomena. In his own words, "vegetation is regarded as a physical mechanism by means of which water is transported from the soil to the atmosphere; it is the machinery of evaporation as the cloud is the machinery of precipitation."

This stressing of vegetation as the best measure of climate is plainly of geographic value; the "natural vegetation" of a region is at least in some measure a measure of its economic potential for settlement, especially for farming economies. But we may wonder whether or not a comprehensive classification should take account of that other great branch of climatology allied to the field of environmental physiology. The direct effect of climate on man's health and comfort has never been reduced to a set of simple parameters analogous to those of the classifications described in this essay. Most of us are familiar with Griffith Taylor's simple but effective "climographs" and "hythergrams"; today the environmental physiologists are seeking, under the stress of military necessity, measures of the climatic stresses of the human frame of the same type, though based on experimental evidence. Windchill, for example, has been reduced to a numerical measure of outstanding value in assessing the direct effect of low temperatures on the human frame (Siple and Passel, 1945).

In any ultimate synthesis it seems inevitable that this little explored field must be somehow incorporated. The geographer cannot ignore a specification of the direct environmental influence of climate; the latter does not work wholly on his activities, but also on man himself.

REFERENCES

Ackerman, E. A., 1941. The Köppen classification of climates in North America. *Geographical Review 31*, 105–111.

Blair, T. A., 1942. *Climatology*. Prentice-Hall.

de Candolle, A., 1875. Les groupes physiologiques dans le régne végétal. *Rev. Scientifique, Ser. 2, 16*, 364–372.

de Martonne, E., 1927. Regions of interior basin drainage. *Geographical Review 17*, 397–414.

Drude, O., 1887. Berghaus' *Physikalischer Atlas* (Part 5). Gotha.

Gorczynski, W., 1935. Uber die Klassification der Klimate. *Gerl. Beitr. zur Geophys. 44(2)*, 199–210.

Gorczynski, W., 1942. Climatic types of California according to the decimal scheme of world climates. *Bull. Amer. Meteorol. Soc. 23*, 161–165 and 272–279.

Gorczynski, W., 1945. Comparison of climate of U.S. and Europe. *Polish Inst. of Arts and Sciences*, New York. 239–247.

Haurwitz, B. and Austin, J. A., 1944. *Climatology*. McGraw-Hill.

Jones, S. B., 1932. Classifications of North American climates. *Economic Geography 8*, 205–208.

Kendrew, W. G., 1922. *Climate*. Oxford.

Kendrew, W. G., 1937. *Climates of the Continents* (3rd ed.). Oxford.

Köppen, W., 1900. Versuch einer Klassifikation der Klimate. *Geog. Zeit. 6*, 593–611.

Köppen, W. and Geiger, R., eds., 1936. Das geographische System der Klimate. *Handbüch der Klimatologie 1, C.*

Lang, R., 1920. *Verwitterung und Bodenbildung als Einfuhrung in die Bodenkunde*. Stuttgart.

Leighly, J. B., 1926. Graphic studies in climatology. *Univ. California, Publications in Geography 2(3)*, 55–71.

Linnser, C., Review of his scientific work by Abbe, C., 1905. *U.S. Weather Bureau Bull. 36*, 211–233.

Meyer, A., 1926. Uber einige Zusammenhange zwischen Klima und Boden in Europa. *Chemie der Erde 2*, 209–347.

Penman, H. L., 1940. Meteorological and soil factors affecting evaporation from fallow soils. *Quart. Journ. Royal Meteorol. Soc. 66*, 401–410.

Penman, H. L., 1948. Natural evaporation from open water, bare soil and grass. *Proc. Royal Soc. A, 193*, 120–145.

Russell, R. J., 1926. Climates of California. *Univ. California, Publications in Geography 2(4)*, 73–84.

Russell, R. J., 1931–32. Dry climates of the United States. *Univ. California, Publications in Geography 5(1)*, 1–41 and 245–274.

Russell, R. J., 1934. Climatic years. *Geographical Review 24*, 92–103.

Russell, R. J., 1945. Climates of Texas. *Annals*, Association of American Geographers *35*, 37–52.

Sanderson, M., 1948. Drought in the Canadian Northwest. *Geographical Review 38*, 289–299.

Schimper, A. F., 1903. *Plant Geography*. Translated by Fisher, W. R., Oxford.

Siple, P. A. and Passel, C. F., 1945. Measurements of dry atmospheric cooling in sub-freezing temperatures. *Proc. Amer. Phil. Soc. 89(1)*, 177–199.

Szymkiewicz, D., 1925. Etudes climatologiques. *Acta Societatis Botanicae Poloniae 2(4)*.

Thornthwaite, C. W., 1931. The climates of North America according to a new classification. *Geographical Review 21*, 633–635.

Thornthwaite, C. W., 1933. The climates of the earth. *Geographical Review 23*, 433–440.

Thornthwaite, C. W., 1943. Problems in the classification of climates. *Geographical Review 33*, 233–255.

Thornthwaite, C. W., 1948. An approach toward a rational classification of climate. *Geographical Review 38*, 55–94.

Thornthwaite, C. W. and Holzman, B., 1937. The determination of evaporation from land and water surfaces. *Mon. Wea. Rev. 67*, 4–11.

Transeau, E. N., 1905. Forest centres of North America. *Amer. Naturalist 39*, 875–889.

Trewartha, G. T., 1943. *Introduction to Weather and Climate*. McGraw-Hill.

Van Royen, W., 1927. The climatic regions of North America. *Mon. Wea. Rev. 55*, 313–319.

G. R. McBoyle

10 Climatic classification of Australia by computer

In the last seventy years geographers have been presented with many attempts at climatic classification. Those that have had greatest acclaim have been the classifications of Köppen and Thornthwaite. Köppen and the many modifiers of his classification depended directly on phytogeography using vegetation as a single factor representing the complexity of elements whose amalgam is climate. Using vegetation maps a set of climatic figures could be derived to demarcate the boundaries of vegetation and other landscape elements, for instance the $-3°$ C isotherm for the coldest month indicates the equatorward limit of permafrost. Thornthwaite's main classification based on potential evapotranspiration also relied on information gleaned from vegetation and soil maps so in both his and Köppen's classifications external factors related to, but not wholly identifiable with, climate are employed. Ideally what is required is a classification having reference only to climatic data, and one which avoids any subjective judgment in delimiting the boundaries between different climatic types.

The use of factor analysis on non-physical data for the purposes of delimiting economic regions has been employed by Berry (1967), Berry and Ray (1966) and others. Steiner (1965) foresaw its use in climate to provide "a genuine and rational climatic classification . . . based on climatic elements only and not . . . on external factors." The following regionalization of Australia's climate has been obtained through factor analysis using solely climatic elements, and the result is a classification which is easy to map and interpret on both large and small scales, giving a general yet useful end-product which is an efficient arrangement of the data in as simple a form as possible.

Australia was chosen for study because it is a large, compact continent with generally low relief. Since Australia lies in the subsidence and divergence zone of the subtropical anticyclones much of the country is occupied by dry climates, yet because of its position athwart the Tropic of Capricorn it comes under the control of different wind systems as the pressure cells move north and south bringing different moisture conditions from season to season on the peripheries of the continent. Australia labors under many disabilities because of the relationship of heat and moisture within her lands and since these factors control much of the country's economy they have been used as the main criteria in this classification.

Starting with twenty variables from each of the sixty-six stations marked in Figure 1 a matrix of simple correlations between the input variables was obtained (Table 1) and using this as a starting point a sequence of several processes—the factor analysis—is carried out reducing the complex matrix to a smaller number of "Composite variables" which are sets of indices that summarize the original matrix, that is the original twenty climatic variables.

The climatic data for the twenty variables for each of the stations were obtained from the Meteorological Office Bulletin MO 617f, Part VI, Australasia and the South Pacific Ocean (1962). The variables used in the analysis were:

1	Mean annual temperature, ° F	T_A
2	Average daily mean temperature, ° F, January	T_1
3	Average daily maximum temperature, ° F, January	$T_{1,DMAX}$
4	Average daily minimum temperature, ° F, January	$T_{1,DMIN}$
5	Average daily mean temperature, ° F, July	T_7
6	Average daily maximum temperature, ° F, July	$T_{7,DMAX}$
7	Average daily minimum temperature, ° F, July	$T_{7,DMIN}$

Reprinted, with minor editorial modification, by permission of the author and editor, from *Australian Geographical Studies*, Vol. IX, 1971, pp. 1–14.

ari-ble No.	1	2	3	4	5	6	7	8	9	10	11	12	13	14	15	16	17	18	19	20
1	1.000																			
2	0.880	1.000																		
3	0.699	0.857	1.000																	
4	0.874	0.848	0.498	1.000																
5	0.897	0.595	0.405	0.727	1.000															
6	0.930	0.740	0.679	0.672	0.903	1.000														
7	0.553	0.218	−0.121	0.593	0.780	0.446	1.000													
8	−0.374	−0.683	−0.751	−0.326	0.035	−0.269	0.409	1.000												
9	−0.024	−0.426	−0.562	−0.039	0.380	0.098	0.604	0.911	1.000											
0	−0.663	−0.810	−0.792	−0.556	−0.333	−0.603	0.121	0.885	0.629	1.000										
1	0.163	−0.171	−0.287	0.108	0.464	0.239	0.558	0.653	0.749	0.421	1.000									
2	0.614	0.305	0.125	0.497	0.782	0.632	0.684	0.274	0.525	−0.029	0.815	1.000								
3	−0.651	−0.761	−0.693	−0.559	−0.383	−0.552	−0.030	0.666	0.453	0.738	0.322	−0.243	1.000							
4	−0.532	−0.791	−0.734	−0.536	−0.173	−0.401	0.157	0.812	0.670	0.799	0.628	0.161	0.793	1.000						
5	0.206	0.045	−0.062	0.188	0.288	0.206	0.299	0.168	0.292	0.001	0.392	0.414	0.015	0.184	1.000					
6	−0.772	−0.884	−0.747	−0.723	−0.483	−0.664	−0.092	0.689	0.419	0.842	0.234	−0.284	0.889	0.867	−0.032	1.000				
7	−0.033	0.438	0.493	0.124	−0.461	−0.193	−0.632	−0.789	−0.892	−0.518	−0.705	−0.536	−0.411	−0.678	−0.272	−0.435	1.000			
8	−0.384	−0.404	−0.380	−0.333	−0.225	−0.342	0.025	0.306	0.134	0.414	0.028	−0.314	0.722	0.392	−0.141	0.618	−0.196	1.000		
9	−0.506	−0.251	0.078	−0.498	−0.618	−0.528	−0.555	−0.156	−0.419	0.175	−0.395	−0.510	0.116	0.010	−0.279	0.274	0.413	0.084	1.000	
0	−0.643	−0.606	−0.470	−0.599	−0.493	−0.598	−0.185	0.333	0.048	0.576	−0.166	−0.495	0.610	0.380	−0.270	0.719	−0.121	0.751	0.372	1.000

able 1 Matrix of simple correlations between input variables.

8 Mean annual relative humidity (afternoon), percent — H_A
9 Average relative humidity (afternoon), percent, January — H_1
10 Average relative humidity (afternoon), percent, July — H_7
11 Mean annual precipitation, inches — P_A
12 Average precipitation, inches, January — P_1
13 Average precipitation, inches, July — P_7
14 Mean annual number of raindays (with 0.01 inches or more) — $P_{A,0.01}$
15 Average number of raindays, January — $P_{1,0.01}$
16 Average number of raindays, July — $P_{7,0.01}$
17 Temperature range, ° F, (January—July) — T_R
18 Precipitation ratio (July/January) — P_R
19 Humidity ratio (July/January) — H_R
20 Rainday ratio (July/January) — P_{AR}

The data derived from the Meteorological Office bulletin were subjected to a factor analysis. The lower half of the matrix of simple correlations between the twenty input variables obtained from the computer program is already given in Table 1. Having evaluated all components with eigenvalues equal to or greater than one, the computer output resulted in an unrotated factor solution with three factors explaining 86.79% of the total variation. In order to simplify explanation of the factor structure a Varimax rotation is applied. "From a mathematical viewpoint" says King (1969) "the rotation of factors can be viewed as a consequence of the indeterminacy of the solution." In other words a rotation is applied if there are many intermediate values between 0 and ±1 and if the values are spread between two or more factors. Thus a rotation attempts to elucidate the situation by eliminating the intermediate values and giving a result that has most of the factor loadings either of high or low values. "The most commonly used of these (rotations) is the Varimax routine, which by a series of orthogonal transformations of pairs of factors, seeks to simplify the columns of the factor loading matrix" (King, 1969). The factor scores of the rotated solution, the eigenvalues, and the percentage explanation of the total variance (or communality) of each of the three factors is given in Table 2.

Factor I explaining 48.2% of the total variance is the most important. High loadings on this factor are annual and July humidity, July precipitation, and annual and July rainday averages (positive) as well as annual temperature, January mean, daily maximum and minimum temperatures (negative). Therefore factor I appears to be a general index of cool moist conditions with special emphasis on winter moisture and summer coolness. If the factor scores of each of the sixty-six stations are mapped for factor I and isolines drawn (Figure 2) the role of coastal areas is brought out (high positive figures), particularly the area of southeast Australia, including Tasmania, which is generally considered similar to the West European climatic type because of latitude and insularity. The high negative areas in the northwest indicate regions where the opposite conditions are found—hot, dry conditions. The factor indicates that stations with a high annual number of raindays and high winter humidity also generally tend to have a high annual humidity, high winter precipitation, high number of winter raindays and low annual (particularly summer) temperatures; in

			I	II	III
Factor Number			I	II	III
Eigenvalue			9.026	6.040	1.188
Percentage of Total Variance or Communality			48.2	32.3	6.3
	Variable	Percentage Communality over three factors			
1	T_A	98.66	*−0.757*	0.631	−0.124
2	T_1	95.26	*−0.939*	0.236	−0.125
3	$T_{1,DMAX}$	77.20	*−0.867*	−0.014	−0.143
4	$T_{1,DMIN}$	79.84	*−0.701*	0.549	−0.071
5	T_7	96.65	−0.423	*0.887*	−0.035
6	$T_{7,DMAX}$	85.20	−0.634	0.656	−0.143
7	$T_{7,DMIN}$	81.21	−0.026	*0.886*	0.164
8	H_A	91.74	*0.843*	0.441	0.108
9	H_1	90.24	0.623	*0.717*	0.005
10	H_7	85.64	*0.901*	0.066	0.200
11	P_A	81.13	0.452	*0.756*	−0.189
12	P_1	87.32	−0.032	*0.853*	−0.380
13	P_7	82.80	*0.770*	−0.022	0.485
14	$P_{A,0.01}$	87.78	*0.901*	0.233	0.105
15	$P_{1,0.01}$	19.06	0.078	0.384	−0.193
16	$P_{7,0.01}$	92.99	*0.867*	−0.133	0.401
17	T_R	85.90	−0.561	*−0.731*	−0.100
18	P_R	79.68	0.318	−0.046	*0.833*
19	H_R	49.00	0.144	−0.685	0.024
20	P_{AR}	77.98	0.469	−0.331	0.671

High factor loadings are italicized (over ± 0.70).

Table 2 Matrix of rotated factor loadings.

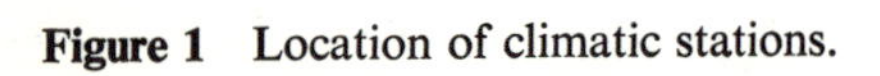

Figure 1 Location of climatic stations.

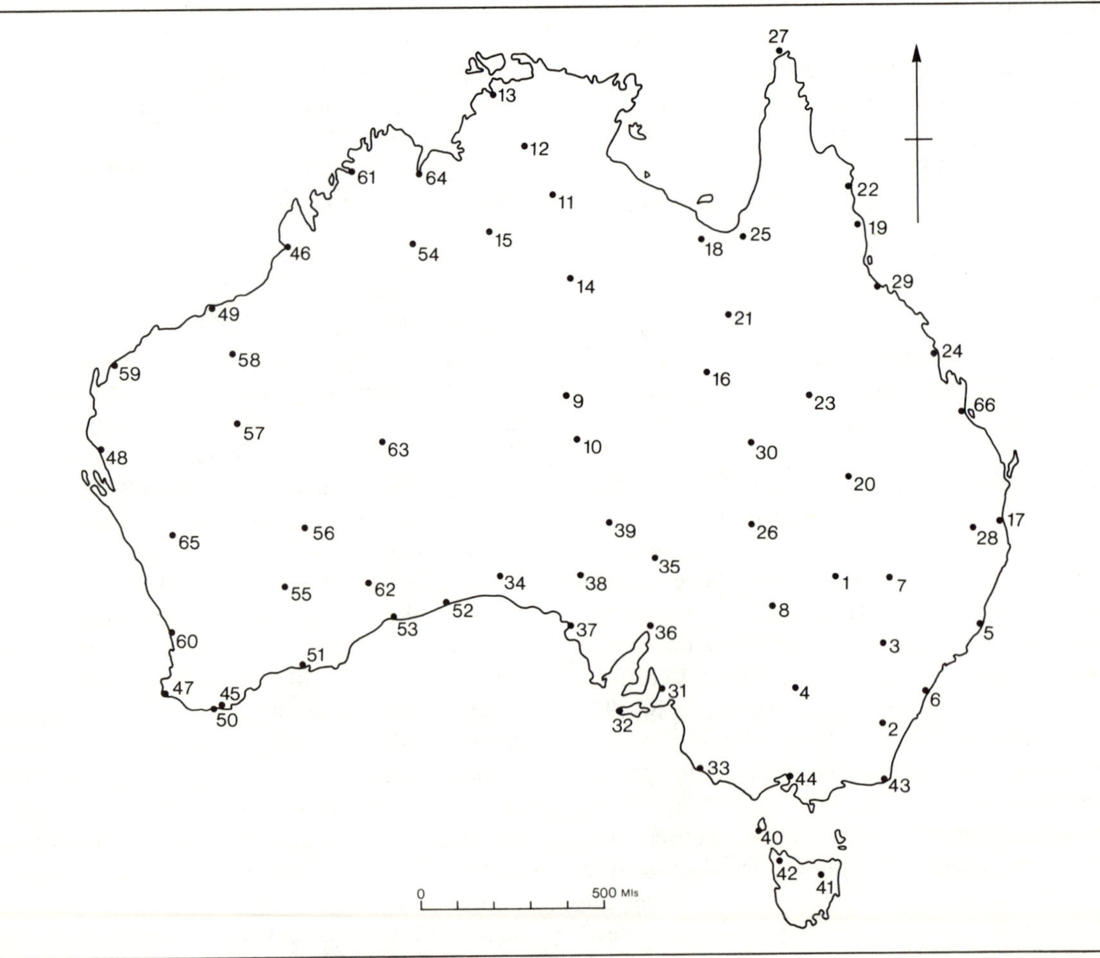

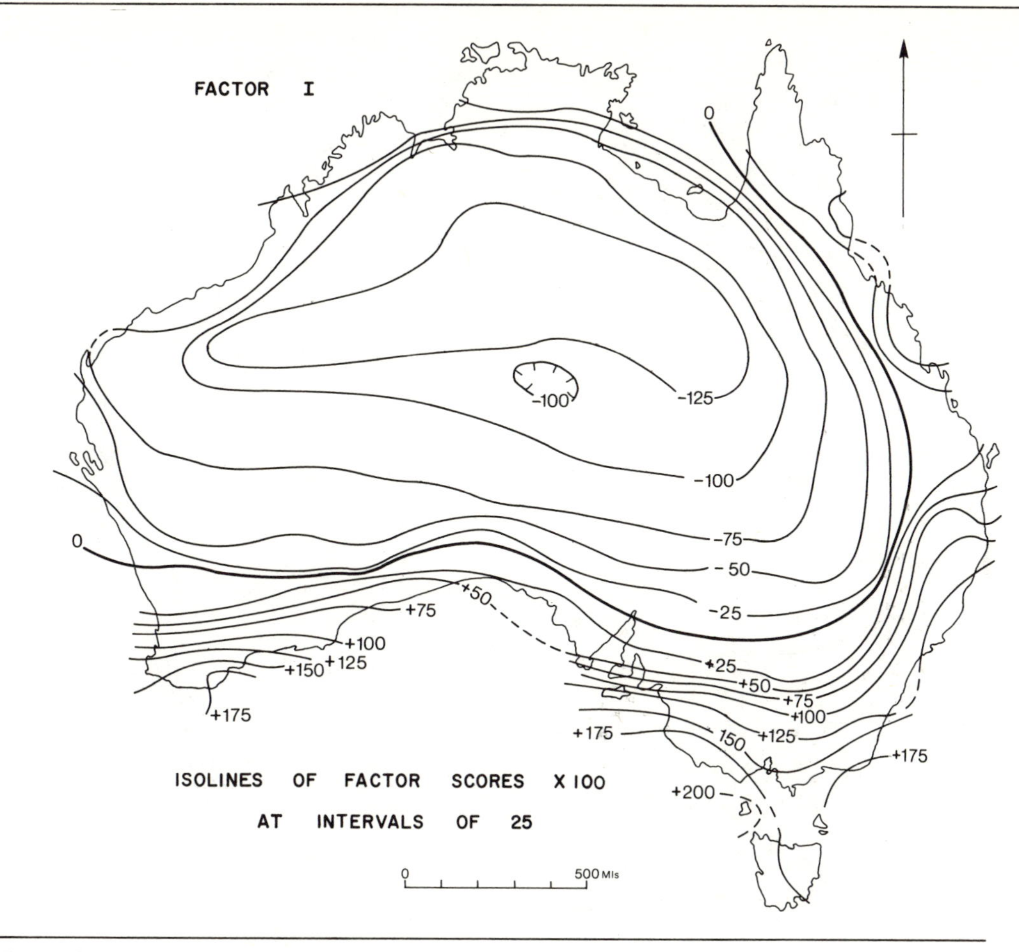

Figure 2 Factor I—"cool and moist index."

other words an all year round moist situation with a winter maximum together with lowered temperatures throughout the year, particularly in summer. In this factor the southeast counters the internal northwest.

The second factor explaining 32.3% of the total variance has been plotted from scores of individual stations and isolines drawn in Figure 3. July daily mean and minimum temperatures, together with annual and January precipitation and January humidity have high positive loadings and the temperature range has a high negative loading on this factor. Other loadings of less importance are the positive loadings of annual temperature and July daily maximum temperature and the negative value for the humidity ratio. Here the situation arises where if there are high winter temperatures in conjunction with summer rains then a high annual rainfall, high summer humidities, little temperature and humidity differences between summer and winter may be expected. Figure 3 is a map of conditions of overall high temperatures and precipitation with a moisture maximum in summer. This is the "tropical summer rain" belt of Australia associated with the monsoon or "Wet" in the northern part of the continent, and the area affected by the rain-bearing southeast trade winds along Australia's east coast. The map could be called a "tropical summer wet index".

Finally, factor III accounts for 6.3% of the total variance. Although the rainday ratio does not reach the required level of 0.7, its value of +0.67 makes it so outstanding compared to the other values that it is considered along with the most important loading, the precipitation ratio, as being the major control of factor III. Figure 4 illustrates the great variation between the southwest and the east of the continent. Factor III indicates mainly winter rainfall and raindays in agreement with the rainy westerly wind belt of the winter season in the Mediterranean areas of Australia, around Perth and Adelaide.

The three factors, "cool and moist", "tropical summer wet", and "Mediterranean" have been used to describe the climate of Australia where originally it was indicated by some combination, in groups or in toto, of the twenty original variables. These new factors have been obtained almost wholly objectively; only the choice of original variables depends on the individual's selection. Values for "cool and moist", "tropical

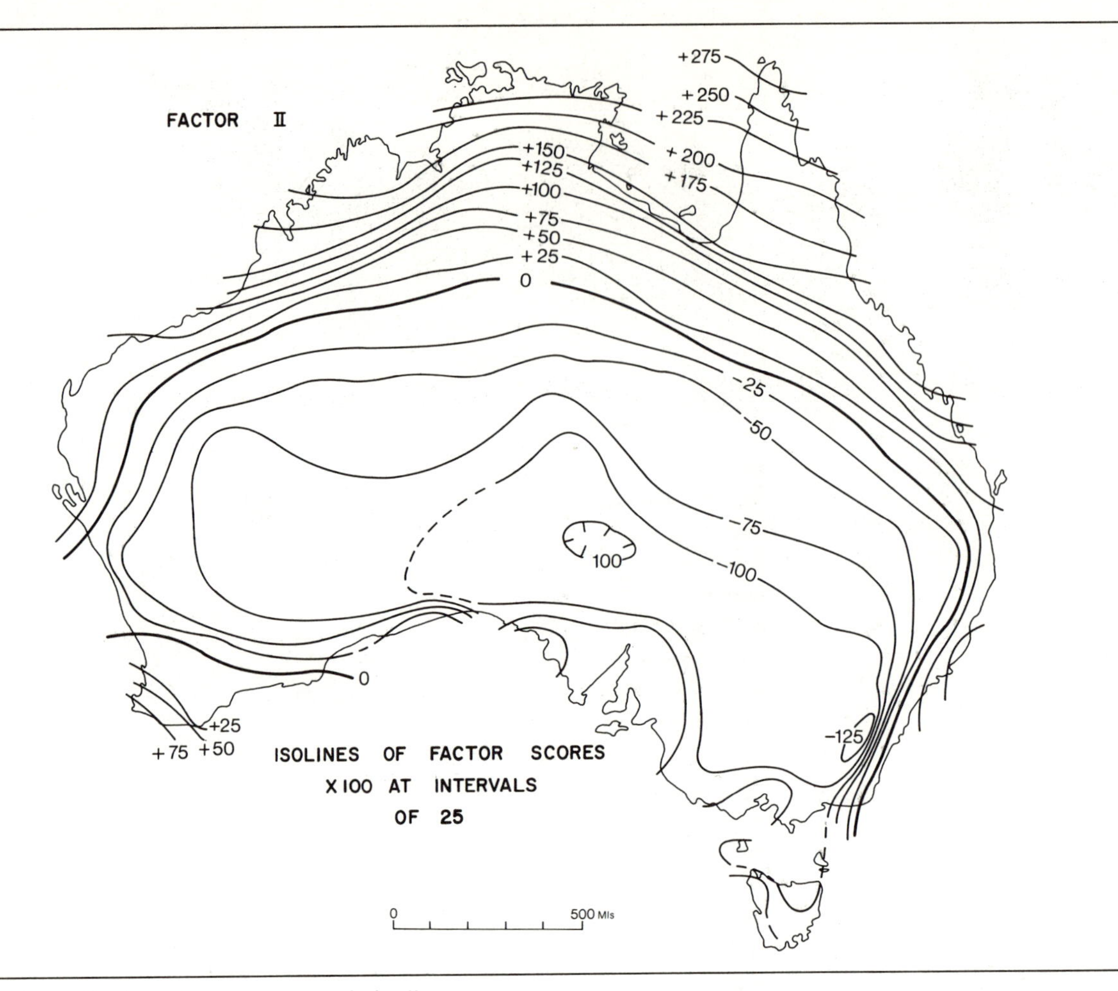

Figure 3 Factor II—"tropical summer wet index."

summer wet" and "Mediterranean" may be obtained from the following equations derived from the factor loadings.

"cool and moist" = $-0.76\ T_A - 0.94\ T_1 - 0.87\ T_{1,DMAX} - 0.70\ T_{1,DMIN} + 0.84\ H_A + 0.90\ H_7 + 0.77\ P_7 + 0.87\ P_{7,0.01}$

"tropical summer wet" = $0.89\ T_7 + 0.89\ T_{7,DMIN} + 0.72\ H_1 + 0.76\ P_A + 0.85\ P_1 - 0.73\ T_R$

and "Mediterranean" = $0.83\ P_R + 0.67\ P_{AR}$

Using these equations, stations whose data were not available for the original compilation may be given scores thus allowing additions to be made to the maps.

If the loadings of the three factors derived from the twenty variables of the original sixty-six stations are used as input variables an attempt may be made to detect homogeneous regions using a distance statistic as a similarity measure. This is a step-wise process and, by a series of discrete steps, factor loadings which are close to one another on a three-dimensional graph may be grouped, since only three factors are being used. Berry (1967) notes that the result is "a complete linkage tree," which "proceeds from n outermost branches, through (n − 1) to (n − 2) to i and (i − 1) to 4, 3, 2 and finally the main trunk 1." It must be remembered that there is no guarantee with this method that the grouping procedure will yield contiguous regions. By means of a group analysis program a linkage tree was obtained which shows the reduction from the detail of sixty-six climatic stations to one group, Australia itself (Figure 5).

From Figure 5, maps can be drawn showing the sixty-six climatic stations grouped into 10 regions (Figure 6 [p.117]), 14 regions (Figure 7 [p.117]), and 28 (Figure 8 [p.118]). The analysis thus gives not only regions of climatic types but also outlines rational boundaries not influenced by an external factor such as vegetation or soil type. The boundaries may be ranked into a boundary hierarchy where the first split into two regions yields the boundary of the first order, the second split into three regions the boundary of the second order and so on. In this manner two ways of examining the regions for climatic analysis are obtained—(a) either from inside the region viewing the boundaries on the horizon and considering the homogeneity within, or (b) viewing

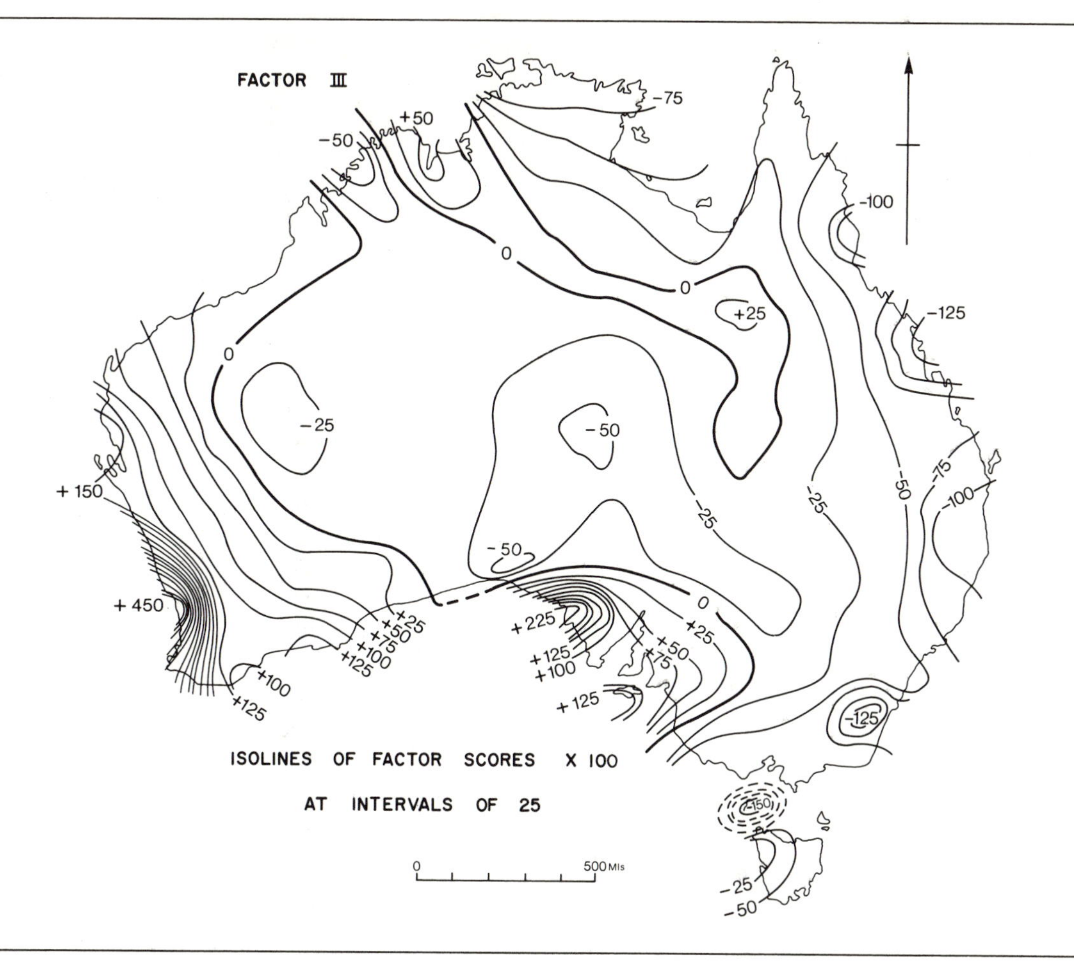

Figure 4 Factor III—"Mediterranean index."

differing climates from a position astride the ranked boundaries.

From the first order boundary obtained for Australia, two regions are determined which are significantly different with respect to annual and winter temperatures, temperature range and humidity. It may be said that the first order boundary is a division of the "tropical summer wet" in the north from the remainder of the continent. The second order boundary divides off the Mediterranean belt of southwest Australia and Spencer Gulf while the third order boundary demarcates the extreme areas of "cool and moist" of factor I, mainly the southeast of the country. Thus differing precipitation variabilities of the continental periphery have been delimited from the dry core of the interior.

Although Figure 6 and a Köppen-Geiger classification for Australia (Figure 9 [p.118]) show a moderate correlation, certain contrasts are evident:

1. The arid area (1) of Figure 6 encompasses most of the semiarid area of Köppen's classification in south and east Australia.

2. Figure 6 has a separate climatic region (7) between the semiarid belt of the north (6) and the northern tropical zone (4).

3. The belt of Mediterranean climate obtained by factor analysis is divided into sub-sections (2, 5, 8, 10) because of climatic differences within the zone.

4. The Cf climate of Köppen is divided into two zones (3, 9) in Figure 6. These are comparable with Köppen Cfa and Cfb.

It appears that the factor analysis has differentiated the climates of the continental periphery to a greater extent than does Köppen, yet his classification has greater insight into variations in the arid zone than the computer technique. Because of these differences Figure 7 was investigated to determine if some climatic variations within the arid zone could be differentiated. A belt composed of two regions, 12 and 14, now may be compared favorably with the southern area of semiarid climate by Köppen's classification. The remaining eastern area of semiarid climate was not differentiated at this level yet it is visible on Figure 8 which depended on twenty-eight groupings. The main reason for these differences is that the computer technique obtained its

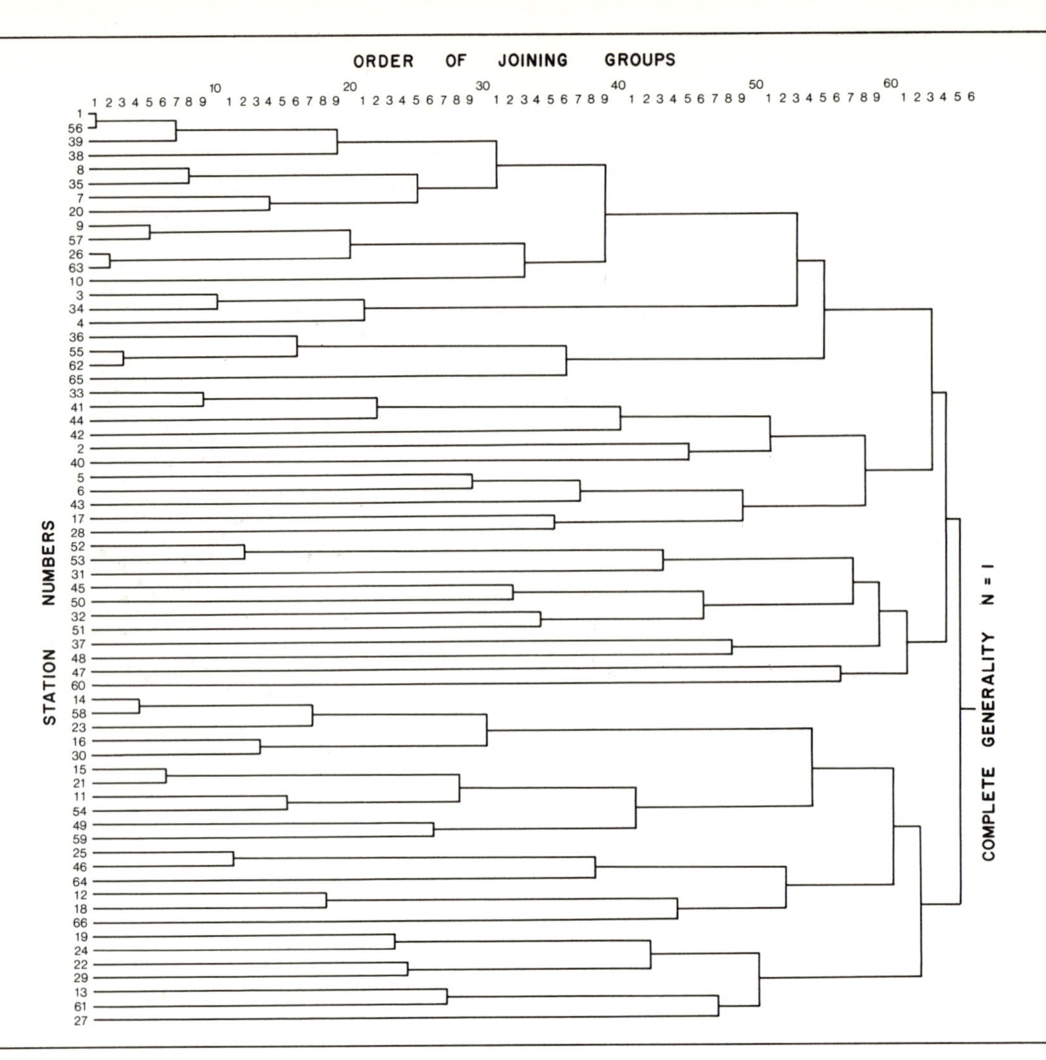

Figure 5 Linkage tree of climatic stations.

regions by numerical hierarchical method whereas Köppen was looking for a different hierarchy of climatic variations and was not attempting to rank by magnitude of change.

The climatic regionalization of Australia that has been obtained differs from those obtained by other methods. Firstly it is dependent solely on climatic elements. Secondly the boundaries indicate where the most rapid change of climate occurs, the first order boundary indicating the most rapid climatic change, the second order the second most rapid and so on. Thirdly the computer method allows an ease in changing the groupings from a global context to local variations. The grouping of the climatic stations into, say, three or four groups gives the world-wide trend from north to south of tropical, desert, Mediterranean and West European climatic types whereas a grouping of fourteen regions is ideal for a climatic study of Australia by itself and a regrouping of twenty-eight may be of use for detailed local work.

REFERENCES

Berry, B. J. L. and Ray, D. M., 1966. Multivariate socio-economic regionalization: a pilot study in central Canada. In *Regional Statistical Studies*, ed. T. Rymes and S. Ostry. University of Toronto Press, Toronto.

Berry, B. J. L., 1967. The mathematics of economic regionalization. *Proc. Brno Conf. on Economic Regionalization*, Czech. Acad. Sciences, Brno.

Blüthgen, J., 1964. *Allgemeine Klimageographie*. Berlin.

King, L. J., 1969. *Statistical Analysis in Geography*. Prentice-Hall, Engelwood Cliffs.

Meteorological Office, 1962. Australasia and the South Pacific Ocean. *Meteorological Office Bulletin MO 617f, Part VI.*

Steiner, D., 1965. A multivariate statistical approach to climatic regionalization and classification. *Tijdschr. Kon. Ned. Aardr. Gen. 82*, 329–347.

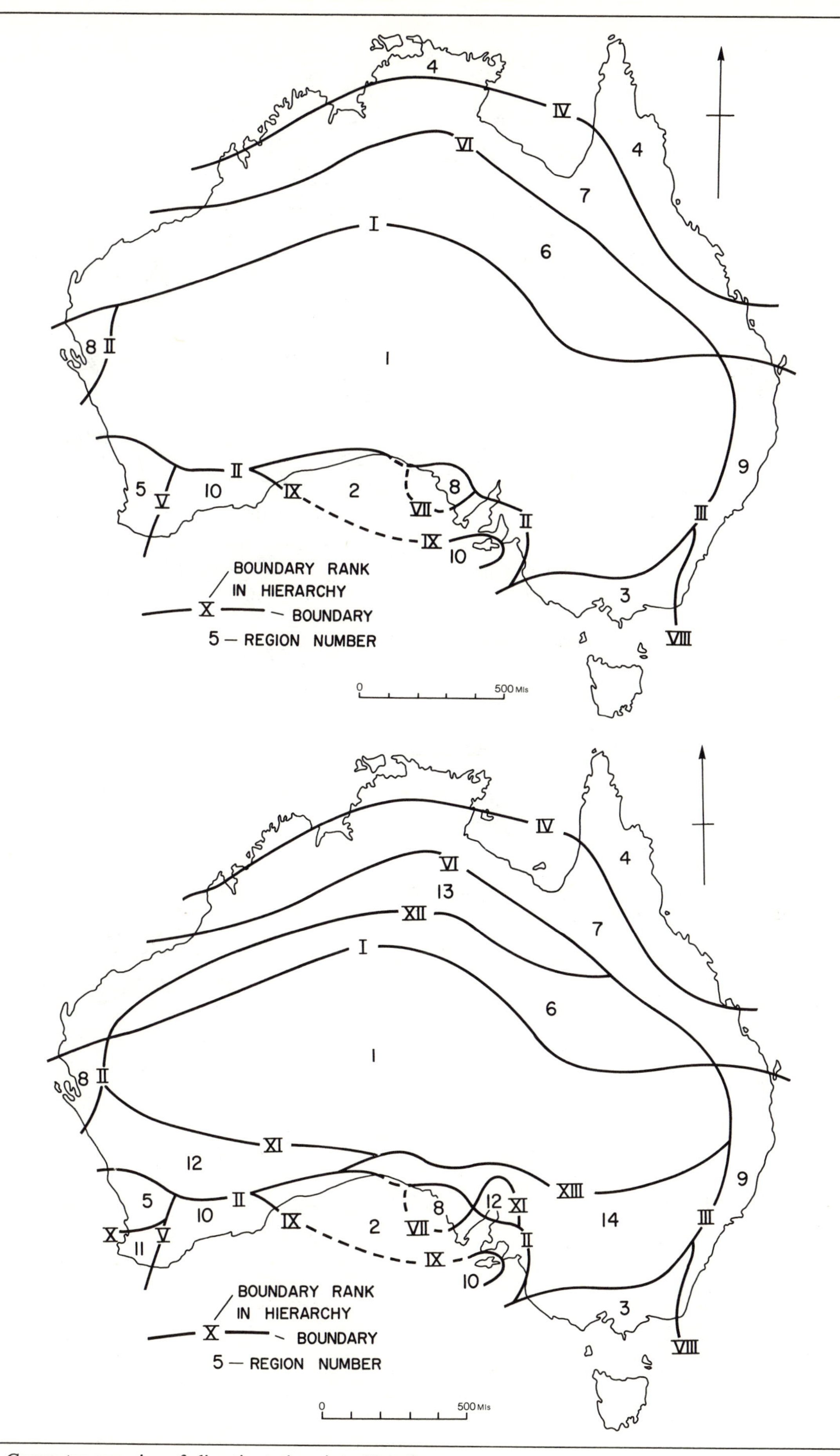

Figure 6 Computer grouping of climatic stations into 10 regions.

Figure 7 Computer grouping of climatic stations into 14 regions.

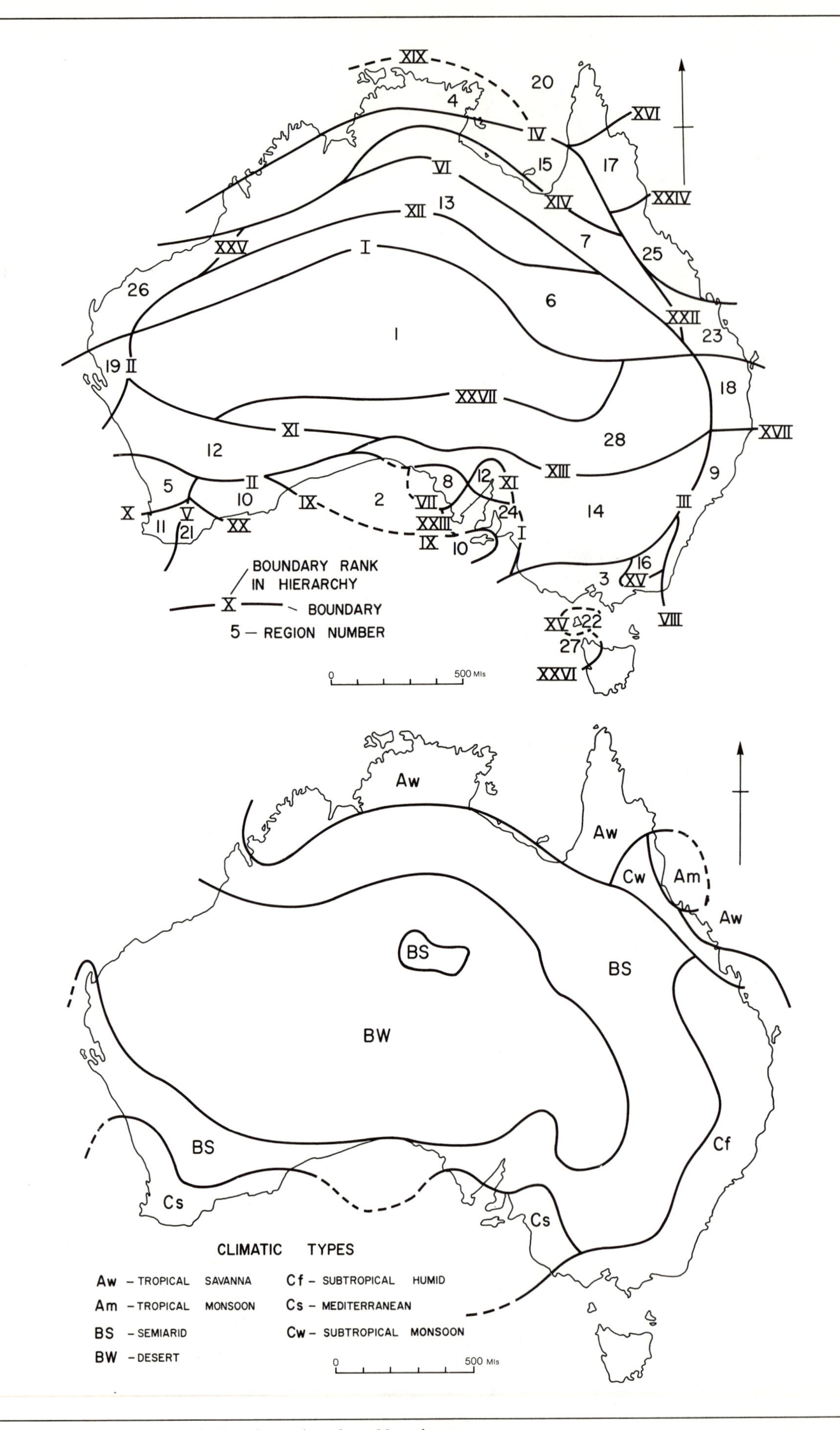

Figure 8 Computer grouping of climatic stations into 28 regions.

Figure 9 Climatic classification of Australia according to Köppen-Geiger from Blüthgen (1964).

Werner H. Terjung

11 Annual physioclimatic stresses and regimes in the United States

Geographical and climatological literature abounds with references to the stress, temperateness, or intemperateness of climates. Most of these generalizations are qualitative, subjective statements, not based on concise criteria defining the constituent elements or describing how they could be mapped. (For an exception, see Bailey, 1964.)

This study proposes a quantitative approach to the problem of climatic stress by introducing three annual indices: annual Cumulative Stress (CS), which measures the cumulative relative variations and departures from the human state of comfort; Proportional Cumulative Stress (PCS), which measures on an annual basis the percentage of heat stress in relation to cold stress; and Annual Physioclimatic Regime (APR), which attempts to synthesize the annual march of physiological climates, their regionalization, and their degree of fluctuation and frequency. The last classification also indicates the possible sources and causes of the various stresses as exhibited in the other indices. All these indices are applied to the United States, including Alaska and Hawaii.

Earlier, this writer introduced a classification of physiological climates that attempted to place man at the center of attention (Terjung, 1966a). This scheme integrated the major climatic elements that influence man physiologically and psychologically (temperature maxima and minima, humidities, radiation, and air movements) and combined a Comfort Index and a Wind Effect Index on seasonal maps. This study laid the foundation for the indices introduced here.

SUMMARY OF THE COMFORT INDEX (CI)

A Comfort Index, integrating the psychophysiological sensations of the "average" person with respect to temperature (as measured by the dry-bulb thermometer) and humidity (as expressed by relative humidity), was attempted by superimposing on a psychrometric chart (containing Effective Temperatures (ET), wet- and dry-bulb temperatures, and relative humidities) a system of categories of comfort, expressed in subjective terms. The divisions between the categories (Figure 1) were determined after a series of existing researches in the field of human comfort had been consulted (Table 1). The results of some of these investigations were plotted on the chart. Winter and summer comfort sensations were not differentiated. Psychological, or subjective, data (for example, "Comfort Zone," Table 1) were preferred when available, since man appears to be subject to numerous cultural idiosyncrasies in regard to his "feelings"—an assumption emphasized by the fact that many physiological data (for example, "Thermal Neutrality," Table 1) did not completely agree with subjective impressions. For instance, most of the physiological ranges are greater than the psychological and seem to extend more into the warmer parts of the comfort continuum, whereas "subjective man" prefers cooler conditions.

The clustering of points and lines in the *comfort zone* (0) indicates general agreement on its delimitation (*A'ABCDD'*). More than 70% of persons questioned in climate chambers considered *AB* (along 64° ET) to be still comfortable in winter; about 90% considered *CD* (along 72° ET) comfortable in summer. The slanting of *AA'* (along 60° wet-bulb) and *DD'* (68° wet-bulb) was based on the consideration that most investigators agree that a relative humidity higher than 70% cannot be regarded as optimum. Only in the cooler part of the continuum (for example, *MN*) does high humidity not bring a warming sensation. From about 45° F down man subjectively feels the opposite (though no physiological evidence has ever been established); in other words, impressions of wet-cold are perceived as more uncomfortable.

Reprinted, with minor editorial modification, by permission of the author and editor, from the *Geographical Review*, Vol. 57, 1967, pp. 225–240, copyright by the American Geographical Society of New York.

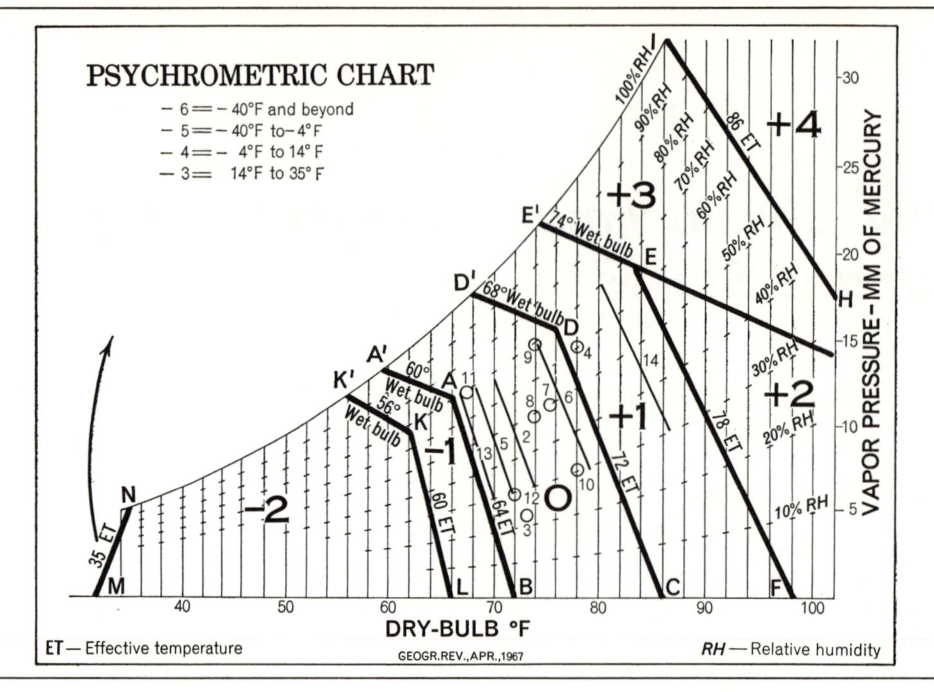

Figure 1 Psychrometric chart. Refer to Table 1 for numerals (1 to 14). Symbols of Comfort Index: −6, ultra cold; −5, extremely cold; −4, very cold; −3, cold; −2, keen; −1, cool; 0, mild; +1, warm; +2, hot; +3, sultry; +4, extremely hot. (After Terjung, 1966a.)

Investigator	ET° Optimum line	ET° Range	Operative range (° F)	Remarks
		Comfort Zone		
Yaglou and Houghton	66 (5)	63–71		Winter, at rest, clothed
Yaglou and Drinker	71 (6)	66–75		Men, summer, at rest clothed
Harris			76 (50% RH) (7)	Summer, clothed
			74 (50% RH) (8)	Winter, clothed
Gordon			74 (70% RH) (9)	Summer, clothed
			78 (30% RH) (10)	Summer, clothed
			68 (70% RH) (11)	Winter, clothed
			72 (30% RH) (12)	Winter, clothed
ASHRAE Lab.	71 (1)			Summer, clothed
	67.5 (2)			Winter, clothed
			73 (25% RH) (3) to 77 (60% RH) (4)	Entire year, slightly active
		Thermal Neutrality		
DuBois and Hardy		64.8–76 (13)(14)		Men, clothed

Table 1 Selected research in human comfort.
(Sources: ASHRAE, 1959; Terjung, 1966a)
Numbers in parentheses refer to Figure 1 and are plotted thereon as lines and circles.
ET, Effective Temperature. RH, Relative Humidity.

Table 2 Comfort index at low temperatures.
(Sources: Lee and Lemons, 1949; Burton and Edholm, 1955)

Dry-bulb temperature F°	Designation	Symbol	Clothing layers needed (approx.)
32 to 14	Cold	−3	4–4.5
14 to −4	Very cold	−4	4.5–5.5
−4 to −40	Extremely cold	−5	5.5–7
Lower than −40	Ultra cold	−6	More than 7

Extremely hot	Sultry	Hot	Warm
$+4/+3 = EH_1$	$+3/+3 = S_1$	$+2/+2 = H_1$	$+1/+1 = W_1$
$+4/+2 = EH_2$	$+3/+2 = S_2$	$+2/+1 = H_2$	$+1/0 = W_2$
$+4/+1 = EH_3$	$+3/+1 = S_3$	$+2/0 = H_3$	$+1/-1 = W_3$
etc.	etc.	etc.	etc.
Mild	**Cool**	**Keen**	**Cold**
$0/0 = M_1$	$-1/-1 = C_1$	$-2/-2 = K_1$	$-3/-3 = CD_1$
$0/-1 = M_2$	$-1/-2 = C_2$	$-2/-3 = K_2$	$-3/-4 = CD_2$
$0/-2 = M_3$	$-1/-3 = C_3$	$-2/-4 = K_3$	$-3/-5 = CD_3$
etc.	etc.	etc.	etc.
Very cold	**Extremely cold**	**Ultra cold**	
$-4/-4 = VC_1$	$-5/-5 = EC_1$	$-6/-6 = UC$	
$-4/-5 = VC_2$	$-5/-6 = EC_2$		

Table 3 Day and night combination of Comfort Index.
(Source: Terjung, 1966a)
Capital letter(s) indicates day condition; subscript indicates diurnal variability. First numeral refers to daytime conditions, second numeral to nighttime conditions.

The *warm* zone (+1), within the confines of *D'DCFEE'*, was delimited along *EF* (78° ET) because less than 10% considered this still comfortable in summer. The slanting of *EE'* was based on the same consideration as before.

The line *HI* (86° ET) is generally regarded as the upper limit for good health and practically the limit of efficient outdoor labor for extended periods; thus the part beyond this line has been designated *extremely hot* (+4). The remaining part, between +1 and +4, was divided into *sultry* (+3) and *hot* (+2) in view of the different sensations man experiences in a moist-hot or dry-hot environment.

The part of the continuum perceived as *cool* (−1), within *K'KLBAA'*, was delimited along *KL* because nobody still felt comfortable below this line in winter. Generally, one to two standard clothing layers (Lee and Lemons, 1949) become necessary for comfort (standing person at about 80 kcal/m^{2}hr, or about 1.5 mets metabolism). The zone bounded by *NMLKK'* (−2) is termed *keen*; about two to four clothing layers become necessary here.

Since at these low dry-bulb temperatures humidity is of little consequence (unless applied to the human body from without), starting at about 32° F (or about 35° ET, along *MN*) a different scheme of classification was adopted, which largely considers the physiological reactions of man in relation to the number of clothing layers needed to keep him at thermal neutrality (Lee and Lemons, 1949). Table 2 presents this part of the Comfort Index in summarized form.

A complete Comfort Index is determined by applying temperature and humidity data for daytime and nighttime (for a particular time period) and combining the two (Table 3) into a single expression. To derive a monthly Comfort Index, for example, mean daily maxima and minima for temperature and relative humidity should be utilized.

ANNUAL CUMULATIVE STRESS (CS)

Only the Comfort Index was used in determining the proposed indices (Terjung, 1966b), since temperature and relative humidity always prevail and are more accessible as data. Each Comfort Index (for daytime and nighttime) was given a weight (Table 4) based on its degree of

Table 4 Cumulative stress index.

COMFORT INDEX	D	N	D²	N²	CS	PCS
EH_3	+4	+1	16	1	17	100
EH_4	+4	0	16	0	16	100
S_1	+3	+3	9	9	18	100
S_3	+3	+1	9	1	10	100
S_4	+3	0	9	0	9	100
S_5	+3	−1	9	1	10	90
H_2	+2	+1	4	1	5	100
H_3	+2	0	4	0	4	100
H_4	+2	−1	4	1	5	80
H_5	+2	−2	4	4	8	50
W_1	+1	+1	1	1	2	100
W_2	+1	0	1	0	1	100
W_3	+1	−1	1	1	2	50
W_4	+1	−2	1	4	5	20
M_1	0	0	0	0	0	—
M_2	0	−1	0	1	1	0
M_3	0	−2	0	4	4	0
M_4	0	−3	0	9	9	0
C_1	−1	−1	1	1	2	0
C_2	−1	−2	1	4	5	0
C_3	−1	−3	1	9	10	0
K_1	−2	−2	4	4	8	0
K_2	−2	−3	4	9	13	0
K_3	−2	−4	4	16	20	0
CD_1	−3	−3	9	9	18	0
CD_2	−3	−4	9	16	25	0
VC_1	−4	−4	16	16	32	0
VC_2	−4	−5	16	25	41	0
EC_1	−5	−5	25	25	50	0
UC	−6	−6	36	36	72	0

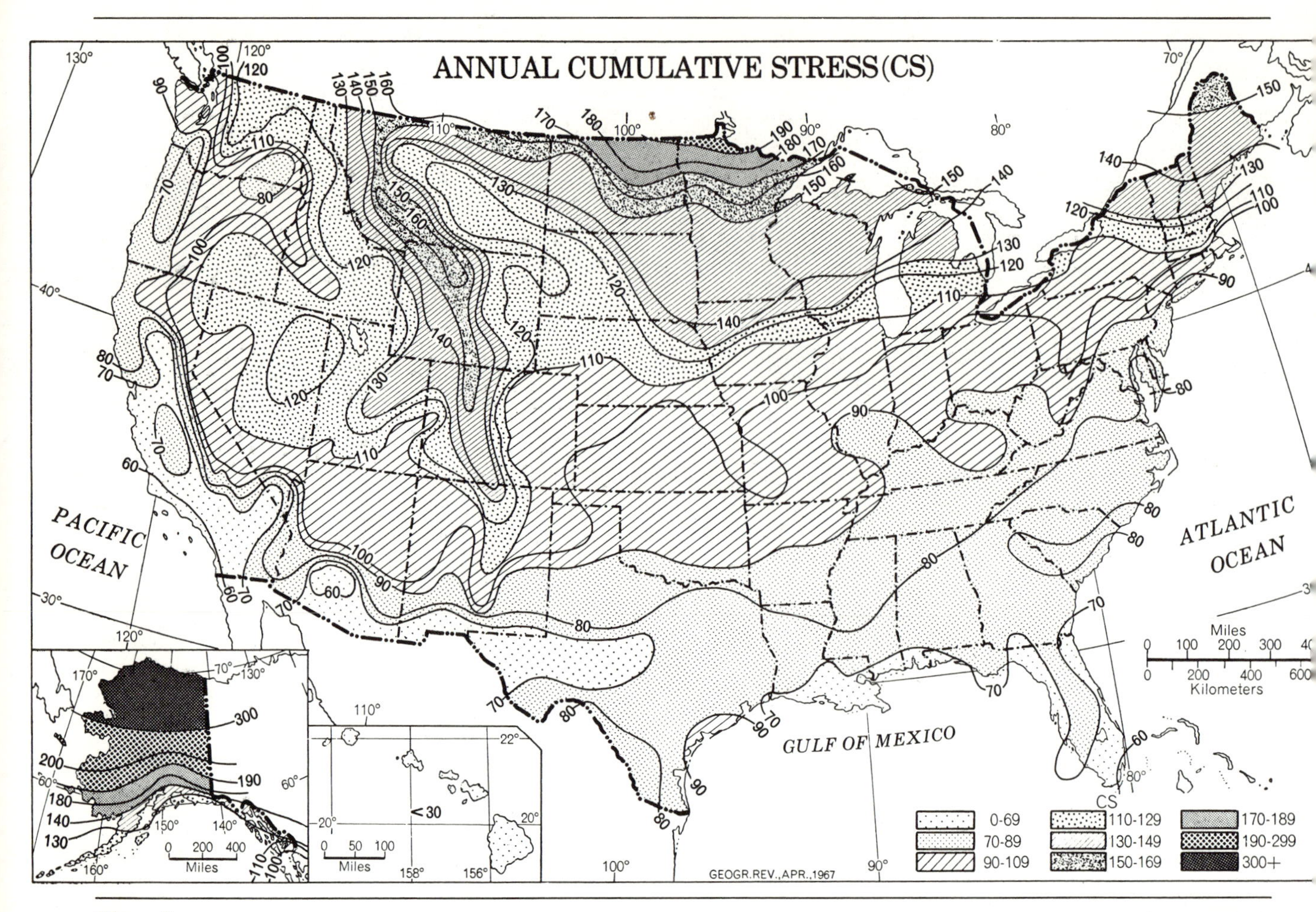

Figure 2

deviation from conditions comfortable for the "average" person. The following expression is introduced:

$$CS = \sum_{1}^{12} (D^2 + N^2),$$

where CS is the annual Cumulative Stress index, D the weight of the respective daytime Comfort Index, and N the weight of the respective nighttime Comfort Index. The annual Cumulative Stress for a given station is the sum of the CS indices for all twelve months. D and N were squared because most physiological *strains* resulting from climatic stress seem to approach exponentially shaped curves of intensity. The reader must be cautioned, however, that this index cannot predict certain physiological reactions to climatic stress, since each physiological effect or displacement follows its own curve in stressful situations. Thus the CS index can provide only a rough measure of relative stress, whether the cumulative stress is based on heat, cold, or a combination of both. Since it is centered on a comfort zone (0 in the Comfort Index nomogram), the index should also provide a measure of temperateness in relation to that zone.

As can be seen from Table 4, least stress (CS 0) could occur only in a somewhat unlikely place that would exhibit an M_1 Comfort Index for all twelve months. A large CS thus indicates high cold or heat stress, or a combination of both, as happens frequently in the middle latitudes.

This index was applied on a monthly basis to 260 first-order stations (U.S. Weather Bureau; Gt. Brit. Meteorological Office, 1958) in the United States (see Table 5 for selected stations), entailing about 6,500 computations. The results were plotted on a map, 1:10,000,000, and isolines of annual CS were drawn. The interval between isolines was arbitrarily chosen at CS 10. At this map scale complexities created by higher altitudes (especially in the Rocky Mountains) could be dealt with only in a highly generalized manner.

The arrangement of isolines on the map (Figure 2) is striking. The eastern two-thirds of the country is latitudinally aligned from south to north with respect to increasing stress; the western mountainous regions exhibit a complex configuration, with high CS values along the Rockies; and the Pacific coast is meridionally aligned. The Hawaiian Islands, a coastal strip in Southern California, and a sliver of land near Miami

STATION	J	F	M	A	M	J	J	A	S	O	N	D	CS	PCS	APR
Birmingham, Ala.	K_1	K_1	C_2	M_3	W_4	S_4	S_4	S_4	H_3	M_3	C_2	K_1	78	41	$K3^{2211}S3$
Barrow, Alaska	EC_1	EC_1	EC_1	VC_2	CD_2	K_2	K_1	K_1	K_2	CD_2	VC_2	EC_1	374	0	$EC4^{22}K4$
Juneau, Alaska	CD_1	CD_1	K_2	K_2	K_1	C_2	C_2	C_2	K_1	K_1	K_2	CD_1	132	0	$K6_3{}^3$
Phoenix, Ariz.	C_2	M_3	M_3	W_4	H_5	H_4	H_3	H_3	H_3	W_4	M_3	C_2	57	39	$H5_{232}$
Little Rock, Ark.	K_2	K_1	C_2	M_3	W_3	S_4	S_4	S_4	H_3	M_3	K_1	K_2	88	36	$K4^{12113}$
Blythe, Calif.	C_2	M_3	M_3	W_4	H_5	H_3	EH_3	EH_4	H_3	H_5	M_3	C_2	84	59	$H4_{132}{}^{02}$
Los Angeles, Calif.	C_2	C_2	M_3	M_3	M_3	M_3	W_4	W_3	W_3	M_3	M_3	C_2	48	6	$M6_3{}^3$
Denver, Colo.	K_2	K_2	K_2	K_1	M_3	W_4	W_4	W_4	M_3	C_2	K_2	K_2	101	3	$K6^{123}$
Bridgeport, Conn.	K_2	K_2	K_2	K_1	C_2	M_3	W_2	W_2	M_3	C_2	K_1	K_2	88	2	$K6^{222}$
Washington, D.C.	K_2	K_2	K_1	C_2	M_3	W_2	S_4	H_3	W_3	M_3	K_1	K_2	84	18	$K5^{12211}$
Miami, Fla.	M_2	M_2	M_1	W_2	W_1	S_3	S_3	S_1	S_3	W_1	W_2	M_1	56	96	$M4W4^0S4$
Tallahassee, Fla.	C_2	M_3	M_3	W_4	H_4	S_4	S_3	S_3	S_4	W_4	M_3	C_2	75	59	$S4_{1232}$
Atlanta, Ga.	K_1	K_1	C_2	M_3	W_4	S_4	S_4	S_4	W_2	M_3	C_2	K_1	75	38	$K3^{222}S3$
Honolulu, Hawaii	W_2	W_2	W_2	W_2	W_2	W_2	W_2	W_1	W_2	W_2	W_2	W_2	13	100	W12
Boise, Idaho	K_2	K_2	K_2	K_1	M_3	M_3	W_4	W_4	M_3	C_2	K_2	K_2	100	2	$K6^{132}$
Chicago, Ill.	CD_1	K_2	K_2	K_1	M_3	W_4	W_3	W_3	M_3	C_2	K_2	K_2	100	3	$K5_1{}^{123}$
Indianapolis, Ind.	K_2	K_2	K_2	K_1	M_3	W_3	S_4	W_2	W_4	C_2	K_2	K_2	99	12	$K6^{11301}$
Des Moines, Iowa	CD_2	K_2	K_2	K_1	M_3	W_3	H_3	W_2	M_3	C_2	K_2	K_2	105	6	$K5_1{}^{1221}$
Dodge City, Kans.	K_2	K_2	K_2	C_2	M_3	W_3	S_4	H_3	W_4	M_3	K_2	K_2	98	15	$K5^{12211}$
Louisville, Ky.	K_2	K_2	K_1	C_2	M_3	H_3	S_4	S_4	W_4	M_3	K_1	K_2	95	24	$K5^{12112}$
New Orleans, La.	C_2	M_3	M_3	W_3	W_2	S_3	S_3	S_3	S_3	W_2	M_3	C_2	66	65	$S4_{0332}$
Portland, Maine	CD_2	CD_2	K_2	K_2	C_2	M_3	W_4	W_4	M_3	K_1	K_2	K_2	133	2	$K5_2{}^{122}$
Detroit, Mich.	CD_1	CD_1	K_2	K_1	M_3	W_4	W_3	W_3	M_3	K_1	K_2	K_2	108	3	$K5_2{}^{023}$
Minneapolis, Minn.	CD_2	CD_2	K_2	K_1	M_3	W_4	H_3	W_3	M_3	K_1	K_2	CD_2	136	4	$K4_3{}^{0221}$
Vicksburg, Miss.	K_1	K_1	M_3	M_3	W_2	S_3	S_3	S_3	S_4	M_3	C_2	K_1	81	49	$S4_{01313}$
St. Louis, Mo.	K_2	K_2	K_1	C_2	M_3	W_2	S_4	H_3	W_3	M_3	K_1	K_2	84	18	$K5^{12211}$
Butte, Mont.	CD_2	CD_2	K_2	K_2	K_2	M_3	M_3	M_3	C_3	K_2	K_2	CD_2	162	0	$K5_3{}^{13}$
Omaha, Nebr.	CD_2	K_2	K_2	C_2	M_3	W_3	S_4	H_3	W_4	C_2	K_2	K_2	111	14	$K4_1{}^{21211}$
Reno, Nev.	K_2	K_2	K_2	C_3	M_3	M_3	W_4	W_4	W_4	M_4	K_2	K_2	107	3	$K5^{133}$
Atlantic City, N.J.	K_2	K_2	K_2	K_1	C_2	W_3	W_2	W_2	M_2	C_2	K_1	K_2	83	4	$K6^{213}$
Albuquerque, N.Mex.	K_2	K_2	K_2	C_2	M_3	W_4	H_3	H_4	W_4	M_3	K_2	K_2	97	10	$K5^{1222}$
New York, N.Y.	K_2	K_2	K_2	K_1	M_3	W_3	W_2	W_2	M_3	C_2	K_1	K_2	85	4	$K6^{123}$
Charlotte, N.C.	K_2	K_2	K_1	M_3	W_4	H_3	S_4	S_4	W_3	M_3	K_1	K_2	92	26	$K5^{02212}$
Bismarck, N.Dak.	CD_2	CD_2	K_2	K_2	C_2	M_3	H_4	W_4	M_3	K_2	K_2	CD_2	150	3	$K4_3{}^{1211}$
Akron, Ohio	K_2	K_2	K_2	K_1	M_3	W_4	W_3	W_3	M_3	C_2	K_2	K_2	95	3	$K6^{123}$
Tulsa, Okla.	K_2	K_2	K_1	M_3	W_4	S_4	S_4	S_4	W_3	M_3	K_1	K_2	97	30	$K5^{02203}$
Portland, Oreg.	K_1	K_1	K_1	C_2	M_3	M_3	W_4	W_4	M_3	C_2	K_1	K_1	72	3	$K5^{232}$
Philadelphia, Pa.	K_2	K_2	K_2	K_1	M_3	W_3	H_3	W_2	M_3	C_2	K_1	K_2	88	7	$K6^{1221}$
Memphis, Tenn.	K_2	K_1	K_1	M_3	W_3	S_4	S_4	S_4	W_2	M_3	K_1	K_2	88	33	$K5^{02203}$
Brownsville, Tex.	M_3	M_3	W_3	W_2	S_3	S_3	S_3	S_3	S_3	H_4	W_4	M_3	75	76	$S5_{133}$
Dallas, Tex.	K_1	K_1	M_3	M_3	W_2	S_3	S_3	S_3	H_3	W_4	C_2	K_1	77	47	$K3^{1221}S3$
Salt Lake City, Utah	K_2	K_2	K_2	K_1	M_3	W_4	H_5	H_5	M_3	C_2	K_2	K_2	107	8	$K6^{1212}$
Burlington, Vt.	CD_2	CD_2	K_2	K_2	C_2	M_3	W_4	W_4	M_3	K_1	K_2	CD_1	138	1	$K4_3{}^{122}$
Richmond, Va.	K_2	K_2	K_1	M_3	M_3	W_2	S_4	H_3	W_3	M_3	K_1	K_2	83	18	$K5^{03211}$
Seattle, Wash.	K_2	K_2	K_1	K_1	C_2	M_3	M_3	M_3	M_3	K_1	K_1	K_2	92	0	$K7^{14}$
Madison, Wis.	CD_2	CD_2	K_2	K_1	M_3	W_4	H_4	W_4	M_3	K_1	K_2	CD_1	133	5	$K4_3{}^{0221}$
Lander, Wyo.	CD_2	K_3	K_2	K_2	C_2	M_3	W_4	W_4	M_3	K_2	K_2	CD_2	145	1	$K5_2{}^{122}$

Table 5 Physioclimatic data for selected stations.

have the lowest values; northern and central Alaska and northern Minnesota and North Dakota have the highest. If the map extended to lower latitudes an increase in CS could again be expected in certain areas (Terjung, 1966 b). For instance, Port Sudan, on the Red Sea, has a CS of 124, and Brazzaville, on the Congo (subject to many months of moist tropical conditions), has CS 92.

The reader might object to the inclusion among the lower CS values of large parts of the South and Southeast that exhibit oppressive conditions in summer. This is a fallacy that results from generalizing about annual conditions from the extreme seasons only. The South's low values derive mainly from its rather mild winters and transitional seasons (Figure 4), whereas the North has no such compensation. Considering their latitude, the coastal regions of southern Alaska exhibit remarkably low values; the Aleutians, for instance, show values similar to those of sections of the Middle West.[1]

All these configurations largely dispel the myth of the "temperate climate" found mainly in the middle

[1] In the Aleutians, however, the Wind Effect Index shows wind to be more detrimental in its chilling effects.

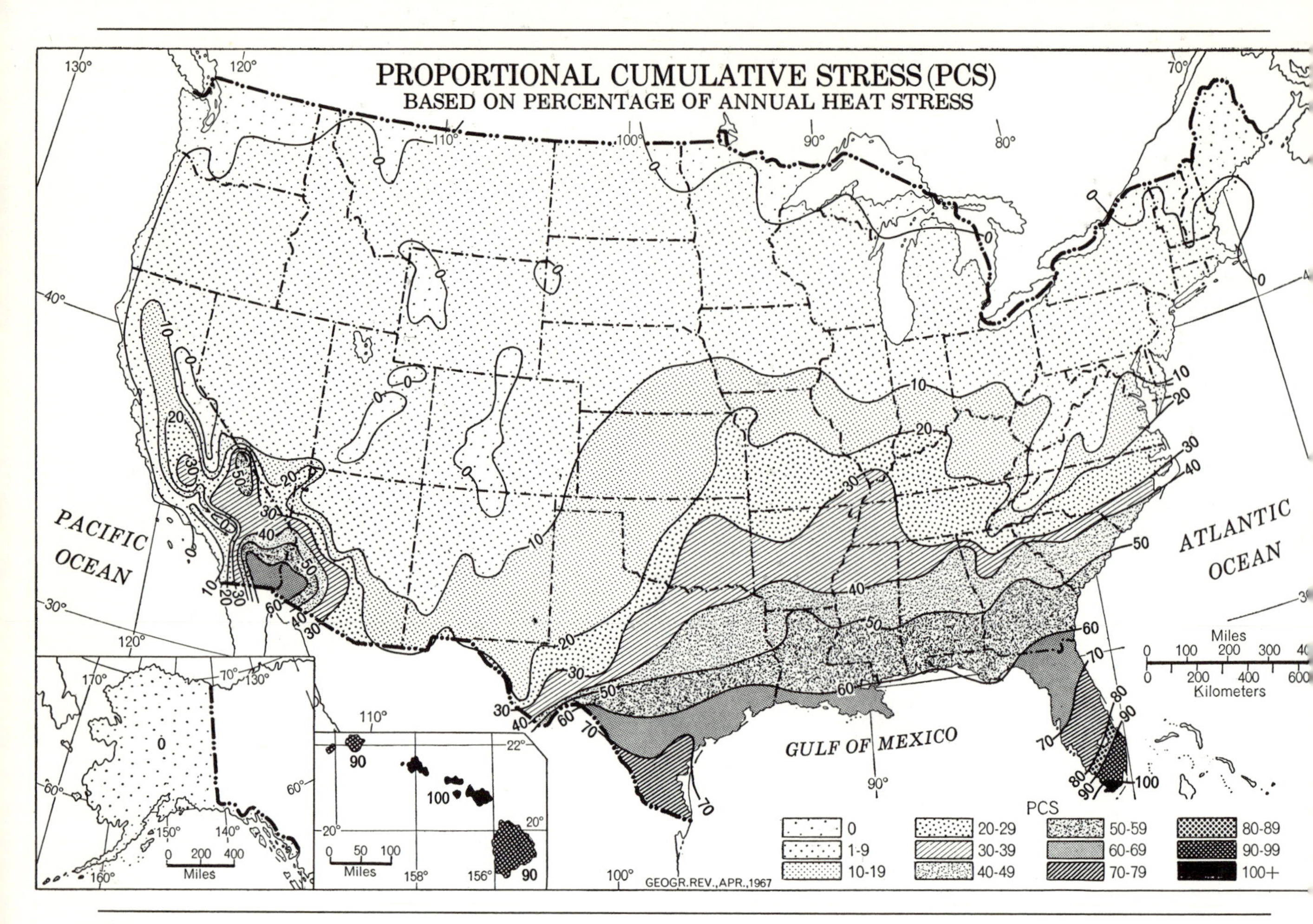

Figure 3

latitudes. In fact, Africa has larger temperate regions than North America. The Middle West, for instance, has CS values comparable with those of areas popularly considered "unfit for the white man." The reader is invited to compare Bailey's map (1964) of temperateness in the conterminous United States with the map presented here. The similarity is remarkable despite Bailey's use of a different approach to derive his M index of temperateness.[2]

What significance to geography has an index such as annual Cumulative Stress? Does man recognize the distribution of climatic stress? A glance at a population-density map of the United States finds no particular correlation, though areas of high CS generally tend to have a low population density. However, one must remember that the average person has only recently achieved an economic living standard that gives him some degree of choice other than purely economic reasons in selecting a location. A new stage of migration, based on amenities, and especially on climate, has barely begun. When the map is considered in relation to rates and areas of in-migration (Current Population Reports, 1962), a pattern results which appears to indicate that man is becoming cognizant of his climatic environment and is beginning to "do something about the weather." Thus most of the areas in the United States with a low CS have one of the higher in-migration rates. Nevertheless, the problem remains that man is not a good judge of what constitutes amenable climates or, particularly, where they are located. Local chambers of commerce are not usually the best guides, catering to myths, irrational notions, and sometimes even misinformation. Figure 4 is offered as a somewhat more reliable guide in selecting, for example, recreation and retirement climates.

[2] However, this writer would be inclined to disagree with Bailey's grouping of certain low-latitude regions on his world map.

PROPORTIONAL CUMULATIVE STRESS (PCS)

One of the shortcomings of the CS index is that it cannot show the constituent factors which caused a particular value—heat or cold or both. Thus an additional index is introduced, which uses the following expression:

$$\mathrm{PCS} = \frac{\Sigma_1^{12}\,(\mathrm{D}_h^2 + \mathrm{N}_h^2)}{\Sigma_1^{12}\,\mathrm{CS}} \cdot 100,$$

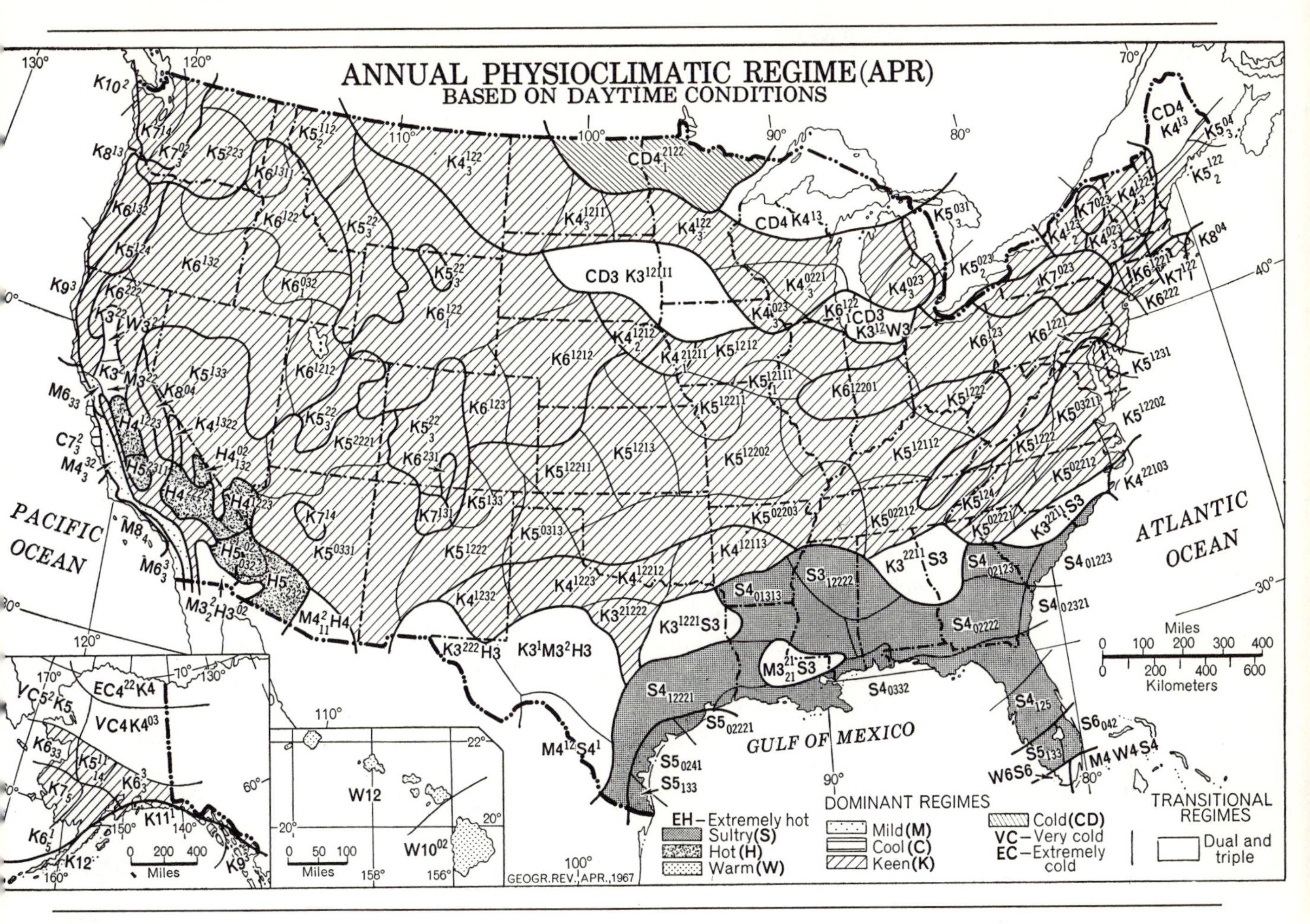

Figure 4

where h is the heat stress in the Comfort Index. Obviously, the remaining percentage must be caused by cold stress. Thus a value of PCS 40 means that 40% of the CS is caused by heat stress and 60% by cold stress. A value of 100 indicates that climatic stress is due entirely to heat, a value of 0 indicates the opposite (see Table 4). The reader is cautioned not to interpret high PCS values as necessarily high general (CS) stress also, since the PCS value portrayed could have derived, especially in the United States, from a very low CS. Thus a PCS value merely points out, on a yearly basis, what proportion of stress felt by man, if any, is due to heat.

The foregoing concept has been applied in Figure 3. Except for the Pacific littoral, the pattern is almost the perfect inverse of Figure 2: regions of high CS appear now with low PCS values and vice versa. Vast sections of the country exhibit values which indicate that heat is not a real problem except for brief periods (see also Figure 4 and Table 5) and that the major environmental stress derives from cold. This applies especially to the areas of high population concentrations. Even large areas of the South, the Southeast, and the western deserts, which often carry the connotation of "hot," trace their CS to rather small parts of PCS. Only in southern Florida and the Hawaiian Islands is cold stress of little consequence. The opposite applies to most of the northern half of the country, much of the Southwest, and the Pacific littoral. The Pacific coastal strip reaches as far south as Long Beach, California.

If this system were extended globally, three basic belts could be expected: those owing their CS entirely to heat, those owing it entirely to cold, and the transition belt due to both influences. Considered in that way, the United States is a perfect example of a "transition climate" and offers a possible criterion for "middle latitude climates."

ANNUAL PHYSIOCLIMATIC REGIME (APR)

An index that measures relative cumulative stress and/or degree of temperateness does not indicate the sources of such stress, and the PCS index only divides them into heat or cold effects; both, however, are necessary for a more nearly complete appraisal of physioclimatic effects.

Another way of indicating annual variation of physiological climate, or in fact any climate, is a portrayal of the spatial distributions of the monthly (or other time unit) types of climate, which results in an areal arrangement of the annual regime. This will supply not only a measure of what and how much is where but also a measure of the amplitude of annual variability. Further, it will supply a tool to regionalize physiological climates on an annual basis. The proposed APR index is based on daytime conditions only, to keep complexities to a manageable level at the map scale employed. In a micro study the classification could be easily modified to accommodate both day and night, and even to add Wind Effect Index variations. For the present purpose it can be simply expressed as

$$APR = \sum_{1}^{12} CI_d$$

where CI_d is the monthly daytime Comfort Index.

Two basic types of regimes result: Dominant Regimes, where one physioclimatic type occurs more frequently than others; and Transitional Regimes, where several types vie with one another. Each basic type has major subdivisions. Dominant Regimes can be extremely hot (EH), sultry (S), hot (H), warm (W), mild (M), cool (C), keen (K), cold (CD), very cold (VC), extremely cold (EC), and ultra cold (UC), depending on the physioclimatic type of the dominant month (several of these do not appear in the United States). Each of the Dominant Regimes is also subdivided, according to the arrangement and number of the other "type-months." Thus we can have a continuum from a regime dominated all twelve months by the same type-month (for example, twelve months of K) to a regime where two different type-months occur the same number of times. This latter condition would indicate the beginning of a Transitional Regime. Thus various intensities of a regime have to be reckoned with. Dominant Regimes that have ten to twelve occurrences of the same type-month, "keen" for example, are considered to be *extremely* keen, those with eight to nine are considered *very* keen, those with five to seven are simply *keen*, and, finally, those with one to four are considered *slightly* keen. Under certain local conditions, as in the United States, divisions within the subtypes might become desirable. This applies especially to that part of the classification where the number of occurrences of the dominant type-month becomes very low in relation to the other type-months. Thus a climatic regime that has a wide range of type-months in the year (for example, in the Middle West) is hard to classify, since it partakes of a whole series of type-months and often just barely falls into any category. But this very fact points up its variability.

Transitional Regimes are subdivided into *dual*, *triple*, and *quadruple* (the last does not occur in the United States), according to the grouping of equal-number type-months. Each subdivision can have many further subdivisions; for example, dual regimes of mild-warm (six months of M, six of W) or warm-sultry.

The foregoing considerations need to be symbolized by some logical system, which enables analysis of an APR classification at a glance. To achieve this goal, a system of letters, numerals, subscripts, and superscripts is introduced. First, a continuum of warmth or cold is developed thus: EH–S–H–W–M–C–K–CD–VC–EC–UC. Second, superscripts are considered warmer type-months, and subscripts colder type-months, in relation to the dominant month (or months). The sequence and arrangement of numerals and letters indicate position in the hot–cold continuum; the numeral following a letter indicates how often that particular physioclimatic type occurs.

For example, a region experiencing five months of K climates (the dominant month), two of C, one of M, and four of CD would be symbolized as $K5_{4}^{21}$, where K indicates the dominant physioclimatic type, and 5 the intensity (frequency) of that type. Setting up this part determines automatically the positions of the other type-months. The numeral 2 in the superscript indicates two months of C conditions, since according to the continuum C comes after K going toward the warmer end of the continuum. The numeral 1 indicates one month of M conditions, since M appears next. The subscript 4 indicates four months of CD conditions, since CD comes after K going toward the colder end of the continuum. When there is no continuous sequence in an APR region, 0 is substituted for the gap. For instance, four months of K, one of C, two of M, two of W, and three of S (skipping H) appears as $K4^{12203}$. An area that has all twelve months dominated by the same climatic condition, for example twelve months of keen, would appear as K12.

The Transitional Regimes are treated similarly. The same continuum applies, except that the coldest type-month of the dominants always comes first. For instance, six months of W and six of S (a dual Transitional Regime) result in W6S6; three of K, one of C, three of M, two of W, and three of H (a triple Transitional Regime) appear as $K3^{1}M3^{2}H3$. Thus subscripts and superscripts are used in the same way.

For clarity, and because of scale limitations, many small, slightly different local APR's were combined with similar larger sections. Micro studies would probably uncover many smaller subdivisions and/or pockets of different regimes, especially in mountainous areas.

From Figure 4 and Table 5 one again perceives the transitional nature of most climatic regimes in the conterminous United States. Except for the Pacific coast, not a single "core regime" (a regime with a dominant type-month of more than seven occurrences at least) is shown. Such regimes appear only in Alaska (K12) and Hawaii (W12). Most striking is the vast extent of a large variety of K regimes, which range mainly from slightly keen to keen. Only along the Pacific coast do very keen and extremely keen regimes appear. Because of the importance of the different K regimes in the United States, an attempt has been made to subdivide them further than the other APR's, by using heavier boundary lines. It is important to realize that the less often a dominant type-month occurs in an APR, the more "noise" is introduced; that is, the system becomes less and less a pure regime with respect to, for example, its "keenness." This can be readily seen from the superscripts and subscripts; many of these K regimes are just one type-month removed from being Transitional Regimes. For instance, a $K5_{3}^{22}$ regime (Rockies) is quite different from a $K3^{21222}$ regime (central Texas), which even contains two months of sultry conditions, whereas the first regime has three months of cold conditions and never reaches warm conditions. The larger the number of digits in its superscripts and subscripts, the more annually variable an APR becomes. As was to be expected, this intense annual amplitude is especially well represented in the Middle West. In general, areas west of 105° W have increasingly less variable K regimes. It is suspected that globally K regimes are most characteristic of the middle latitudes and probably occupy the lion's share of the zone of PCS values below 100 and above 0 (Figure 3). Depending on the point of view or the particular problem, the various K regimes could probably be subdivided in many other ways. A basic division could be that between K regimes which have subscripts (colder) and those which have not (warmer). The former occur in a wide arc extending from the Rockies through Iowa to coastal northern New England. The K regimes could also be divided according to the number of superscript digits. For example, those with five digits have from one to several sultry months; those with three digits reach only warm conditions. The first occur as far west as Kansas; the second obtain especially in the Great Basin.

Slightly cold regimes and their related dual Transitional Regimes are centered on the North and Northeast. Cool regimes occur in a narrow zone along the Pacific coast of north-central California, and various subdivisions of mild regimes are found south and east of this. Farther inland slightly hot or hot regimes are generally encountered in the Central Valley and in the Mojave and Sonoran Deserts. In the Sonoran Desert and Death Valley the superscript indicates also two months of extremely hot (EH) conditions, the most detrimental of all physiological climates at the heat end of the continuum. The Southeast and South exhibit various combinations of slightly sultry and sultry regimes, with the greatest intensity in the Florida peninsula.

In general, six physioclimatic regimes occur in the conterminous United States, virtually all of them in the lower end of the respective type classification; this results in great variety among the many subtypes. Alaska and Hawaii exhibit greater extremeness in number of type-months. Few of the mapped patterns coincide with those of other climatic classifications, an indication that man does not necessarily perceive climatic influence as thus portrayed. He cannot, for instance, "feel" the "Mediterranean" climate of a classification using yearly averages and parameters that are of little consequence to human physiology and psychological impressions of comfort. The Mediterranean climate as it impinges on man in the United States is made up of at least four different basic regimes—K,

C, M, and H—with their subdivisions. The argument could be extended to most of the other conventional climatic types. Thus it seems that by ignoring man in appraisals of the "climate" of an area an important sector has been too frequently slighted.

Many research needs exist, especially studies that might lead to indices utilizing various physioclimatic stress lines in relation to isolines of, for example, health, nutrition, economic and social behavior, and clothing and shelter needs, in deriving, eventually, systems that measure total environmental stress based on cultural and physical variables. Such indices might eventually lead to the appraisal of the potential or "energy" of regions in relation to human occupance.

REFERENCES

ASHRAE, 1959. *Heating, Ventilating, Air-Conditioning Guide 37*. American Society of Heating, Refrigeration, and Air-Conditioning Engineers, New York.

Bailey, H. P., 1964. Toward a unified concept of the temperate climate. *Geographical Review 54*, 516–545.

Burton, A. G. and Edholm, O. G., 1955. *Man in a Cold Environment*. Arnold, London.

Gt. Brit. Meteorological Office, 1958. *Tables of Temperature, Relative Humidity, and Precipitation for the World, Part I, North America, Greenland, and the North Pacific Ocean*. London.

Lee, D. H. K. and Lemons, H., 1949. Clothing for global man. *Geographical Review 39*, 181–213.

Terjung, W. H., 1966a. Physiologic climates of the conterminous United States: A bioclimatic classification based on man. *Annals*, Association of American Geographers *56*, 141–179.

Terjung, W. H., 1966b. *Physiological Climates of Africa*. Unpublished Ph.D. thesis, Univ. California, Los Angeles.

U.S. Bureau of the Census, 1962. Estimates of the population of states and selected outlying areas: July 1, 1962. *Current Population Repts. Ser. P-25, 272*.

U.S. Weather Bureau. Climates of the states. *Climatography of the United States*, Nos. 60-1 to 60-51. Washington.

SUGGESTED READING FOR PART 4

Bryson, R. A., 1966. Air masses, streamlines, and the boreal forest. *Geogr. Bull. 8(3)*, 228–269.

Chang, J.-H., 1959. An evaluation of the 1948 Thornthwaite classification. *Annals*, American Association of Geographers *49*, 24–30.

Maunder, W. J., 1962. A human classification of climate. *Weather 17*, 3–12.

Terjung, W. H., 1968. Bi-monthly physiological climates and annual stresses and regimes of Africa. *Geografiska Annaler 50A*, 173–192.

Terjung, W. H., 1968. Some thoughts on recreation geography in Alaska from a physio-climatic viewpoint. *The Californian Geographer 9*, 27–39.

Wallén, C. C., 1967. Aridity definitions and their applicability. *Geografiska Annaler 49A*, 367–384.

Part Five

Climatic Change

Beckinsale's critique on the modern theories of climatic change sets the perspective for Part 5. He examines the relationship between theory and empirical results, finding the two to be frequently at odds. The article encompasses such questions as whether the CO_2 theory seems valid to explain major climatic changes, and whether an ice age is brought on by an increase or decrease in global air temperatures.

When and what was the Little Ice Age? How many major climatic epochs have there been since the Pleistocene era? Lamb's explanation for the meteorological changes throughout this period is based on an understanding of the changes in the intensity of the circumpolar vortex and its latitudinal shifts. Furthermore, he has been able to reconstruct surface pressure maps back to 1750, and has developed indices of summer wetness and winter coldness back to 800 AD by content analyses of lay materials.

In the third article Namias demonstrates clearly how meteorological conditions form only one part of the system of the biosphere. He attributes the cool climate experienced in the eastern U.S.A. in the 1960s directly to the incursion of polar air following upper air trajectory shifts, themselves triggered by air-sea interaction conditions over the North Pacific sea body. But wherein lay the cause of the oceanic warming? The answer has yet to be found, but could it perhaps lie latent in the first article of this part?

R. P. Beckinsale

12 Climatic change: a critique of modern theories

The study of palaeoclimatology and changes of climate involves academic disciplines ranging alphabetically from astronomy to zoology, with geology and meteorology dominant. Meteorologists must decide on the rôle of meteorological processes. They find present weather observations helpful in interpreting climatic fluctuations during historic times but as yet these observations lack real length. Climatological experts as a whole take a global or hemispherical point of view in spite of occasional signs of environmental prejudice which persuades a few of them to allow a small polar tail to wag the vast tropical dog.

Geologists have the advantage of being able to investigate existing ice-sheets and glaciated landscapes but in addition have access to ancient landforms and deposits as a guide to palaeoclimates. With the aid of varves, pollen analysis, deep-sea sedimentation, C^{14} datings and various other isotopic measurements, they have obtained a fairly detailed time-scale for at least the last main ice-advance and -retreat and a general chronology for much of geological history.

Unfortunately the problem of climatic change cannot be divorced from an outstanding problem of geology and geophysics—the nature of orogenesis (mountain-building) and of epeirogenesis (major earth-movements or continent-building). The eustatism of Suess, whereby sea-level rose or fell largely because of changes in the capacity of the ocean-basins, no longer dominates geological attitudes to diastrophism. It has been replaced by a chaos of ideas, among which the dominant trend is a shift towards a less catastrophic and more uniformitarianistic[1] form of landmass-controlled diastrophism (Chorley, 1963). With notable exceptions, geophysicists and geologists seem increasingly opposed to universal catastrophes such as abrupt or well-defined mountain-building phases. Mountains are considered to arise gradually over a long period of time and each continent or even each major sub-division of a continent tends to have its own orogenic and epeirogenic time-scale. Within these individual time-scales the possibility of global and zonal coincidences of low and of high relief is not excluded.

Fortunately for the climatologist, the reality of vertical isostatic movements in the Earth's crust is generally admitted by geophysicists however much they may disagree on the processes and mechanisms involved, whether sub-crustal "currents" or phase-change with depth and pressure and so on. The latter disagreement, however, brings a difficulty in palaeoclimatology as it involves the failure of geophysicists to agree upon the reality and extent of continental drift or the migration of land-masses in lateral directions. As global temperatures are arranged in a zonal or systematic latitudinal pattern, the presence, for example, of coal seams in Antarctica poses serious (but not insuperable) difficulties if the continents or/and the poles have not shifted laterally. In this respect it happens that even if continental drift has occurred on a large scale, allowance must still be made for the existence of hot as well as of cold periods in the past (Figure 1). The recent publication of Nairn (1961) and Schwarzbach (1963) makes further discussion here on warm climates unnecessary. Each has an extensive bibliography.

SOME BASAL GEOLOGICAL FINDINGS ON PALAEOCLIMATOLOGY

For simplicity we will concern ourselves mainly with climatic variations in temperature, on which the geological evidence at present available indicates:

(1) Throughout decipherable geological time, long periods of warm climates have existed and have been

[1] That is by processes which are still at work today and so able to be investigated and measured by modern scientists.

Reprinted, with minor editorial modification, by permission of the author and the editors, from *Essays in Geography for Austin Miller*, J. B. Whittow and P. D. Wood (editors), University of Reading, 1965, pp. 1–38.

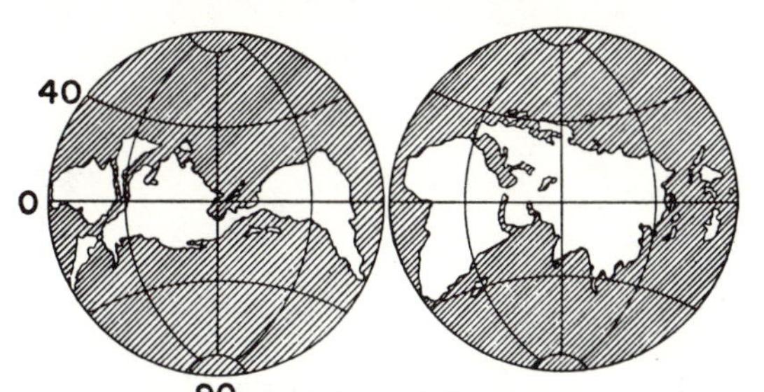

Figure 1 Popular explanation of hot and cold climates on continents based on lateral shift of land-masses. (After Lyell, 1872.)

interrupted at wide intervals by colder periods with extensive glaciers and ice-sheets.

(2) Modern isotopic measurements appear to show that some distinctly warm geological periods experienced appreciable climatic oscillations. Thus, the later Jurassic was warm with much wider tropical and subtropical zones than exist today whereas the succeeding Cretaceous was cooler than the later Jurassic and probably experienced two temperature cycles with a thermal amplitude of 4° or 6° C and a time-scale of about 20 or 30 million years.

(3) Palaeotemperature measurements also indicate that in spite of slight cooling at the end of the Cretaceous period, the Tertiary climate remained warmer than today although during its course of 60 million years or so temperatures dropped appreciably. According to laboratory assessments, Miocene temperatures were about 4° C lower than those of the preceding Oligocene which is generally reckoned as being a few degrees cooler than the Eocene. The overall Tertiary drop of temperature is variously reckoned by geologists as being about 8° C. The findings, however, do not entirely exclude the possibility of an oscillation of a few degrees with a 20 or 30 million year periodicity.

(4) The isotopic analysis of deep-sea sediment cores has also revealed at least three significant features of late Pliocene and Pleistocene climates.

(*a*) During the last few hundred thousand years of the nonglacial Pliocene period the climate gradually became cooler and at the same time experienced fluctuations of about the same periodicity as those that occurred later in the Pleistocene but of a much smaller amplitude.

(*b*) Not less than about 800,000 years and not more than 1,500,000 years BP, a relatively abrupt change of temperature marked the onset of the great continental ice-sheets. There is disagreement on the abruptness of this change. Some geologists postulate a span of 6,000 years or even less between the nonglacial Pliocene and the time of the great continental glaciations (Ericson, Ewing, Wollin, 1963) while others consider the change to be more gradual (Riedel, Bramlette, Parker, 1963). There is no doubt, however, that the change was appreciable.

(*c*) The post-Pliocene or Pleistocene epoch underwent several major temperature oscillations with a periodicity of a few ten thousand years. The amplitude in equatorial and tropical regions of the ocean was about 6° C and the periodicity about 41,000 years (Emiliani, 1956).

(5) A large body of geological evidence, in addition to variations in deep-sea sediments, indicates that the major Pleistocene ice-advances and -retreats were not simple and continuous but had superimposed upon them lesser oscillations of a secondary amplitude and a period of about 10,000–20,000 years.

(6) Many small or "secular" fluctuations of climate (that is with a time-scale of hundreds of years only) occurred at least during the latest of these lesser oscillations and continued to the present day. This detailed recent climatic chronology, since about 45,000 BP, can be based upon laboratory data (especially C^{14} and pollen analysis) and since about 1700 on some form of meteorological data. There seems no reason why fluctuations of various amplitudes and time-scales should not be a continuing feature throughout geological time.

The geological evidence outlined above suggests the reality of world climatic changes from warm to cold, of warm climates with cooler phases, and of cool climates with colder phases. The total amplitude of the changes may well be under 20° C for decipherable geological history or about 10° above or below the norm, whatever that may be (Oligocene?).[2] The seasonal regional changes will however be appreciably greater than this especially in poleward latitudes in continental interiors with a long low-sun season.

Yet, while it is necessary to assume that warm (i.e., warmer than today) is normal in a time sense and that warm and cold climates both undergo oscillations, it seems equally necessary to base any detailed theory of climatic change on knowledge of late Pliocene and post-Pliocene times, as by shortening the investigation to the last 2,000,000 years the possibility of large scale continental drift is virtually eliminated. To be able to assume the broad present continental spacing or continental-oceanic horizontal pattern is a great advantage although vertical changes, such as mountain-building, will be active upon it.

GENERAL REQUIREMENTS OF ANY ACCEPTABLE THEORY OF QUATERNARY CLIMATIC CHANGE

Any satisfactory hypothesis of climatic change during the Quaternary epoch must explain:

(*a*) the co-existence of vast, thick ice-sheets around or near both poles and extending over the adjacent oceans;

(*b*) large repetitive advances and retreats of the ice-line and snow-line on mountains, lowlands and oceans, more or less contemporaneously throughout the northern hemisphere and probably almost contemporaneously between the southern and northern hemispheres; and

(*c*) the presence of vast areas of permafrost and of periglacial land-forms in regions far equatorward of the ice-sheets.

The obvious solution is to assume major and minor decreases in global air temperatures sufficient for these purposes and incidentally sufficient also to lower evaporation in semi-arid sub-tropical climates.

The main objection to such a simple explanation is that in high latitudes cold normally decreases precipitation, especially in continental interiors with strong anticyclonic airmasses in the winter half-year. As the thickness and extent of the Pleistocene ice-sheets required a large accumulation of snow, it has been suggested that ice-ages were caused by an increase of solar radiation, and so of precipitation, during a time of general lowering of temperatures. This theory in various forms became popular among mid-twentieth-century meteorologists. It may be crudely expressed as follows: As global temperatures rose, snowfall increased steadily on mountainous lands in high latitudes and exceeded snow-melt so that ice-sheets accumulated, but beyond a certain temperature or insolation, melting began to outstrip accumulation and the ice-sheets retreated. Later, when insolation began to fall again, at a certain temperature snowfall again exceeded melting and ice-sheets reappeared but in this waning phase of the insolation the decreasing temperature gradually lessened the snowfall and ice-sheets shrivelled through lack of snow-accumulation. The fact that a cooling mechanism was used to remove ice-sheets did not impair the popularity of this meteorological theory (Simpson, 1934, 1957; Bell, 1953), which is shown in a greatly simplified form in Figure 2.

However, after 1955 the solar-radiation increase hypothesis became less tenable when analyses of deep-sea sediment cores and other laboratory measurements indicated that global temperatures were very low during times of most extensive glaciation. Moreover, we hope to show later that other meteorological arguments can be put forward to explain the large ice-accumulation required.

MAIN THEORIES ON CLIMATIC CHANGE

As the main theories on climatic change have been summarized in Brooks (1949), Shapley (1953), Flint (1957), Wright (1961), and Schwarzbach (1963), only new theories since about 1950 and modern adaptations of older theories will be stressed here.

The more acceptable hypotheses are usually classified according to the genetic physical properties involved into four main groups, depending respectively on changes in: solar radiation; atmospheric transparency; earth's geometry; terrestrial geography.

[2] It seems impossible to suggest world average temperatures with any accuracy. Today the world average air temperature is 15° C (59° F). This probably fell to about 9° C (48° F) during the last main ice advance. The normal in geological time might be about 19° C (66° F).

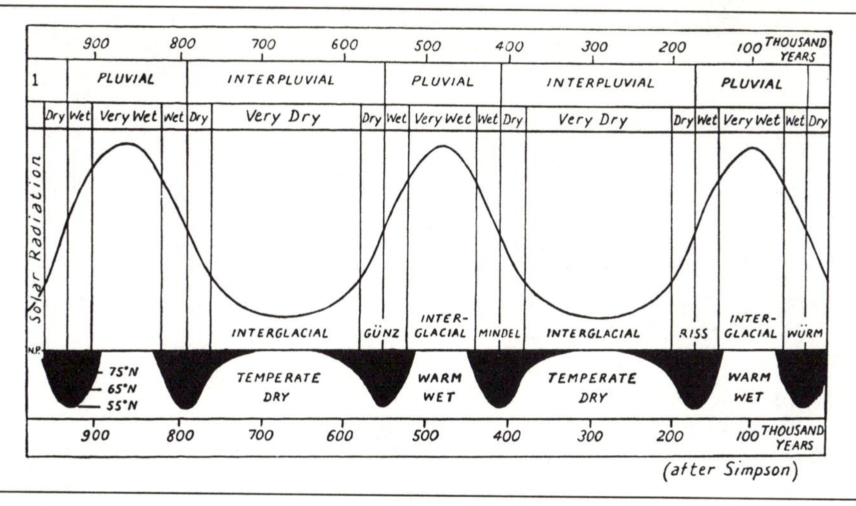

Figure 2 A simplification of Simpson's (1957) radiation curve and its effect in polar latitudes on ice-advances and ice-retreats (shown in black) and on rainfall in tropical latitudes (shown in 1).

A. Theories Postulating Changes in Solar Radiation

These theories depend mainly on variations in the quantity of solar emission or radiation, and in the quality or wave-length of that radiation.

(1) *Short-term fluctuations of solar radiation*
The existence of periods of sunspot maxima and sunspot minima has long been recognized and cycles of 11 years, 22 years, 44 years, and so on are now generally admitted. Although coincidences of sunspot activity and surface weather are often striking at least for the longer periodicities, they cannot be proved to be causal. Many of them are "vague statistical similarities, which usually turn out to have a very low level of significance" (Tucker, 1964). However, much stress is laid today on variations in the intensity and latitudinal position of the main components of the general atmospheric circulation, which lead to changes in the rate of exchange of air between the tropics and polar areas. It is often suggested that some of these variations show associations with solar activity, and that if they were maintained for a sufficiently long period either on the positive or on the negative side of the mean they could acquire a considerable magnitude. The difficulty, however, is that the time-scales available in the Pleistocene seem decidedly short for such slow accumulative effects.

(2) *Short-term fluctuations in the nature of solar radiation*
Several meteorological hypotheses on climatic change involve a change in quality rather than in quantity of solar radiation. Short-term variations are known to occur in the solar spectrum, especially in the ultra-violet range and in the more complex cosmic radiation, expressed variously as X-rays, cosmic corpuscles, fluxes of highly energetic photons and so on, which increase during solar flares. Solar ultra-violet variability, especially in the 0.2 micron to 0.3 micron wavelength band, appears to influence the "photochemical equilibrium concentrations of ozone" in the upper stratosphere. Apparently also the relation of some of this selective emission to the earth's geomagnetic field might especially affect the polar regions. Yet it seems in fact that some recent strong solar flares have had very little effect on surface weather and that the earth's main ice-caps, particularly in Antarctica retain their glacial equilibrium and show an indifference to these spasmodic emissions. On the other hand, it has been argued that "irregular solar activity of the selective variety controls the entire complex of climatic cycles, including the major Pleistocene glacial–interglacial cycles" (Willett, 1953). Details of these short-term variations in the nature and quantity of solar radiation, and of their possible weather correlations, will be found in the symposium issued by UNESCO (1963).

In regard to all theories concerning short-term variations of solar radiation it must be noticed that they receive no significant help from modern measurements of the "solar constant" (insolation at the upper edge of the atmosphere) which appears (from observations made at the Earth's surface and involving problematical adjustment to conditions outside the atmosphere) to have remained almost constant at 1.95 cal cm^{-2} min^{-1} during the last few decades. However, it is often pointed out that a possible slight increase could reach some magnitude over a long period of time, especially if an accumulative or "feed-back" mechanism can be devised.

(3) *Medium-term fluctuations in solar radiation*
The meteorological theory of Simpson, already noticed, required fluctuations in solar radiation of a periodicity of about 400,000 years. This was modified later by Willett (1949), and Bell, to incorporate four solar

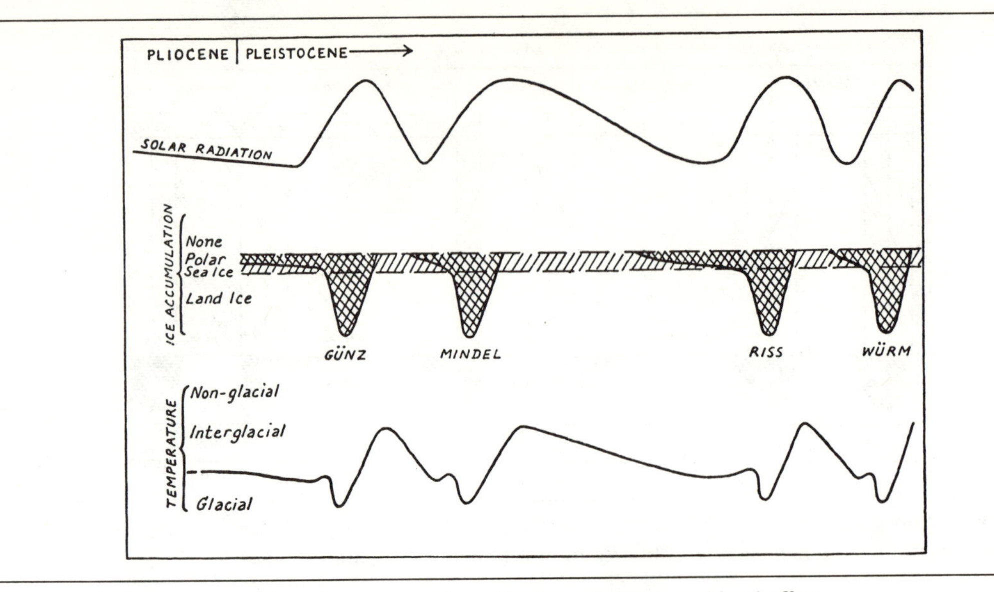

Figure 3 Solar variation curve and ice-advances and ice-retreats and their combined effect on air temperatures in polar latitudes as devised by Bell (1953).

radiation maxima during the Pleistocene with a periodicity of 200,000 years or less (Figure 3).

In 1957, Simpson re-issued his scheme and, while retaining his basal mechanism of 400,000 year cycles of solar-radiation increase, assumed three complete radiation oscillations since the onset of the Pleistocene and attempted to fit glacial and inter-glacial phases into them (Figure 2).

Recently, Kraus (1960) has pointed out *inter alia* that Simpson's suggestion that "surface air temperatures in all seasons at the Equator were 6° C higher than today at the maximum of a glacial phase . . . is in contradiction to the oceanographic evidence which indicates substantially colder ocean surface temperatures in the tropics at that time." Moreover, it appears that there are more plausible explanations to obtain the increased activity of the general atmospheric circulation postulated by Simpson, and these could be "associated with a lower mean temperature of the air—a condition shown to be almost certainly essential for the establishment of a glacial phase" (see Theories 7 and 9).

There seems to have been no notable theory based on decrease of solar radiation since that by Viete in 1950, which achieved the necessary precipitation by means of a steeper temperature gradient between the polar areas and the tropics which increased the atmosphere circulation and storminess in temperate latitudes. However, as Lamb points out, this idea is so much an inevitable implication of modern understanding of the general circulation of the atmosphere that the majority of meteorologists now believe in it in some form.

(4) *Long-term cyclical or periodic variations in solar emission*

The possibility of the sun being a variable star was suggested in the 1880's. This possible considerable variability in the sun's emission of heat was imputed to "fluctuations in the rate of shrinkage of its diameter, brought about by the unequal struggle between the diminishing amount of heat in the interior and the increasing force of the gravitation of its particles, and by the changes in the enveloping atmosphere (chromosphere) of the sun, which, like an enswathing blanket, arrests a large portion of the radiant heat from the nucleus, and is itself evidently subject to violent movements, some of which seem to carry it down to the sun's interior. Unknown electrical forces may also combine to add an element of variability" (Wright, 1893).

A recent theory on the sun's variability was formulated mathematically by Öpik (1953) who considers that fluctuations in energy generation in the sun's interior lead to peripheral expansion and a decrease of interior temperature. This self-regulating process may be accompanied by several minor reactory cycles before the sun recovers stability. The major occurrence appears to be at intervals of about 250 million years and to have a period of a few million years, but within it, shorter-term variations, if they exist, cannot be deduced by calculation. A fluctuation of solar emission of 9% from the present (average global temperature about 15° C) would reduce average global temperatures to 5° C; a rise of 9% would raise it to the "normal" global average of 22° C (Öpik, 1958).[3]

(5) *Long-term aperiodic increases in solar radiation*

The germ of the idea of aperiodic variations in the sun's heat dates back at least to Newton. Its modern adaptation suggests that at wide intervals of time (every 100 million years or so) the sun passes through nebulosities or clouds of cosmic dust or interstellar matter. During

[3] By normal is meant the average over all geological ages.

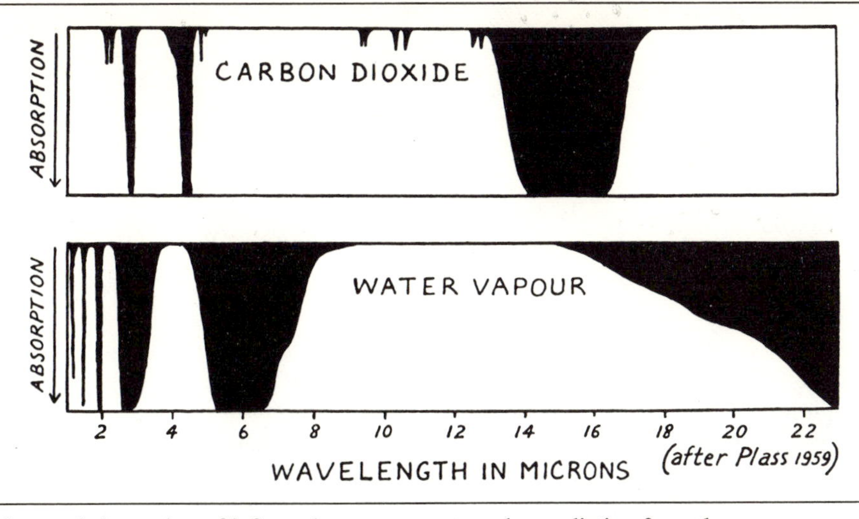

Figure 4 Spectral charts of absorption of infra-red waves, or outward re-radiation from the earth, by carbon dioxide and water vapor. The wide absorption band of carbon dioxide at 13 to 17 micron wavelengths coincides with wavelengths at which re-radiation from the earth is most intense and at which there is a gap or window in infra-red absorption by water vapor. (From Plass, Carbon dioxide and climate. Copyright © 1959 by Scientific American, Inc. All rights reserved.)

the passage, which is of the order of a million years, the cosmic particles are absorbed by the sun and their kinetic energy converted into heat. Thereby solar radiation is increased, mostly in the ultra-violet regions of the electro-magnetic spectrum. The more enthusiastic supporters of this idea suggest that "there is now incontrovertible astronomical evidence . . . that the radiation reaching the Earth from the sun will in the past have been subject to irregularities of increase . . . the probability is strong that an adequate first cause of the major climatic variations lies here" (Hoyle and Lyttleton, 1950). However, it should be noticed that recent ice-advances appear to have been associated with a decrease of global temperatures and so presumably also of solar radiation.

(6) *Long-term aperiodic decreases in solar radiation*
The inverse of the above theory was put forward in 1837 by Poisson, a French physicist and mathematician, who assumed that the Earth had at times moved through "cold regions" in space. It was soon suggested that these cold regions could, among other factors, have been caused by a relative lack of showers of meteorites, asteroids and other minor astronomical bodies falling upon the sun's surface. "While it is impossible absolutely to disprove this hypothesis, it labors under the difficulty of having little positive evidence in its favor" (Wright, 1893).

B. Theories Based on Variations in Atmospheric Transparency

The chief constituents of the atmosphere involved in this group of theories are carbon dioxide (CO_2), volcanic dust, ozone (O_3) and water vapor (H_2O).

(7) *Variations in carbon dioxide content of atmosphere*
Changes in the CO_2 content of the lower atmosphere were advocated by Tyndall in 1861 and later, for example, by Chamberlin (1899) in his famous "attempt to frame a working hypothesis of the cause of glacial periods on an atmospheric basis". Today it is known that CO_2 is mainly transparent to short-wave solar radiation but absorbs strongly in the long-wave terrestrial radiation band from 13 microns to 17 microns and that an increase of it would, other things remaining equal, raise air temperatures and vice versa (Figure 4). The problem is to assess the probability of and effectiveness of CO_2 changes in the atmosphere where at present it forms just over 0.03% of the total gaseous content.

The probability of large changes in the CO_2 content of the atmosphere is doubted by most scientists because the Earth's carbon cycle is controlled largely by CO_2 absorption by the oceans which form a vast reservoir for carbon compounds. As absorption of CO_2 by the hydrosphere is mutually adjusted to the CO_2 pressure in the atmosphere, in order to double the CO_2 content of the atmosphere much more than double the amount present in it would have to be added.

There is, of course, the probability of time-lags in the Earth's carbon cycle. A short lag (of the order of a few years) may occur due to slowness of horizontal exchanges, for example of transporting into the southern hemisphere an excess of atmospheric CO_2 in the northern hemisphere. A longer lag may occur due to the great depth and volume of the oceans which might require "a thousand years or more for the balance between calcium carbonate, calcium bicarbonate and atmospheric carbon dioxide to be struck" (Matthews,

1959). Others postulate up to 50,000 years for the achievement of equilibrium conditions in the atmospheric-hydrospheric carbon cycle.

The idea of the effectiveness of CO_2 as a factor in climatic change was revived recently because a slight increase in world temperatures appeared to be due "to back radiation from the carbon dioxide produced by fossil-fuel combustion" (Callendar, 1961). The effect of CO_2 on back-radiation (that is downward infra-red flux) and so on surface temperatures was calculated by Plass (1956) who concluded that if the CO_2 content of the atmosphere were doubled or halved the Earth's surface temperatures would change by about $+3.0°$ C or $-3.2°$ C. Plass went much further and evolved the first modern theory of climatic change based on CO_2 variations. According to his hypothesis, after the Earth's carbon dioxide supply has decreased slightly,

(*a*) Global temperatures fall and ice-sheets form, so reducing the volume of the oceans by between 5 and 10%;

(*b*) As these ice-sheets contain relatively little carbonate matter, they indirectly increase the CO_2 concentration in the reduced oceans which in their turn eventually, by transference, increase the CO_2 content of the atmosphere;

(*c*) Thereupon the atmosphere warms, the ice-sheets melt and the increase in oceanic volumes upsets the hydrospheric-atmospheric CO_2 equilibrium. When this equilibrium is again achieved the CO_2 cycle returns to (*a*) and begins anew, the periodicity being about 50,000 years.

It will be noticed that the primary cause leading to an unbalance of the Earth's carbon cycle is unknown and that the hypothesis is largely based on a secondary mechanism dependent upon variations in terrestrial geography. It seems that "no geological factors are known which could have influenced the CO_2 balance so considerably as to provide a possible explanation of the fundamental characteristics of palaeoclimates" (Schwarzbach, 1963).

Recently, Kondratiev and Niilisk (1960) re-calculated infra-red radiation effects and showed that if other feedback mechanisms, such as water-vapor, are also considered, the thermal influence of CO_2 is much less than supposed. Thus the CO_2 theory in any form seems inadequate as a cause of major climatic changes. But if, as is generally admitted, the recent slight increase in CO_2 concentration has contributed slightly to the relatively recent warming trend of surface temperatures (Godson, 1963) there seems no great difficulty in devising terrestrial geographical changes in the past which may have had, through upsets in the Earth's carbon cycle, relatively short-term negative or positive thermal effects of a small amplitude. The danger seems to be that because the influence of CO_2 variations is relatively small the importance of more potent mechanisms, such as water-vapor, cloud and the snow-cover albedo, will tend to cause it to be ignored. However, Kraus (1960) thought that a low value of the CO_2 mixing ratio would cause a high tropopause, more intense infra-red cooling and hence heavier rainfall in the tropics and relatively cold surface temperatures at high latitudes. He thought CO_2 a stronger climatic-change influence than was ozone.

(8) *Variations due to volcanic aerosols*

The correlation between volcanic eruptions and decrease in surface air-temperatures due to the screening effect of dust-veils in the lower stratosphere seems established by the marked decrease of direct solar radiation that has occurred at many observatories after volcanic explosions. However, today it has been shown that, unless the vulcanism is widespread and persistent, this screening effect is regional or zonal rather than hemispheric or global (Gentilli, 1948) and is decidedly ephemeral, being restricted at most to within 10 years of any eruption. The great eruption on the island of Bali on 17 March 1963 caused a sharp drop of about 5% in the direct solar radiation at Pretoria, South Africa, between 20 and 22 April. But the diffuse sky radiation increased distinctly. "In both cases the difference amounted to about 4 cal cm^{-2} hr^{-1} on a horizontal plane. However, there was no remarkable change in the total or global radiation. This new state continued in May and June" (Burdecki, 1964). It might be, however, that in climates cloudier than Pretoria's much more of the diffuse sky-radiation would fail to reach ground-level.

The volcanic-dust theory faces many other difficulties. The matter ejected by geysers and violent volcanoes consists largely or partly of water-vapor and carbon dioxide, and in fact excessive vulcanism is occasionally postulated as a means of upsetting the balance of the Earth's carbon cycle. Many volcanoes spend most of their operative years beneath the sea (Zapffe, 1954) and many, particularly as happened in Tertiary times, are not excessively explosive. Yet a period of active orogenesis and epeirogenesis would probably be associated with increased vulcanism which might be sufficiently sustained or frequent to have an accumulative global effect.

The volcanic-veil theory is, as stated above, based essentially on the violent ejection of fine dust into the stratosphere where it would interfere with the transmission of solar radiation to the lower atmosphere. The dust would settle within 1 to 10 years of ejection and its movement would tend to have a slow net poleward drift. The quantity required is not great and Humphreys (1940) calculated that the amount needed to cut solar radiation by 20%, "if discharged by volcanoes each year for 100,000 years would make a layer of dust only one fiftieth of an inch thick" (Wexler, 1952). This may satisfactorily explain the frequent absence of detectable ash-layers in deep-sea sediment cores and in deep ice-cap bores but in fact it virtually reduces the dust-veil theory to impotency. A solar radiation reduction of 10% would be ample for our purposes; 10,000 years or 15,000 years would cover an ice-advance; the dust-screen then needed according to the above calculation would be so fine that diffuse radiation from its particles would presumably be mainly downward. In brief, it appears that volcanic aerosols cause appreciable spasmodic decreases in regional and zonal air temperatures but, even allowing for slight repercussions on world atmospheric circulations, they seem quite incapable of achieving the status of a major factor in climatic change. They may, however, be important in the secondary and minor fluctuations of climate if eruptions are frequent enough for cumulative effects to be built up.

(9) *Climatic changes due to variations in the ozone layer*
The relative concentration of ozone in the upper stratosphere at 30 to 50 km has a very important screening effect on solar radiation. It absorbs about 5% of the total radiation and so deprives the troposphere or surface atmosphere of most of the ultra-violet radiation. Ozone absorbs strongly also in a "narrow band centered about 10 microns, located in the important 'window' region of the water-vapor spectrum" (Wexler, 1953). Consequently, changes in the concentration and height of the ozone greatly affect incoming solar radiation (by short-wave absorption) and slightly affect outgoing or terrestrial radiation by infra-red (10μ) absorption.

Attempts to correlate sunspot cycles with variations in total atmospheric ozone have been criticized by meteorologists who consider that total ozone amounts are determined by "an equilibrium of ultra-violet-dependent photochemical reactions" (Mitchell, Dobson, Normand, 1962). It seems that possible absolute variations in the ozone layer must be associated with variations in solar radiation (theories 1–6). The latest suggestions (Godson, 1963) are that during periods of increased solar activity, solar radiation is enhanced mainly near the 0.2 micron wavelengths (that is towards the shortest wavelengths). This increases the concentration of ozone and so leads to greater absorption of ultra-violet in the ozonosphere. However, the interactions between this increased absorption and surface air conditions are extremely complicated, particularly as some form of stratospheric circulation is involved, but a general increase in surface air temperatures might be expected in summer and an increase in high latitudes in the winter half-year.

It will be noticed that this ozone-absorption variation is a secondary product of variations or changes in solar emission, including the possible accretion by the sun of "interstellar matter" or cosmic dust. In view of Godson's suggestions it seems that the inverse condition (decreased ozone concentrations) would lead to lower global temperatures because of lessened absorption and in spite of lessened screening effect.

Kraus, as already noticed, preferred to substitute for Simpson's solar-radiation cycles a scheme whereby with more or less constant direct solar radiation the air temperatures fluctuate because of changes in "the infra-red cooling rate" of the atmosphere due to changes in either the O_3 or the CO_2 mixing ratio or in both. "A decrease in the stratospheric ozone mixing ratio would tend to lower the temperature and raise the height of the tropical tropopause"; this might increase tropical rainfall and by reducing "the amount of heat available for export to extra-tropical latitudes . . . would be conducive

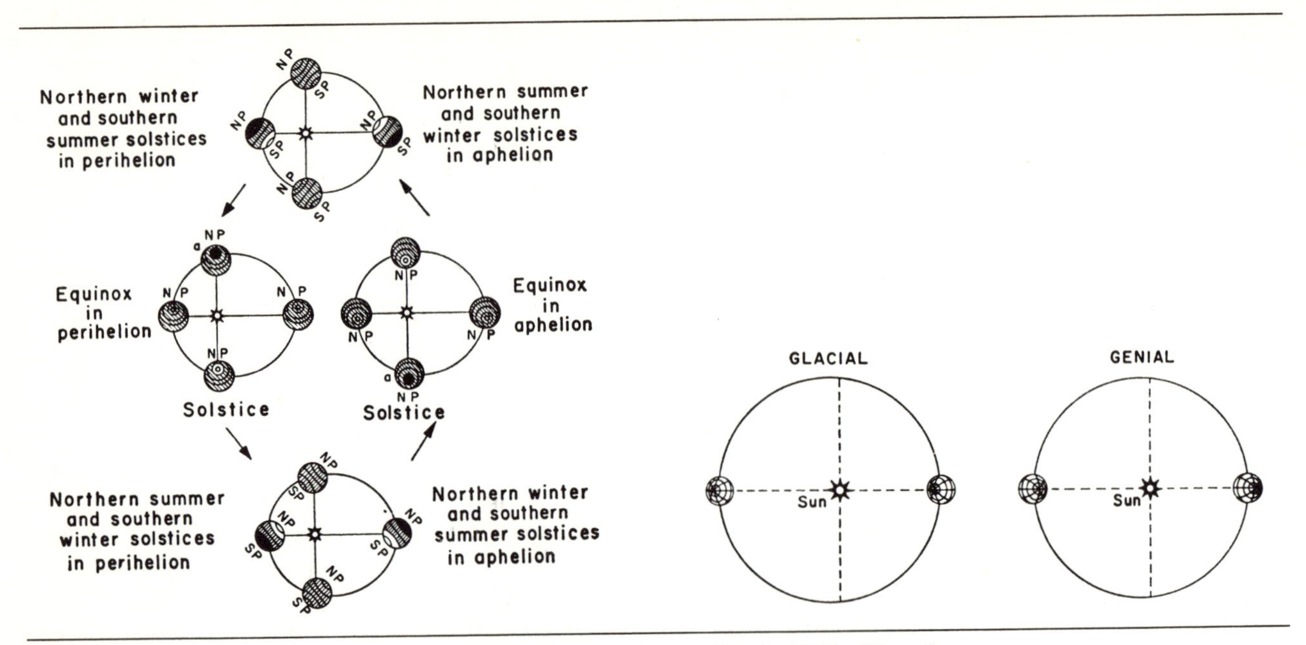

Figure 5 Diagram to illustrate the precession of the equinoxes (Lyell, 1872). Here the difference between aphelion (94,500,000 miles) and perihelion (91,500,000 miles) is greatly exaggerated. This exaggeration still persists in modern textbooks.

Figure 6 Diagram to illustrate the combined effect of eccentricity of the Earth's orbit and the precession of the equinoxes (Ball, 1891). On left, maximum eccentricity coincides with winter in aphelion in northern hemisphere giving glacial conditions there; on right about 10,500 years later, maximum eccentricity with winter in perihelion in northern hemisphere gives genial climates there. Maximum eccentricity was assumed to be considerable and rarely achieved.

to colder conditions there". Thus variations in the ozone concentration might cause relatively short climatic fluctuations with minor amplitude. It seems rather more doubtful whether they could account by themselves for the sequence of glacial and interglacial stages (1960).

(10) *Variations in water vapor content of the atmosphere* Attempts to base climatic change on variations in the water vapor content of the atmosphere, as distinct from temperature-induced changes, date back at least to the 1880's. The latest suggests that "any widespread application of an atmospheric chemical or physical process which would hinder the mechanism of precipitation" would result in major changes of climate. The author, however, admits that "there may be better ways of making a glacial period" (Workman, 1962).

C. Theories Based on Changes in the Earth's Geometry

In 1688, Hooke suggested that warm palaeoclimates might be due to changes in the inclination of the Earth's axis. Known variations in the orbital position of the Earth's equinoxes (equator) and solstices (tropics) later evoked the climatic-change theory of Adhémar in 1842 and its elaboration by Croll (1875) and subsequently by many others. Today many variables in the Earth's geometry are recognized but by far the three chief are:

(i) the eccentricity of the Earth's orbit, with a periodicity of about 92,000 years and a mainly tropical climatic effect;

(ii) the precession of the equinoxes, with a periodicity of about 22,000 years (Figure 5); and

(iii) variations in the obliquity of the plane of the ecliptic, with a significant periodicity of about 41,000 years, and a strongly polar climatic effect. The obliquity, today about $23\frac{1}{2}°$, has in the past varied periodically from $21\frac{1}{2}°$ to $24\frac{1}{2}°$ (Figure 6).

Taken either in isolation or combined, as is usual today, we shall label them theories 11 to 14. Their combined or total direct effect on insolation (theory 14) was calculated by Milankovitch (1941) and the resultant thermal curves have, in various forms, been correlated, with due allowance for the time-lag involved in terrestrial mechanisms, with the nature of deep-sea sediment cores by Emiliani (Figure 7) and others and with recent changes of sea-level by Fairbridge (1963) and others. It seems that the direct correlation between geometrical curves and ice-advances postulated by Zeuner (1945) and others, lacks accuracy, because of non-allowance for time-lag mechanisms. The main criticisms of the planetary geometry climatic theory have been summarized by Flint (1957).

Recently Woerkom (1953) supplied certain refine-

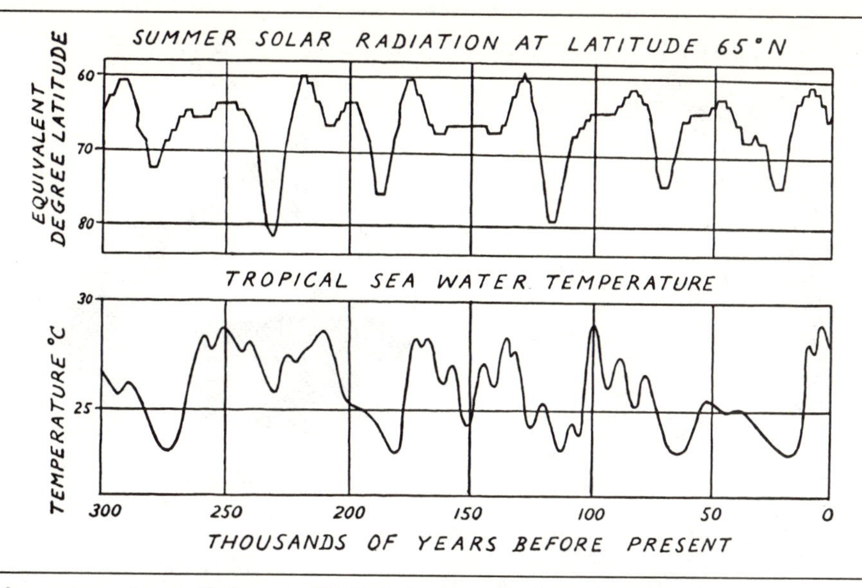

Figure 7 Curves of summer solar radiation at 65° N latitude and of tropical seawater temperature. The solar radiation is expressed as apparent shifts in latitude; thus a poleward extension of the curve indicates a relative drop in summer solar radiation to the amount experienced today at the latitude indicated. (After Emiliani (1958) and Woerkom (1953).)

ments to Milankovitch's calculations and concluded that although the climatic or insolation effects of the combined geometrical variations are "insufficient to explain the periods of glaciation, nevertheless they must have had their consequences". Most meteorologists agree that the direct effect on global temperatures would be only in the magnitude of 1° to 2° C yet this is the only certain factor we have so far discussed that has a periodicity comparable with that of the ice-advances and -retreats during the later Quaternary period. Irrespective of its magnitude it must be included at least as a minor cause in all theories of climatic change.

One of the most striking uses of the climatic effect of the Earth's geometrical variation has been made recently by Emiliani and Geiss (1957), in a theory to which we will refer again later. They show that the combination of the three main geometrical variations causes a variation in semi-annual insolation with an approximate average periodicity of about 41,000 years in middle and high latitudes and 21,000 years in low latitudes. Although the yearly insolation remains nearly constant, summer and winter insolation vary in an opposite way and probably the summer insolation is more critical than that of winter. These astronomical or geometrical effects are considered to form the trigger-action which sets off glaciation which itself then soon begins to dominate the climatic change. The authors tend to play down the geometrical variations because their effects have not been recognized in times of warm climates. But it seems highly probable that these 40,000-year periodicities will never be strongly reflected in biological and geological conditions when the global climate is warm. Only when climates reach a stage when vegetation existence is threatened and critical physical changes such as freezing occur are such small variations likely to instigate great changes. A large increase in the Earth's albedo due to expansion of ice-sheets and snow-cover forms a powerful adjunct to the geometrical-insolation effect.

D. Theories Based Mainly on Changes in Terrestrial Geography

Theories of climatic change based on alterations of the Earth's surface are excessively complex as the combinations, known and possible, are numerous. Yet probably the geological or geomorphological theories are far from being fully developed. Only the most recent and more important will be summarized here.

(15)*Changes in the horizontal arrangement of landmasses*
Continental drift and similar horizontal movements of landmasses will not explain the wide variations with time of climates during the 800,000 or 1,500,000 years of the Quaternary epoch. Yet the postulation of increased climatic continentality, so well argued by Brooks, is not entirely irrelevant to Pleistocene problems. When the whole of a polar region poleward of about 50° or 55° latitude, ocean and continent alike, is covered with ice or snow, winter continentality becomes excessive while summer continentality remains relatively unimportant. This effect, however, is subsidiary to the primary cause of the frigidity.

(16) *Shifts in the position of the polar axis*
By omitting the influence of horizontal or lateral migrations of continents, it might be thought that the possibility of influence by shifts in the polar axis was

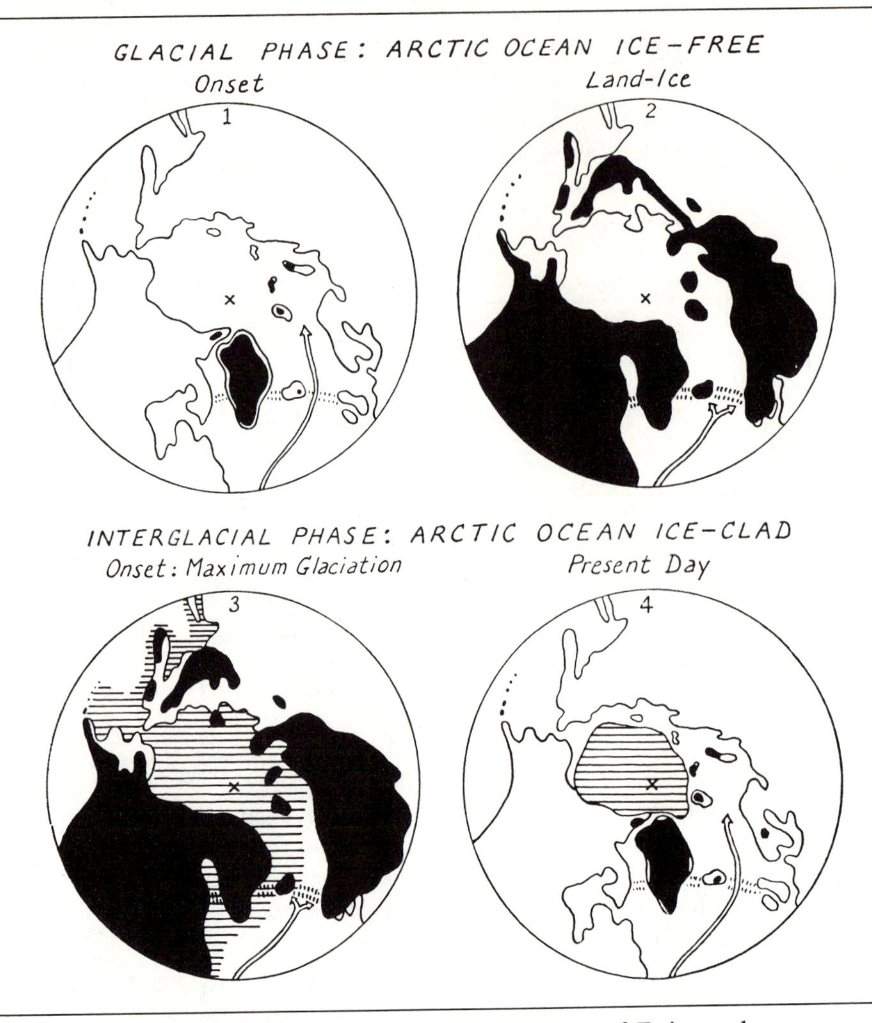

Figure 8 Diagrammatic interpretation of four stages in the glacial theory of Ewing and Donn. Land ice is shown in black; sea ice by horizontal shading; the warm North Atlantic surface drift by arrows; and the Davis-Iceland-Faroes submarine ridge by hachures. (From *Climates of the Past* by Martin Schwarzbach, © 1963, Litton Educational Publishing, Inc., reprinted by permission of Van Nostrand Reinhold Company.)

also automatically eliminated but geophysicists are not so easily ignored. As early as 1908, Simroth postulated a "pendulum oscillation" of the Poles with resultant changes in the oceans and in global climates. Recent literature is copious and stimulatingly self-contradictory. For example, in 1962 Jardetzky postulated a mobile interior to the Earth that allows a solid outer shell to glide over an internal yielding layer so causing polar-shift, while Lamar and Kern (1962) suggest that "phase changes in the upper mantle may provide a mechanism to prevent polar wandering". Yet these geophysical disagreements concern us, as in 1956–59 Ewing and Donn formulated a theory of ice-ages which included a mechanism to explain the periodic ice-advances and -retreats in the Pleistocene (Figure 8).

They suggest that the Quaternary glacial began when the Poles shifted into their present thermally-isolated locations. Thus the North Pole migrated into an ice-free Arctic Ocean warmed by water-exchange with the Atlantic, thereby providing the ideal meteorological conditions for large snowfalls and ice-accumulation on lands near the Arctic sea. On these peripheral land-masses, glaciers expanded and increased the Earth's albedo sufficiently to lower mean global temperatures and to encourage the further expansion of ice-sheets. Ultimately so much water was locked up in the land-ice that sea-level fell and the exchange of water over the Arctic–Atlantic (submarine) sill was considerably reduced. This reduction led to warming of the Atlantic oceanic circulation whereas the Arctic froze and thereby reduced Polar precipitation to such an extent that the glaciers dwindled and sea-level rose. Thereupon the influx of warm water into the polar sea increased, the Arctic ice-pack itself melted and the whole cycle (snow accumulation—ice advance—drop in sea-level, etc.) began anew. Thus the peculiar, largely-enclosed nature of the Arctic Ocean is thought to have provided an essential element of the mechanism required for periodic ice-advances and -retreats. The direct part of the mechanism is the small amount of relatively warm ocean water that passes poleward over the Faeroe–Icelandic submarine sill; the trigger mechanism for the climatic change is, of course, the shift of the Poles, which presumably has ensured that the periodic advance and retreat of ice will continue until the Poles migrate elsewhere.

This interesting scheme demonstrates two of the main problems confronting many modern theories on climatic change: namely, the need to account for thick ice-accumulations, and the difficulty of raising regional geographical mechanisms into global factors. A regional viewpoint immediately draws attention to global or antipodean complications. Thus Antarctica, much of which lies today at over 8,000 feet, would presumably have generated vast ice-sheets before the peripheral Arctic lands had achieved minor glaciers. Evidence from deep-sea sediment cores supports the view that "a secular deterioration of climate, with superimposed temperature oscillations leading to glaciation in Antarctica, took place during the late Pliocene" (Ericson, Ewing, and Wollin, 1963). In this case, the south polar ice-sheets presumably had lowered sea-level by 30 meters or even by 50 meters before the Arctic came into play. The surface water of the poleward-trending branch of the North Atlantic Drift thus becomes a less powerful mechanism to control the large and deep Arctic Ocean.

Similarly the need to explain ice-accumulation and ice-movements in territories now free of ice, such as northern Canada, often leads to geographical difficulties. Ewing and Donn mention the "unexplained glacial conditions which have continued in Greenland" and which contrast very sharply with the present ice-free condition in Northern Canada at the same latitudes. The Greenland ice survival is imputed to feeding from moist Atlantic airmasses, and the north Canadian ice-free condition to lack of precipitation from dry airmasses from off the frozen Arctic Sea. But a simple geographical survey reveals that the Greenland ice-sheet, although preserved in a saucer-shaped topography, is buttressed by mountains 6,000 feet and in many areas over 7,000 feet high. Much of it is in fact *above* the snow-line. Were northern Canada at these altitudes would it not eventually accumulate an ice-cap? As will now be shown, increase in the height of landmasses forms the main basis of some of the most popular twentieth-century hypotheses on climatic change.

(17) *The catastrophic-orogenic climatic theory*

Theories of climatic change based on changes of landmass altitude and of relief take many forms. Some include ingenious horizontal and vertical re-arrangements of mountain barriers and submarine oceanic sills, including for example a direct ocean connection between the Atlantic and Pacific across the isthmus of Panama. A popular recent theory is based on the belief that each main period of orogenesis, such as the Tertiary Alpine uplift, was followed after a lag of a million years or so by an ice-age. The sudden mountain-building phase was considered to chill global temperatures mainly by increasing the loss of terrestrial radiation direct into space ("holes in the glass-house"). The outward radiation losses were increased by reflection from the extended snow-cover and from the upper surface of the extra cloud-cover. The total cooling due to the various orographic influences was calculated by Brooks to range from $2\frac{1}{2}°$ C up to 6° C, but he also emphasized that orogenesis cannot be dissociated from interference with atmospheric and oceanic circulations.

(18) *Uniformitarian-orogenic climatic theories*

The meteorological influences discussed above apply equally to the uniformitarian approach to continent- and mountain-building. This considers elevation of landmasses to be the outcome between wide sub-crustal epeirogenic forces, more localized orogeneses and the opposing forces of denudation. The factors include isostatic depression due to increased crustal load, i.e., by ice-sheets, and isostatic uplift due to decreased crustal load. In geological history there appear to have been periods of strong oceanogenesis or weak orogenesis and these phases of extensive oceans and low landmasses seem to have been warm.

On the other hand other geological periods have witnessed long spells of relatively active mountain-building. Thus in the Tertiary period during a span of 20 or 30 million years the general altitude of the continents was raised from about 300 to 800 meters. Local orogenic tracts were raised by thousands of meters and were kept elevated because sub-crustal uplift exceeded erosion as is happening for example in many highland areas today (Schumm, 1963) but these orogeneses have a long time-scale and need not be synchronous throughout the world. Moreover, the elevation of landmasses in the critical zones poleward of about 45° latitude will be of most significance in theories on climatic change. Flint (1957) shows that all the highlands related to existing and Quaternary glaciers underwent considerable uplift in late Pliocene and early Pleistocene times. He goes on to suggest that the Pleistocene ice-sheets can be satisfactorily explained by the present relief pattern and present general atmospheric circulation if a sufficient lowering of temperature occurs. Only when land uplifts were "unusually high and widely distributed, especially in regions traversed by westerly winds, could small fluctuations of solar energy succeed in reducing temperatures enough to bring about the building of great glaciers". On this argument it appears that an increase of altitudinal and relief influence that had been in progress for about 30 million years, and presumably was partly or largely responsible for a lowering of bottom seawater temperatures by about 8° C since the mid-Oligocene (Emiliani, 1955), reached critical proportions just before and during the Quaternary era.

(19) *Insolation-topographic theory*

In 1957, Emiliani and Geiss developed an elaborate climatic-change hypothesis similar in essentials to the above. Glaciation is explained as a result of topographic uplift being influenced by a varying astronomical cause, combined with a varying heat exchange between the oceans and the continental ice-sheets and with certain time-lag effects due to plastic deformation of the ice, crustal warping, heat absorption by ablation and so on. The astronomical cause could be either fluctuations in solar emission, which lacks observable basis, or variations in summer insolation which do occur and appear to have a phase (40,000 years in tropics and 21,000 in polar areas) in agreement with much recent geological evidence. The latter astronomical cause is preferred but "the question is open to further investigation" and the terrestrial factors should apply to either.

Emphasis is laid on the importance of snow-cover and ice-caps which having once reached a certain size expand mainly because of their own effect on albedo, atmospheric circulations, regional temperatures and sea-level. Inevitably the problem arises of how to accomplish the eventual removal of the ice-sheets during interglacial phases. This is done almost independently of the geometrical cause by means of isostatic depression of the Earth's crust under the ice-sheets coupled with flattening of the ice-caps due partly to evaporation exceeding precipitation over their anticyclonic centers and partly to thinning under expansion due to the ice's own weight.

These various topographic climatic theories, with their many complications, have encouraged much modern research. The topographic factor is now known to overcome at least partly the need to produce an appreciable snowfall in a period when reduced temperatures will tend to decrease precipitation in cold latitudes. Small altitudinal changes can cause a striking increase in precipitation in cooler latitudes where topographic obstructions enhance other precipitation-forming processes. Given sufficiently high topographic barriers in, for example Scotland and Scandinavia, there seems no difficulty in accumulating tremendous ice-sheets within a few millennia. In very cold areas it has always been assumed that relief had a negligible influence on precipitation but Rubin and Giovinetto (1962) show that increase in slope or heightened relief combines with atmospheric features to produce an appreciable increase in snow accumulation even in the glacial climate of central West Antarctica.

The complicated relations between mountain barriers and air-flow and precipitation amounts have evoked as yet only a meager literature but enough has been written to emphasize that rapid ice-cap growth would at a fairly early stage become mainly peripheral and that heavy snow-accumulations on windward and warmer margins may have adverse effects on leeward slopes where lee-wave eddies or descending airflow may at times cause evaporation. Some of the anomalies of ice-movement and of the peculiar tracts that appear to have had little ice or snow may perhaps be explained by "foehnization."[4] Also, it seems certain that not enough attention

[4] That is by the effect of airflow warmed by descent and so capable of appreciable evaporation.

has been paid to peripheral growth on the warmer edges of the ice-sheet where moist airmass invasions are possible. Thus Leighley (1949) considers that an anticyclonic airflow over an ice-sheet in north-eastern North America would help to extend the ice westwards whereas a similar type of air circulation over a north-western European ice-sheet would hamper rather than aid its growth towards the continental interior.

Hypotheses based partly or largely on topography and land altitude have been aided by modern measurements of isostatic rebound and changes of sea-level. The concept of isostatic depression due to the weight of ice-sheets was used by Upham in the 1890's as an important mechanism in an orographic climatic-change theory. Today the maximum amount of regional depression is generally reckoned at between one-third and one-quarter of the ice-thickness and some form of peripheral upwarping is associated with it. The time-lag between excessive crustal load and sub-crustal adjustment is probably of the order of ten thousand years. In Fennoscandia the maximum isostatic rise since the final ice disappearance about 6,800 BC is estimated at 250 meters. Theoretically a further 200 meters rise should occur before isostatic equilibrium is completed but the early rapid uplift (maximum up to 50 cm/yr at head of Gulf of Bothnia) has now slowed down to a maximum of less than 1 cm/yr. The total recovery time has been calculated at 18,000 years. Recently Farrand (1962) has drawn rebound contours for the Laurentian Shield which also show a decreasing rate of uplift from the time of deglaciation (about 12,000–8,000 BP) to the present. Abundant geological evidence demonstrates uplifts of 150–220 meters in arctic Canada and some raised shorelines are 300 meters above present sea-level.

Thus a significant minor topographical periodicity must be introduced into glacial phases. As the ice-sheet accumulates, after 10,000 years or so isostatic depression begins and eventually reaches about one-quarter of the ice-thickness, which itself is largely dependent on the diameter of the ice-sheet except in mountain-enclosed basins. The time comes when the widening of the ice-sheet and its tendency to create relatively dry anticyclonic airmasses, reduces the central snow accumulation until it is no greater or less than the isostatic depression. At the same time peripheral upwarping may slightly complicate the general regime of the ice-cap. A few thousand years after the ice-sheet begins to dwindle the isostatic rebound commences and eventually outstrips the eustatic rise of sea-level due to ice-melt. But for many millennia the land mass locally is far lower than at the onset of glaciation. Such isostatic mechanisms, involving up to 500 meters vertical depression, must be especially important in the regimes of ice-sheets superimposed on convex or flat topographies which favor ice dispersion but their total vertical effect will be greatest in concave topographies, such as Greenland, where they encourage ice-accumulation and longer ice-survival.

Post-1950 research has also greatly assisted one other aspect of the topographic effect of ice-sheets—their lowering of global sea-level. Sonic seismic soundings indicate that the main ice-caps, especially in Antarctica are much thicker than hitherto supposed. These soundings and other geological evidence have allowed the following re-estimations of the total ice-volume and so of changes of sea-level involved: (Donn, Farrand, and Ewing, 1962).

Present: Area 14.9 km^2 or 10% land surface; volume 34 million cubic km; equivalent sea-level lowering 86.5 meters or 284 feet.

Pleistocene maximum ice-advance: Area 45 million km^2 or 30% present land surface; volume 98 million cubic km; equivalent sea-level lowering 246 meters or 807 feet. The estimates, especially of the past, are rough but there seems no doubt that at the maximum ice-advances the sea-level fell about 500 feet below its present stand.

Thus today the topographic influence on temperature decrease can be expressed more strongly. Apart from the important increase in the mean elevation of the continents by about 500 meters in Cenozoic times and the active orogeneses of Pleistocene times, we can state another significant internal mechanism for altitude variations. During ice-advances, snow-accumulation increased continental heights regionally by 1,000 or 1,500 meters or more for the period before isostatic depression and ice-dispersion began actively to flatten the ice-sheets.[5] Simultaneously sea-level dropped by up

[5] For the probable rates of growth and shrinkage of ice sheets *see* Weertman (1962, 1964).

to 246 meters below its mid-Pliocene stand. The mean altitude increase for all continents exceeded 600 meters, with a temperature equivalent decrease of about 3° C or about 1° C for the whole surface of the globe. These amplitudes are small but as land-ice is concentrated poleward of 50° latitude this ice-sea vertical-change mechanism assumes a mean maximum magnitude exceeding 1,200 meters, or of over 6° C temperature equivalent decrease, for polar landmasses generally. This condition, however, is a gross underestimate of the total climatic effect which includes the great increase in albedo due to extensions of ice-sheets and snow-cover. The fact seems to be that there is little difficulty in assuming a tremendous ice-advance once some factor has raised land-heights or lowered regional temperatures sufficiently to allow ice-sheets to *expand* for a few millennia. The possibility that such global temperature changes could be accomplished by minor changes in the present-day atmospheric circulation if sustained over a long period forms the basis for the last climatic-change theory discussed here.

(20) *Built-in meteorological theories*
This ultra-uniformitarian approach depends largely on the evidence of complicated variations of climate within the period of recorded observations (Lamb and Johnson, 1959). It rests securely on the contention that small thermal and circulation changes can if extended over long periods have profound and wide-scale effects particularly if associated with "feed-back mechanisms," such as the ocean. Thus an extremely small annual deficit or surplus in the heat-storage of the tropical seas would eventually have a large effect on world temperatures. Mathematically these ideas are perfectly tenable but the prolonged time-scale required seems unfavorable.

Of the built-in climatic-change theory Godson (1963) writes "one is tempted to regard climatic change as a small residual of already small residuals, and therefore beyond the range of quantitative prediction or even qualitative explanation" but in the light of considerable short-term fluctuations in the general circulation of the atmosphere "we need search only for factors that could modify the energy input of the general circulation over a large area and inquire whether these factors could persist for long periods of time". Godson, however, represents the modern ultra-meteorological viewpoint. He rules out topography as a variable parameter in the Pleistocene and includes solar ultra-violet mechanisms as internal. From one such internal mechanism, changes in snow- and ice-albedo, he develops "a self-sustaining mechanism, capable of producing widespread glaciation". In this connection it is interesting to notice that Brooks also stressed the strong lowering effect on air temperatures of increased ice-cap albedo. He estimated that if a sea-covered pole was chilled from just above to just below its freezing point (−2° C), the ultimate result would be an ice-cap of 25° latitude or 1,500 miles radius and a final drop of temperature at the pole of −27° C (Schwarzbach, 1963). It is generally recognized that this is an overestimate but the real difficulty with the use of snow-albedo and most other "built-in" mechanisms is that they are secondary factors. As Sutcliffe (1963) affirms, "it would seem impossible on the face of it for the internal variability of the atmosphere–ocean system to account for the return of permanent ice-fields when these had been absent for millions of years, and other factors, terrestrial or extra-terrestrial, must then be introduced".

Elementary Theory of an Ice-age

The factors, primary and secondary, available for use in a climatic-change theory may be crudely summarized as:

Long-term variations

(1) Fluctuations in solar emission, due to atomic reactions of sun: at intervals of perhaps 250 million years, with periodicity of few million years. Perhaps shorter periodicities.

(2) Fluctuations in solar emission due to changes in sun's cosmic environment (cosmic dust, etc.).

(3) Earth's continental–oceanic crustal vertical movements (epeirogenesis; orogenesis; oceanogenesis), which combine with

(4) Regional denudation or geomorphic cycles, the duration of which is variously estimated at 15 million to over 110 million years regionally. Probably 30 to 50 million years is common for some coincidence of wide regional peneplanation and over 100 million years for significant global coincidence of wide regional planations.

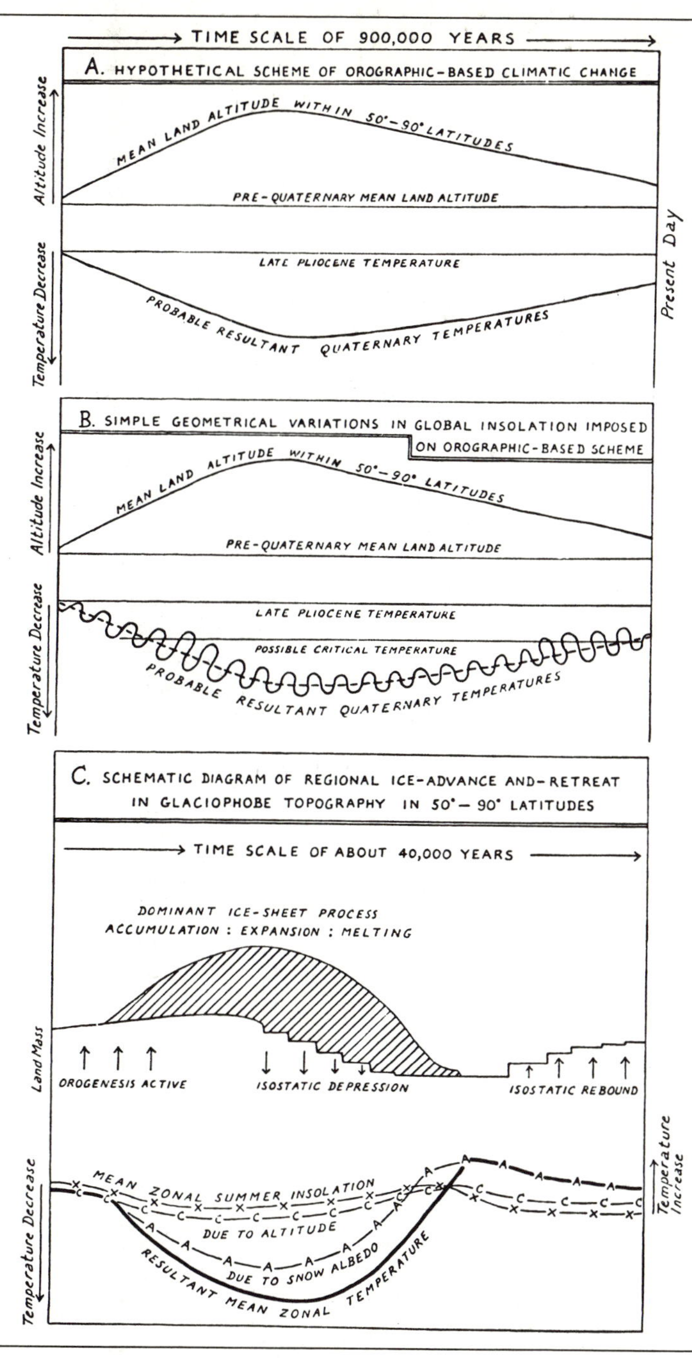

Figure 9 Schematic diagram to illustrate the probable effect on temperatures in 50°–90° latitude of: *A*, Changes in mean altitude of zonal land-masses; *B*, Proven minor climate oscillations imposed on *A*; and *C*, Of a single major ice-advance and retreat affected by the combined influence of changes in summer zonal insolation and in mean altitude of land-surface. Shading denotes ice-cap. (Compiled by R. P. Beckinsale.)

Short-term variations: observed qualitatively or quantitatively

(5) Geometrical, solar-insolation variations with significant periodicities of about 20,000, 40,000, and 90,000 years.

(6) Variations in carbon dioxide content of atmosphere (dependent on 3, 4, and 7) with a periodicity of 50,000 years.

(7) Isostatic and eustatic mechanisms dependent on ice-sheet formation (itself dependent on 1–6) with a periodicity of 30,000–50,000 years.

(8) Volcanic dust veils. Spasmodic and of doubtful periodicity, dependent on 3.

In general it seems necessary to accept that a short-term oscillation in solar-radiation effect due to changes in the Earth's geometry, is permanently imposed upon

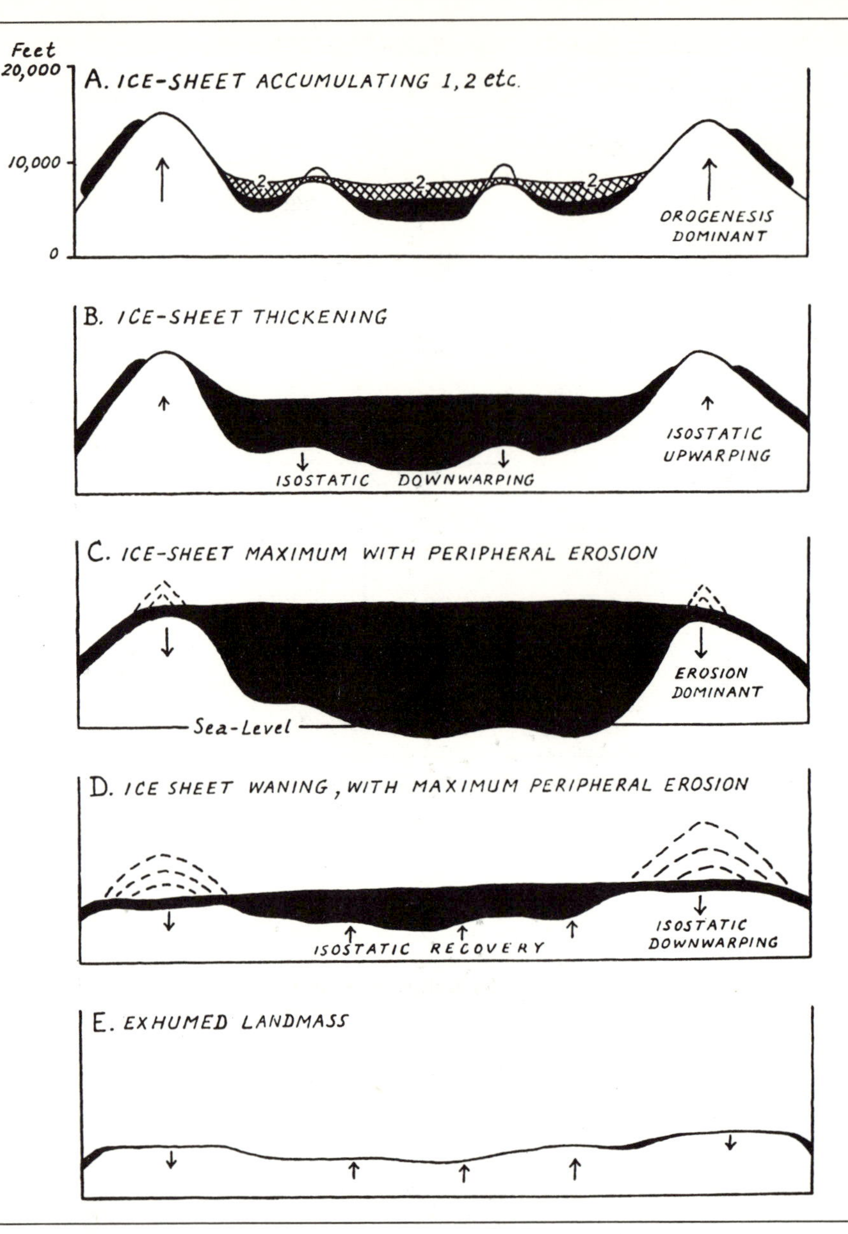

Figure 10 Diagram of possible ice-advance and ice-retreat in a polar glaciophile topography. In *C* and *D* prolonged erosion breaches and lowers the peripheral mountains so facilitating ice-drainage and the reduction of altitude and concavity. Time scale of the order of over 1 million years. (Compiled by R. P. Beckinsale.)

variable long-term oscillations in landmass altitude. When epeirogenesis and orogenesis dominate strongly over denudation, global temperatures fall; when oceanogenesis and low relief prevail global temperatures rise. But a great coincidence of lofty or mountainous relief in high latitudes can act as the equivalent of a large mean general rise in land relief and is probably essential for the formation of an ice-age. These two more or less "permanent" factors, geometrical and orographic, may or may not be superimposed on long-term oscillations in solar emission. In any event, when the general thermal swing is in warm stages, the short-term oscillations of solar-insolation are unimportant; when the swing is in cool or critical phases from a meteorological (and biological) point of view the short-term cycles become significant and act as trigger mechanisms. Similarly, as stated above, orographic effects may vary in potency, for example, according to their zonal or latitudinal location, and all active orogenic epochs are not associated with ice-ages.

Thus, if possible variations in solar emission are ignored, the main proven cause of an ice-age may be crudely expressed as in Figure 9*A*; the climatic oscillations might be generalized as in Figure 9*B*; and the broad details of a single ice-advance and -retreat in a

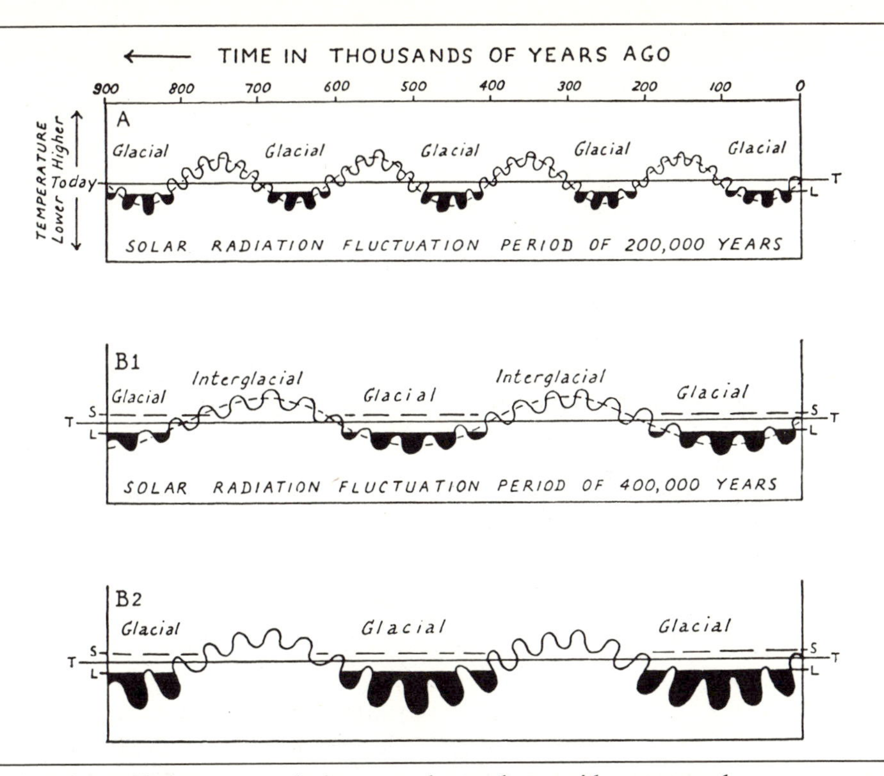

Figure 11 Schematic solar radiation curves during a cool-sun phase, with superposed variations in summer insolation at 65° N. *A* shows a solar radiation fluctuation period of 200,000 years and a summer zonal insolation variation period of about 20,000 years. In *B* the periods are 400,000 years and about 40,000 years respectively. In *B*2 the curve represents the resultant zonal air temperatures due to all periodic cosmic and geographic influences. S, marks critical temperature for formation of polar sea-ice; T, present-day temperature; L, critical temperature for ice-advance on land. It should be noticed that for the sake of simplicity the possibility of absolute changes in mean land-altitude (as in Figure 9, *A* and *B*) and in mean solar radiation has been ignored here. (Compiled by R. P. Beckinsale.)

glaciophobe (flat or convex) topography in latitudes 50°–90° as in Figure 9*C*. An ice-advance and -retreat in a glaciophile, or concave, topography would tend, other conditions being equal, to occupy a longer time-scale and would complicate the chronology of ice-retreats. Such a scheme for a lofty concave polar topography is generalized in Figure 10.

However, it will be readily apparent that the above schemes quite fail to explain the probable existence of *warm* interglacials during an ice-age. The geological evidence seems definitely to have established that times of maximum ice-advance were cold climatically and that at least some interglacials were warm. It happens that there are several theories which explain the possibility of ice-advances and of warm interglacials during increase of solar radiation but few or none which attempt simply to correlate ice-advances with solar-emission decrease and interglacials with solar-emission increase. Whereas Öpik's hypothesis of reduction in solar emission every 250 million years supplies an adequate primary cause, as an alternative or complement to topographic change, for an ice-age, his idea of minor solar-cycles during this cool-sun phase lacks, as yet, scientific basis. But if some such form of minor, short-term variations in solar emission do not occur, reduction of terrestrial relief by erosion is too slow a process to provide an alternative factor.

There seems no alternative but to accept the probability of short-term variations in solar radiation and to try to adapt a scheme such as that by Simpson (1957) or by Willett and Bell so as to correlate all ice-advances with increasing cold and interglacials with warmth. A problem then is whether to associate each solar-emission phase with two ice-advances or with one; whether the solar-phases should have a periodicity of approximately 400,000 years or 200,000 years. No doubt, the former would suit better a time-scale of 1,500,000 years. The theoretical solar-radiation scheme proposed in Figure 11 would be representative of the complexities of glacial and pluvial phases. The combined geometrical and terrestrial factors, with their secondary meteorological reactions would presumably act upon the minor oscillations shown and the climatic effects would be excessively complex.

The position of present time in the climatic-change cycle is not known, except that the present lies within a cool phase. Simpson (1959) avers that geological evidence shows clearly that "the Earth has recently passed out of a glacial into an interglacial period so that the present day is at the division between two phases. The existing small ice-cap at latitude 75° N will decrease and disappear during a cold-dry interglacial". But the fact is that the ice-cap is on the sea and probably increased cold would merely strengthen it. The present climate may be on either a rising or a falling curve of solar-emission. A small increase of present global temperatures would probably melt the Arctic Sea ice-pack but it would need either a large climatic amelioration or a long time-span to remove the thick ice-sheets of lofty Antarctica.

REFERENCES

Bell, B., 1953. Solar variation as an explanation of climatic change. In *Climatic Change*, ed. H. Shapley. Cambridge, Massachusetts. 123–136.

Brooks, C. E. P., 1949. *Climate through the Ages*. McGraw-Hill. 395 pp.

Burdecki, F., 1964. Meteorological phenomena after volcanic eruptions. *Weather 19*, 113–114.

Callendar, G. S., 1961. Temperature fluctuations and trends over the earth. *Quart. Journ. Royal Meteorol. Soc. 87*, 1–12.

Chamberlin, T. C., 1899. An attempt to frame a working hypothesis of the cause of glacial periods on an atmospheric basis. *J. Geol. 7*, 545–584.

Chorley, R. J., 1963. Diastrophic background to twentieth-century geomorphological thought. *Bull. Geol. Soc. Amer. 74*, 953–970.

Croll, J., 1875. *Climates and Time in their Geological Relations*. London.

Donn, W. L., Farrand, W. R., and Ewing, M., 1962. Pleistocene ice volumes and sea-level lowering. *J. Geol. 70*, 206–214.

Emiliani, C., 1955. Pleistocene temperatures. *J. Geol. 63*, 538–578.

Emiliani, C., 1956. Oligocene and Miocene temperatures of the equatorial and sub-tropical Atlantic Ocean. *J. Geol. 64*, 281–288.

Emiliani, C., 1958. Ancient temperatures. *Scient. Amer. 198*, 54–63.

Emiliani, C. and Geiss, J., 1957. On glaciations and their causes. *Geologische Rundschau 46*, 576–601.

Ericson, D. B., Ewing, M., and Wollin, G., 1963. Pliocene-Pleistocene boundary in deep-sea sediments. *Science 139*, 727–737.

Ewing, M. and Donn, W. L., 1956–59. A theory of Ice Ages. *Science 123*, 1061–1066; *126*, 1157–1162; *129*, 463–465.

Fairbridge, R. W., 1963. Mean sea level related to solar radiation during the last 20,000 years. In UNESCO, *Changes of Climate*. Arid Zone Research XX, 229–242.

Farrand, W. R., 1962. Postglacial uplift in North America. *Amer. Journ. Sci. 260*, 181–199.

Flint, R. F., 1957. *Glacial and Pleistocene Geology*. New York, 553 pp.

Gentilli, J., 1948. Present-day volcanicity and climatic change. *Geol. Mag. 85*, 172–175.

Godson, W. L., 1963. The influence of the variability of solar and terrestrial radiation on climatic conditions. In UNESCO, *Changes of Climate*. Arid Zone Research XX, 323–331.

Hoyle, F. and Lyttleton, R. A., 1950. Variations in solar radiation and the cause of ice ages. *J. Glaciol. 1*, 453–455.

Humphreys, W. J., 1940. *Physics of the Air* (3rd ed.). McGraw-Hill. 516 pp.

Jardetzky, W. S., 1962. Aperiodic pole shift and deformation of the Earth's crust. *J. Geophys. Res. 67(11)*, 4461–4472.

Kondratiev, K. Y. and Niilisk, H. I., 1960. On the question of carbon dioxide heat radiation in the atmosphere. *Geofisica pura e applicata 46*, 216–230.

Kraus, E. B., 1960. Synoptic and dynamic aspects of climatic change. *Quart. Journ. Royal Meteorol. Soc. 86*, 1–15.

Lamar, D. L. and Kern, J. W., 1962. Stability of the Earth's axis of rotation and phase changes. *J. Geophys. Res. 67(11)*, 4479–4484.

Lamb, H. H. and Johnson, A. I., 1959. Climatic variation and observed changes in the general circulation. *Geografiska Annaler 41*, 94–134.

Leighley, J., 1949. On continentality and glaciation. *Geografiska Annaler 31*, 133–145.

Lyell, C., 1875. *Principles of Geology* (12th ed.). London.

Matthews, M. A., 1959. The Earth's carbon cycle. *New Scient. 6(2)*, 644–646.

Milankovitch, M., 1941. *Canon der Erdbestrahlung* Kgl. Serb. Akad. Belgrade.

Mitchell, J. M., jr., Dobson, G. M. B., and Normand, C. W. B., 1962. On ozone. *J. Geophys. Res. 67*, 4093–4095.

Nairn, A. E. M. (ed.), 1961. *Descriptive Palaeoclimatology.* New York. 380 pp.

Öpik, E. J., 1953. A climatological and astronomical interpretation of the ice ages and of past variations of terrestrial climate. *Contributions from Armagh Observatory 9.*

Öpik, E. J., 1958. Climate and the changing sun. *Scient. Amer. 198*, 85–92.

Plass, G. N., 1956. The carbon dioxide theory of climatic change. *Tellus 8*, 140–156.

Plass, G. N., 1959. Carbon dioxide and climate. *Scient. Amer. 201*, 41–47.

Riedel, W. R., Bramlette, M. M., and Parker, F. L., 1963. "Pliocene-Pleistocene" boundary in deep-sea sediments. *Science 140*, 1238–1240.

Rubin, M. J. and Giovinetto, M. B., 1962. Snow accumulation in central West Antarctica as related to atmospheric and topographic factors. *J. Geophys. Res. 67*, 5163–5170.

Schumm, S., 1963. The disparity between present rates of denudation and orogeny. *U.S. Geol. Surv. Prof. Paper 454-H.*

Schwarzbach, M., 1963. *Climates of the Past.* Van Nostrand Co. 328 pp.

Shapley, H. (ed.), 1953. *Climatic Change.* Cambridge, Massachusetts. 318 pp.

Simpson, G. C., 1934. World climate during the quaternary period. *Quart. Journ. Royal Meteorol. Soc. 60*, 425–478.

Simpson, G. C., 1957. Further studies in world climate. *Quart. Journ. Royal Meteorol. Soc. 83*, 459–481.

Simpson, G. C., 1959. World temperatures during the Pleistocene. *Quart. Journ. Royal Meteorol. Soc. 85*, 332–349.

Sutcliffe, R. C., 1963. Theories of recent changes of climate. In UNESCO, *Changes of Climate.* Arid Zone Research XX, 277–280.

Tucker, G. B., 1964. Solar influences on the weather. *Weather 19*, 302–312.

UNESCO, 1963. *Changes of Climate.* Arid Zone Research XX. 488 pp.

Viete, G., 1950. Über die allgemeine atmosphärische Zirkulation während der diluvialen Vereisungs-perioden. *Tellus 2*, 102–115.

Weertman, J., 1962. Stability of ice-age ice caps. *U.S. Army Corps Engineers, C Region's Research Report 97*, 12 pp.

Weertman, J., 1964. Rate of growth or shrinkage of non-equilibrium ice sheets. *J. Glaciol. 5(38)*, 145–158.

Wexler, H., 1952. Volcanoes and world climate. *Scient. Amer. 186*, 74–80.

Wexler, H., 1953. Radiation balance of the earth as a factor in climatic change. In *Climatic Change*, ed. H. Shapley, Cambridge, Massachusetts. 73–105.

Willett, H. C., 1949. Long period fluctuations in the general circulation of the atmosphere. *J. Meteor. 6*, 34–50.

Willett, H. C., 1953. Atmospheric and oceanic circulation as factors in glacial-interglacial changes in climate. In *Climatic Change*, ed. H. Shapley, Cambridge, Massachusetts. 51–71.

Woerkom, A. J. J. Van, 1953. The astronomical theory of climatic changes. In *Climatic Change*, ed. H. Shapley, Cambridge, Massachusetts. 147–157.

Workman, E. J., 1962. The problem of weather modification. *Science 138*, 407–412.

Wright, G. F., 1893. *Man and the Glacial Period.* New York. 385 pp.

Wright, H. E., jr., 1961. Late Pleistocene climate of Europe; A review. *Bull. Geol. Soc. Amer. 72*, 933–983.

Zapffe, C. A., 1954. A new theory for the great ice ages. *Physics Today 7(10)*, 14–17.

Zeuner, F. E., 1945. *The Pleistocene Period.* Ray Soc., London. 447 pp.

H. H. Lamb

13 On the nature of certain climatic epochs which differed from the modern (1900-39) normal

INTRODUCTORY

Climatic changes during the last five or six thousand years appear, at some times, to have allowed sufficient vegetation for primitive men and animals to travel across what are now deserts in north Africa, central Asia and northern Mexico-south-western United States of America, and, at other times, to have cut off these routes of trade and migration. Within the last thousand years various categories of evidence suggest that there was, at first, rather little permanent ice on the Arctic seas, and later, such a great extension of this ice that grain growing was for centuries impossible in Iceland and the total evacuation of that country was considered; at the same time, the cod fishery almost disappeared even from the Faroe Islands. In the worst decades Scotland, as well as parts of Scandinavia and Iceland, experienced famine; upland farms and villages in England and Germany may have been abandoned partly for this reason.

The last hundred years, or rather more, have seen a significant warming over most of the world, particularly the Arctic. The limits of open water and of the cod fishery (cod population effectively checked by the 2° C isotherm of water temperature), as well as those of very diverse biological species, have been displaced poleward in the Northern Hemisphere. Glaciers have shrunk. During about the same period, the levels of water bodies and the discharges of rivers in the lower latitudes (in both the arid and equatorial zones) have generally gone down materially, e.g., the Great Salt Lake of Utah, the Caspian Sea, the East African lakes and the Nile. Corresponding changes of mean annual rainfall have been reported, amounting in some places to 30–40%; though India appears to have been more or less unaffected, and China and Japan show more complex variations (with rainfall minima about 1900 and 1940). Most of the trends of the past hundred years have halted or show signs of reversal since 1940, though in China in 1960 extreme aridity was again reported.

HISTORICAL SURVEY

At least four climatic epochs since the Ice Age seem likely to repay more study by meteorologists than they have so far received: (*a*) the post-glacial climatic optimum (warm period culminating between about 5000 and 3000 BC); (*b*) the colder climatic epoch of the early Iron Age (culminating between about 900 and 450 BC); (*c*) the secondary climatic optimum in the early Middle Ages (broadly around AD 1000–1200 or rather longer); (*d*) the Little Ice Age (cold climate very marked between about AD 1430 and 1850).

The first step must be to build up knowledge of the world climatic patterns prevailing during these epochs. The following outlines can already be discerned:

Post-Glacial Optimum. Warm Epoch (circa 5000 to 3000 BC)

The distribution and general extent of land ice was probably not very materially different from now, and the world sea level similar to today.

The extent of ice on land, decreasing throughout the warm epoch, may have reached its minimum around 2000–1500 BC; i.e., after the main part of the warm epoch was over.

By soon after 4000 BC the world sea level had risen to about its present level (Godwin, Suggate, and Willis, 1958), or possibly a few meters above (Brooks, 1949; Fairbridge, 1961). The rise over the previous 10–12,000 years from its minimum stand in the Ice Age can be primarily attributed to reduction of the ice sheets on land to somewhere near their present extent. Isostatic effects, however, altered the geography near the former

Reprinted, with minor editorial modification, by permission of the author and publishers, from *The Changing Climate*, H. H. Lamb, Methuen & Co., Ltd., 1966, pp. 58–112.

ice sheets; there were extensive submerged lands and shallow seas for some thousands of years (see, for example, development of the Baltic and North Sea between about 8800 and 5000 BC in Zeuner, 1958). Because of this, and because of rather higher ocean temperatures than now (1° C in the tropical Atlantic (Emiliani, 1955) and probably several degrees in the Arctic (Brooks, 1949)), the present general sea level could have been attained at a time when the land ice was still slightly more extensive than now.

The Arctic Ocean, with much open water, was probably ice-free at least in summer, though not the channels of the Canadian Archipelago (Brooks, 1949).

Fossil marine fauna (molluscs and edible mussels) and evidence of past vegetation and bog growth indicate much higher sea and air temperature than now quite generally in high latitudes north and south, e.g., at Spitsbergen (Brooks, 1949; Schwarzbach, 1959). Camel remains found in Alaska and tiger in the New Siberian Islands, approx. 75° N are attributed to this period (Alissow, Drosdow, and Rubinstein, 1956) but probably imply temperate, not tropical conditions.

In the (present) temperate zone of the Northern Hemisphere temperatures were higher than now and there is evidence of a northward anomaly, especially before 5000–4000 BC, of the atmospheric circulation.

The vegetation belts were displaced poleward and to greater heights above sea level than now. In Europe the summer temperatures can be estimated as prevailing 2–3° C higher than now (Schwarzbach, 1950; Godwin, 1956), in North America rather less above present levels (Schwarzbach, 1961). Winter temperatures, though possibly conditioned by anticyclones, can never have attained the severity now reached in air streams coming over the continents from the polar ice. The snow line was 300 meters above the present level in central Europe (Schwarzbach, 1961). Annual mean temperatures in Europe were about 2° C higher than now (West, 1960). Since the fringe of the dry-climate landscape of the steppe reached Leningrad and the Volga Basin (Alissow et al., 1956; see also Buchinsky, 1957), at least in the period up to about 4000 BC, it seems safe to assume that the subpolar depressions and the axis of the main anticyclone belt were generally displaced north at that time in the European sector, perhaps by as much as 10° of latitude.

Later in the warm period there were considerable and long-lasting variations of rainfall, and it has been suggested that the climate of Europe became generally wetter after the development of an enlarged Baltic, perhaps as early as 5000 BC. Depressions presumably tended to pass on more southerly tracks than formerly, gradually coming nearer to their present general latitudes. Milder, more oceanic winters would help maintain high annual mean temperatures.

In the Sahara and deserts of the Near East from about 5000–2400 BC there was an appreciably moister climate than now (Butzer, 1957a, 1958).

The evidence, archaeological and zoological, is especially convincing as regards the early part of this period. Desiccation of the climate, possibly beginning well before 2400 BC, might have a somewhat delayed effect because of the higher water table and presumably more extensive oases.

The high pressure belt presumed generally well north of the Mediterranean around 5000 BC and after. [sic] Africa and the Near East could come under the influence of broadened trade-wind and equatorial zones and more widespread summer (monsoon) rains. With this pattern winter rains in the latitude of the Mediterranean may also have occurred.

Gradually, the effects of the high pressure belt returning to lower latitudes seem to have been increasingly felt and, with increasing aridity in the region, by 3000 BC some of the animal migrations were cut off.

In Hawaii also the "climatic optimum" period gives evidence of greater rainfall than now in the (widened?) trade-wind zone.

In the southern temperate zone, a moister epoch, apparently rather warmer than now, was experienced though the temperature anomalies seem to have been less than in Europe. The evidence is largely from the extent and distribution of forest species in southernmost South America and New Zealand (Auer, 1958, 1960; Cranwell and von Post, 1936; Schwarzbach, 1950). Firm conclusions regarding the prevailing latitudes covered by the subtropical high pressure and west wind belts must await further work in other areas. (Work on the poor floras of the smaller islands in the southern temperate zone has, however, been able to throw no light on this matter, since no changes are apparent there.) It seems clear that in the earlier post-glacial period between about 7000 and 5000 BC Tierra del Fuego had had a rather dry, anticyclonic climate with prevailing west winds and limited forest extent on the west side only; New Zealand had a distribution more similar to that of today. During the warm-climate period which followed the forest spread in New Zealand warmth-loving species seem to have gained and wind directions appear to have been rather more variable than earlier (expanded equatoria [sic] and trade-wind zones?). The dates depend sufficiently on radio-carbon tests and can be confidently accepted for comparisons with the Northern Hemisphere (Auer, 1958).

Antarctica also experienced a warm period after the main Ice Age (when the Antarctic ice sheet had been several hundreds of meters thicker than now). Ahlmann (1944) tentatively, but doubtless rightly, identified this with the time of the post-glacial warm epoch in the northern and southern temperate zones.

The dating of the warmest epoch in the Antarctic as contemporaneous with that in the other climatic zones may be accepted as required by the results of radio-carbon dating of changes of world sea level (Godwin et al., 1958) and of the climatic changes in Tierra del Fuego, near 55° S.

In the Wohlthat Mountains, near 72° S 10° E, temperatures were so many degrees higher than now that there were considerable streams of running water and fluvial erosion of the landscape; between 600 and 1,890 meters above sea level lakes were formed which have subsequently frozen solid and remain as "fossils" (Ahlmann, 1944).

Flohn (1952) estimates that annual temperatures in this epoch were 2°–3° C higher than now in the Antarctic and in Tierra del Fuego and also on the Himalayan Mountains.

Post-Glacial Climatic Revertence. Early Iron Age Cold Epoch (circa 900–450 BC)

Brooks (1949) suggested that the sharp worsening of the climate of Europe, which came with floods and storms and advances of the Alpine glaciers between about 1400 and 500 BC, following a much more gradual decline of temperature for centuries previously, implies and coincided with rather sudden re-formation of the "permanent" pack-ice cover on the Arctic Ocean north of 75°–80° N.

Regarding the recession in the temperate zone, most of the present glaciers in the Rocky Mountains south of 50° N were formed about this time, certainly after 2000 BC (Matthes, 1939). In the Austrian Alps glaciers advanced to near the same limits which they regained, or more generally passed, around AD 1600 (Firbas and Losert, 1949). In Europe generally the most impressive feature is the evidence of a sharp increase of wetness.

There was a widespread re-growth of bogs after a much drier period (Godwin, 1954). This change ("recurrence surface" or *Grenzhorizont*) is still the most conspicious feature of peat bog sections all over northern Europe from Ireland to Germany and Scandinavia, represented by a change of color from the black lower peat to the lighter upper layers. Lakeside dwellings in central Europe were flooded and abandoned (Brooks, 1949), and ancient tracks across the increasingly marshy lowlands in England and elsewhere in northern Europe were adapted to changed conditions or abandoned too (Godwin and Willis, 1959).

In Russia forest had spread farther south than in the warm epoch, advancing for instance along the lower Dnieper, and the species represented (beech, hornbeam, fir rather than oak) indicate some lowering of the summer temperatures. It is not clear whether any important dry period occurred in Russia, as it did farther west in Europe, between the end of the warmest epoch and the sharper Iron Age recession (Buchinsky, 1957).

In the Mediterranean and North Africa the climate seems to have been drier than during the climatic optimum, but not so dry as today. Rainfall was apparently not quite so rare as now in summer. Roman agricultural writers (e.g., Saserna) noted that around 100 BC the vine and the olive were spreading north in Italy to districts where the weather was formerly too severe: from this it would appear that the preceding centuries had had a cooler climate also in the Mediterranean.

The southern temperate and Antarctic zones had also entered a colder period by around 500–300 BC. In southern New Zealand the evidence is consistent with prevailing westerly winds, but in Tierra del Fuego spread of the forest to cover the whole island on both sides of the watershed indicates quite frequently easterly winds and hence depression tracks on the whole in lower latitudes than now (Auer, 1960; Cranwell and von Post, 1936).

Secondary Climatic Optimum (circa AD 1000 to 1200)

This epoch appears to show most of the same characteristics as the post-glacial epoch both in the Northern Hemisphere and in the Antarctic, only in less degree, perhaps because of its shorter duration.

The Arctic pack ice had melted so far back that appearances of drift ice in waters near Iceland and Greenland south of 70° N were rare in the 800s and 900s and apparently unknown between 1020 and 1200, when a rapid increase of frequency began. This evidence hardly supports Brooks' suggestion that the Arctic Ocean again became ice-free during this epoch, though "permanent" ice was probably limited to inner Arctic areas north of 80° N and possibly not including the Canadian Archipelago (to judge from occasional exploits there by the Old Norse Greenland colonists).[1] From the evidence of early Norse burials and plant roots in ground now permanently frozen in southern Greenland, annual mean temperatures there must have been 2°–4° C above present values. It seems probable that sea temperatures in the northernmost Atlantic were up by a similar amount.

In western and middle Europe vineyards extended generally 4°–5° latitude farther north and 100–200 meters higher above sea level than at present (Lamb, 1959). Estimates of the upper limits of the forests and of tree species on the Alps and more northern hills in central Europe range from 70 to 200 meters above where they now stand (Gams, 1937; Firbas and Losert, 1949). These figures suggest mean summer temperatures about 1° C, or a little more, above those now normal.

[1] Glaciological and other studies of Arctic ice islands, and their presumed growth when formerly part of the Ellesmere Land ice shelf, have not so far been reduced to an agreed time scale (Crary, 1960; Stoiber *et al.*, 1960). It seems most probable, however, that the ablation period in progress in the early 1950s began only about 40 years ago and that the total age of the ice is 620 years or less, implying growth during the Little Ice Age epoch and that net ablation prevailed before that, during the secondary climatic epoch, in the Canadian Archipelago.

In North America archaeological studies in the upper Mississippi valley (approx. 45° N) suggest a warm dry epoch, followed by a change to cooler, wetter conditions after AD 1300 (Griffin, 1961).

In lower latitudes Brooks (1949) names this as a wet period in central America (Yucatan) and probably in Indo-China (Cambodia). There is evidence of greater rainfall and larger rivers in the Mediterranean and the Near East (Butzer, 1958). There is some evidence of a moister period in the Sahara from 1200 or earlier, lasting until 1550 (Brooks, 1949).

In southernmost South America the forest was receding rapidly to western aspects only, indicating a drier climate than in the previous epoch and more predominant westerly winds.

On the coast of east Antarctica, at Cape Hallett, a great modern penguin rookery appears from radiocarbon tests to have been first colonized between about AD 400 and 700, presumably during a phase of improving climate, and to have been occupied ever since (Harrington and McKellar, 1958). This tends to confirm the earlier assumption of explorers of the Bunger Oasis in east Antarctica of a period of marked climatic improvement about a thousand years ago, since which there has been only a modest reversion.

The Little Ice Age (circa AD 1430–1850)

There is manifold evidence of a colder climate than now from most parts of the Northern Hemisphere.

The Arctic pack ice underwent a great expansion, especially affecting Greenland and Iceland, and by 1780–1820 sea temperatures in the North Atlantic everywhere north of 50° N appear to have been 1°–3° C below present values (Lamb and Johnson, 1959). Indirect evidence suggests that these (or even slightly lower) water temperatures were already reached by the 1600s.

Decline of the forests at the higher levels in central Europe between about 1300 and 1600 evidently had some catastrophic stages, especially after 1500, and Firbas and Losert (1949) believe this may have been the time of principal change of vegetation character at levels above 1,000 meters since the post-glacial climatic optimum. (In Iceland the relict woodland surviving from the climatic optimum virtually disappeared early in this epoch, doubtless partly by human agency.) Also near the Atlantic coast of Scotland eye-witness reports of the time (Cromertie, 1712) suggest widespread dying off of woods in the more exposed localities, presumably because of miserable summers and increased damage from salt-spray. In Europe around 50° N it seems, however, that the prevailing summer temperatures were mostly about their present level (in the 1700s slightly above), though the winters were generally more severe.

Some notably severe winters also affected the Mediterranean. Glaciers advanced generally in Europe and Asia Minor, as well as in North America, and snow lay for months on the high mountains in Ethiopia where it is now unknown.

The Caspian Sea rose and maintained a high level until 1800. Records of the behavior of the Nile suggest that this was a time of abundant precipitation in Ethiopia but very low levels of the White Nile, which is fed by rainfall in the equatorial belt: the equatorial rains were evidently either weak or displaced south.

The evidence generally points to an equatorward shift of the prevailing depression tracks in the Northern Hemisphere and more prominent polar anticyclones.

The Southern Hemisphere seems largely to have escaped this cold epoch until 1800 or after, though by then temperatures were possibly somewhat lower than today in some parts of the southern temperate zone. Between 1760 and 1830 the fringe of the Antarctic sea ice appears to have been generally a little south of its present position and the southern temperate rain-belt apparently also displaced south.

Recession followed after 1800–30 until 1900 or later. The rain zone and depression tracks moved north and there were great advances of the glaciers in the Andes and South Georgia, as well as some extraordinarily bad years for sea ice on the southern oceans (Aurrousseau, 1958; Findlay, 1884). This recession in the southern temperate and sub-Antarctic zones was, however, out of phase with the trend by then going on in the Northern Hemisphere. Since 1900 the temperature of the southern temperate zone as a whole may have been rising like that of other zones (Callendar, 1961); in the sub-Antarctic the rising trend began even later and may be out of phase with the latest trend in the Northern Hemisphere (Willett, 1960; Mitchell, 1961).

Summary of the Historical Survey

The first three epochs appear to demonstrate respectively: (*a*) climatic zones displaced towards high latitudes, equatorial/monsoon rain belt widened; (*b*) climatic zones displaced towards low latitudes, equatorial belt narrowed; (*c*) as (*a*) but in less degree.

These variations fit with the sequence of contractions and expansions of the circumpolar vortex proposed by Willett (1949), with minor modifications of date and extra detail thanks to recent additions to knowledge.

The last epoch appears to present a different pattern, with an equatorward shift of the climatic zones in the Northern Hemisphere accompanying a (probably smaller) poleward shift in the Southern Hemisphere, followed by return movements in both hemispheres during the nineteenth century.

Climatic fluctuations in the present equatorial, arid, temperate and polar zones have been such as accord with these changes.

The amplitude of the temperature fluctuations appears to have been greatest in high latitudes, so that the meridional gradient of surface temperature, at least between 50° and 70° N must have been materially less in the warm epochs than in the cold ones. It appears, however, that the temperatures on mountains even in low latitudes were raised 2°–3° C in the warm epochs[2] and lowered by 1°–2° C below present values during the Little Ice Age. This means that the possibility remains that the meridional (Equator-Pole) gradient of upper air temperature was actually less during the cold epochs, the surface temperatures in high latitudes being particularly unrepresentative at such times because of frequent strong inversions (Lamb and Johnson, 1959).

Climatic trends in the Far East bear a rather complicated and partly inverse relationship to those elsewhere, e.g., evidence of generally late cool springs in Japan AD 1000–1200, milder winters between about 1700 and 1900 than before or since (Arakawa, 1956, 1957; Lamb and Johnson, 1959). From study of variations within the last two centuries Yamamoto (1956) describes how this inverse relationship tends to come about: (*a*) during periods of strengthened zonal circulation of

[2] From Himalayan evidence, quoted by Flohn (1952). Evidence from other regions desirable.

the atmosphere (strengthened Siberian anticyclone and weak polar anticyclone in the Asian sector) Japan experiences more cold air from the continental interior in winter (and some Arctic outbreaks in the rear of depressions); (*b*) during periods of weaker zonal circulation (Siberian anticyclones weaker or displaced northwest, polar anticyclones covering more of northern Asia) the main cold air mass in winter streams west over central Asia towards Europe, and Japan comes more under the influence of Pacific anticyclones and mild oceanic air. However, farther north, the Okhotsk Sea develops more ice at such times, under the influence of the polar anticyclones, and this produces a tendency for poor summers in Japan.

Superimposed upon, and running through, the bold climatic phases and trends here defined many workers have found evidence of apparent periodic oscillations, particularly affecting rainfall but presumably also affecting the latitudes of the most frequent depression tracks over Europe and Asia.

A periodicity of 180–200 years has been suggested by variations in the Baltic ice and levels of the Caspian Sea (Betin, 1957), one of 400 years from Chinese data (Link and Linkova, 1959), one of about 600 years from recurrent regeneration layers ("Recurrence surfaces" or *Grenzhorizonte*) in the peat bogs in many parts of temperate Europe and a periodicity of 1,700–2,000 years from variations in the rivers, lakes and inland seas in European Russia and in central Asia.

There seems no doubt, however, that the four epochs described earlier are those which represent the greatest departures from present day conditions. If periodic variations of any agency influencing the Earth and its atmosphere be involved, it nevertheless seems that in these epochs some special conjunction of circumstances —possibly including external circumstances such as volcanic dust—must have come into play.

It may be significant that the two cold epochs more or less coincide with III and IV of the post-glacial world-wide waves of volcanic activity (I-IV) identified by Auer (1958, 1959). Radio-carbon dates of these volcanic phases are: I—around 7000 BC; II—3000–250 BC; III—around 500–0 BC; IV—around AD 1500–1800. Moreover III was marked by volcanic activity over both hemispheres, e.g., Kamchatka, Iceland, Andes and Tristan da Cunha (see also Thorarinsson et al., 1959); whereas IV seems to have been largely in the Northern Hemisphere and equatorial zones, apart from the Southern Hemisphere eruptions of 1835 and certain later years. Thus the different climatic distributions deduced for these epochs correspond well with a difference that may reasonably be presumed in the geographical distribution of dust veils in the high atmosphere.

METEOROLOGICAL INVESTIGATIONS OF CHANGES IN THE GENERAL ATMOSPHERIC CIRCULATION

By Surface Pressure Maps Back to 1750

Monthly m.s.l. [mean sea level] pressure charts have been constructed in the British Meteorological Office, covering as much of the world as possible, for each January and each July back to the earliest years for which usable observation data could be found. The internal consistency checks provided by a network of observing stations, some in each area having long series of observations, have made it possible, after tests for probable error of the isobars, to reconstruct the pressure field for a worthwhile area of Europe back to 1750. Pressure distribution over the North Atlantic Ocean from as early as 1790 can be established within a tolerable error margin (standard error ± 2.5 mb in January, ± 1.0 mb in July) by using 40-year means. The method, sources of data and tests used have been published elsewhere (Lamb and Johnson, 1959).

Figures 1 and 2 show the 40-year average m.s.l. pressure in January and July respectively for the earliest possible period over the Atlantic Ocean (1790–1829) compared with the modern normal charts based upon the *Historical Daily Weather Maps* (1900–39). Solid lines are used for the isobars where the pressure values can be regarded as known within the error margins mentioned above, broken lines are used where the general pattern of the pressure field is reasonably certain but the values are not good enough for useful measurements to be based upon them.

January

Average pressure gradients have increased in January from the period around 1800 to the present century (Figures 1 (*a*) and (*b*)).

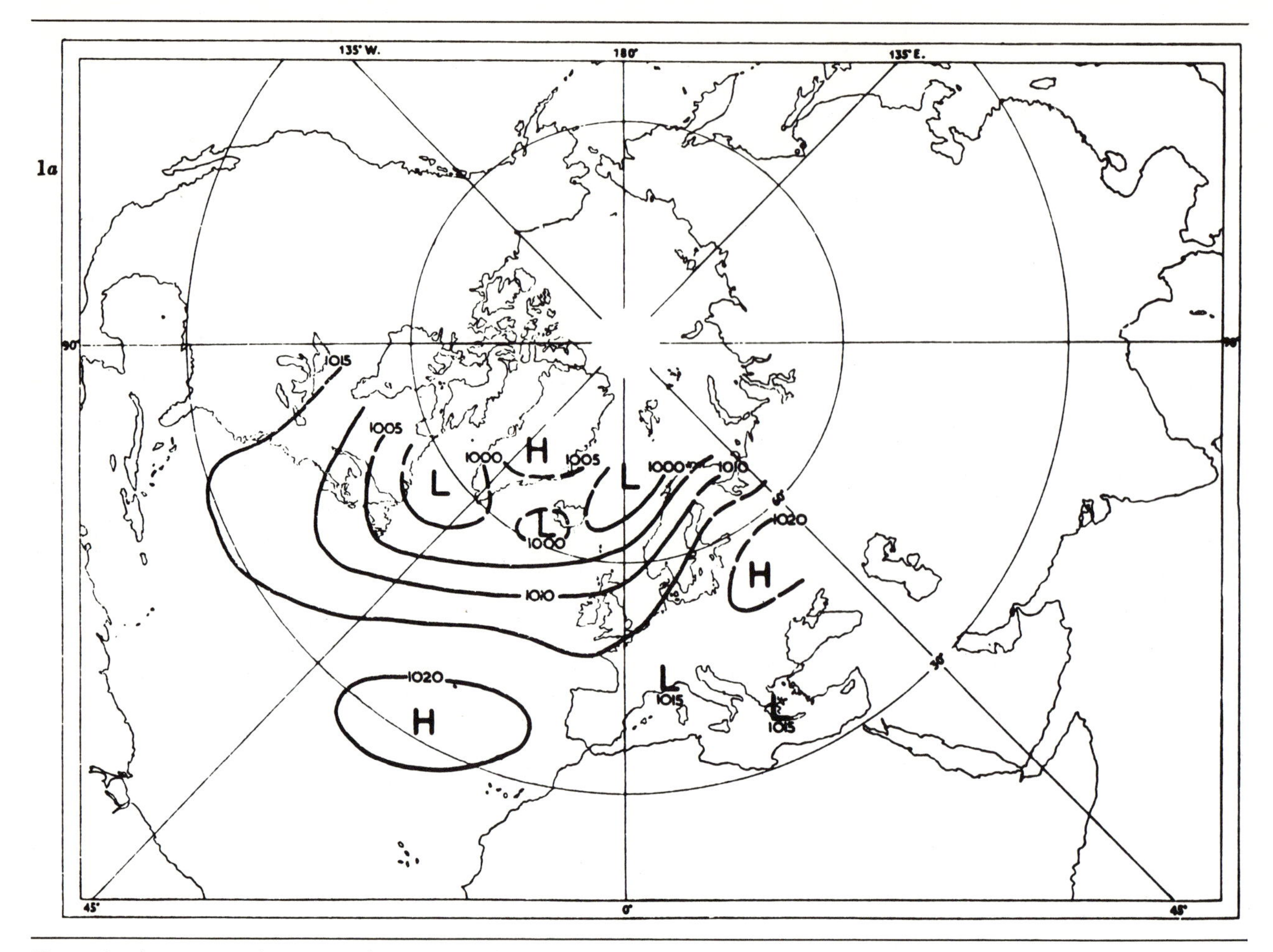

Figure 1 Average m.s.l. pressure for January: (a) 1790–1829; (b) 1900–39.

In Figure 3 a limited selection of pressure difference indices of circulation intensity in January are presented. From these, and others not shown, it is seen that the increase of intensity of the zonal circulation since the middle or earlier part of the nineteenth century is a world-wide phenomenon, affecting at least the westerlies and trade-wind zone in the North Atlantic and the southern westerlies. The peak intensity of the zonal circulation so far was reached about 1930 in the North Atlantic and 1910 in the Southern Hemisphere. Over the North Atlantic the increase of the zonal circulation from around 1800 to 1930 appears to amount to 5–10%. Changes in the extent of Arctic ice (for which the North Atlantic is virtually the only outlet) appear, however, to amplify circulation changes in this sector.

It is an advantage of the map method that indices of circulation vigor can be measured at points where the main air streams are best and most regularly developed. Comparison of numerous correlation coefficients between measures of the same air stream taken at different points have demonstrated that such indices are the most representative (Lamb and Johnson, 1959). The Northern Hemisphere indices in Figure 3 have been chosen in this way, but, for the earliest years, and in the Southern Hemisphere, it is still necessary to make measurements only near where there happen to be observing stations. The monthly mean pressure gradient between 50° and 60° N for westerlies over the British Isles, used in Figure 3 (*c*) to cover a longer period than any of the indices in Figure 3 (*a*) or (*b*), is an index linked with the overall pressure range between Azores maximum and Iceland minimum in January by a correlation coefficient of about +0.6 (1800–79, +0.62; 1880–1958, +0.56—both statistically significant beyond the 0.1% level). It is reproduced here because it shows the increase of vigor of the westerlies from a generally lower level in the eighteenth century.

The period around 1790–1830 was evidently one of rather notable minimum strength of the zonal circulation in January, though there had been another minimum earlier, probably before 1750.

The variations in the mean strength of the North Atlantic westerlies nicely parallel the changes of mean January temperature in Britain and central Europe. The mean January temperature of central England (Manley, 1953) between about 1740 and 1850, was for instance, over 1.5° C lower than between 1900 and 1940. Since this type of temperature trend is typical for the

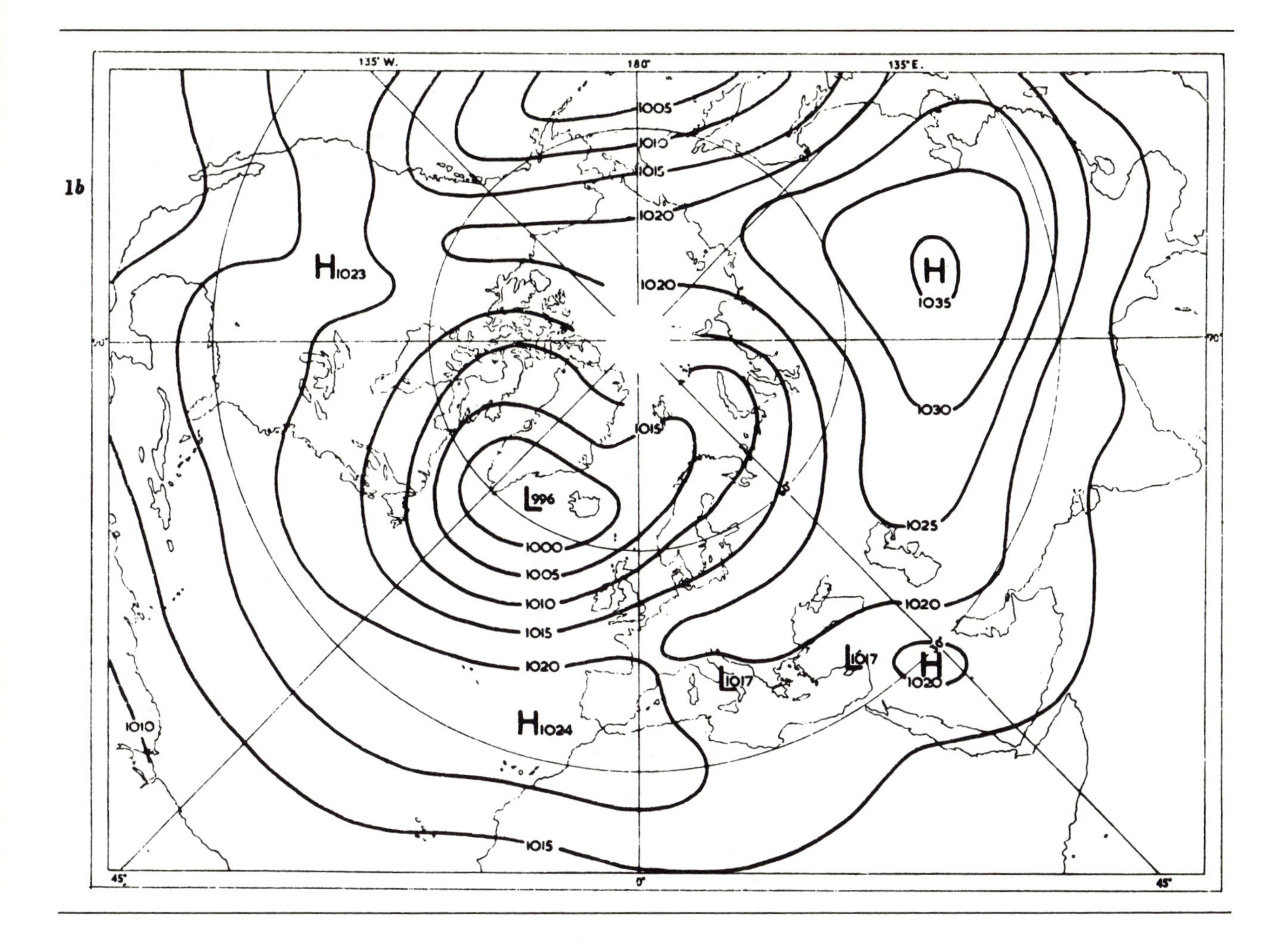

colder half of the year, it is reasonable to suppose that other months have also partaken in the change of strength of the North Atlantic westerlies.

The mean meridional component, which is much weaker than the mean zonal component, of the circulation (meridional component measured by pressure differences along 55° N) does not appear to have shared the increasing trend of the zonal circulation in January, but seems rather to show a long-term fluctuation with a possible period of around 60 to 70 years.

The importance of considering the meridional circulation and the zonal circulation separately in connexion with the recent warming of the Arctic has been further demonstrated by Petterssen (1949) and by Wallén (1950, 1953) who discovered that between 1890 and 1950 the frequency of northerly meridional situations decreased and that of southerly meridional situations increased markedly in the region of the Norwegian Sea and Scandinavia; since 1950 northerly meridional situations have once more become prominent. This discovery might be related to changes of wave-length (and displacement of the preferred positions for troughs and ridges) in the upper westerlies of which we shall give further evidence later.

Variations of the meridional component of the circulation, especially the winter southerlies over the North Sea, must be linked in some way with the occurrence of blocking anticylones, which is also subject to variations possibly with periods of 60–90 years. Since no persistent trend is found in the meridional component, the generally colder winters of the period around 1800 and earlier are mostly to be explained by weakness of the mean westerly flow towards Europe rather than by blocking, which was only really prominent at long intervals both in the epochs of weak and of stronger westerlies.

July

Changes in the July circulation from 1790–1829 to 1900–39 (Figure 2) are much less obvious than those in January, though Europe seems to have had a weaker and more anticyclonic pressure field in the earlier period.

The changes are better seen from the trends of the intensity indices shown in Figure 4. (Following the principle of measuring pressure gradients where the main air streams are best and most regularly developed, the most representative indices of the July circulation are not found at the same points as those used for

2a

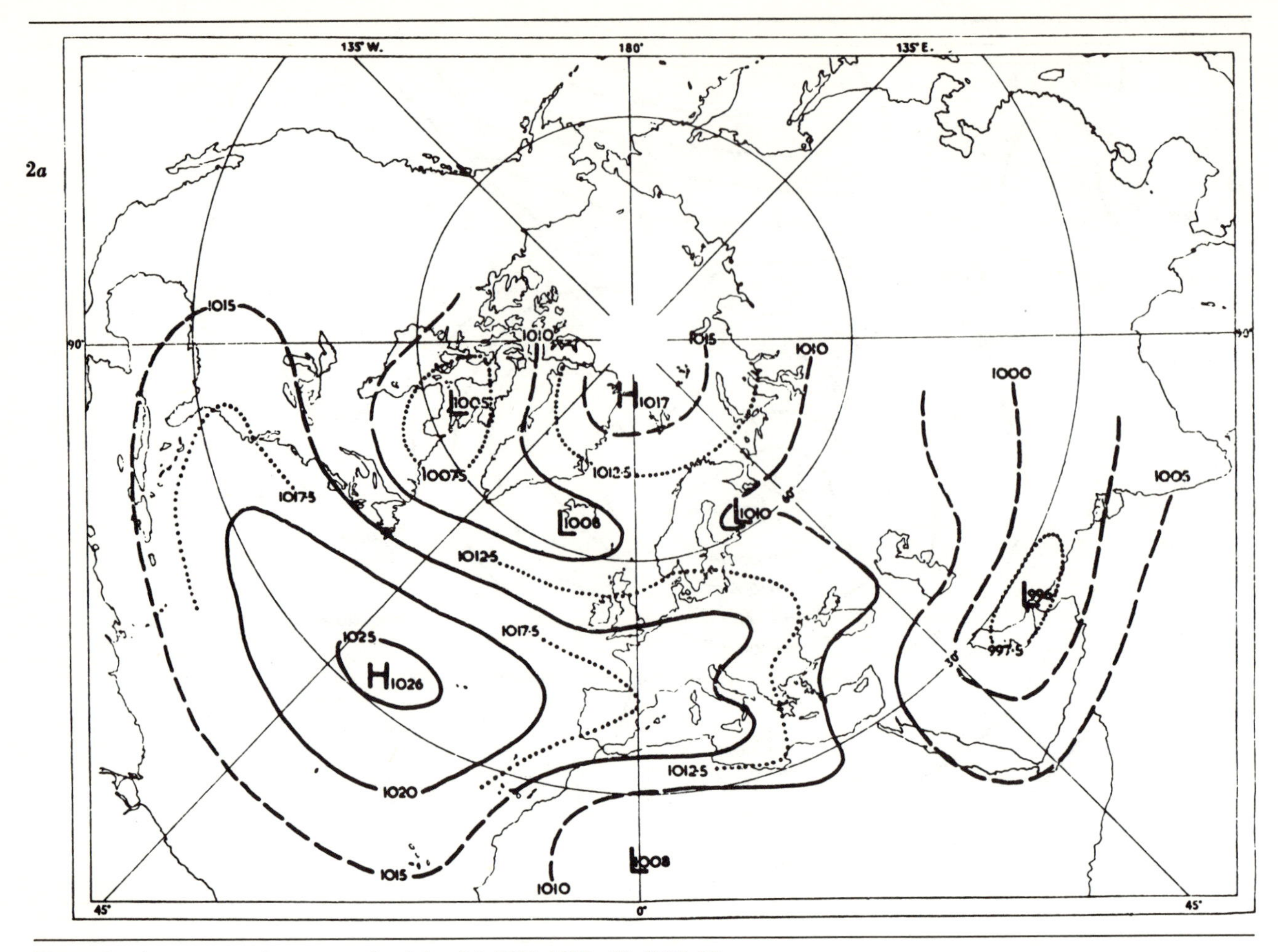

Figure 2 Average m.s.l. pressure for July: (a) 1790–1829; (b) 1900–39.

January.) The general nature of these curves is an increase of intensity from low values around the middle of the nineteenth century to high values in the twentieth century. A major peak intensity of such indices generally seems to have been passed between 1900 and 1950—mostly in the decades 1900–09 or 1920–39—but again appears earlier in the Southern Hemisphere around 1890–1920 (1897–1906 New Zealand, 1908–17 Chile).

The rather stronger July circulation in the Northern Hemisphere before 1800–50 seems to have shown different patterns from those prevailing in the present century. Hints of rather lower pressure in the Labrador region (east of the North American cold trough) and of higher pressure over central and northern Europe are developments that might be expected to accompany the persistence of a cold surface (with some permafrost and remnants of ice and snow) in somewhat lower latitudes than now occurs.

The mean meridional components of the circulation, illustrated in Figure 4 (*c*) by the pressure difference between 0° and 10° E at 55° N (northerly wind component over the North Sea), in July appear to show a rising trend. This is apparent also over the China Sea and is probably involved in the increasing gradient for southwesterly winds over the Newfoundland Banks (Figure 4 (*c*)). The North Sea index rises from the 1790s to the present century; if there are any real oscillations superimposed, the period seems to be 70–90 years—between peaks around 1820 and 1900 and between minima in the 1780–90s, the 1850s and 1940s—and the rising trend may be still continuing.

Summary

Increases of strength of the zonal circulation have been found in January and July over widely separated parts of the globe, apparently being quite general from around the middle of the last century and culminating around 1930 in the Northern Hemisphere and 1900–10 in the Southern Hemisphere. The trend of some indices suggests that an overall increase of energy may have been going on since well before 1850 in July as well as January. Possibly the rather stronger summer circulation over the North Atlantic and neighboring longitudes in the late 1700s than in the mid-1880s should be regarded as a local difference associated with the presence of some persistent snow and ice in rather lower latitudes before, say, 1830 than at any time since. Increased strength of the general circulation has probably occurred

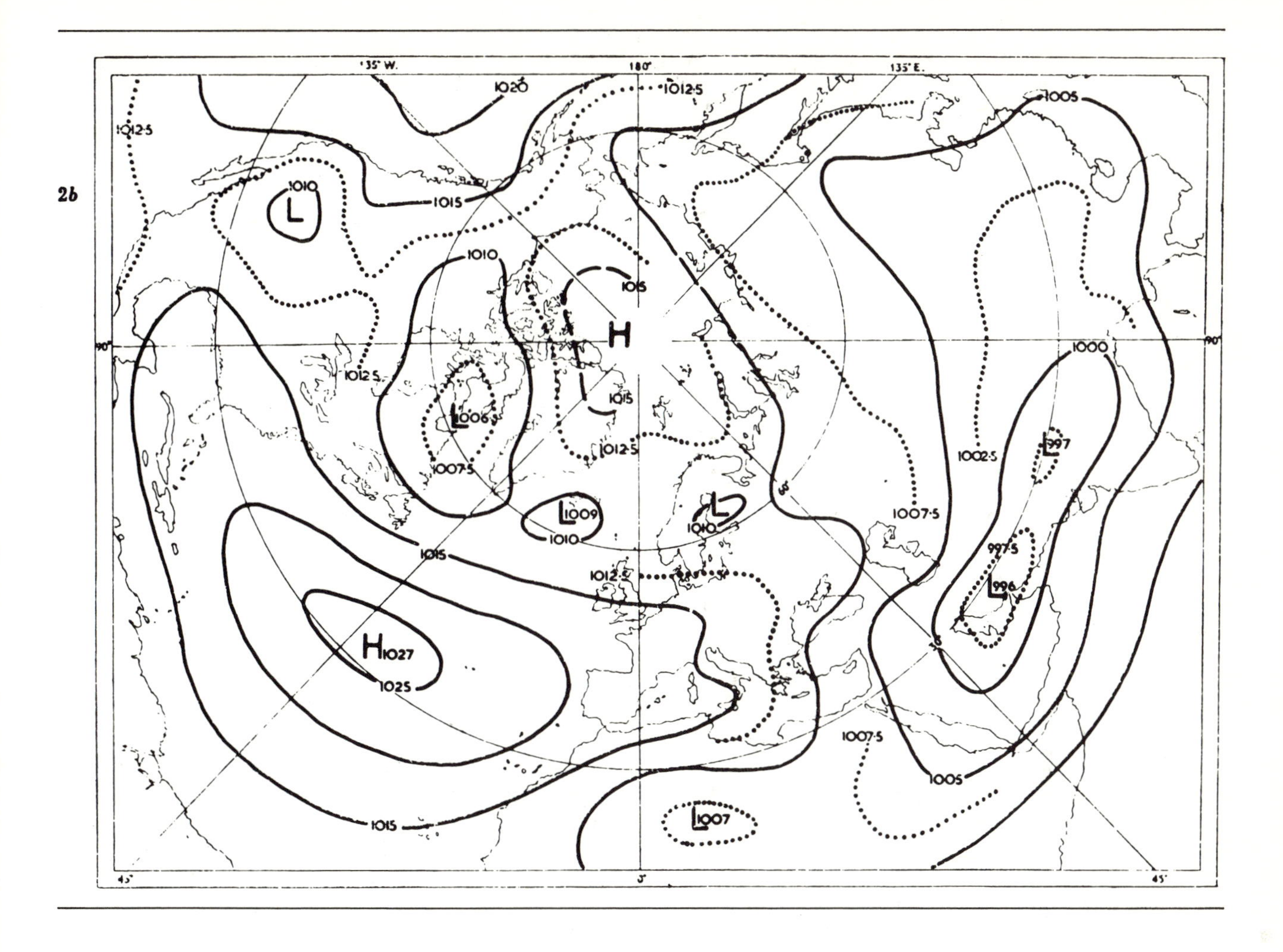

in most months of the year. The increase from *circa* 1800 to 1930 in the strength of the zonal circulation over the North Atlantic in January amounts to between 5 and 10%, but in general is probably less than this.

A general rise in the winter temperatures in Europe accompanied the increased vigor and prevalence of the westerlies. Summer temperatures showed a much smaller, and not statistically significant, fall.

Changes of average latitude of the main zones of the atmospheric circulation have also been studied. They seem to show a general equatorward displacement, especially of the sub-polar depression track, in winter and summer, by 2–4° of latitude in those periods and regions where sea ice has attained an abnormally great extent—e.g., around 1800 in the North Atlantic and around 1900 near Chile. A progressive equatorward trend of the northern and southern high pressure belts in the Atlantic sector in January during the present century is so far unexplained.[3]

A tendency for increased ice on the Arctic seas in this sector since about 1940 may be a consequence of this. In July the trend seems to have been the other way till about 1930. Over the Indian Ocean and Australian sectors of the Southern Hemisphere the main pressure belts have all moved 2°–4° south in January and north in July during the present century,[4] displacements whose trend seems to follow that of the intensity of the (monsoon) pressure gradients developed over Asia, especially the winter northerlies and summer southerlies over eastern Asia, passing its extreme point and reversing broadly around 1930.

An Important and Severe Stage in the Climatic Change since 1800

Regarding the recent period of climatic recovery from the Little Ice Age and strengthening atmospheric

[3] Willett (1961) has lately reported on an index of solar activity which shows a highly significant negative correlation with the latitude (but no relation with the intensity) of the strongest upper westerlies across North America in January over the period 1900–60. A smaller negative correlation coefficient between relative sun-spot number and the latitude of the westerlies was also obtained. Hence the extremely disturbed sun in recent sun-spot cycles may have some bearing on this equatorward trend of the pressure and wind zones in the North Atlantic.

[4] The January trend affecting summer rainfall in Australia was first reported by Kraus (1954), who does not, however, seem to have been aware of the reverse tendency in the southern winter.

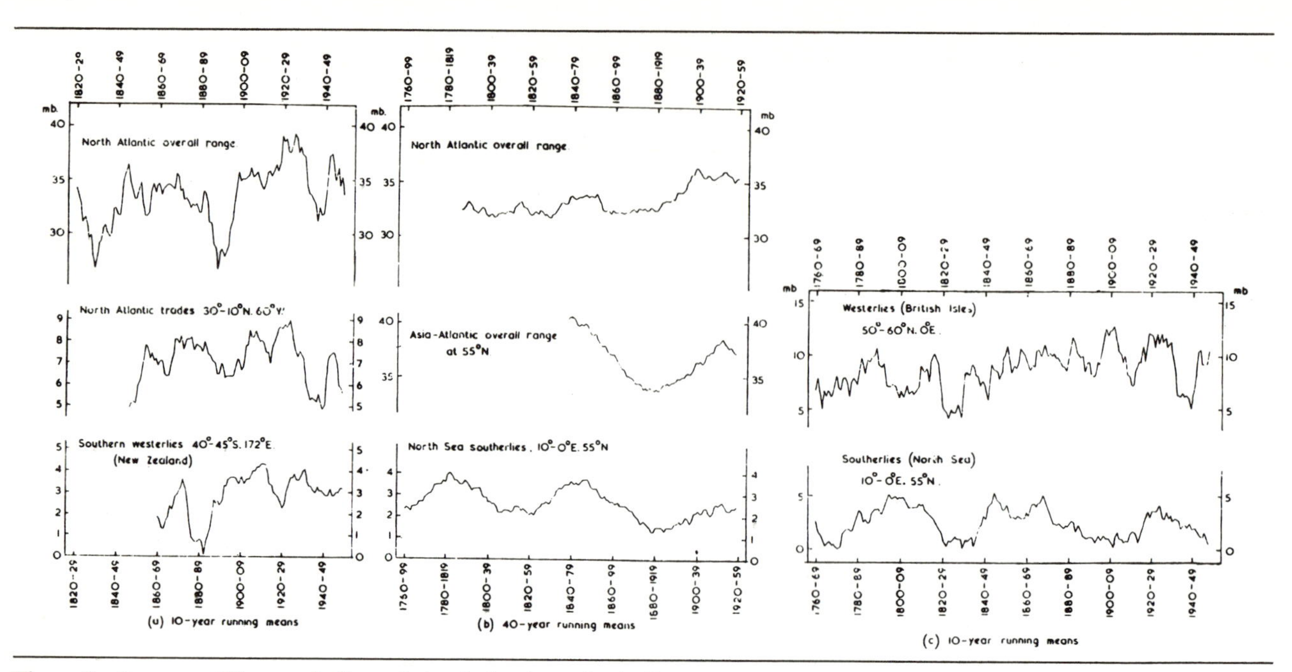

Figure 3 Pressure difference indices of circulation intensity in January.

circulation, with warming of most latitude zones and increasing aridity in the desert zone (Butzer, 1957b), the particularly strong Northern Hemisphere circulation of the 1920s and 1930s and of the decade 1900–09, at least as regards the sector between North America and Europe, has already attracted the attention of meteorologists (Wagner, 1940; Scherhag, 1950; Lamb and Johnson, 1959). The peculiar characteristics of the 1830s do not, however, seem to have received notice, perhaps because it has not previously been possible to present circulation maps. The mean pressure distributions for January and July for each decade since 1750 can now be presented and those for 1830–39 are shown here in Figure 5.

This decade stands out from the series 1750 to date in several respects:

1 Highest pressure generally over Europe and the Mediterranean.
2 Most northerly position of the high pressure belt in this sector.
3 Lowest pressure in Gibraltar, evidently with easterly winds and probably a still greater low pressure anomaly to the south and south-west.
4 Very high temperatures at Gibraltar (unhomogenized series but positive anomalies of 2°–4° C in both winter and summer months appear common in the 1830s).
5 Fragmentary reports of an abnormal climatic regime at Madeira, tending to warmth and summer drought, and at Malta, where the persistent (summer and winter) drought raised public alarm, particularly between 1838 and 1841—incidentally leading to the institution of regular rainfall measurements.

It would probably be worth while to devote effort to establishing fuller information about the climate (and displacements of the normal climatic zones) during this period, particularly over North Africa and perhaps generally over as much of the world as possible.

At present, it seems relevant to point to the following circumstances:

1 A general upward trend of intensity of the zonal circulation was probably already well under way, and may have been so since before 1750 (see British Isles curve in Figure 3 (*c*)). This is also suggested by a study of the North Atlantic trade winds from 1827 onwards, using ships' logs, by Privett and Francis (1959): it appears, moreover, that the 1830s were a decade of strong trades, somewhat above the smoothed trend.
2 The 1830s more or less coincided with pronounced peaks in summer and winter (1820–33 and 1838 in January, 1830–39 in July) of Scandinavian blocking anticyclones—just possibly a periodic phenomenon with periodicities of 60–90 years associated with solar events (Scherhag, 1960). There were some exceptionally high monthly mean pressures in the January of this period at Trondheim and St. Petersburg (though St. Petersburg also had a few very low ones), with exceptionally strong, and prevalent, westerlies to the north and easterlies to the south.
3 Iceland and Greenland records suggest that enormous quantities of Arctic ice were broken up during the 1830s and drifted away into the Atlantic, followed by a period 1840–54 when ice was hardly ever seen near Iceland and when the permanent ice had retreated greatly both east and west of Greenland.

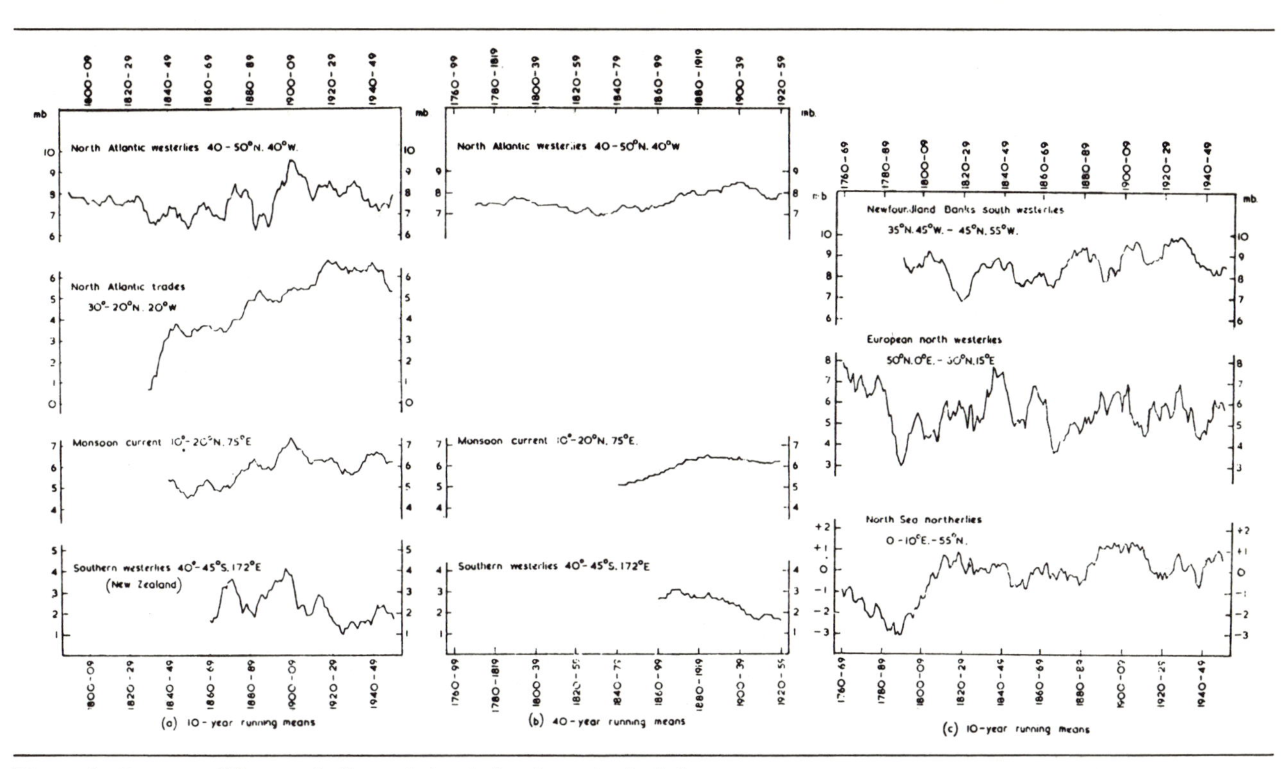

Figure 4 Pressure difference indices of circulation intensity in July.

4 Sydney and Adelaide had long dry periods about this time—Sydney from the 1820s to the early 1840s, Adelaide through the 1840s and possibly earlier, implying prevalence of far southern positions of the anticyclone belt and depression tracks in that sector of the Southern Hemisphere. In Chile increasing rainfall in the Santiago district from 1820 onwards (after a long very dry epoch) has been taken as a sign of the temperate and subpolar depression belt returning northwards.

5 The Antarctic sea ice seems also to have been more erratic in the 1830s and after than in the preceding decades. In 1832, 1840–44, and 1855, enormous amounts of ice broke away into latitudes between 60° and 40° S but seem to have left the higher latitudes unusually ice-free.

The Arctic ice never regained the extent which it apparently had before 1840. The circulation maps also make it doubtful whether any truly analogous patterns have occurred in recent years with those before 1840—a point which may be significant for long-range forecasting.

Consideration of the Probable Upper Air Circulation in the Period Around 1800

Figures 6 and 7 display the longitudes of lowest and highest m.s.l. pressure over the North Atlantic in January and July from 1790 to the present time. The latitudes used were chosen (*a*) to make use of the most reliable portions of the isobaric charts and (*b*) to be where the identifiable features are related to the prevalent cyclogenetic and anticyclogenetic effects east of upper troughs and ridges respectively.

The wave length L of a standing wave is taken to be expressed by the Rossby formula

$$U = \frac{\beta L^2}{4\pi^2}$$

where U is the prevailing zonal velocity of the upper westerlies and β is the rate of change of the Coriolis parameter with latitude at the latitude of the mainstream of the westerlies.

Both Figures 6 and 7 show that the longitude spacing of surface systems generally increased and decreased at times when the indices of circulation intensity were increasing and decreasing respectively. This is particularly clear with the spacing between the West Atlantic trough and mid-Atlantic ridge which are closest to the mainstream of the upper westerlies: the spacing of these surface features should correspond to about half a wave length. This result appears to give qualitative confirmation of the changes of circulation intensity found from the surface pressure gradients. But both in January and July the change of wave length is surely too great (implying at 25 to 30% change of intensity) to be accounted for wholly in this way. It seems therefore that some slight change of latitude of the strongest upper westerlies must also have taken place; in

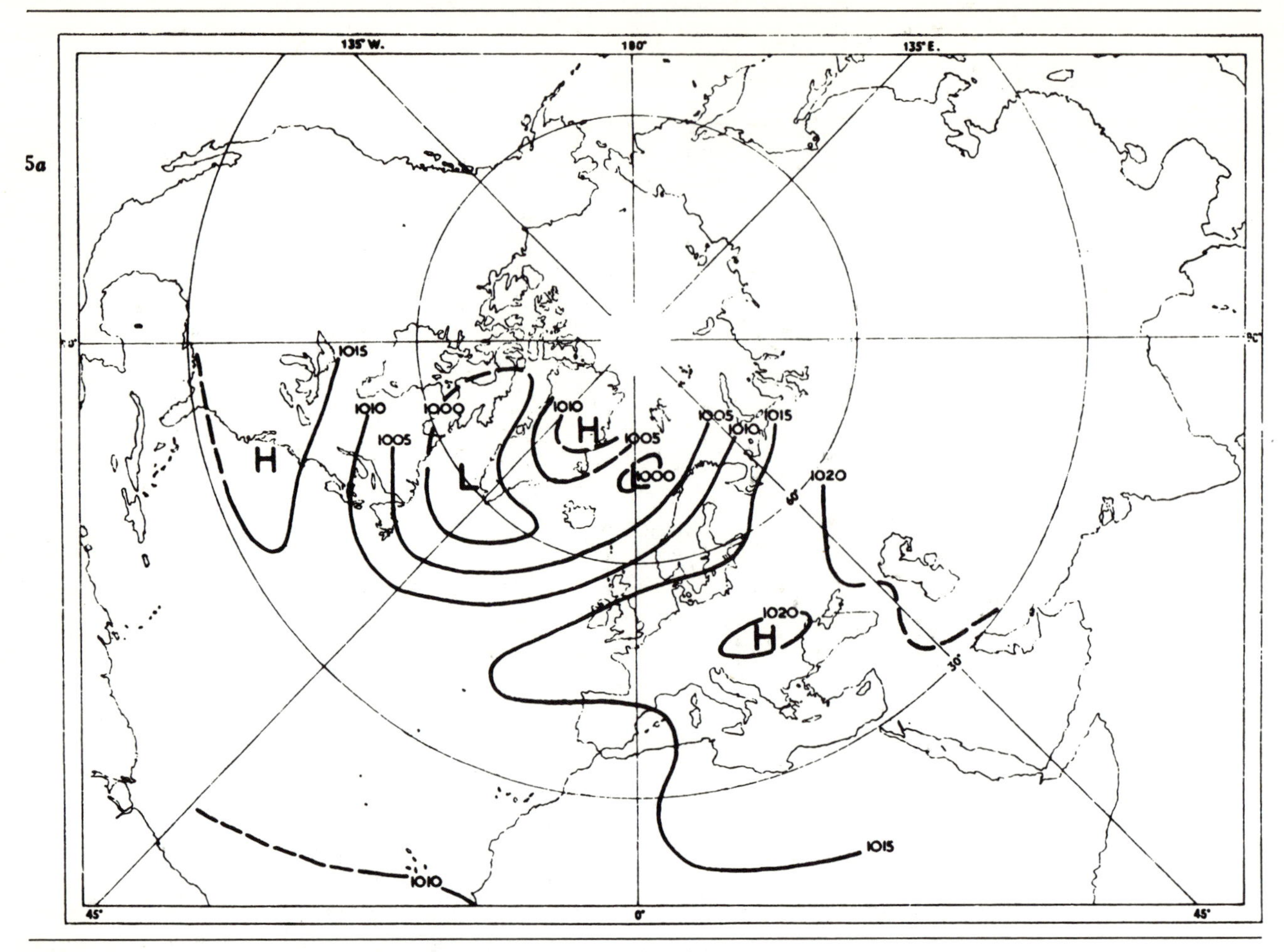

Figure 5 Average m.s.l. pressure for 1830–39: (a) January; (b) July.

particular, a poleward displacement, of a few degrees of latitude from the times of weak circulation around 1800–50 to the epoch of strongest circulation in the present century seems to be implied.

The consistency of these trends suggests that the surface patterns, and in particular the longitudes of certain features, may be used judiciously to indicate probable changes in the main tropospheric westerlies aloft. This principle is the basis of the experimental investigations described in this and the following section.

A likely first approximation to the 1,000–500 mb thickness patterns prevailing over the North Atlantic in January and July around AD 1800 was obtained (Figure 8) by considering the average sea-surface temperatures observed between 1780 and 1820 (Lamb and Johnson, 1959) and shifting the mean thickness isopleths from their modern positions by the same amount as the sea-surface isotherms. Broadly this preserved nearly saturated adiabatic equilibrium with the sea-surface temperature, except for the highest thicknesses in January and the lowest thicknesses in July where vertical stability prevails over some parts of the Atlantic—these exceptions were, however, at the edges of the region over which surface pressures could be mapped.

Over the neighboring land regions some slight equatorward displacement of the thickness values in the period around 1800 would be implied if a join-up with those over the ocean is to be achieved. Such a displacement might have been produced either by the dynamic effects of modified or southward-displaced upper westerlies or, more directly, by a small reduction of the effective insolation.

The latter suggestion gains *a priori* plausibility because Wexler (1956), in an attempt to explore the possibility of explaining Ice Age phenomena (precipitation pattern and atmospheric circulation) in terms of a persistent 20% reduction of the insolation due to volcanic dust, obtained estimated anomalies of the 700 millibar and deduced surface pressure patterns which amounted to a remarkable qualitative prediction of the pressure charts subsequently constructed by Lamb and Johnson (1959) from actual data for the period around AD 1800. In particular, Wexler predicted as features of the January pattern an elongated polar trough and intensified frontal activity along the Atlantic

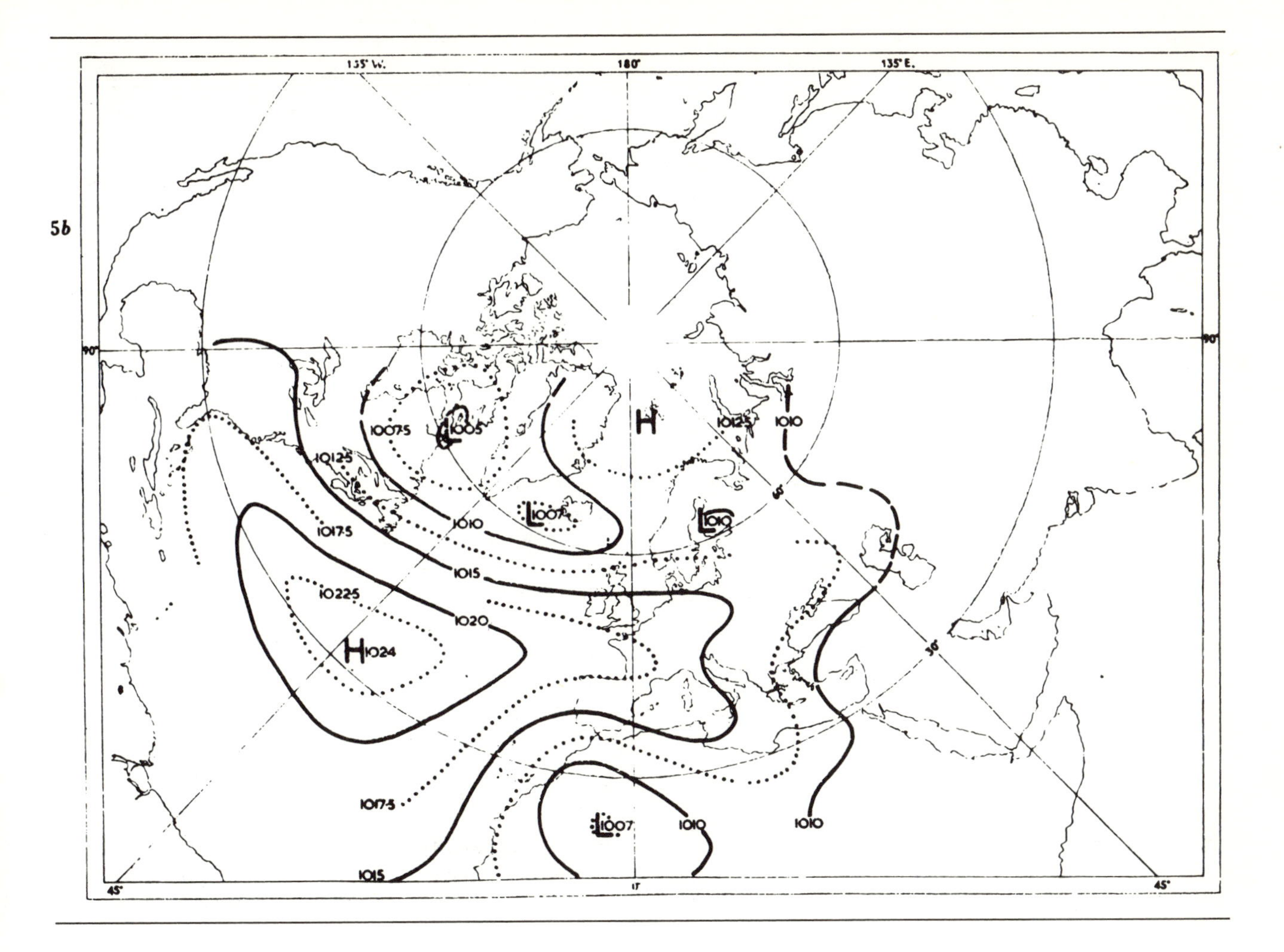

coast of the United States, higher pressure over Greenland and more prominent cold northerly flow over the Norwegian Sea and North Sea affecting the coasts of Scandinavia and Britain.

Wexler moved the modern normal insolation isopleths by distances sufficient to reduce the amount of insolation by 20%. Over the relatively cloud-free snow-covered interior of the continent he assumed the 1,000–700 millibar thickness values to be controlled by the insolation and moved the thickness lines by the same amount, so as to coincide with the same insolation.

Applying Wexler's argument in reverse to the 1,000–500 millibar thickness distribution around AD 1800, we see in Figure 8 (*a*) and (*b*) that a reasonable join-up with the thickness isopleth positions over the North Atlantic could be achieved if the isopleths over North America (and continental Eurasia and Africa) were shifted by an amount to correspond with a 5% reduction of insolation or rather less—possibly with any reduction between 0 and 5% or a little over. (This seems not unreasonable in view of the 5 to 10% reduction of circulation vigor over the North Atlantic in winter and the smaller reductions found at other seasons and places.) The only regions which did not fit were: (*a*) the coastal region near Newfoundland and Labrador, where a greater reduction of mean thickness over the sea (and neighboring land) may reasonably be attributed to colder water and more ice; (*b*) western and central Europe, where greater reductions of thickness seem to be required and may perhaps be taken as dynamic, i.e., linked with the other evidence of a shorter wave length and more western position of the quasi-permanent trough in the upper westerlies.

Figures 9 and 10 show the 500 millibar patterns derived by using these thickness patterns and 1,000 millibar contours corresponding to the actual pressure fields.

Since sea temperatures 1780–1820 were available for the North and South Atlantic Ocean, it seemed worth while to apply a similar experiment to the great ocean region as a whole, excluding areas disturbed by strong water currents, strong upwelling and zones of intense cloudiness (as in the equatorial and subpolar rain-belts). The zones of strong water currents and of upwelling are only fringe regions of the ocean but their delineation is unavoidably somewhat uncertain and arbitrary. Over

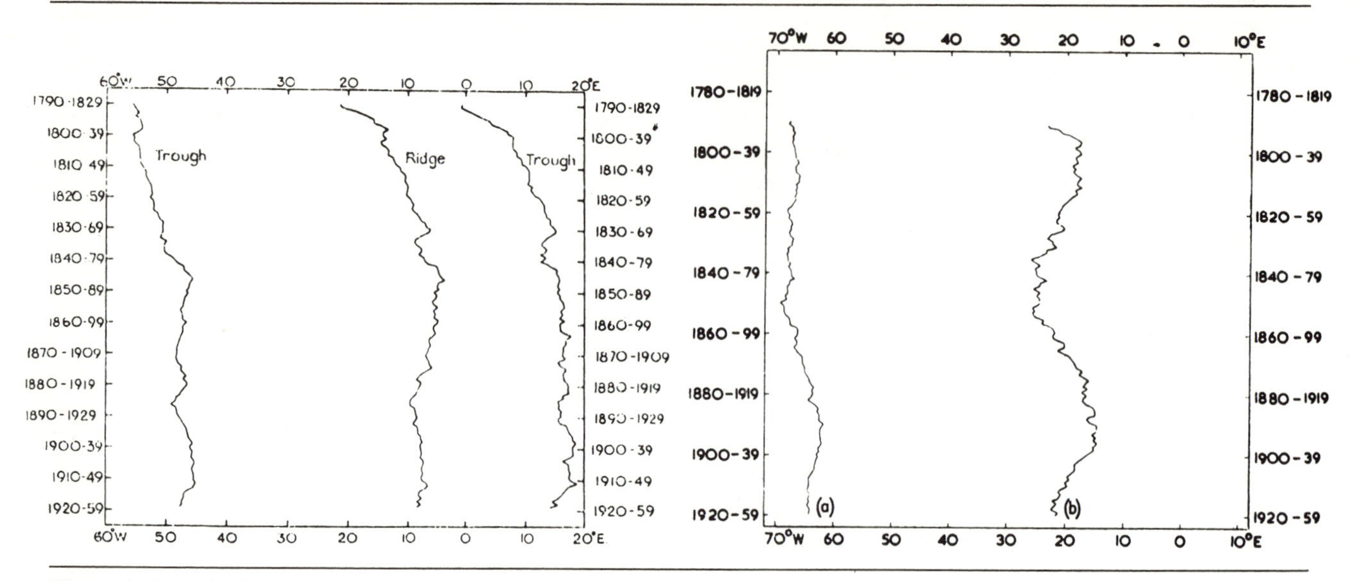

Figure 6 Longitudes of the semi-permanent surface pressure troughs and ridges at 45° N in the Atlantic sector in January (40-year running means).

Figure 7 Longitudes of the semi-permanent surface pressure troughs and ridges in the Atlantic sector in July (40-year running means): (a) longitude of trough at 55° N; (b) longitude of ridge at 55° N.

the broad remaining regions of the ocean, where both positive and negative anomalies occurred, it seems likely that the general (overall average) level of surface-water temperature and of 1,000–500 millibar thickness is controlled by the amount of insolation actually penetrating the atmosphere to the surface. The results of trying the obviously crude assumption that this was the whole explanation of shifts of sea temperature (and thickness) isopleths since 1800 were: (*a*) over the North Atlantic between 30° and 55° N, including the fringes where cold currents were probably expanded—an indicated reduction of insolation in 1780–1820 by 7 or 8% (almost certainly too large); (*b*) over the North Atlantic between 30° and 55° N, omitting the fringes—a reduction of 3 to 4%; (*c*) over the North and South Atlantic between 50° N and 40° S, omitting the fringes and the equatorial rain-belt—a reduction of 0.7 to 1.5%.

The result under (*c*) is possibly too small because the most extensive region of colder surface north of 50° N is excluded and apparently other cold areas near 40° S. We may guess that anomalies in the great Pacific Ocean were smaller, though on the whole in the same sense.

Summing up it seems therefore that there is a case for supposing a small reduction of, say, 1 to 2% in the average available insolation over the surface of the globe during the epoch 1780 to the 1820s compared with the present century. Over the Northern Hemisphere the reduction may have been rather over 2%. We notice, however, that the greatest displacement of the isopleths was over the North Atlantic Ocean, so that the increased Arctic ice and spread of cold water north of 50° N was playing perhaps the most immediately important part in modifying the atmospheric circulation pattern at that time. It seems further to be implied by the 500 millibar changes in Figures 9 and 10 that the circulation, although perhaps generally rather weaker around 1800, was locally strengthened where the cold surface over the continents and the Labrador current reached rather lower latitudes than today. Other features indicated were a shortened wave length[5] and large amplitude waves.

An Experiment in the Systematic Treatment of Documentary Weather Records since AD 800

Efforts are being made to extend our pictures of the monthly mean atmospheric circulation over Europe still farther back. Several decades before 1750 can be covered by numerous well kept registers of wind and weather —e.g., on board navy ships berthed for a month or more in various harbors from the Baltic and Iceland to the Mediterranean (and possibly farther afield). Assembly of this data takes time and could perhaps profitably

[5] Betin (1957) has noticed an interesting variation of climatic behavior in Europe which seems likely to be related to changing wave length and most frequent ridge and trough positions. Over the eighty years of decreasing ice between about 1870 and 1950 there was a negative correlation coefficient between Baltic ice and the level of the Caspian Sea, which is fed by rainfall in the Volga Basin, whereas the longer term trends of both since 1550 run parallel. Presumably between 1870 and 1950 cyclonic or anticyclonic conditions over the Baltic tended to cover the Volga Basin too, whereas for a long period previously this was not the case.

be pursued in different countries. It is known that there were drastic climatic vagaries in the decades that can be covered, particularly the 1690s.

A wealth of manuscript information from still earlier times regarding the character of particular months and seasons exists in state, local, monastic, manorial and personal accounts and chronicles. Compilations have been made by meteorologists and others in many countries, so that by now it is possible to attempt numerical assessment of various phenomena, though much care is needed and special techniques have to be devised to watch and allow for changes in the fullness of reporting.

As a first attempt at systematic use of this material to reveal something about the changes in the prevailing atmospheric circulation from the warm climate period of the early Middle Ages right through the Little Ice Age to the present day, the reports from different places in Europe between 45° and 55° N and between Ireland and Russia were used. This is a particularly suitable region for study because it is always affected by the behavior of the mainstream of the zonal circulation aloft and because reports are most abundant there. One can commonly compare contemporary events in the Mediterranean, Scandinavia and elsewhere. Immediate objects in view were to establish the climatic sequence in Europe more firmly, and in more detail, and to discover how far different longitudes underwent similar experiences.

The most reliable surface weather indications in the early manuscripts relevant to this study were thought to be:

1 Severity or mildness of the weather prevailing in the main winter months of December, January, and February. The effects upon landscape, transport, and the agricultural economy are likely to have been reported in all important cases. It should be possible to identify confidently the persistent spells: mild winter by rains, flooding and thunderstorms even in continental regions, also by early or out-of-season flowering of plants; severe winters by frozen rivers, lakes and seaways, and by many sorts of privation and damage.

2 Raininess or drought in summer. Again the effects upon the landscape and upon agriculture are reasonably sure to have achieved mention in all outstanding cases. Wet summers produce flooding and ruined crops, though highly colored accounts of individual thunderstorms may occur in otherwise good summers. Dry summers are known by parched ground and dwindling rivers, whilst the grain crops are usually good; forest fires are also particularly liable to occur. The rain character of a summer is surer of faithful recording than the temperature, since an oppressive heat wave might well be the only recorded reference to temperature in an otherwise poor summer.

Since long spells of weather of set character persisting from July to August are one of the most prominent features (Lamb, 1953) of the climate of temperate Europe (early recognized in the Saint Swithin and Seven Sleeper legends), and since the circulation patterns involved presumably correspond to the quasi-stationary pattern attained at the climax of the summer heating of the hemisphere in the individual years, July and August only were used as the time unit for summer.

The compilations available included: Buchinsky's (1957) for the Russian plain (references especially to the Ukraine, the Moscow region and Poland); Hennig's (1904), chiefly covering central Europe and Italy, but also including references to other places between Ireland and Poland; Easton's (1928) regarding winters in western (and central) Europe between Scandinavia and the Mediterranean; Vanderlinden's (1924) for Belgium; an unpublished collection kindly made available by Schove for the British Isles, including use of Britton's chronology of early time (1937) and original sources. Amplifying evidence from vine harvests in Luxembourg (Lahr, 1950), Baden (Muller, 1953) and ice on the Baltic (Betin, 1959) was also considered.

Crude numerical indices were then defined and applied to the years in groups of not less than a decade (thus eliminating uncertainties in early times regarding the exact year of a particular occurrence). Data for some groups of years appear full and self-consistent from quite early times—e.g., remarkable warmth and dryness of central European summers between AD 988 and 1000—but a complete sequence of decade characteristics can hardly begin before AD 1100. So as not to make too great demands in terms of approximately uniform standards of reporting only the simple indication of excess of mild or cold, wet or dry for individual decades is here attempted (Figure 11). Half-century means of the following indices appeared, however, to give reliable

8a

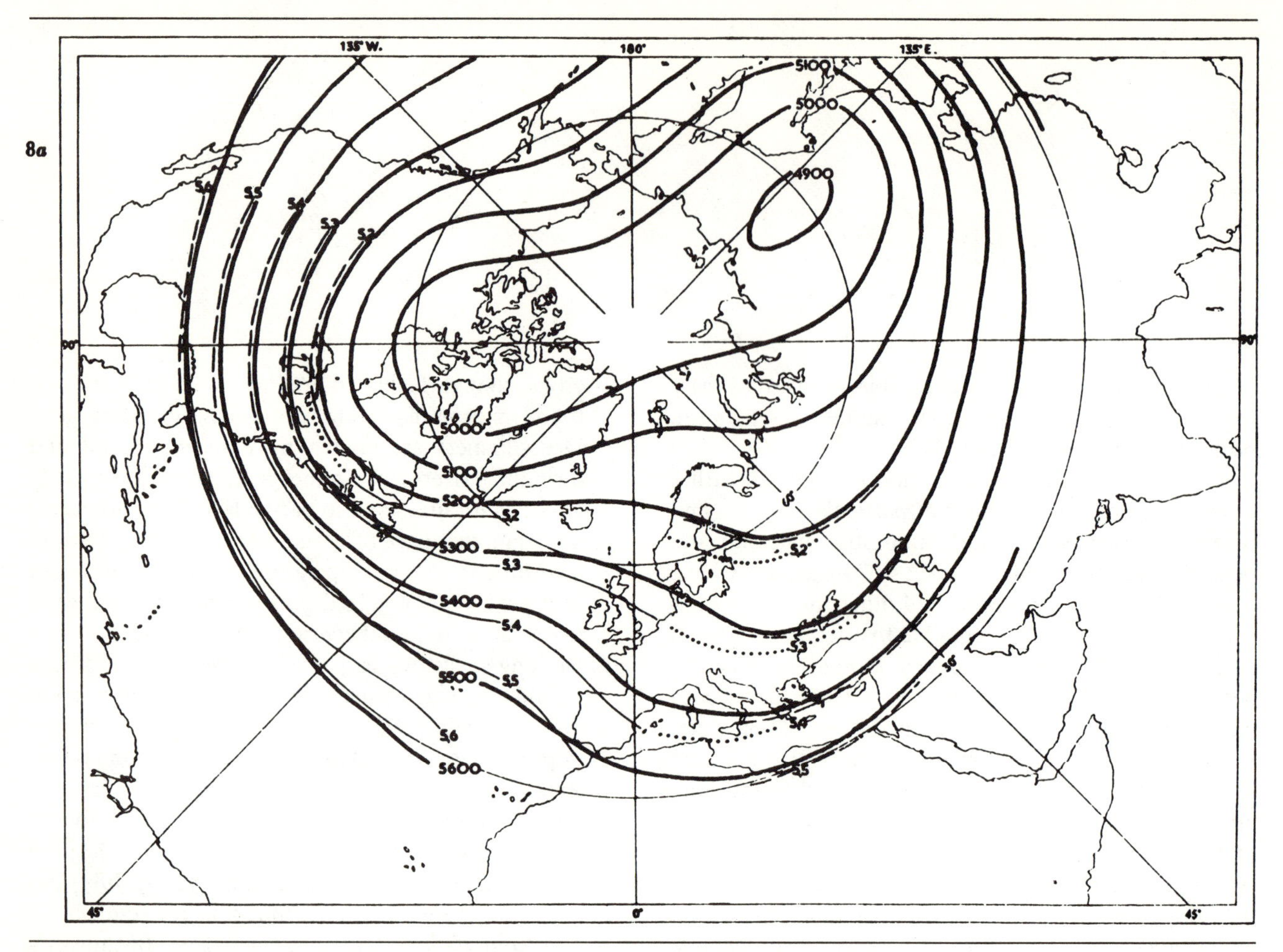

Figure 8 Average 1,000–500 mb thickness: (a) January; (b) July. Bold lines averages for recent years: (a) 1950–58; (b) 1949–57. Narrow lines suggested for 1780–1820 from sea temperatures. Broken lines suggested for about 5 per cent reduction of insolation.

numerical values from an earlier date, and have been tentatively extended back to AD 800.

1 Winter severity index. The excess number of unmistakably mild or cold winter months (December, January, and February only) over months of unmistakably opposite character per decade—excess of cold months counted negative. (Unremarkable decades score about 0. Extreme decade values of the index in Europe range from about +10 to −20.)

2 Summer wetness index. Each month (July and August only) with material evidence of drought counted 0, unremarkable months 0.5, months with material evidence of frequent rains and wetness counted 1. (Unremarkable decades score about 10. Extreme decade values of the index in Europe range from about 4 to 17.)

For the epoch since regular weather observations became available, these were preferred for identifying notably wet or dry, mild or severe months.

The earliest and longest records here used are a daily weather register from Hesse 1621–50 (Lenke, 1960), central England temperatures from 1680 to 1952 (Manley, 1953, and private communication), rainfall in England and Wales from 1727 (Nicholas and Glasspoole, 1931) and in Holland from 1735 (Labrijn, 1945).

The problem of welding the earlier and later series of months considered notable into a single series was tackled by studying overlap periods of several decades for which both instrument measurements and descriptive chronicles were to hand. The overall numbers of months marked as displaying noteworthy anomalies remained satisfactorily steady between 1100 and 1550 at about one-third of all the winter months and 40% of the Julys and Augusts. (The frequency of months noted in the Russian chronicles was an exception, being fairly steadily about half that for more western longitudes and tending to concentrate on months of disastrous severity: index values obtained from this source were therefore doubled to assimilate them to the others.) Rather higher frequency of months noted as extreme between about 1550 and 1700 was thought more likely due to the peculiar climate of that time than to relaxed criteria of extreme weather. The criteria adopted in using the instrument records of the period since 1680 to 1800 therefore were: (*a*) winter months counted as mild or severe if the temperature anomaly exceeded the

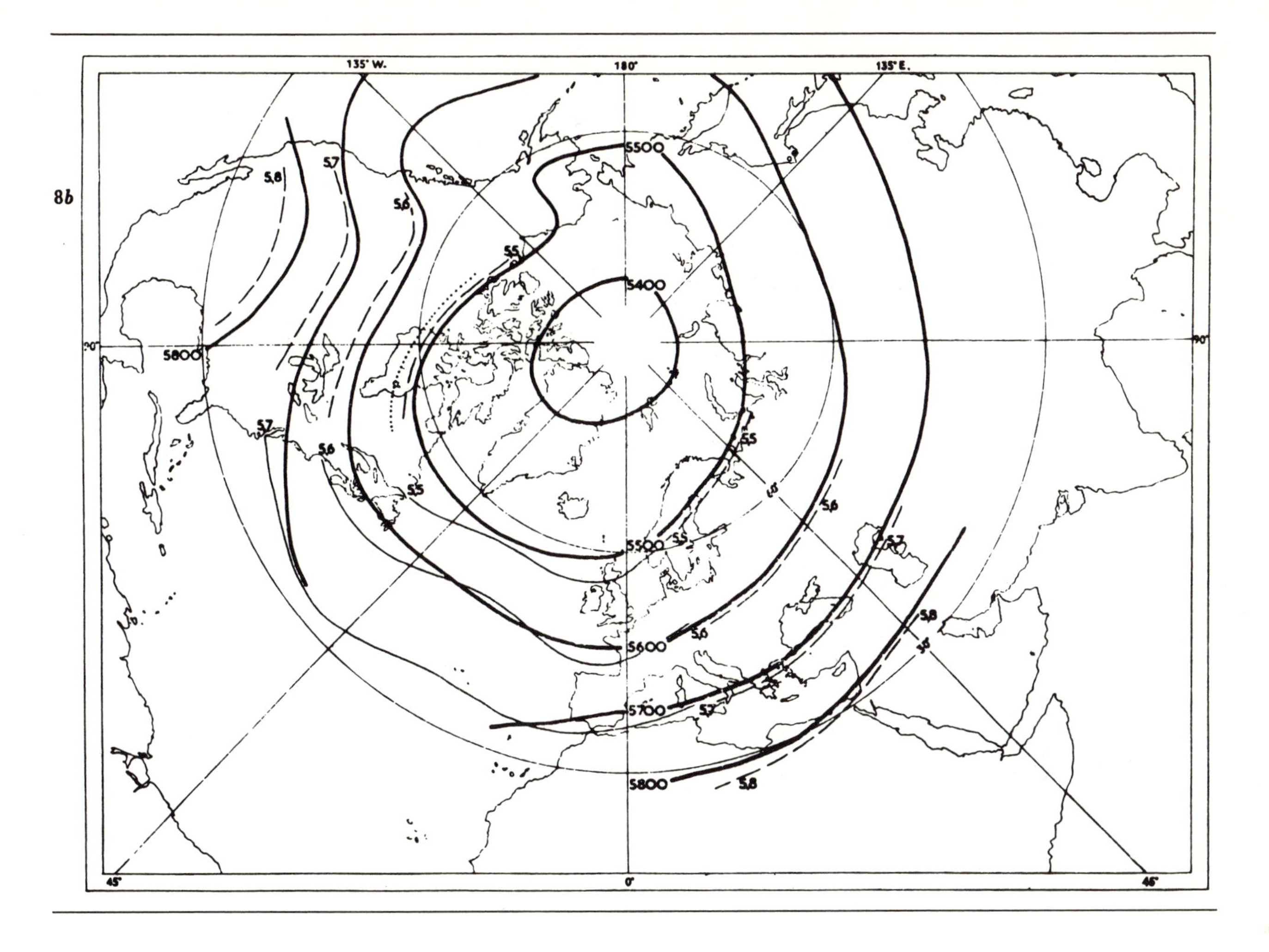

standard deviation from the longest period mean; (*b*) Julys and Augusts counted as wet or dry if the rainfall measured was within the highest or lowest quintile.

The possibility of a slight "change of zero" in the 1700s or early 1800s must, however, still be borne in mind. With this reservation Figure 11 displays the course of the summer and winter climate of Europe in different longitudes near 50° N, by decades, from 1100 to 1959.

All longitudes are not always affected by the same anomalies at the same time. Correlation coefficients between the winter index values in Britain and Germany and in Britain and Russia 1100–1750 were respectively +0.45 and +0.31: these both appear statistically significant at the 1% level, but in both cases coefficients with reversed sign were found in some centuries during this period.

Comparisons of the winter severity index in Europe with an indicator of winter character in Japan—the freezing dates of Lake Suwa (Arakawa, 1954)—produced no significant correlation coefficients.

There seems (Figure 11) to have been more tendency for like character of the winters, and of the summers, in all European longitudes between 1150 and 1250—in the case of the winters as late as 1350 or after—and since 1850 than at most other times. These have been the periods of most predominance of mild winters, presumably with westerly winds sweeping far across Europe, and good summers. The predominance of mild winters, greatest between 1150 and 1300, but also noteworthy in certain earlier and later periods (Figure 12), was punctuated by individual decades with cold winters everywhere in Europe near 50° N. There were roughly half-century intervals between the decades with most mild winters, also between those with most cold winters—perhaps another suggestion of a periodicity in the occurrence of blocking anticyclones over North Europe.

Between 1550 and 1700 all hints of this oscillation are lost owing to the heavy preponderance of cold winters, especially in Russia and Britain. This distribution suggests that the westerlies were weak and that northerly windstreams over the Russian plain and over the Norwegian Sea were important—the latter would tend to explain the very rapid worsening of the

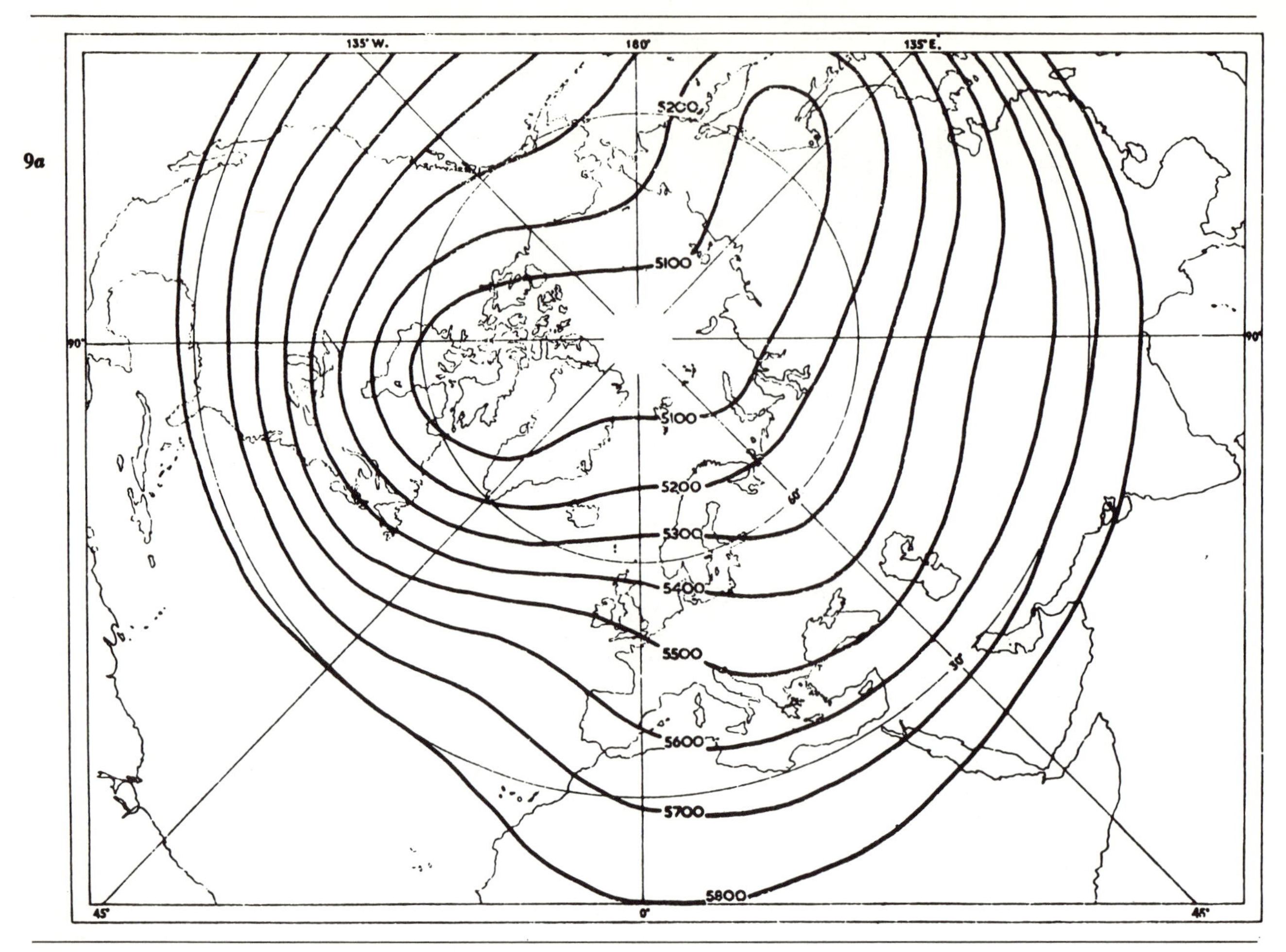

Figure 9 Average 500 mb contours for January: (a) 1949–58; (b) 1780–1830. (Based on Figures 1(a) and 8(a).)

ice situation in Iceland and Greenland waters, especially after 1550. Preponderance of wet summers (Julys and Augusts) in all European longitudes near 50° N between 1550 and 1700 is also noticeable.

Examining the smoothed trends of summer wetness and winter severity indices presented by the running five-decade (half-century) average values in Figure 12, one apparently discovers a general westward progress across Europe of a region of maximum summer wetness between 1250 and 1400–1500 and a corresponding eastward progress during the climatic recovery between about 1700 and 1900. Also shown is a general westward retreat from about 1200 onwards of the predominance of mild winters and a return movement (more clearly seen from the isopleths for predominance of cold winters withdrawing east) between 1700 and 1900. There are complexities of detail which tend to obscure these trends, but summer and winter sequences—the one of temperature, the other of a different element—appear to run parallel during both the climatic decline preceding the Little Ice Age and the recovery afterwards. Moreover the fact that any semblance of an orderly progression should be revealed by such primitive data may seem surprising and possibly justifies one in disregarding complexities.

Viewed in this light, the epochs 1550–1700 and about 1000–1200 appear as times of standstill, when the trend was halted and whatever fluctuations went on affected all European longitudes alike. These are perhaps properly regarded as the culminating periods respectively of the Little Ice Age and of the warm climate (secondary optimum) before it.

Figure 12 even suggests a still earlier period, between 800 or earlier and 900, of eastward spread across Europe of a region of predominantly mild winters and dry summers. Six to ten notably severe winters seem to have occurred rather earlier, between 764 and 860, mostly still more remarkable in the eastern Mediterranean (ice on the Adriatic, the Bosphorus, Dardanelles and Nile); there was apparently only one more similar occurrence afterwards—in 1011—until the 1600s.

Severe winter weather in Europe is related to a westward and southward shift (or expansion) of the cold trough in the upper westerlies normally found nowadays over eastern Europe; mild winter weather is similarly closely related to the East Atlantic warm

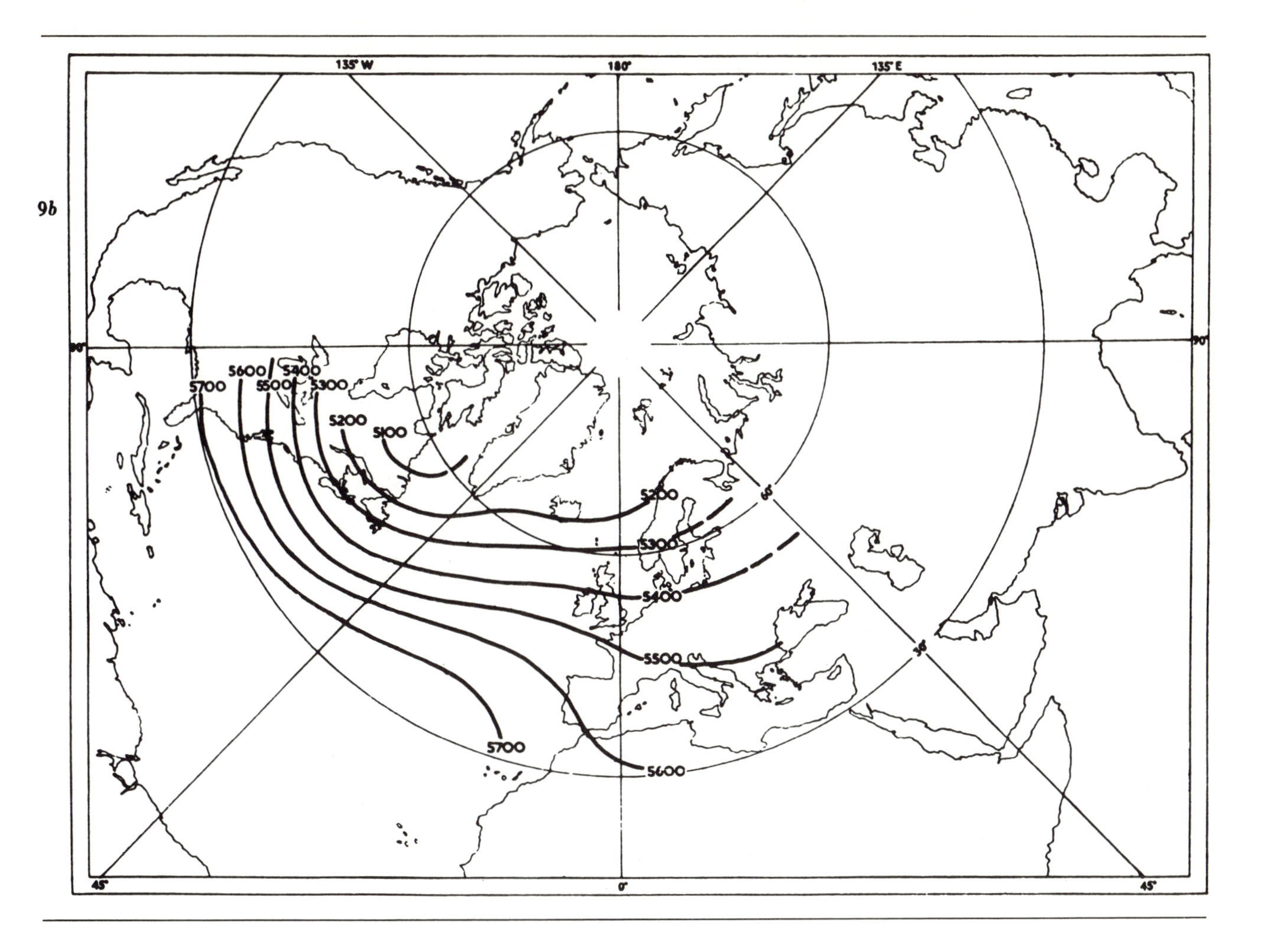

ridge (Figure 9). Wetness of the summers depends not only on proximity to the depression track (around 62° N in this sector nowadays), but tends to be most pronounced in the region of the upper cold trough normally discernible over Europe and immediately east thereof (Figure 10).

Westward and eastward movements of the features displayed in Figures 11 and 12 before and after the Little Ice Age are such as might accompany respectively shortening and lengthening of the waves in the upper westerlies downstream from a more nearly anchored ridge in the vicinity of the Rocky Mountains.

The experiment was therefore tried of measuring the apparent longitudinal displacements of the features identified on Figures 11 and 12, with the following results:

Winter: in 1550–1700 maximum winter mildness some 20°–30° longitude west of today's and in 1150–1300 once more about today's position.

Summer: in 1550–1700 greatest predominance of wet Julys and Augusts 15°–30° longitude farther west than in recent times; before 1300 farther east than nowadays.

Table 1 summarizes measurements, and directly derived estimates, of parameters relating to the circulation over the North Atlantic sector.

The latitude of the strongest flow at the 500 millibar level is taken as the middle of the zone of closest packed isopleths in the neighborhood of the point of inflexion (between trough and ridge) in mid-Atlantic. The latitude of the depression track is taken over the eastern Atlantic between about 20° W and 10° E, in order to be related to the 500 millibar measurement yet more indicative of weather in temperate Europe.

The longitude of the European upper trough is measured in 45°–55° N, the same latitude zone to which the summer wetness and winter severity indices (Figures 11 and 12) apply. The estimates of wave length, and of change of wave length from the modern normal, allow for a smaller sympathetic westward or eastward movement of the cold trough near the Atlantic margin of North America.

These measurements appear to admit the solutions in terms of change of latitude and/or intensity of the main circulation features which are outlined in the following paragraphs:

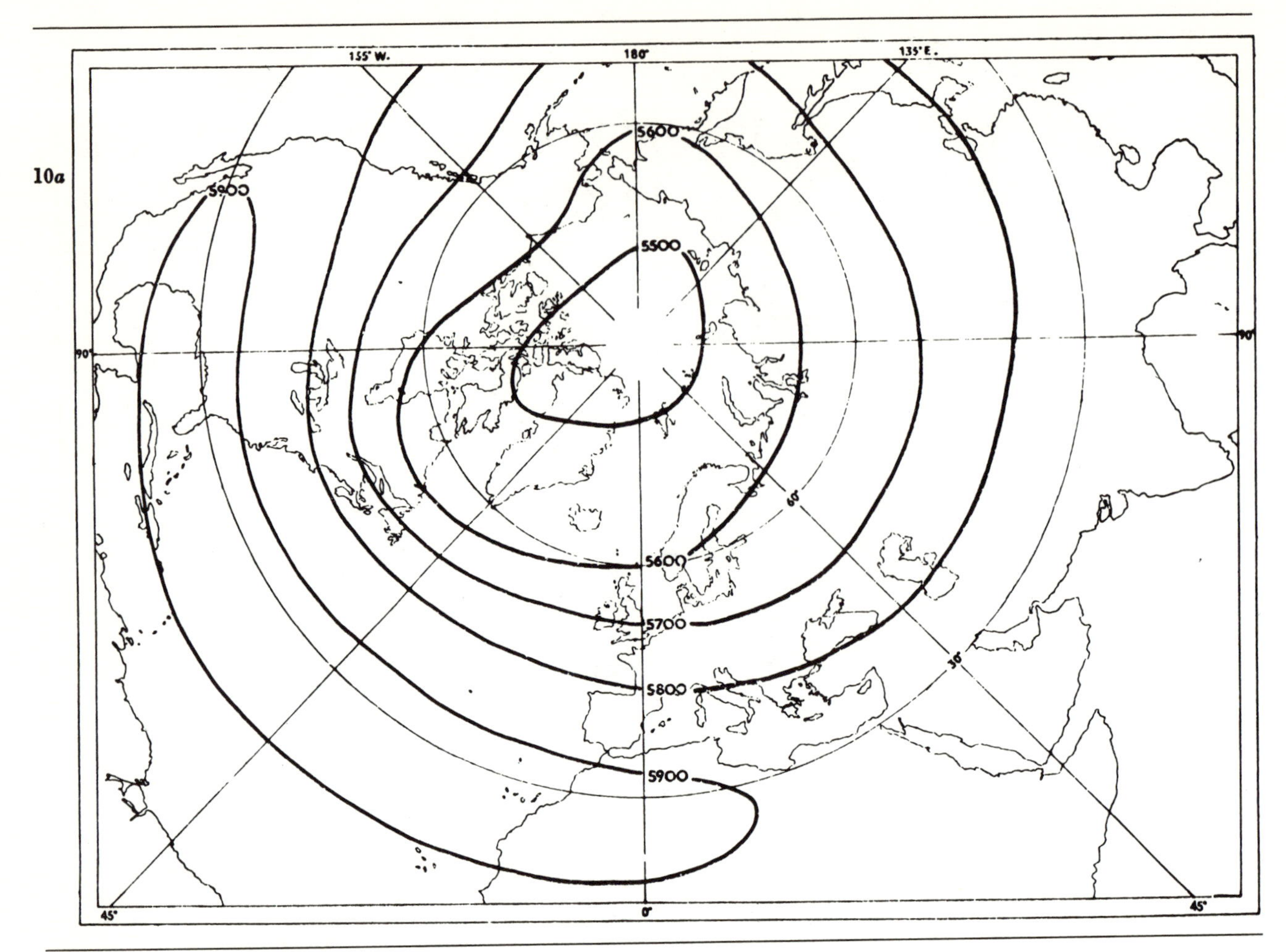

Figure 10 Average 500 mb contours for July: (a) 1949–58; (b) 1780–1830. (Based on Figures 2(a) and 8(b).)

Table 1 Key parameters of the atmospheric circulation, North Atlantic sector

	Modern normal 1900–39 (1949–58 500 mb values)	Extreme years in the early 1940s (partly after Scherhag)	Little Ice Age culminating period 1550–1700	Early Middle Age warm epoch 1000–1200
Summer (July)				
Latitude of strongest 500-mb flow	48° N	50°–52° N		
Departure from modern normal	—	+3°		
Longitude of European trough at 500 mb	10°–20° E	15°–20° E	0°–10° W[a]	
Departure from modern normal	—	Slight +	−15° to 30°	+10
Wave length change implied (° longitude)	—	+6°	(−10° to 20°)[b]	(+5° to 10°)
Wave length (° longitude)	78°	84°	(65°–70°)	(83°–88°)
Latitude of depression track	62° N	65° N		
Departure from modern normal	—	+3°		
Winter (January)				
Latitude of strongest 500-mb flow	60° N	55°–60° N		
Departure from modern normal	—	−3°		
Longitude of European trough at 500 mb	40°–50° E	25°–33° E	10°–20° E[a]	
Departure from modern normal	—	−15°	−20° to 30°	Slight
Wave length change implied (° longitude)	—	−10° to 15°	(−20°)	
Wave length (° longitude)	120°–130°	110°	(100°–110°)	
Latitude of depression track	68° N	65° N		
Departure from normal	—	−3°		

[a] Values read off the charts for the period around 1800.
[b] Values in parentheses are derived estimates; all other values measured from the charts and diagrams.

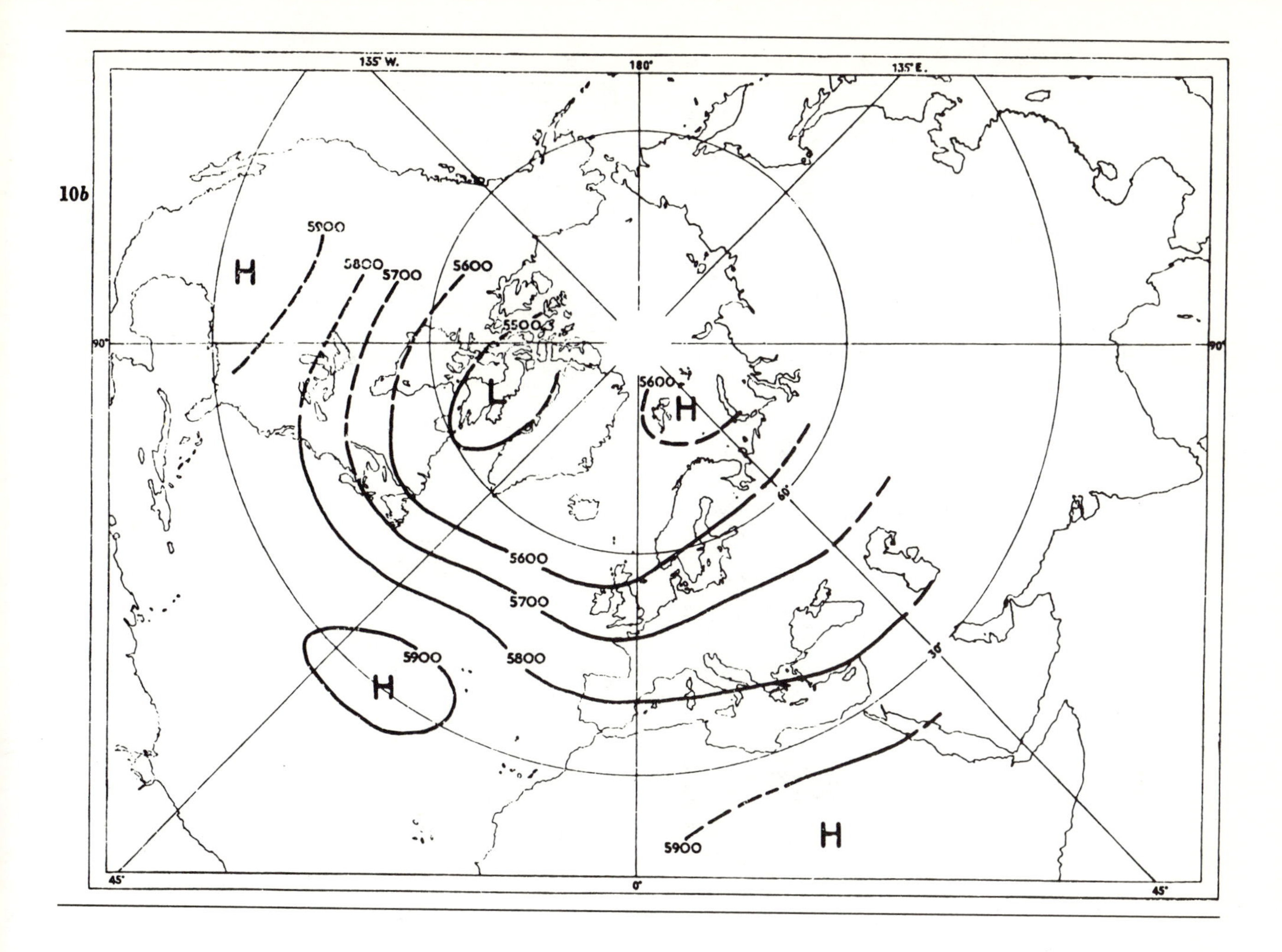

LITTLE ICE AGE CULMINATING PERIOD 1550–1700

Summer

Either flow weakened by about 30% or strongest flow shifted south by 5° or more. The likeliest solution, following the indications regarding intensity changes and latitude shifts on the charts around 1800 and earlier, including fragmentary charts for the 1690s, appears to be southward shift of the main flow by 4°–5°, with only trivial, if any, weakening of the summer circulation in the North Atlantic sector.

Main depression track was 57°–60° N over the eastern Atlantic sector, possibly rather south of 57° N over the ocean itself, but reaching 60° N over Russia.

There should have been also a change in prevailing wave number, five to six waves having been normal around the hemisphere, possibly at times a good fit with a five-wave pattern which might become correspondingly firmly established. (The modern normal wave length is intermediate between that required for a four- and a five-wave pattern.)

Using the indications of Figure 10 (*b*), we may judge the likeliest positions for summer cold troughs in middle latitudes in a five-wave pattern during the Little Ice Age epoch as (*a*) near 60°–70° W; (*b*) 0°–10° W; (*c*) 60°–70° E; (*d*) 130°–140° E; (*e*) 130° W.

Winter

Either flow weakened by about 30% or strongest flow shifted south by 5° or more. The likeliest solution seems to be that the flow was in general weakened over the North Atlantic by 5–10% in winter and the mainstream shifted south by 3°–5°.

Main depression track was 63°–65° N. It is obvious from the charts around 1800 that this position of the main depression track allowed for a scatter with many more depressions than now entering the Mediterranean.

There should have been a change of wave-number around the hemisphere from the twentieth-century predominance of three-wave patterns at the times of most vigorous winter circulation, especially in January, to one where four-wave patterns were commoner, often even at the peak intensity of the winter circulation. (This is a feature which has tended to reappear in recent

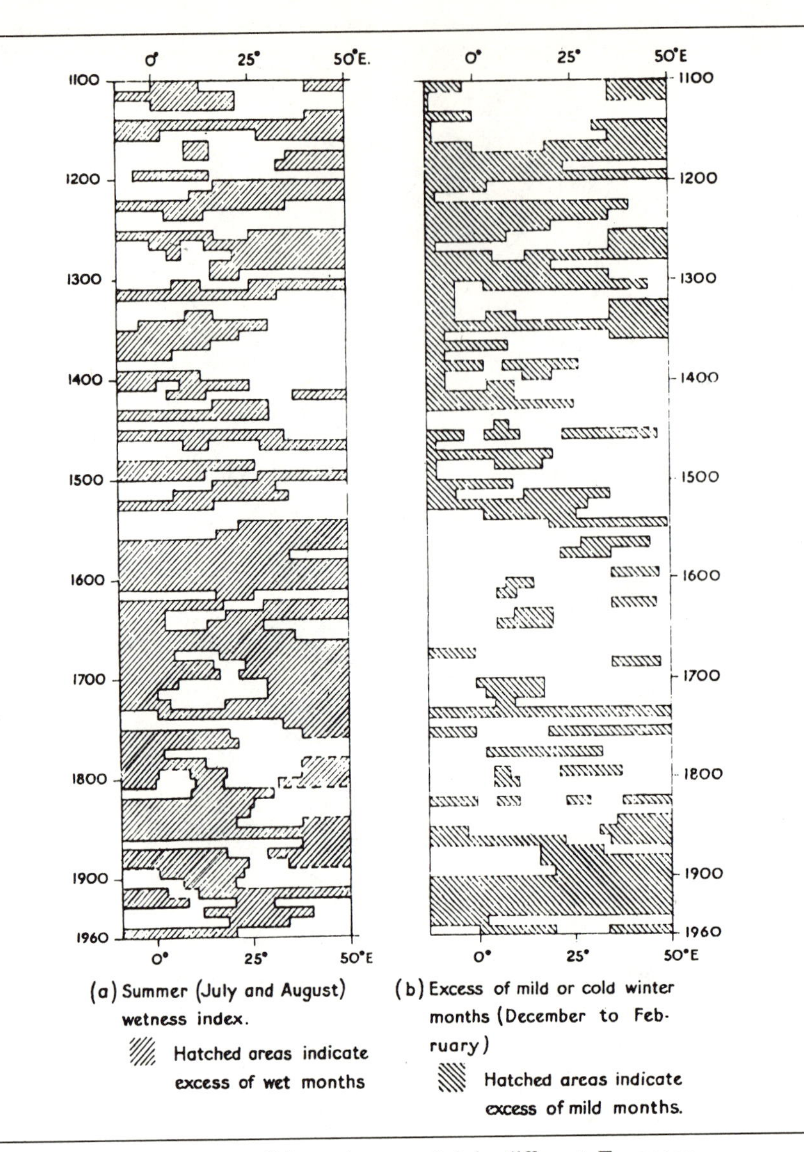

Figure 11 Summer wetness index and winter mildness (or severity) in different European longitudes near 50° N by decades from 1100 to 1959.

years, in the weaker and southward displaced circulations in several winters of the 1940s and 1950s.)

Using the indications of Figure 9 (*b*), we may judge the likeliest positions for winter cold troughs in middle latitudes in the Little Ice Age epoch as: (*a*) 80°–90° W; (*b*) 10°–20° E; (*c*) 110°–130° E; (*d*) more variable in position, but probably tending to be near enough to North America at times to reduce the Rocky Mountains warm ridge to insignificance. (This is a suggestion which might explain an apparent tendency even in the nineteenth century for severe winters to occur simultaneously in western and eastern North America and Europe.)

EARLY MIDDLE AGES WARM EPOCH 1000–1200

Summer

Either flow 30% stronger than now or strongest flow shifted north by 3° to 5° from the modern normal. The likeliest solution seems again to be that the main anomaly was one of latitude position.

Main depression track over the eastern Atlantic was 65°–67° N. Presumably there was considerable similarity with 1959 and the best summers of the 1930s and 1940s in Europe (1933–35, 1941–43, 1945, 1947, 1949); the marked aridity of the 1100s in Europe probably means that the depression track was then a degree or two still farther north, passing near the east coast of Greenland and generally near 70° N in the European sector. This would be consistent with the suggestion of little floating ice on the Arctic seas south of 80° N and an anticyclone belt over Europe, whilst Mediterranean summer droughts should have experienced more breaks than nowadays.

Winter

The evidence suggests a more northerly position of the depression track than now, perhaps allowing the same

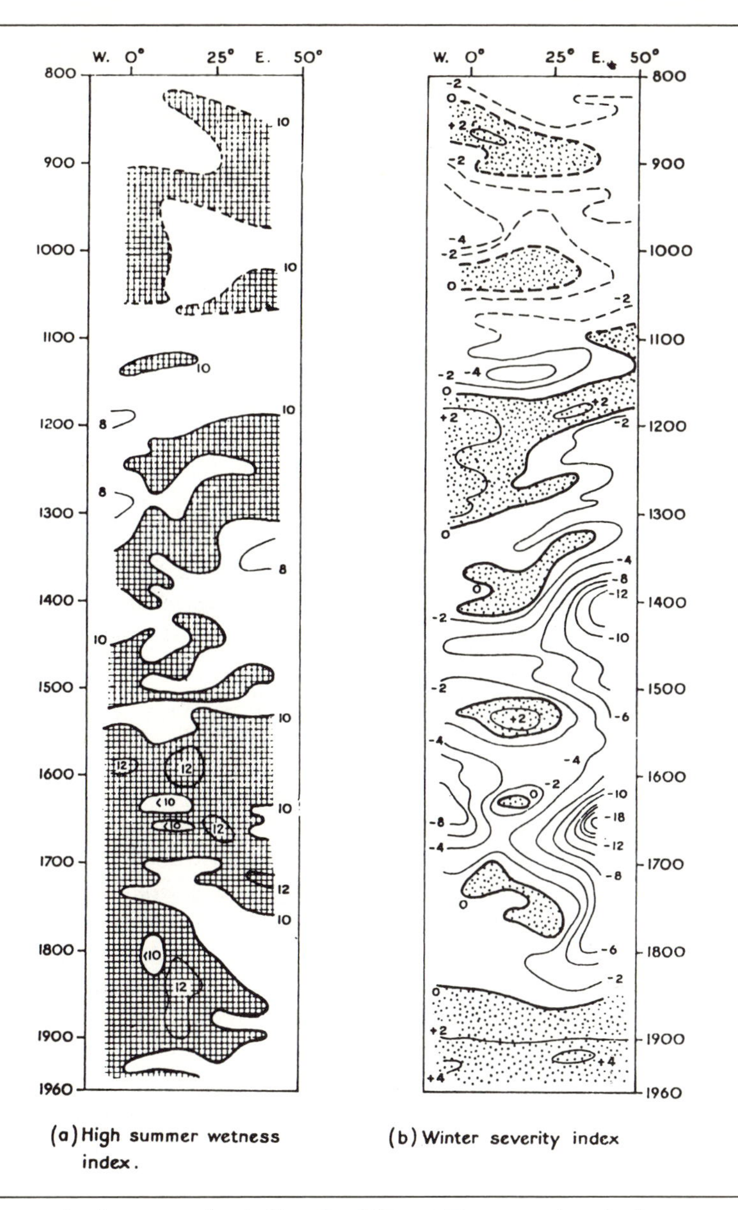

Figure 12 Summer wetness and winter severity indices in different European longitudes near 50° N (overlapping half-century means) from 800 to 1959. Cross hatching indicates excess of wet Julys and Augusts. Dots indicate excess of mild Decembers, Januarys, and Februarys.

wave length as nowadays with a rather weaker atmospheric circulation. If we assume that a depression track in the Barents Sea was common owing to the less extent of ice, there may have been periods when a two-wave pattern was attained at the climax of the winter circulation vigor. The cold belt across Europe near 50° N (Figures 11 and 12) in the winters of the 1100s would then be most readily attributed to easterly winds with a long land track, as would also occur in other years and decades when blocking anticyclones were prominent. This pattern would probably imply more rain-giving depressions in the Mediterranean than normal in the winters of the twentieth century, but also less extreme cold than at those times when northerly outbreaks over Russia and the Norwegian-Greenland Sea were commoner (e.g., around AD 800 and 1600).

Botanical studies, which already suggest an important increase of total rainfall in northern Europe around 1200–1300 and a drier time between 1700 and 1800 (anticyclonic summers and winters in central eastern Europe indicated by the charts of 1750–1800), may be able to add more firm indications of the prevailing temperatures, rainfall and latitudes of depression tracks in different centuries. Further evidence might also come from similar studies of documentary records of seasonal weather character in the latitudes of the Mediterranean and northern Europe and in the Far East, whilst archaeological studies in the Americas may have a part to play.

TENTATIVE GENERAL CONCLUSIONS

The climates of the different epochs here discussed are amenable to interpretation largely in terms of (*a*) intensity changes and (*b*) latitude shifts of the main limbs of the zonal circulation, accompanied by appropriate changes of wave length and trough positions in the belt of westerlies. The latitude and longitude changes must have had direct consequences in modifying climates in every latitude zone and have perhaps been more conspicuous than the changes of circulation strength.

There is some evidence for supposing that both the circulation in general and the radiation available were weakened during the Little Ice Age AD 1430–1850 by a few per cent in the Northern Hemisphere and perhaps by about 1–2% over the world, most probably due to volcanic dust. Correspondingly, it might be reasonable to suppose that the circulation in general, and (more doubtfully) the radiation available, were slightly above their twentieth-century strength in the early Middle Ages (especially 1000–1200), though it would be safer to assume that they were not far different from modern values. Possibly the chief difference between the modern situation and that of the early Middle Ages warm epoch and of the major post-glacial climatic optimum around 5000–3000 BC lies in the duration of these warm epochs and the sea temperatures consequently attained in the Atlantic and Arctic.

As the energy of the circulation increased (at least between 1800 and recent decades) the Northern Hemisphere circulation seems in general to have shifted poleward, perhaps mainly controlled by the displacement of the main thermal gradient accompanying the shrinking Arctic ice and winter snow cover on land.

It is clear that the extended Arctic ice and cold water of the period around AD 1800 and earlier modified the Atlantic sector circulation, causing peculiarly great latitude shifts there and even local strengthening of the circulation and thermal gradients near the southern extremities of the extended cold troughs—in an epoch when over wider regions the energy was reduced.

Correspondingly, in the early Middle Ages warm epoch, when the energy generally available may be supposed to have been greater (if only because of higher sea temperatures), the circulation seems to have been not always stronger than now (and possibly sometimes weaker) in northern Europe. The likeliest reason for this seems to be the remoteness at that time of the Arctic ice limit. Indeed, the patterns of that time may have some bearing upon what would happen in various latitudes if the Arctic ice were artificially disposed of.

When the general atmospheric circulation increases in energy the strongest (most disturbing) effects should be expected in those sectors where a broad quasi-permanent ice or cold-water surface has protruded farthest towards low latitudes. At times very strong circulations and very abnormal patterns might then be produced and become very variable from year to year if the protrusion of Arctic ice were to break up or shrink rapidly. The peculiar climatic course of the 1830s in the North Atlantic and neighboring regions should probably be viewed in this light. The decade was one of extreme variations between persistent blocking and intense zonal circulations, most strongly developed in unusually high latitudes, so that the low values of some of the intensity indices in that decade may be misleading —indices taken in what are at most other times the best positions being just then unrepresentative.

Wexler has pointed out that any long-period changes in the insolation available should produce a much quicker response over the great land masses than in the oceans and regions of quasi-permanent ice. This seems to be supported by our study of the state of the general circulation around 1800, when a trend towards increasing energy may (according to some indications) have been already under way for about a century, and we observe a peculiarly great equatorward displacement of the circulation over the North Atlantic Ocean.

Peculiar instability of the climate in the Atlantic-European sector at various times between 1250 and 1550, with harsh alternations of wet and dry, warm and cold years in various (especially eastern) parts of Europe in the 1300s and (especially) 1400s, may perhaps be regarded as a phenomenon of the trend towards colder climate and increasing areas of snow and ice surface crudely corresponding to the vicissitudes of the 1830s during the recovery trend.

Clearly general trends of circulation strength must be judged by measures of the circulation in many different parts of the world and allowance made for passing phases in which misleading effects arise in regions affected by persistent ice or cold water—especially the North Atlantic.

To judge the position of the most recent decades in relation to longer-term trends it may be necessary to consider whether the appearance of rhythmic changes of wave position, wave length and circulation intensity affecting Europe—rises to maxima around 1100 and 1900–30 and a minimum around 1650—represents some long-period oscillation. So far the evidence seems against this, since the fluctuation was probably asymmetric in time—the decrease spread over 300–500 years, the recovery nearly complete in 150–200 years—and the Southern Hemisphere was probably not affected. Nevertheless there is a good deal of evidence that the general circulation has for the time being fallen away from its maximum intensity around 1900–30—evidently not due to volcanic dust—and the trend of the North Atlantic circulation towards lower latitudes in winter (and probably other seasons) needs watching and explaining. In this connexion the case made out for solar weather relationships by Willett (1949, 1961) and Baur (1956, 1958) cannot be ignored.

REFERENCES

Ahlmann, H. W., 1944. Nutidens Antarktis och istidens Skandinavien. *Geol. Fören. Stockh. Föhr. 66*, 635–654.

Alissow, B. P., Drosdow, O. A., and Rubinstein, E. S., 1956. *Lehrbuch der Klimatologie*. Deutscher Verlag der Wissenschaften, Berlin.

Arakawa, H., 1954. Five centuries of freezing dates of Lake Suwa in central Japan. *Arch. Met., Wien, B. 6*, 152–166.

Arakawa, H., 1956. Climatic change as revealed by the blooming dates of the cherry blossoms at Kyoto. *J. Meteor. 13*, 599–600.

Arakawa, H., 1957. Climatic change as revealed by data from the Far East. *Weather 12*, 46–51.

Auer, V., 1958. *The Pleistocene of Fuego-Patagonia. Part II: History of the flora and vegetation.* Suom. Tiedeakat., Geologia-Geographica. Helsinki.

Auer, V., 1960. The Quaternary history of Fuego-Patagonia. *Proc. Royal Soc., B, 152*, 507–516.

Aurrousseau, M., 1958. Surface temperatures of the Australian seas. *J. and Proc. Royal Soc. N.S.W. (Sydney) 92(IV)*, 104–114.

Baur, F., 1956. *Physikalisch-statistische Regeln als Grundlagen für Wetterund Witterungs-Vorhersagen 1.* Akademische Verlagsgesellschaft, Frankfurt am Main.

Baur, F., 1958. *Physikalisch-statistische Regeln als Grundlagen für Wetterund Witterungs-Vorhersagen 2.* Akademische Verlagsgesellschaft, Frankfurt am Main.

Betin, V. V., 1957. Ice conditions in the Baltic and its approaches and their long-term variations. *Gosudarst. Okeanogr. Inst. Trudy 41*, 54–125.

Betin, V. V., 1959. Variations in the state of the ice on the Baltic Sea and Danish Sound. *Gosudarst. Okeanogr. Inst. Trudy 37*, 3–13.

Britton, C. E., 1937. A meteorological chronology to AD 1450. *Geophys. Mem. 70.* Meteorological Office, London.

Brooks, C. E. P., 1949. *Climate through the Ages.* (2nd ed.), London, Ernest Benn.

Buchinsky, I. E., 1957. *The Past Climate of the Russian Plain.* (2nd ed.), Leningrad, Gidrometeoizdat.

Butzer, K. W., 1957a. Mediterranean pluvials and the general circulation of the Pleistocene. *Geografiska Annaler 39*, 48–53.

Butzer, K. W., 1957b. The recent climatic fluctuation in the lower latitudes and the general circulation of the Pleistocene. *Geografiska Annaler 39*, 105–113.

Butzer, K. W., 1958. Sudien zum vor-und frühgeschichtlichen Landschaftswandel der Sahara. *Abh. math.-nat. Kl. Akad. Wiss. Mainz 1.*

Callendar, G. S., 1961. Temperature fluctuations and trends over the earth. *Quart. Journ. Royal Meteorol. Soc. 87*, 1–12.

Cranwell, L. M. and Von Post, L., 1936. Post-Pleistocene pollen diagrams from the Southern Hemisphere. 1: New Zealand. *Geografiska Annaler 18*, 308–347.

Crary, A. P., 1960. Arctic ice island and ice shelf studies, *Scientific studies at Fletcher's Ice Island, T-3: 1952–1955. Geophysics Research Papers (63), Vol. III*, 1–37. Boston Air Force Cambridge Research Center. (AFCRC-TR-59-232(3)ASTIA document no. AD-216815.)

Cromertie, G., Earl of, 1712. An account of the mosses in Scotland. *Phil. Trans. Royal Soc. 27*, 296–301.

Easton, C., 1928. *Les hivers dans L'Europe occidentale.* Leyden, E. J. Brill.

Emiliani, C., 1955. Pleistocene temperatures. *J. Geol. 63*, 538–579.

Fairbridge, R. W., 1961. Eustatic changes in sea level. In *Physics and Chemistry of the Earth 4.* Pergamon. 99–185.

Findlay, A. G., 1884. *Directory for the South Pacific.* (5th ed.), Richard Holmes Laurie, London.

Firbas, F. and Losert, H., 1949. Untersuchungen über die Entstehung der heutigen Waldstufen in den Sudeten. *Planta 36*, 478–506.

Flohn, H., 1952. Allgemeine Zirkulation und Paläoklimatologie. *Geol. Rundschau 40*, 153–179.

Gams, H., 1937. Aus der Geschichte der Alpenwalder. *Zeit. des deutsch. und österreich. Alpenvereins 68*, 157–170.

Godwin, H., 1954. Recurrence surfaces. *Danmarks Geol. Undersögelse II: Raekke 80.*

Godwin, H., 1956. *History of the British Flora.* Cambridge, University Press.

Godwin, H., Suggate, R. P., and Willis, E. H., 1958. Radiocarbon dating of the eustatic rise in ocean level. *Nature 181*, 1518–1519.

Godwin, H. and Willis, E. H., 1959. Radio-carbon dating of pre-historic wooden trackways. *Nature 184*, 490–491.

Griffin, J. B., 1961. Some correlations of climatic and cultural change in eastern North American prehistory. *New York Academy of Sciences, Symposium on solar variations, climatic changes and related geophysical problems.*

Harrington, H. J. and McKellar, I. C., 1958. A radio-carbon date for penguin colonization of Cape Hallett, Antarctica. *N.Z.J. Geol. Geophys. 1*, 571–576.

Hennig, R., 1904. Katalog bemerkenswerter Witterungsereignisse von den ältesten Zeiten bis zum Jahre 1800. *Abh. preuss. meteorol. Inst. 2(4).*

Kraus, E. B., 1954. Secular changes in the rainfall regime of southeast Australia. *Quart. Journ. Royal Meteorol. Soc. 80*, 591–601.

Labrijn, A., 1945. Het klimaat von Nederland gedurende de laatste twee en een halve eeuw. *Kon. Nederl. Meteorol. Inst., Meded. Verh. 49(102).*

Lahr, E., 1950. *Un siècle d'observations météorologiques appliquées à l'étude du climat luxembourgeois.* Bourg-Bourger Verlag, Luxembourg.

Lamb, H. H., 1953. British weather around the year. *Weather 8*, 131–136 and 176–182.

Lamb, H. H., 1959. Our changing climate, past and present. *Weather 14*, 299–318.

Lamb, H. H. and Johnson, A. I., 1959. Climatic variation and observed changes in the general circulation. *Geografiska Annaler 41*, 94–134.

Lenke, W., 1960. Klimadaten 1621–1650 nach Beobachtungen des Landgrafen Herman IV von Hessen. *Ber. deutsch Wetterdienstes 63.*

Link, F. and Linkova, Z., 1959. Méthodes astronomiques dans la climatologie historique. *Studia geoph. geod. (Prague) 3*, 43–61.

Manley, G., 1953. The mean temperature of central England, 1698–1952. *Quart. Journ. Royal Meteorol. Soc. 79*, 242–261.

Matthes, F. E., 1939. Report of committee on glaciers. *Trans. Amer. Geophys. Un. 1*, 518–520.

Mitchell, J. M., 1961. Recent secular changes of global temperature. *New York Academy of Sciences, Symposium on solar variations, climatic changes and related geophysical problems.*

Müller, K., 1953. *Geschichte des badischen Weinbaus.* Laar in Baden, von Moritz Schauenburg.

Nicholas, F. J. and Glasspoole, H., 1931. General monthly rainfall over England and Wales, 1727 to 1931. *Brit. Rainf.* 299–306.

Petterssen, S., 1949. Changes in the general circulation associated with the recent climatic variation. *Geografiska Annaler 36*, 212–221.

Privett, D. W. and Francis, J. R. D., 1959. The movement of sailing ships as a climatological tool. *The Mariner's Mirror 45*, 292–300.

Scherhag, R., 1950. Die Schwankungen der allgemeinen Zirkulation in den letzten Jahrzehnten. *Ber. Deutsch Wetterdienstes, U.S. Zone 12*, 40–44.

Scherhag, R., 1960. *Einführung in die Klimatologie.* Braunschweig, Westermann.

Schwarzbach, M., 1950. *Das Klima der Vorzeit.* Enke, Stuttgart.

Schwarzbach, M., 1961. The climatic history of Europe and North America. In *Descriptive Palaeo-climatology*, ed. A. E. M. Nairn. Interscience, London. 255–291.

Stoiber, R. E., Lyons, J. B., Elberty, W. T., and McCreahan, R. H., 1960. Petrographic evidence on the source area and age of T-3, *Scientific studies at Fletcher's Ice Island, T-3: 1952–1955. Geophysics Research Papers (63), Vol. III*, 78. Boston Air Force Cambridge Research Center. (AFCRC-TR-59-232(3) ASTIA document no. AD-216815.)

Thorarinsson, S., Einarsson, T., and Kjartansson, G., 1959. On the geology and geomorphology of Iceland. *Geografiska Annaler 41*, 135–169.

Vanderlinden, E., 1924. Chronique des événements météorologiques en Belgique jusqu'en 1834. *Mém. Acad. R. Belg. 2, 5.*

Wagner, A., 1940. *Klima-änderungen und Klimaschwankungen.* Braunschweig, Vieweg.

Wallén, C. C., 1950. Recent variations in the general circulation as related to glacier retreat in northern Scandinavia. *Geofis. pura appl.* (Milan) *18*, 3–6.

Wallén, C. C., 1953. The variability of summer temperature in Sweden and its connection with changes in the general circulation. *Tellus 5*, 157–178.

West, R. G., 1960. The Ice Age. *Adv. Sci.* (London) *16*, 428–440.

Wexler, H., 1956. Variations in insolation, general circulation and climate. *Tellus 8*, 480–494.

Willett, H. C., 1949. Long-period fluctuations of the general circulation of the atmosphere. *J. Meteor. 6*, 34–50.

Willett, H. C., 1960. Temperature trends of the past century. *Centenary Proc. Royal Meteorol. Soc.* 195–206.

Willett, H. C., 1961. Review of "Atlas of planetary solar climate with sun-tide indices of solar radiation and global insolation" (Bollinger). *Bull. Amer. Meteorol. Soc. 42*, 303–304.

Yamamoto, T., 1956. On the climatic change in Japan and its surroundings. *Proc. Eighth Pacific Sci. Congress, 1953, Quezon City, Philippine Islands 2A*, 1113–1128.

Zeuner, F. E., 1958. *Dating the Past.* (4th ed.) Methuen.

Jerome Namias

14 Climatic anomaly over the United States during the 1960's

Climatic fluctuations on all time scales have always received a great deal of attention, and the cooling trend observed in numerous areas of the world during the 1960's has been no exception (Lamb, 1966; Mitchell, 1970; Wahl and Lawson, 1970; Namias, 1969). Speculation as to the cause of the cooling has involved air pollution, volcanic activity, solar variations, and other more bizarre phenomena. Generally omitted from consideration is large-scale and long-term air-sea interaction—perhaps because of the unavailability of reliable long series of oceanic temperature data or because of the tacit assumption that the sea is always a slave to the atmosphere. It seems to me that, in the quest for causes of climatic fluctuations, scientists may be overlooking the most important factor by neglecting this interaction.

Although most hypotheses dealing with short- and long-period climatic changes invoke uniformly acting global mechanisms, it is quite possible that regional mechanisms interacting with each other through atmospheric dynamics can produce hemispheric and even global fluctuations. In the following paragraphs I shall describe and attempt to explain one such regional interaction—namely, between the North Pacific and North America.

The winter weather over the contiguous United States was indeed colder than normal across the eastern two-thirds of the nation, where temperatures averaged from 1° to 4° F (approximately 0.5° to 2° C) below the 1931–60 mean (Figure 1). West of the continental divide, temperatures averaged above normal. During practically all of the ten winters, temperatures averaged below normal in the eastern half of the nation.

These temperature anomalies are easily associated with the prevailing flow pattern of the winds in mid-troposphere, as shown by average decadal height contours at the 700-millibar pressure level and by isopleths of departure from normal (here, normal is defined as an average for the winters of 1947–63) for the ten winters (Figure 2 [p. 182]).

Since the prevailing winds flow along the contours with low heights to the left, this chart shows that the long wave pattern affecting much of the Northern Hemisphere, and especially North America, was appreciably amplified above the normal. With a stronger ridge (northward bulge in the flow) over western North America and a stronger trough (southward bulge) to the east, the more frequent deployment of Arctic air masses into the eastern half of the nation is assured. This is obvious from the isopleths of 700-millibar height anomaly and is verifiable from an objective system (Klein, 1965) for specifying mean temperature anomalies at the earth's surface from 700-millibar height anomalies. In view of this relationship, it seems unlikely that increased air pollution, variation in volcanic activity, or human intervention was the cause of the decadel temperature fluctuation over the United States. If these factors do not cause the U.S. temperature fluctuation, then it is possible that fluctuations elsewhere are caused by direct regional interactions and their further consequences (Namias, 1963).

At the same time that the eastern United States was abnormally *cold*, the sea surface over much of the North Pacific was abnormally *warm*. Figure 3 [p. 183] shows the average departures of sea-surface temperatures in the winters of the 1960's from normals based on a period of more than 40 years prior to 1945 (United States Naval Hydrographic Office, 1944). Whatever the cause of the oceanic warming, it is likely to have produced an aberration in the winter-time atmospheric circulation over the North Pacific—most probably by abnormal excitation of cyclones (Namias, 1963). Once the cyclonic activity increased and the vorticity (or curl of the winds) was transported aloft, the resulting standing (or forced) long-wave central Pacific trough created downstream perturbations in the manner shown in Figure 2.

Reprinted, with minor editorial modification, by permission of the author and *Science*, from *Science*, Vol. 170, pp. 741–743, 13 November 1970. Copyright 1970 by the American Association for the Advancement of Science.

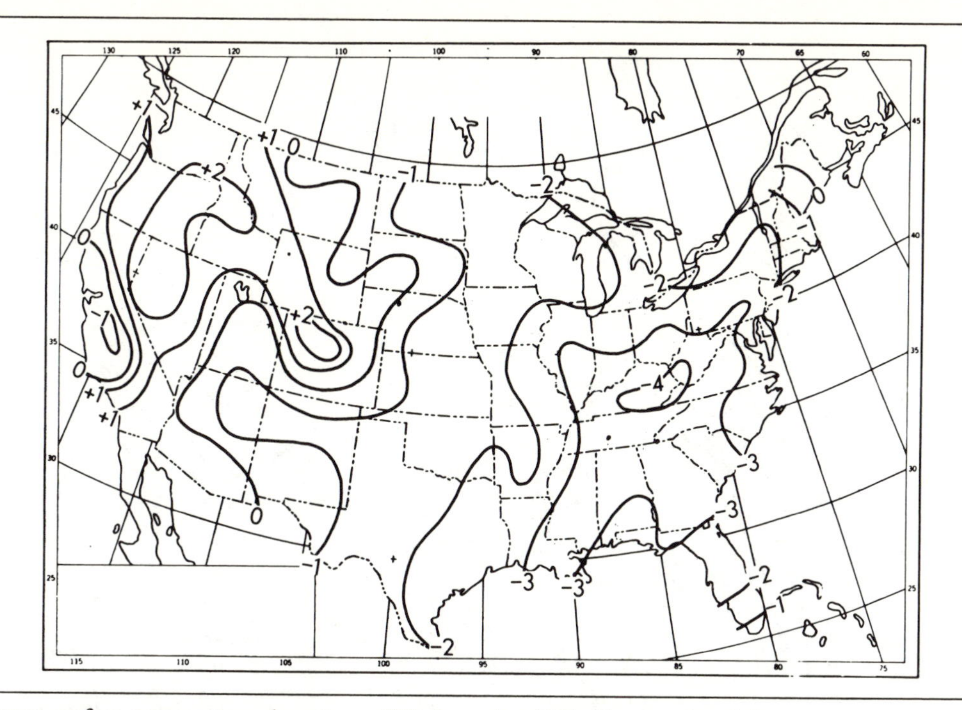

Figure 1 Average surface temperature departures (°F) from the 1931–60 normals of the winters 1960–61 through 1969–70. (December through February are defined as the winter months.)

These perturbations are the well-known long or Rossby waves, and their observed positions agree with theoretical (Rossby, 1939) or empirically derived (O'Connor, 1969) teleconnections.

The *fall* sea-surface temperature anomalies of the last decade (not shown) averaged up to 1° F (approximately 0.5° C) warmer than those of the following winters over much of the central North Pacific. Thus, the winter storms had an unusually warm initial reservoir on which to feed, and some of the anomalous heat was extracted through increased latent and sensible heat losses associated with stronger winds.

There seems to be no strong reason why repetitive conditions such as those described cannot lead to climatic fluctuations of a much longer time scale than a decade. It may be shortsighted to invoke extraterrestrial or man-made activity to explain these fluctuations.

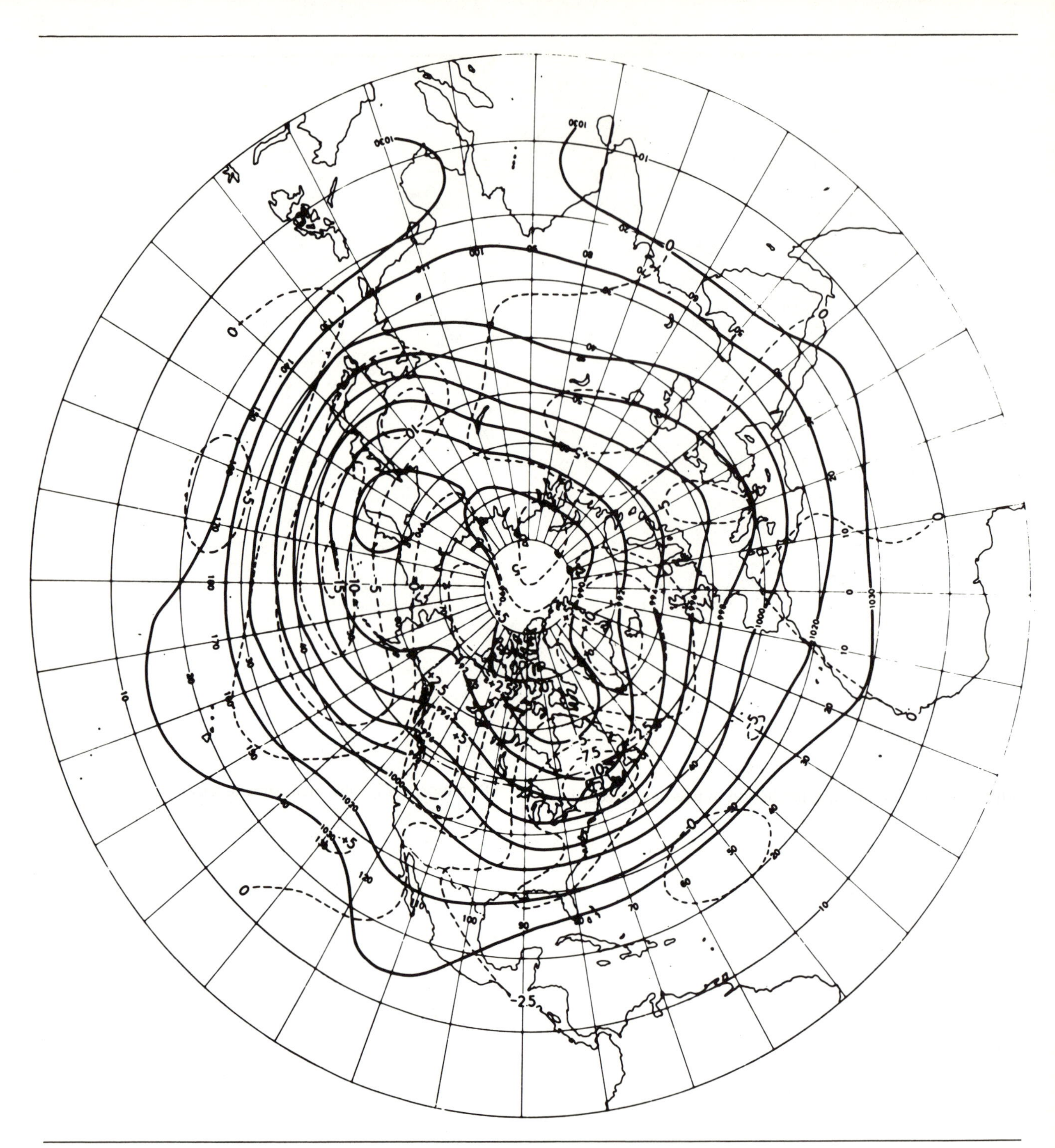

Figure 2 Mean 700-mb height contours (solid lines) and isopleths of departure from normal (broken lines) for the winters of the decade from 1960–61 through 1969–70. Contours and isopleths of anomaly are labeled in tens of feet (multiply by 3.048 for meters).

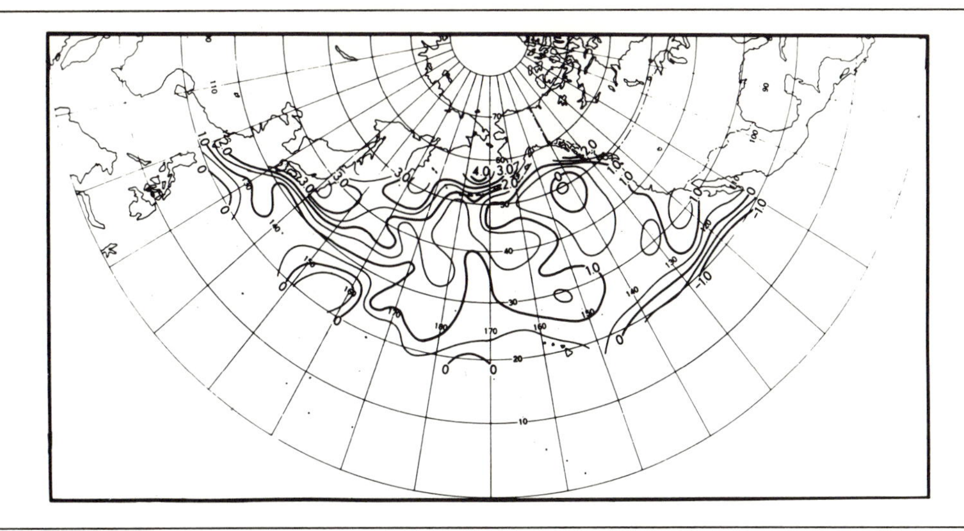

Figure 3 Average departures of sea-surface temperature for the ten winters, 1960–70, from long-term (about 40-year) means ending about 1945. Temperatures were extracted from monthly means published by the Bureau of Commercial Fisheries (1961–70) and from values furnished by the Japanese Meteorological Agency. Isopleths are constructed at 0.5° F (approximately 0.3° C) intervals.

REFERENCES

Bureau of Commercial Fisheries, 1961–70. *California Fisheries Market News Monthly Summary. Part 2.* U.S. Dept. Interior, Bureau of Commercial Fisheries, San Diego.

Klein, W. H., 1965. Application of synoptic climatology and short-range numerical prediction to five-day forecasting. *U.S. Wea. Bur. Res. Pap. 46*, 109 pp.

Lamb, H. H., 1966. Climate in the 1960's: Changes in the world's wind circulation reflected in prevailing temperatures, rainfall patterns and the levels of the African lakes. *Geographical Journal 132*, 183–212.

Mitchell, J. M., jr., 1970. A preliminary evaluation of atmospheric pollution as a cause of the global temperature fluctuation of the past century. In *Global Effects of Environmental Pollution*, ed. S. F. Singer. Dordrecht, Holland. 139–155.

Namias, J., 1963. Large-scale air–sea interactions over the north Pacific from summer 1962 through the subsequent winter. *J. Geophys. Res. 68(22)*, 6171–6186.

Namias, J., 1969. Seasonal interactions between the north Pacific Ocean and the atmosphere during the 1960's. *Mon. Wea. Rev. 97*, 173–192.

O'Connor, J. F., 1969. Hemispheric teleconnections of mean circulation anomalies at 700 mb. *ESSA Tech. Rept. WB 10*, U.S. Dept. Commerce, Washington, D.C.

Rossby, C. G., 1939. Relations between variations in the intensity of the zonal circulation and the displacements of the semi-permanent centers of action. *J. Marine Res. 2*, 38–55.

U.S. Naval Hydrographic Office, 1944. *World Atlas of Sea Surface Temperatures.* H.O. Pub. 225 (2nd ed.), Washington, D.C.

Wahl, E. W. and Lawson, T. I., 1970. The climate of the midnineteenth century United States compared to the current normals. *Mon. Wea. Rev. 98*, 259–265.

SUGGESTED READING FOR PART 5

Gentilli, J., 1971. Climatic fluctuations. In *Climates of Australia and New Zealand*, ed. by J. Gentilli. World Survey of Climatology, Vol. 13, Elsevier Publishing Company. 189–211.

Lamb, H. H., 1969. Climatic fluctuations. In *General Climatology*, ed. by H. Flohn. World Survey of Climatology, Vol. 2, Elsevier Publishing Company. 173–249.

Lamb, H. H., 1970. Climatic variation and our environment today and in the coming years. *Weather 25(10)*, 447–455.

Mitchell, J. M., jr., 1965. Theoretical paleoclimatology. In *The Quaternary of the United States*, ed. by H. E. Wright, jr., and D. G. Frey. Princeton University Press. 881–901.

Mitchell, J. M., jr., 1968. Causes of climatic change. *Meteorological Monographs*. (American Meteorological Society) *8(30)*, 159 pp.

Rasool, S. I. and Hogan, J. S., 1969. Ocean circulation and climatic changes. *Bull. Amer. Meteor. Soc. 50(3)*, 130–134.

Sellers, W. D., 1969. A global climatic model based on the energy balance of the Earth-atmosphere system. *J. Applied Meteor. 8*, 392–400.

Part Six

Weather Modification – Man-controlled

Wyckoff's account of weather modification since 1947 sets the scene for this part. He describes the state of the theory and operational factors to date and provides some answers to queries on dissipation of cold fog, activation of rain, and suppression of hail and hurricanes.

Simpson and Simpson explain the difficulties inherent in hurricane modification: because of the nature and complexity of the phenomenon it seems more expedient to conduct the experiments in the real world. Their analyses of mature hurricanes in 1961 and 1963 were encouraging but not conclusive. Were the observed alterations caused by extraneous natural factors or manmade?

The Los Angeles smog problem is infamous. Its severity has stimulated the suggestion of numerous solutions, more than a few highly amusing. Neiburger makes a rational and thorough examination of the major proposals, convincingly illuminating the impracticality of most of them. He calls for a total systems approach to solving the problem, such that an assessment of the operational and economic outputs be included to determine the feasibility of the proposal. He concludes that control of pollution at the source is still the most expedient measure.

The implications of weather modification on a global level are enormous. Morris provides a much-needed perspective on the legal aspects of weather modification, the laws in vogue, and the decisions of the courts. Here is a subject that will become a crucial issue in international politics. Who has the right to deflect a hurricane from one shore only to deploy its fury on a neighboring country? And within the nation, who has the right to decide that avoidance of a hurricane at the coast is worth the resultant reduction in normal precipitation totals inland?

Peter H. Wyckoff

15 Evaluation of the state of the art

It is now 20 years since the early experiments of Langmuir, Schaefer, and Vonnegut planted the seed of credibility that man could influence the weather. The bright hopes of 1946–1947 have now been tempered with experience, and it has become obvious that the atmosphere is a much more complex system than originally realized. The era of the rainmaker with his pots and pans and secret formula has given way to the sober and deliberate study of the atmosphere by scientists using the latest observational, laboratory, and theoretical techniques. The challenge of the atmosphere in recent years has attracted an increasing number of sound analytical minds from all disciplines of scientific research, and progress is now being made in substituting fact for fiction and scientific logic for inspirational hunches. The scientific assessment for the future implementation of weather modification is no less bright than the dreams of the earlier rainmakers, but it is tempered by judgment and a sober realization of the task remaining ahead.

Weather modification, like any emerging science, has been divided into two main schools of thought. One school of thought is represented by the basic researcher who in the quest for scientific fact, is satisfied only with the logical progression of proven scientific evidence obtained in the laboratory, fully supported by theoretical reasoning, and verified in the atmosphere by a series of critical experiments. It was by such a logical and deductive process that atomic energy was developed and made available to mankind today. It is a proven approach and has demonstrated its efficacy many times in the physical, chemical, and mathematical sciences.

The second school of thought consists of those who are concerned with the realization that the atmosphere is a complex and huge entity containing many variables, some of which are beyond the understanding of man today or in the near future. This group feels that the laboratory approach can never simulate the environment of the earth's atmosphere in a test tube, and that the only approach is to conduct experimentation in the actual atmospheric environment, and then attempt to assess results through statistical evaluation and the hypothecation of theoretical models. This group approaches the problem in much the same way as an engineer who is confronted with a mysterious black box with a multiplicity of electrical terminals but with no way for looking inside. He measures the resistance, capacitance, and inductance across each set of terminals, and then attempts to devise an equivalent wiring diagram of resistors, condensers, and inductors which would reproduce the readings which he has taken on the black box. He finds, to his dismay, that several different configurations will produce the same response as his initial measurements on the terminals of the black box and he cannot determine which one is correct. He must now go to more sophisticated techniques, relying upon hunches and inspiration to decide which ones to try first and in what order. Without a certain amount of insight and experience, he may never solve the problem.

There is, of course, a third position which maintains that in order to make finite progress in a reasonable period of time, both schools of thought must be preserved and fostered, and that both schools of thought may become sterile without cross fertilization taking place between them. The traditional role of the National Science Foundation over the past seven years of its experience in the field of weather modification has been to sponsor and encourage projects following both schools of thought and through personal contacts, sponsored symposia, conferences, and report dissemination to provide the cross fertilization required. The fruits of these efforts are evident in the program of the Bureau of Reclamation for increasing precipitation over catchment basins and reservoirs in the Western States and the plans of the Federal Aviation Agency to incorporate cold fog dissipation as an operational

Reprinted, with minor editorial modification, by permission of the author and editor, from *Human Dimensions of Weather Modification*, W. R. D. Sewell (editor), University of Chicago, Department of Geography Research Paper No. 105, 1966, pp. 27–39.

procedure at commercial airports throughout the nation.

While much remains to be done before the dreams of 20 years ago can become a reality, steady progress has been made in the understanding of atmospheric processes which may chart the course for weather modification.

Fog and Cloud Modification

The modification of fogs or clouds is usually attempted either to dissipate them for the improvement of visibility or to cause more rapid development of the dynamics and liquid water content of the cloud so that natural precipitation may result.

The dissipation of cold fog for the improvement of visibility is already a proven technique using either dry ice or the release of propane gas through expansion nozzles to produce local freezing. It is being used operationally by the U.S. Air Force, by United Airlines, and in several foreign countries, such as Russia and France.

The dissipation of warm fogs, with fog particle temperatures above freezing, however, has not yet been satisfactorily achieved. Dissipation attempts follow three main techniques. One technique attempts to evaporate the fog droplets by the introduction of massive quantities of heat from open flames or infrared burners. This is the basic mechanism used by the British in World War II to clear landing strips for military aircraft, and is known as the FIDO system. It is a successful system provided that sufficient heat can be provided and maintained in the runway area. It is expensive to install and operate, requires large expenditures of fuel, and is localized to the treated runway areas. Unless taxi strips, maintenance areas, loading ramps, etc., are included in the ring of heat, air traffic will bottleneck as soon as it leaves the runway area. Pilots feel uneasy in landing between walls of fire because of the almost certain destruction of the aircraft by fire if it swerves off the runway. Landing problems are also complicated by the turbulence created by thermals at the critical control point just before touch down on the runway.

The first technique also involves the use of fans to draw dry air into the fog from above the thermal inversion. When sufficient dry air is mixed into the fog, the fog particles will evaporate. This has been successfully used in radiation types of fog where the fog layer is shallow and dry air is available a short distance aloft. The U.S. Army has successfully used helicopter downwash to produce this mixing.

The second technique attempts to sweep out the fog droplets by releasing a water spray from an aircraft flying over the fog deck and mechanically driving the fog particles to the ground. This will work successfully if sufficient water can be released by the spray aircraft; however, the quantities of water required have been computed to be logistically unfeasible. Charged sand and concentrated salt brine have also been tried with no major improvement in logistic limitations for practical application.

A third technique attempts to stimulate coagulation of the fog particles into large enough droplets to fall out as a drizzle. The injection of electrical charges into the fog to produce attraction between fog droplets holds some promise for success, but a successful way of introducing charged ions without developing space charges barriers has not yet been developed. The coagulation of fog droplets in sound beams using sirens, whistles, loud speakers, etc., has been successfully demonstrated in laboratory enclosures but in the free atmosphere where standing waves cannot be produced by reflections from solid surfaces such as walls or reflectors, the amount of energy required is prohibitive. The use of finely ground deliquescent salt particles blown into the fog by fans has been successful in some cases by producing droplet growth centers which accumulate liquid water from the saturated water vapor in the air and cause the existing fog droplets to evaporate. The drizzle of brine droplets which results, however, is corrosive to metal and produces stains on grass, runways, and buildings.

There has been sufficient success in the dissipation of warm fog to predict that eventually an economical, clean, and logistically feasible technique can be developed. Perhaps the application of several of the above mentioned techniques applied either simultaneously or in sequence will provide the answer.

In the opposite sense to dissipation, there is often need to preserve a cloud which is in danger of evaporation as the air mass in which it was originally generated is compressed by flowing downward in following the lee side of a mountain peak. Such rain shadow areas are common natural occurrences behind mountain chains. Research has shown that if these clouds are seeded with dry ice or silver iodide when they extend above the freezing level over the mountain peak, ice crystals will be formed which will release the heat of fusion of ice and provide sufficient additional buoyancy to the cloud to prevent dissipation as it passes over to the adjacent valley. Rain may then develop naturally in the cloud and provide moisture in the previous shadow area.

Precipitation and Modification

The problem of artifically stimulating rain to fall from a cloud is largely one of inducing approximately a million or more tiny cloud particles to join together and form a single rain droplet. Most successful attempts at rain stimulation have been applied to clouds where the liquid water cloud droplets are 10° C to 15° C below freezing. The addition of silver iodide or dry ice will produce freezing of the super-cooled droplets, and the ice crystals will grow in size either by the absorption of water vapor from the surrounding moist air or by the capture of adjacent liquid water droplets. When the ice crystals have grown to sufficient size, they will fall out of the cloud by gravity. If the lower atmosphere is below freezing, the crystals of ice will fall as snow, but if the lower atmosphere is warm, the ice will melt into raindrops. The exact mechanism is actually quite complex, and in some cases it appears that the super-cooled cloud droplets combine into relatively large sized liquid water drops before freezing. The mechanism of this agglomeration of liquid water is not well understood, but measurements from aircraft have proven their existence. Various theories such as ice splintering, electrostatic forces, turbulent mixing, etc., have been proposed, and it is probable that all of these mechanisms are valid. Which mechanism may predominate in any given storm type has not yet been determined.

It is estimated that in a coastal storm where moist maritime air is lifted up over a coastal mountain range such as in the northwestern Pacific coast range, approximately 25% to 30% of the available moisture will fall as precipitation by orographic lifting on the upslope of the mountain. Ordinary convective thunderstorms, on the other hand, are formed by thermal lifting of air masses and are estimated to deliver approximately 10% of the available liquid water as precipitation on the ground. While the effect of seeding such storms will vary greatly depending upon the meteorological situation, an average increase of 10% to 15% in the amount of rain reaching the ground seems to be indicated by a review of such attempts made over the past 10 years. It should be noted that this figure applies only to an increase in rainfall from clouds which are already raining or about to rain. There is still no known way to stimulate useful rain from air masses which do not contain a natural reservoir of water vapor or from clouds which are too shallow to contain the necessary liquid water content. Most of the water content of air passing over land masses is of maritime origin acquired from long periods of passage over the oceans, and very little water falling upon land areas is recycled back into the clouds again. Rainfall patterns and drought cycles are produced by variations in the global circulation trajectories of large-scale air masses, and cloud seeding is ineffective if the upper level air masses are dry.

Not all rain falls from clouds which contain ice crystals. In the tropics, rain commonly falls in great quantities from clouds which have not reached the freezing level. The mechanism responsible for the coagulation of the cloud droplets into rain is not known, but it is suspected that electrostatic effects, droplet collisions, cloud dynamics, and natural sea salt nuclei play an important role. One theory is based upon the observation that clouds generated over the ocean generally contain larger drop sizes on the average than equivalent clouds formed over the continent. This is believed to be due to the fact that over the ocean the number of natural condensation nuclei is relatively small and there is less competition between droplets for the amount of water vapor available. Thus, there is plenty of vapor available for all, and they grow to large sizes. The theory further hypothesizes that these nuclei upon which water vapor condenses consist mostly of giant salt nuclei produced by evaporated brine of the ocean spray. Since the vapor pressure of a cloud water particle formed on a salt particle is less than that for pure water, the particle will continue to grow as long as the salt content of the cloud drop is highly concentrated. If the salt nucleus were large enough to begin with, the cloud particle may grow to raindrop size before the dissolved salt concentration is no longer effective in reducing the vapor tension of the drop. Any reduction in the updraft forces of the cloud will then permit the larger and more mature drops to cascade downwards in a brief but drenching shower.

This theory has been the basis of a seeding project now underway in the Virgin Islands to seed warm cumulus maritime clouds with finely pulverized salts. The salt is released from an aircraft into the updraft of a building cumulus deck. After approximately a half an hour or less of ingestion, the newly created brine droplets have grown to raindrop size and fall out after the updraft currents have subsided. Observations to date have indicated excellent results on clouds exceeding 4,000 feet in thickness, but scientific measurements are yet to be made for verification. There is hope that the warm cloud problem may be solved in the near future, but mostly for maritime situations.

Hail Suppression

A hailstorm is essentially a severe storm containing violent updrafts which sustain large supercooled liquid water contents in suspension at temperatures below freezing. Ice crystals formed at lower levels of the storm are driven violently upwards into the supercooled liquid water accumulation zone where they grow by successive collisions with supercooled water droplets. Since the updrafts are turbulent, a growing hailstone may pass up and down through the liquid water accumulation zone many times, building up successive onion-like layers of ice. When the ice particles grow large enough to overcome the updraft forces or are turbulently tossed out of the updraft, a hail fall results. A typical hailstorm usually has many individual cells, and hail has been observed to fall in narrow bands in five or six widely separated areas within a short length of time from a single storm. The exact mechanism of hail formation is still not well understood, and it is suspected that several mechanisms can be responsible for hail formation. The fact that some hailstones are clear ice and others milky in color indicates that different processes of growth were in operation.

Hail modification attempts in the U.S.A., Russia, Switzerland, and Argentina, have utilized seeding by silver iodide. While the exact reason for the observed reduction of hailstone sizes or conversion into raindrops is not well understood, it is hypothesized by some that the silver iodide serves to freeze the liquid water in the accumulation level of the storm and no longer make it available for accretion on other ice particles passing through the zone. The Russians have been especially careful to pinpoint their injection of silver iodide into the collection zone of the storm through the use of artillery shells and rockets directed by radar. Their claims of millions of rubles saved each year in crop damage through pinpoint seeding lends credence to this technique. In Argentina, hail suppression was conducted with ground-based silver iodide generators. Their results indicate that on days that a cold front had just passed, hail damage was reduced by 70% due to seeding. Strangely enough, on non-frontal days, seeding appeared to double the amount of hail damage. It is apparent that different storm structures were involved on the two types of days. If the Russian model is believed, then one might hypothesize that low level seeding produced ice crystals lower in the cloud which were then fed into the liquid accumulation zone and grew to damaging size. It is possible that on days of cold front passage, the accumulation zone received enough of the ground released silver iodide to be effective in converting its liquid water content into ice. In any event, there does appear to be great need to study the dynamics of hailstorm structure and considerable reason to believe that hailstone suppression can be attained in the not too distant future.

An interesting set of experiments have been conducted in Italy and in Kenya. Small rockets containing an explosive charge of TNT are fired into hail-bearing clouds in large numbers by local farmers and ranchers. In Kenya it is reported that hail losses on the one plantation protected by this rocket firing program have been reduced to less than 1% of the damage experienced by 12 nearby plantations which were not protected. The same experience is claimed by the grape growers in Italy. This result is consistent with experiments being conducted in this country and elsewhere on the freezing of supercooled water droplets by shock waves produced by explosions. When balloon-borne explosives are detonated in the center of a cloud of supercooled water droplets, ice crystals are observed to fall from the cloud immediately after the explosion. The use of supersonic military aircraft to project sonic booms into hail clouds may become a subject of investigation for hail suppression studies in the near future.

Lightning Suppression

The actual mechanism for the production of electricity in clouds is not understood at the present time. There are some observations which indicate that electrical fields appear in clouds as soon as ice crystals are formed. This simple explanation, however, does not cover the situations where strong electrical fields have been observed in warm clouds where no ice is present, and theories have been advanced that space charge is carried from the surface layer of air near the earth into the clouds by the convective updraft of the cloud. Obviously again the problem is complex and probably many different mechanisms are at work depending on the meteorological situation. Regardless of the mechanism involved, the reduction of destructive lightning strokes which damage millions of dollars of forest lands and property each year requires attention from the weather modification expert.

Massive seedings of lightning storms with silver iodide by the U.S. Forest Service has indicated that a significant reduction in cloud to ground lightning strokes can be obtained. Laboratory tests indicate that the presence of ice crystals in a volume of air exposed to breakdown electrical potentials will cause electrical breakdown to occur at lower potential gradients than in clear air. There is a possibility that the production of a higher density of ice crystals in lightning-bearing clouds will reduce the potential at which lightning strokes can occur and thereby reduce the average charge lowered to the ground by the stroke. It is also possible that a very high density of ice crystals will drain off sufficient charge internally from the cloud to prevent lightning initiation entirely. Further evidence of the effectiveness of this approach was obtained by the U.S. Army who released millions of tiny metallic needles into a thunderstorm from an aircraft, and observed a significant reduction in electrical field gradient in the vicinity of the charged cloud when the needles were present. Whatever the mechanism may be for charge formation, techniques appear to be emerging for draining off these charges through seeding techniques. The practicable reduction of lightning strokes from clouds does appear to be possible through modification techniques in the reasonably near future.

Severe Storm Modification

Hurricanes

Hurricanes are spawned in the warm tropical waters of the Caribbean by the absorption of warm moist air into the rising convective tower of an active growing cumulus cloud. As the heavily laden moist surface air is sucked into the rising cloud, it is cooled by expansion as it reaches the rarified upper levels of the atmosphere, and the water condenses out in the form of billions of tiny cloud droplets. In condensing the water vapor out of the air, the water releases its heat of condensation to the surrounding cloud mass and causes it to rise with increasing velocity like a hot air balloon. This in turn sucks up more warm moist surface air which also condenses, and the rising tower grows with ever increasing intensity. The high velocity winds ejected from the top of the cloud tower distribute themselves horizontally in radial trajectories from the center in all directions, and because of the rotational forces produced by the turning of the earth, take on a spiral motion like water draining from a bath tub. After traveling horizontally for many miles from the convective center, these winds lose their momentum and fall back to the surface of the warm ocean where they replenish their heat and moisture supply and are sucked back into the core to repeat the cycle all over again. This circulatory system with the core cloud acting as a pump extends for 60 to 100 miles or more in radius for a mature storm, and involves energy of the motion of millions of tons of fast moving air exceeding many thousands of megatons of nuclear energy.

When massive quantities of silver iodide are periodically released over an 8 to 10 hour period into the supercooled portion of the center convective cloud which drives the storm, the cool water droplets are converted into ice. In freezing, the ice liberates its heat of fusion to the surrounding cloud mass and induces a momentary increase in the vertical motion of the cloud. It is hypothesized that this increase in vertical motion forces the rotating wind structure outward to cover a larger area than before. Since the storm is forced to spread out over a larger area but has gained no additional momentum, the average wind velocity is reduced significantly and the storm becomes less damaging. The heat liberated by the freezing of the cold cloud droplets after massive quantities of silver iodide are injected is approximately equivalent to the heat liberated by an atomic bomb, but acts over the period of many hours rather than a few minutes.

The fuel on which the hurricane feeds is the shallow layer of warm water which lies on the surface of the tropical seas. Studies are being made at the present time of the way in which this shallow surface layer of warm water forms on the surface of the ocean. If some means can be found to seal off the surface of the ocean by spreading a thin layer of an oily material over the water, then the supply of moisture upon which the hurricane feeds may be cut off. It may also be possible to produce mixing of the warm surface water with the cooler water just a few feet below it and lower the surface temperature below the critical point. It is also conceivable that artificial cirrus clouds can be maintained by aircraft or rockets high above the danger area to reduce solar heating of the sea surface. In any event, the suppression of hurricanes will never be achieved by brute force methods, since even the most heroic forces controlled by man are only puny compared to nature. It will be necessary to discover the mechanisms by which nature builds up its huge storehouse of energy, and to interfere with a critical ingredient while the process of build up is still in the formative stages.

The research required to achieve this knowledge is well underway, and mathematical models are being studied on high-speed computers to find this weak link in the chain of hurricane formation. The use of silver iodide seeding is not the final answer to hurricane modification, but is a valuable experiment to provide further information on hurricane structure and dynamics. Some day, it is hoped, the key will be found, and hurricanes will be starved into submission in their own breeding ground. Satellites will provide the warning that a convective tower is forming, and prompt aircraft measurements of air temperatures and motions coupled with submarine measurements of ocean temperature at the breeding site will feed data into electronic computers and provide the necessary information required to suppress the process. Until this insight is gained, man will continue to batten down the hatches and place emphasis on early warning to reduce the loss of life and property from these storms.

Tornadoes

The suppression of tornadoes is a problem of even greater difficulty than the hurricane because of their spontaneous appearance with very little warning. The tornado is known to contain a significant quantity of electrical energy, but it is not yet apparent whether the electrical activity is a side effect or a dominant driving force in tornado formation. Since the "will o' the wisp" appearance of tornadoes precludes any actual seeding attempts, the approach to this problem is presently confined to laboratory simulation and theoretical computer modeling. It will probably be many years before sufficient insight can be gained into the tornado mechanism to indicate fruitful modification procedures. There is some hope that the severe storms which spawn the tornado may be amenable to treatment rather than the tornado itself, and a more feasible approach would be to prevent the storm from reaching the proportions of tornado generation. Energies involved in these storms are huge even judged by atmospheric standards, and the immediate approach to the problem is not evident.

Climate Modification

Climate modification is distinguished from weather modification by the span of time covered by the modification activity. If the permanent pattern of weather activity is altered by intervention, whether intentional or not, then climate modification is understood to have taken place.

The unintentional modification of the atmosphere by man is a matter of record due to his pollution of the atmosphere, his modification of the surface climate structure through urbanization, his deforestation of large wooded areas, and his building of dams and diversion of rivers, to mention a few. The modification of the mean average surface air temperature upwards by several degrees in heavily populated regions is a matter of record. Such inadvertent modification is cause for concern as to future consequences, and deserves careful study.

In the strict sense, the term climate modification is usually applied to those activities of man which are performed intentionally to change the climate. To date, studies leading to the possible modification of the large-scale circulations of the atmosphere have been confined to the electronic computer. The primitive equations of motion of the earth's atmosphere have been combined into a mathematical model, and hypothetical cases of climate modification have been tested by mathematically removing the arctic ice pack, removing portions of mountain ranges, changing ocean temperatures, etc. While these mathematical models respond well to large-scale modification inputs, they are not yet sensitive enough to detect small-scale changes such as might be produced by cloud seeding. Until more physical and meteorological cloud mechanisms can be tied down into mathematical terms, it seems unlikely that the computer will be able to predict local seedability of cloud cover. It may, however, be able to predict whether moisture-bearing air masses will pass over a local area and whether they will persist for any length of time.

The possibility of being able to alter the global pattern of atmospheric air masses has been the subject of much speculation. Such heroic measures as melting the arctic ice cap, diverting warm ocean currents, artificially producing widespread cirrus cloud cover by rocket or aircraft seeding, injecting dust or reflecting needles into orbit above the earth, etc., are being studied with respect to their possible effects upon circulation patterns. Preliminary studies have shown that a possibility does exist to alter favorably the climate of a local region of the earth by massive tampering, but that equally less favorable or possibly disastrous results would be evident in other regions of the earth's surface. Actual climate modification attempts cannot be made until the entire consequences on a world-wide basis can be assessed. There is too much at stake to permit irresponsible tampering of the atmosphere at this time on a global scale. It is hoped that the next 5 to 10 years will provide the necessary knowledge and increased computer capacity to permit an accurate assessment of the consequences of some of the more simple and less ambitious aspects of climate modification which can be attempted with safety in the early future.

Summary

In summary, our knowledge of weather modification has progressed greatly in the past 20 years, and with this progress has come an increased respect and appreciation for the complexity of the atmosphere. We feel that we have now come to grips with the problem, and progress will come by small but steady increments throughout the years to come.

The dissipation of cold fog is a matter of operational implementation. Warm fog dissipation has not yet been solved on an economical or logistically feasible basis, but with greatly increased efforts, a solution should emerge in a relatively few years.

Rainfall augmentation using cloud seeding techniques on clouds which are already precipitating or about to precipitate appears to be effective in many cases when properly applied, and an average precipitation increase of 10% to 15% has been observed.

Hail suppression appears to be feasible here in the United States in the very near future. It is claimed that hail suppression is performed in Russia now on an operational basis. An intensive effort in this country on hail suppression research is definitely indicated.

Lightning suppression attempts show hopeful trends which may result in operational techniques in 5 to 10 years.

The suppression of severe storms appears possible, but will require much more research before the real key to the answer can be discovered. It is by no means hopeless, and research should be continued and increased.

The answer to the drought cycle problem will not come from cloud seeding, but will require means for influencing the large-scale motions of moisture-bearing air masses aloft. The present approach through mathematical modeling appears to be fruitful and should be encouraged and increased.

Weather modification has now emerged from the early days of the medicine man approach into a full blown science. Scientific talent has now been attracted to its challenge from all disciplines of physics, chemistry, mathematics, and meteorology, and the science of weather modification is now emerging as a legitimate and respected profession. Its progress in the future will depend upon its acceptance by the public and by the support which the Federal Government is willing to provide.

Robert H. Simpson and Joanne Simpson

16 *Why experiment on tropical hurricanes?*

A series of seeding experiments is being conducted on tropical hurricanes by an interagency program of the U.S. Government called Project Stormfury. The two main agencies carrying out the work are the U.S. Navy and the Environmental Science Services Administration, E.S.S.A. (formerly Weather Bureau). So far, the objectives are primarily scientific, since no avenue has yet been found which promises in the near future to lead to practical modification. This article is mainly confined to the scientific aspects of the program.

So far, seeding experiments have been carried out on two hurricanes, Esther in 1961 and Beulah in 1963. The results were encouraging but inconclusive. The main reason for inconclusive results in hurricane experimentation is the very large natural fluctuations, or high noise level, that these storms undergo. Since hurricanes can develop, collapse, or entirely reverse course in six hours, the consequences of a man-made alteration are very difficult to isolate.

These hurricane experiments require coordination of 10 to 15 specially equipped aircraft to document "before and after" structure. The experiments are costly in terms of money, trained manpower, and frustration. The current series is based on a physical chain of reasoning, from about four or five basic assumptions, some of which are not possible to test directly with existing technology. These are some of the reasons that many meteorologists argue that hurricane experimentation is premature. It has been stated that one should know more about hurricanes, be able to model any experiment numerically using the hydrodynamic equations, and be able to validate the assumptions before field experiments are undertaken.

This has not been the approach of the classical physicist, for whom the early atomic pioneers such as Rutherford and Bohr will serve as familiar examples. These men chose to learn about atomic structure by conducting experiments upon actual atoms—this is our example and motivation in hurricane experimentation. Of course, atoms are small enough to be brought into the laboratory. On the other hand, humans cannot go inside them and observe. Actually, much more was documented about hurricane structure in 1960 than was known about atomic structure in 1900–1910 when the bombardment experiments were pioneered.

However, the cut and try approach, which has sometimes been successful in curing diseases and in certain engineering problems, should not be considered in tackling the hurricane. The large natural fluctuations and small sample alone should preclude any experiment not founded on a predictive physical theory. Furthermore, the tropical storm's huge energy transactions far exceed any energy source that man could bring to bear. A mature hurricane of moderate strength and size releases as much condensation heat energy through its cloud systems in a day as the nuclear fusion energy of about 400 20-megaton hydrogen bombs. Of this, about 3%, or 12 bombs' worth, is converted into energy of winds.

The present Stormfury approach to hurricane experimentation postulates instabilities which might be triggered by relatively small amounts of energy, introduced strategically. The large natural fluctuations, frequently occurring without apparent external forcing, suggest this is a working hypothesis. However, many hurricane experts disagree; they believe that a hurricane is a stable circulation controlled by its boundary conditions. Some possible modification experiments based on the premise of boundary control will be suggested in the concluding section.

The experiment to be described here chooses a mature hurricane in as steady a state as possible. The authors prefer at this time to work with mature storms, in preference to directing experiments toward "nipping in the bud" incipient hurricanes. The reasons are as follows: In the first place, little is known or can be

From *Transactions of the New York Academy of Sciences*, Volume 28, No. 8, pp. 1045–1062, R. H. Simpson and J. Simpson.

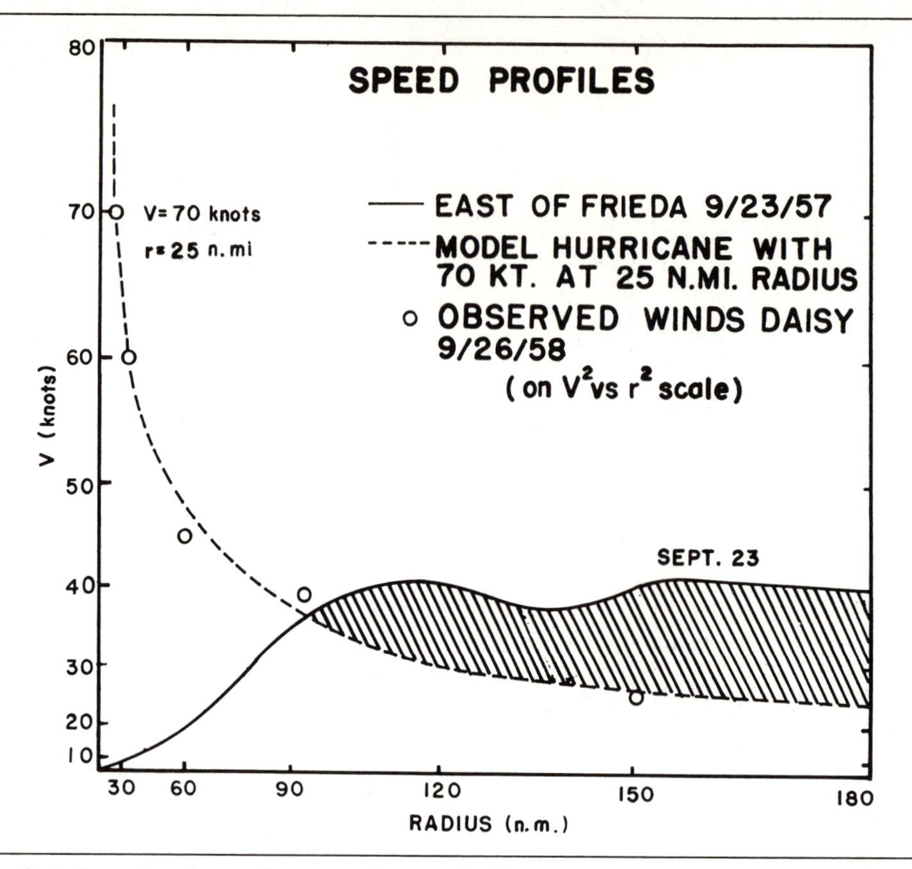

Figure 1 Profiles of wind speed against radius comparing a typical hurricane with a typical tropical storm. Solid curve is tropical storm Frieda (23 September 1957). Dashed curve is hypothetical hurricane with a profile of tangential wind speed given by $V_0r^{1/2}$ = constant. Circles denote winds observed in hurricane Daisy (26 August 1958). The areas beneath the two curves, i.e., the total kinetic energy, are about equal.

predicted about the formation and deepening process, so that analysis of causality in such experiments would be difficult if not futile. About 50 incipient cases are found for every one that deepens into the "tropical storm" category (winds 40–75 miles per hour) and only roughly one-half of those storms reach full hurricane intensity (winds about 75 miles per hour). The unique deepening symptoms, if there are any, remain unknown today. Secondly, tropical storms and hurricanes produce about one-quarter of the annual rainfall in such populated areas as Japan, India, Southeast Asia and the South-eastern United States. Thus from practical considerations, it would not be desirable to "nip these storms in the bud," provided even that we could identify them in the bud and knew how to do so.

Both scientific and practical reasons unite, in our opinion, to concentrate upon the mature hurricane, the structure and operation of which is documented in broad outline. This documentation is due largely to the studies of the National Hurricane Research Laboratory (N.H.R.L.) in Miami, Florida. Their program, using many instrumented aircraft for storm measurements, was established by the U.S. Government (Weather Bureau) in 1956.

One pertinent aspect of the N.H.R.L. findings is that sub-hurricane tropical storms release amounts of energy and rainfall comparable to the full hurricane (Riehl and Gentry, 1958). The main difference is the ring of intense winds in the region of the cloud wall surrounding the eye. This ring of high wind velocities is responsible for the major part of wind and water damage. Figure 1 illustrates this point by comparing the typical radial wind profiles of a full hurricane (Daisy, 1958) with that of a tropical storm (Frieda, 1957). Both storms released about the same total kinetic energy. Fortunately there appears to be little correlation between wind intensity and rainfall amounts produced by tropical storms. Most hurricanes that cross a coastline drop the major portion of their rain at inland stations after the winds diminish below hurricane force; the rate of movement is more decisive than wind intensity in determining the amount of rain that the land stations receive.

Storms like Daisy are highly destructive if they strike populated areas, while ones like Frieda are relatively harmless, since the damage increases roughly as the square of the highest sustained wind. Thus, only 10% reduction in wind speed should reduce the destruction

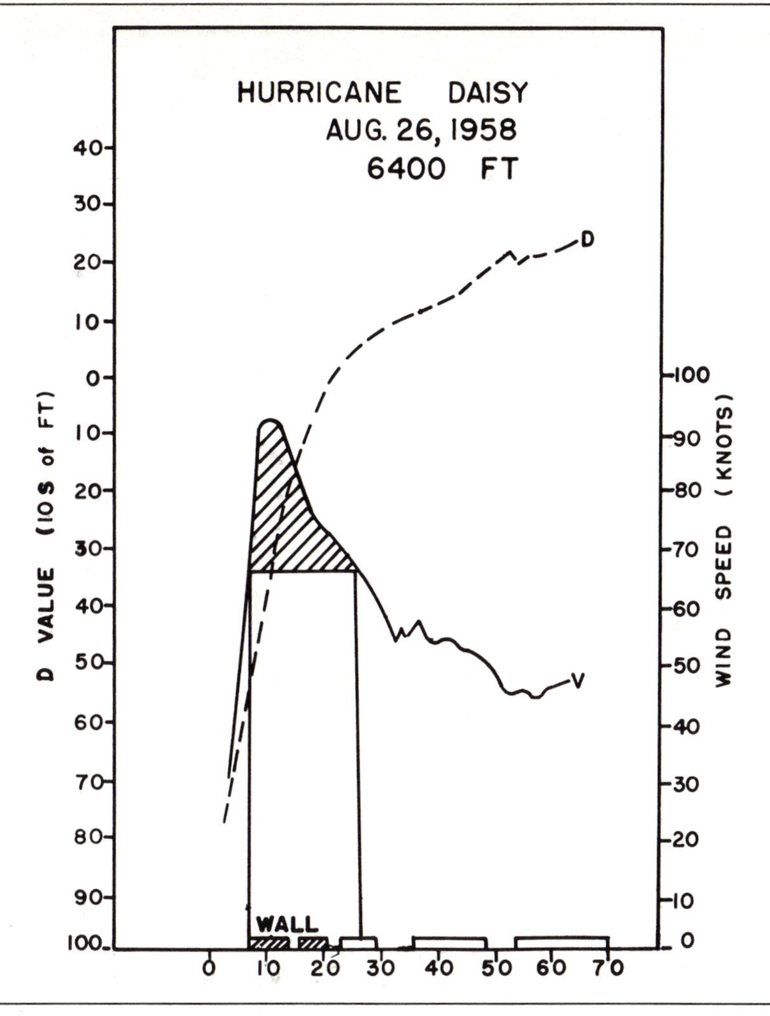

Figure 2 Radial wind and D value profile (radar altitude minus pressure altitude) in hurricane Daisy, 26 August 1958. Location of major cloud bands (from radar) are shown below, with wall cloud shaded. The hatched region under the wind speed curve (radial interval roughly 6–27 miles) shows the narrow ring in which the winds are above hurricane force.

by roughly 20%, and so forth. Bearing in mind the 1.4 billion dollar damage of Betsy in 1965, a 20% reduction in the destruction by a single major storm would support all the hurricane research ever carried out up to the present time, leaving many millions for future programs.

Therefore, the question to which we address our science then is whether a Daisy can be converted into a Frieda by the triggering of an instability. So far, we have been postulating cloud seeding as a possible trigger. Figure 2 shows a closeup of the wind profile of hurricane Daisy. The problem is whether the high winds in the radial interval 10–30 miles from the center can be reduced. This is the wall cloud region shown in Figure 3. The diameter of the wall cloud varies somewhat from storm to storm and sometimes within the life cycle of an individual hurricane, but the essential relations between cloud and wind profiles obtain in the majority of active storms. Can the extreme wind speeds in the wall cloud be significantly reduced by artificial means?

The current Stormfury eyewall hypothesis reasons from cloud seeding to a postulated reduction of maximum wind speeds in the hurricane. The cloud seeding must be done selectively in the wall cloud surrounding the eye; in fact, the silver iodide should be introduced into or just upwind of the most intense convection, called the "chimney cloud" region (Figure 4). The purpose of the silver iodide is to convert supercooled liquid water into ice in this portion of the eyewall, releasing latent heat in the process. The next question is how latent heat release here could cause a reduction of the maximum wind speed of the storm.

To develop this hypothesis (Simpson, Ahrens and Decker, 1963) we use a physical model of how a hurricane engine works, based upon the years of observations carried out by N.H.R.L. This model includes four basic assumptions; three are fairly well confirmed by observations. These are:

(1) Large amounts of supercooled liquid water (roughly 1–4 gm m^{-3}) exist in the hurricane clouds. In

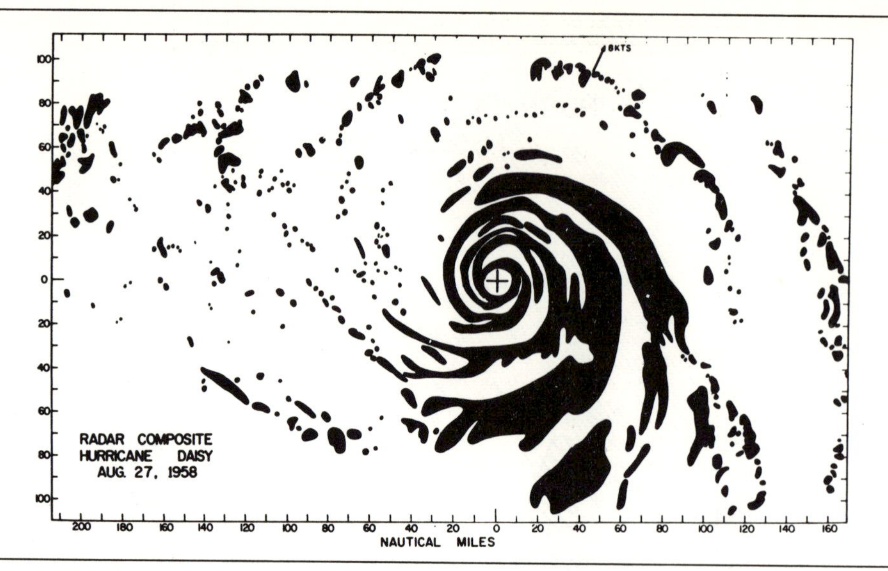

Figure 3 Radar composite of cloud structure in hurricane Daisy on 27 August 1958. This changed very little from the preceding day.

Figure 4 The hurricane model. The primary energy cell (convective chimney) is located in the area enclosed by the broken line.

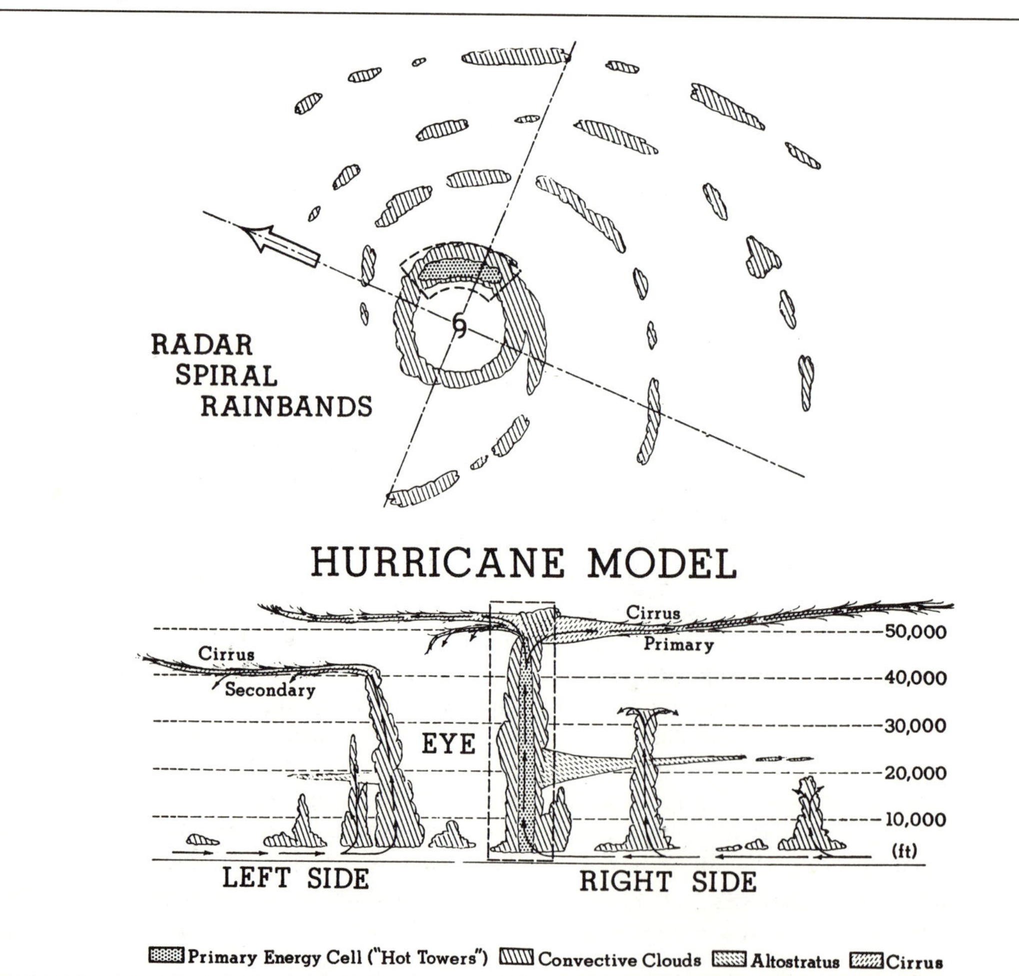

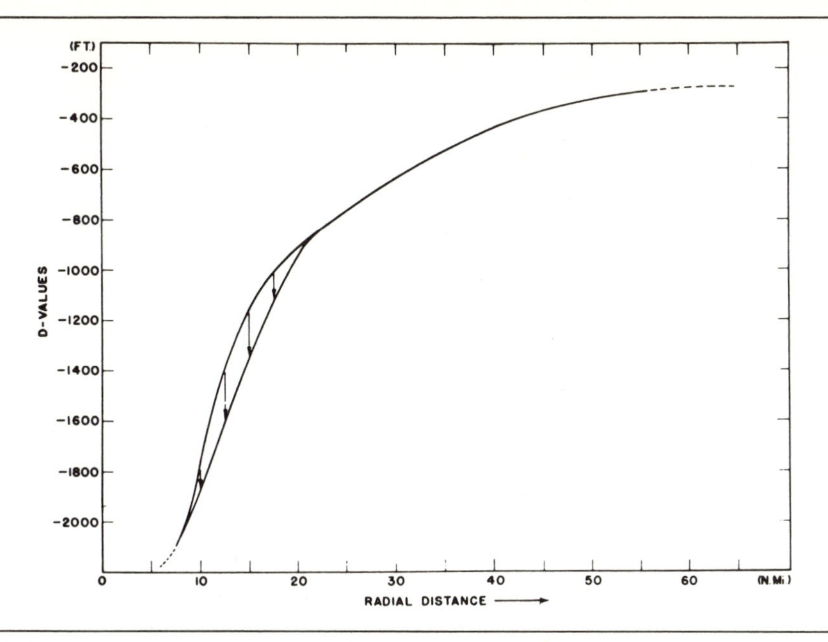

Figure 5 Anticipated change in shape of pressure surfaces due to seeding. D is altimeter correction, or radar altitude minus pressure altitude.

other words, much of the supercooled water does not naturally become converted into ice. Present airborne technology does not offer any immediate way to test this assumption by measurement, although indirect evidence (Simpson, 1963) favors it.

(2) Once frozen as a consequence of seeding, the ice crystals are largely exported by the storm outflow.

(3) An upper boundary of the storm exists, at or near the tropopause, where pressure surfaces are approximately level. This means that inward pressure gradients are produced by relative warming in the troposphere.

(4) Many storms show a radial wind profile which is "dynamically unstable." This means that the wind speed decreases outward from the center rapidly enough so that if a ring of air is given an initial push outward, it may be expected to continue accelerating outward despite the higher pressures toward which it moves.

If these conditions are met, Figure 5 shows that the heat released by the seeding can cause the D values to drop about 200 ft (corresponding to a pressure fall of about 6 millibars) in the eyewall region, thus reducing the slope of the surface pressure profile curve by a calculated 15 to 20%. This is the key link between the changing cloud system and the storm's circulation dynamics. The inward-directed pressure gradient enables the air to keep whirling in a tight circle about the storm center by providing a balance to the centrifugal acceleration. Thus the air approaches the storm center and develops terrific counterclockwise winds (on the northern hemisphere of the rotating earth) before rising in the towering cloud wall near the eye and being ejected aloft. If the inward penetration were reduced, that is, if the eyewall ascent could be moved outward from the center, the counterclockwise speeds achieved would be lessened in proportion, as in the case of the ice skater's rate of rotation which would be slowed if her arms were moved outward.

This is how cloud seeding in the hurricane is expected to work: The reduction in pressure gradient in the eyewall region upsets the original balance of forces that allowed the inflow to penetrate near the storm center. If this gradient is weakened, the eyewall should dissipate and reform farther out. Consequently the main ascent of air would occur at a greater radius and winds should not penetrate the storm core any further than they do in the sub-hurricane tropical storm. Thus the winds could not attain such high speeds as before the modification occurred.

The theory predicts the first few links in the chain quantitatively, up to the percentage change in the pressure gradient. Beyond that, existing models cannot predict the amount or duration of the wind reduction, nor can they tell how far outward the eyewall should migrate or remain. The dynamic-thermodynamic relationships in hurricanes are not yet so well understood, nor are numerical or laboratory models yet within orders of magnitude of the realism or complexity necessary to simulate such experiments. Thus there remain only two open avenues of approach, which we believe should be pursued simultaneously. The first consists of careful measurements in full-scale hurricane experiments. The second involves step-by-step testing of each link in the chain that will lend itself to evaluation by available techniques.

In the second category, laboratory tests on the seeding

methods have been carried out and two sets of field experiments have been completed on tropical cumulus clouds under fair weather conditions. It has been established beyond reasonable doubt that the seeding materials do in fact convert a supercooled cloud to ice, releasing about the predicted amount of heat. Furthermore, the released heat can markedly affect the dynamics of an individual cloud, causing it to grow spectacularly, under certain specifiable initial and environmental conditions (Simpson et al., 1965, 1966; Ruskin, 1967; Simpson, 1967; Averitt and Ruskin, 1967).

For the hurricane case, however, the critical link is that between the cloud thermodynamics and the storm's larger-scale circulation dynamics and there is presently available *no* meaningful way to test this postulate except by a full-scale experiment on a mature hurricane. Therefore, despite the expense and staggering difficulties, a program of experimentation has been undertaken.

A preliminary experiment was run on hurricane Esther in 1961 (Simpson, Ahrens and Decker, 1963). Following the one seeding, marked changes appeared in the eyewall downstream of the seeded region. The maximum wind speed dropped by about 10%. Subsequently, Project Stormfury was officially formed and an operationally successful seeding was achieved on Beulah in 1963. Eight specially equipped aircraft were required. Two were heavily radar-equipped Constellations, three were jet aircraft used as seeders and photographic planes provided by the U.S. Navy, and the remaining three were the (E.S.S.A.) Research Flight Facility's aircraft with radar, cameras, and digitalized recorders for measuring wind, temperature, and some aspects of the cloud structure. The results of this experiment have been reported earlier (Simpson and Malkus, 1963; 1964) so that only highlights are reviewed here in Figures 6–8.

These results are encouraging, but the observed changes following seeding were small. Natural changes of this sort and larger are documented (Ito, 1963) and Project Stormfury observed an example in (unseeded) hurricane Betsy in 1965. Since hurricane changes in both senses appear equally probable, one approach would be to seed a sufficiently large sample of cases so that statistical analysis can separate real and induced changes from the natural noise level. Unfortunately, it might require a hundred cases or more and as many years to accumulate the data.

In addition to the vast expense and logistics, rules governing the conduct of the project provide that only Atlantic storms can be seeded and then only in a restricted region, from which they cannot reach populated areas within 36 hours (Figure 9). About two mature storms in three years cross this region. Thus the sampling objective could not be fulfilled in our lifetime. One alternative is to work on Pacific typhoons, which from some island bases would occur many times more frequently. So far both logistic and political difficulties have delayed serious contemplation of this move. Another alternative is more definitive experiments on the few available individual Atlantic storms. A repetitive seeding experiment was designed and operational plans readied for the 1965 season.

If possible, it was planned to seed the eyewall of a single storm up to five times in succession at two-hour intervals, observing both core and periphery at numerous levels up to 60,000 ft. The experiment would continue for a total of eighteen hours, from four hours before the first seeding to six hours following the final one. If the present Stormfury eyewall hypothesis is worth pursuing further, the repeated eyewall seeding should lead to more pronounced and sustained changes of the sort observed in Esther and Beulah. Direct extrapolation of the simplified reasoning contained in the hypothesis predicts this result qualitatively. However, there are sound grounds for reasoning against this end product, even if the first few steps in the chain of reasoning are verifiable. After a short period, the storm may be able to readjust its structure through processes that we presently cannot foresee. Alternatively, additional instabilities could be activated, or conversely, external factors could so dominate the storm that no apparent effects of the seeding would be detectable.

In any case, execution of the experiment is essential to find out these answers and also to determine other important aspects of hurricane processes. We can hardly fail to learn more about hurricanes and their modification potential after a successful operation of this sort has been effected. In 1965, all aircraft were deployed to Puerto Rico in pursuit of hurricanes Betsy and Elena, the only ones approaching the area, but

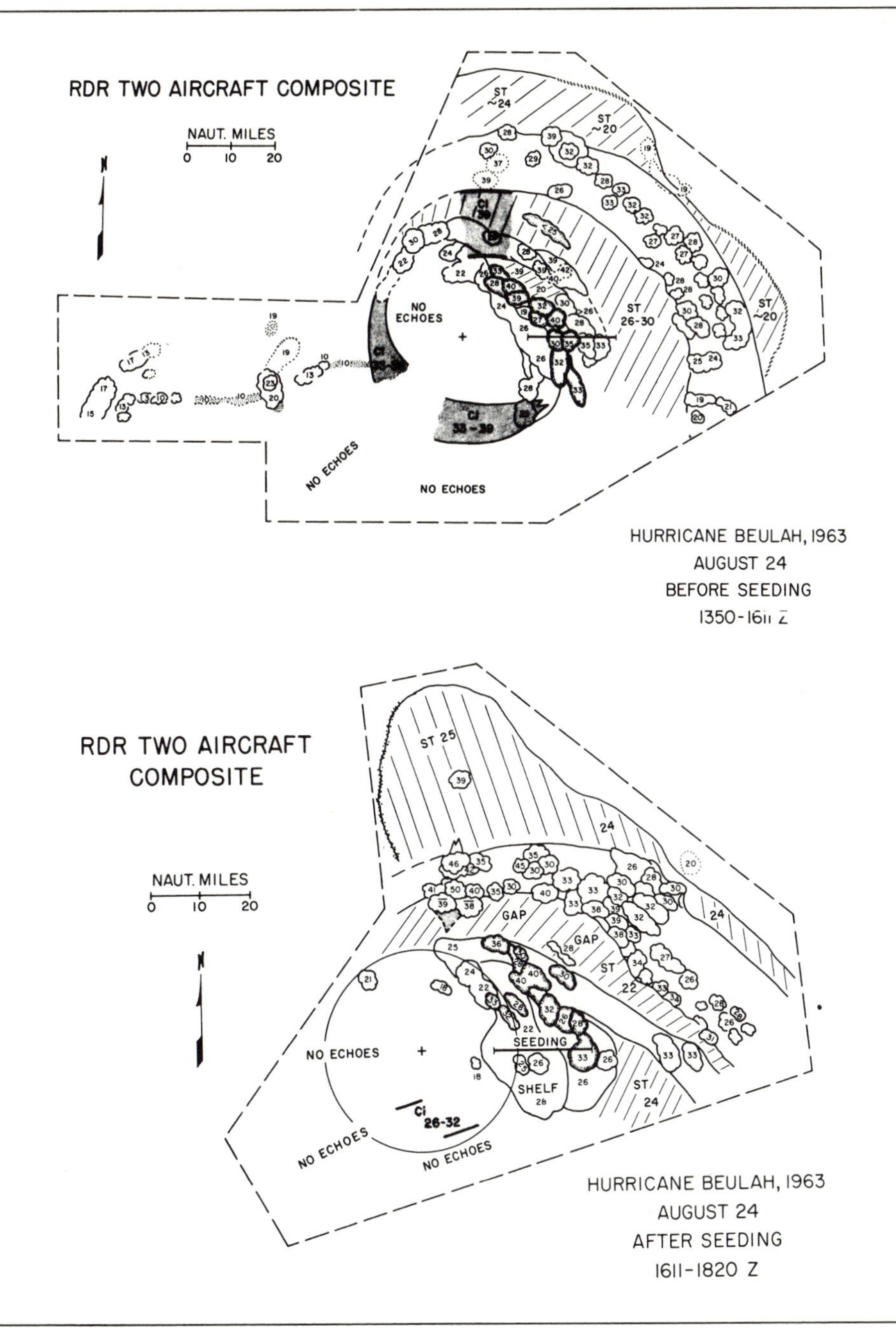

Figure 6 RDR radar composite maps of cloud echoes in hurricane Beulah, 24 August 1963. Numbers indicate echo heights in thousands of feet. A: Before seeding at 1611Z (Greenwich time). B: After seeding.

neither storm was seedable. Preparations are again being readied to carry out one or more of these experiments in 1966.

Cloud seeding may turn out not to be an effective means of hurricane modification. Although less likely, it may even turn out not to be a useful way to explore causal relationships in a purely scientific hurricane experiment. At the very least, however, we have learned that quantitative hurricane experiments are feasible, and this knowledge is available for use with other types of experimental approaches.

However, before mounting any new experimental programs, any one of which probably will be at least as costly and difficult to stage as Stormfury, preliminary research and computations should be undertaken. A scientific experiment cannot be conducted on the basis of a model or technique alone, but rather it requires that the model and experimental technique be properly

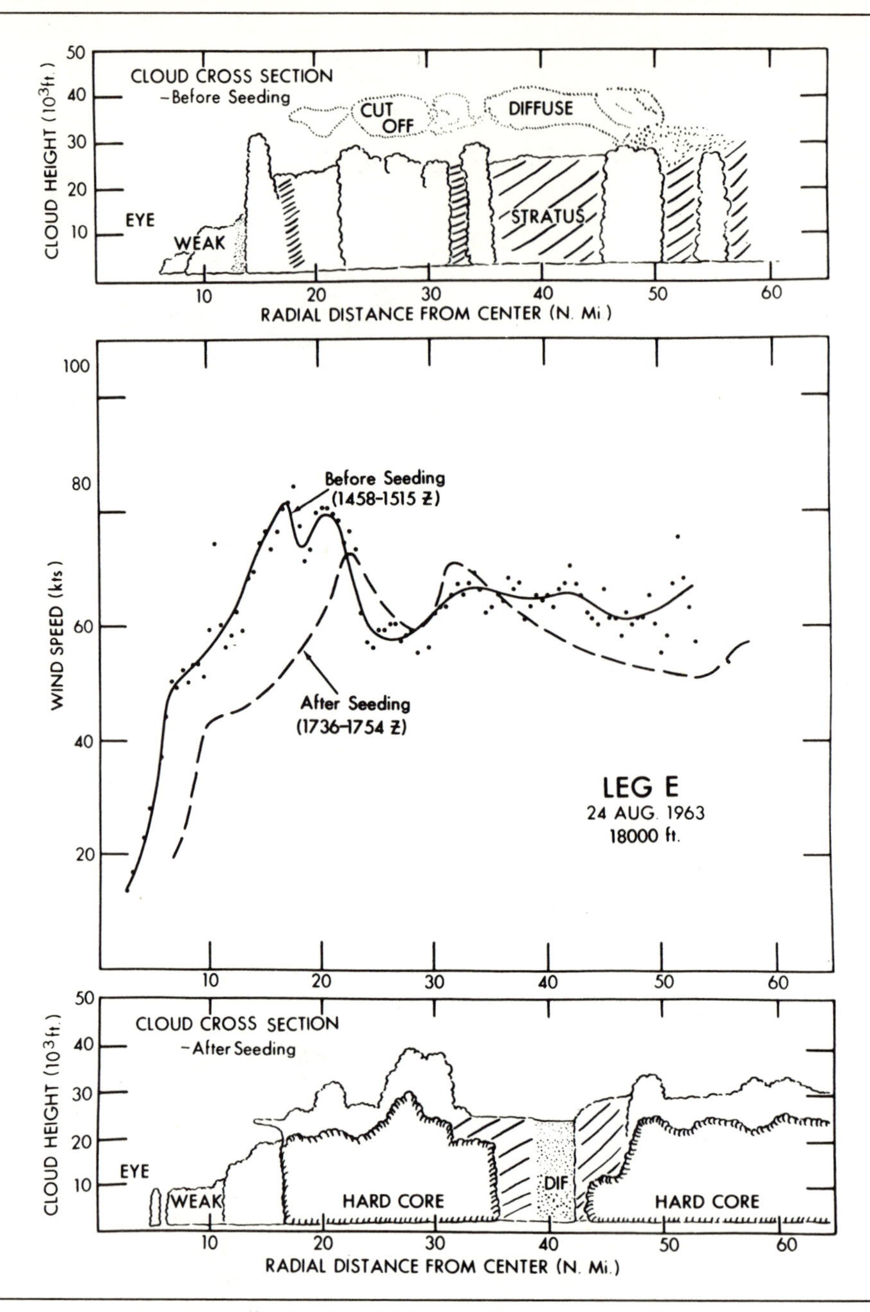

Figure 7 Cloud and wind profiles in hurricane Beulah, 24 August 1963. Above and below: Cloud cross sections (made from RDR radar) before and after seeding. Center: Wind speed. Solid curve before seeding, dashed curve after. Leg E is one of several radial legs flown by the aircraft at 18,000 feet.

mated. There are, however, concepts whose potential for experimentation need careful examination. For instance, other avenues of cloud modification might be considered, such as the prevention of drop coalescence or the cooling of the cloud particles. Chemicals to accomplish these results have been suggested. However, it appears unwise to apply such techniques to hurricanes without construction of hypotheses and preliminary tests on cumulus clouds.

Aside from cloud modification, two other approaches have been suggested. These are directed to opposite parts of the hurricane machinery. Neither requires postulating instabilities in the storm processes, but both are directed toward controlling boundary fluxes. The first concerns the alteration at high levels of the radiative balance in tropical cyclones and the other involves artificial restriction of the critical energy fluxes at the sea-air boundary.

One of the exciting discoveries of the early *Tiros* satellites with infrared radiation sensors was that the

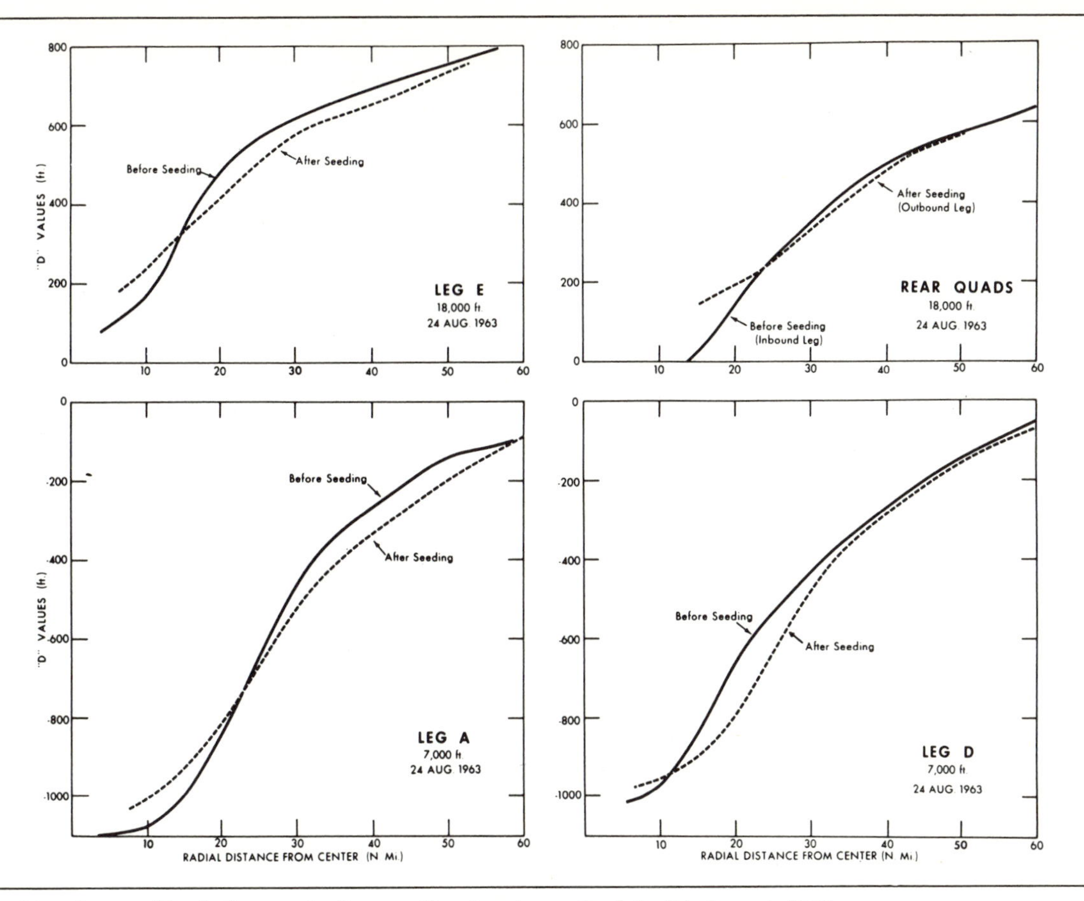

Figure 8 D value profiles before and after seeding hurricane Beulah, 24 August 1963. Compare with Figure 5.

fluxes of long-wave radiation emitted outward to space from hurricane and typhoon cloud systems were much larger than had been anticipated. If this radiation were trapped in the high troposphere, the tropopause should become warmer and lower, presumably lowering the elevation of the vital outflow of air from the storm. Earlier findings (Riehl, 1963) suggest that the higher and colder the tropopause, the higher and colder the outflow and thus the more efficient and intense the hurricane heat engine. Coupling these effects emphasizes the potential importance of any artificial means for upsetting the radiative balance or of lowering the tropopause by selective absorption of radiation near the tropopause level.

Interestingly enough, techniques imposing a greenhouse effect over certain portions of the hurricane cloud system may be easier to apply than it will be to assess the consequences of their application. One approach to this problem is based on research at The Johns Hopkins University (Simpson, 1965), which has produced means of generating very small plastic bubbles in large numbers with diameters as small as several microns. These can be generated and dispersed from jet aircraft. It is possible to impregnate these bubbles with such materials as diatomaceous earths to provide selective absorption of radiation in the infrared range corresponding to the mean radiating temperature of the hurricane cloud system.

One scheme by which an interesting radiation experiment might be initiated is suggested in Figure 10a. At the top of the wind circulation relative to the center of the hurricane is a cusp point which, kinematically, is a point of zero wind speed relative to the moving storm center at the surface. Plastic bubbles inserted at the top of the storm in the vicinity of the cusp point would tend to maintain their position relative to the moving center and, by absorption of infrared radiation, contribute to a progressive warming of the layer. A hypothetical alteration of the sounding in the column beneath the bubbles is illustrated in Figure 10b. It would be worthwhile to initiate a program to assess by computation the possible consequences of such modification on the intensity and movement of the storm system.

Although numerical models of hurricanes have not yet approached the stage where they can predict realistic results of an actual experiment, there is some indication that laboratory models may soon contribute to this objective. Some initial work, as yet unpublished, at the Florida State University (Hadlock, personal

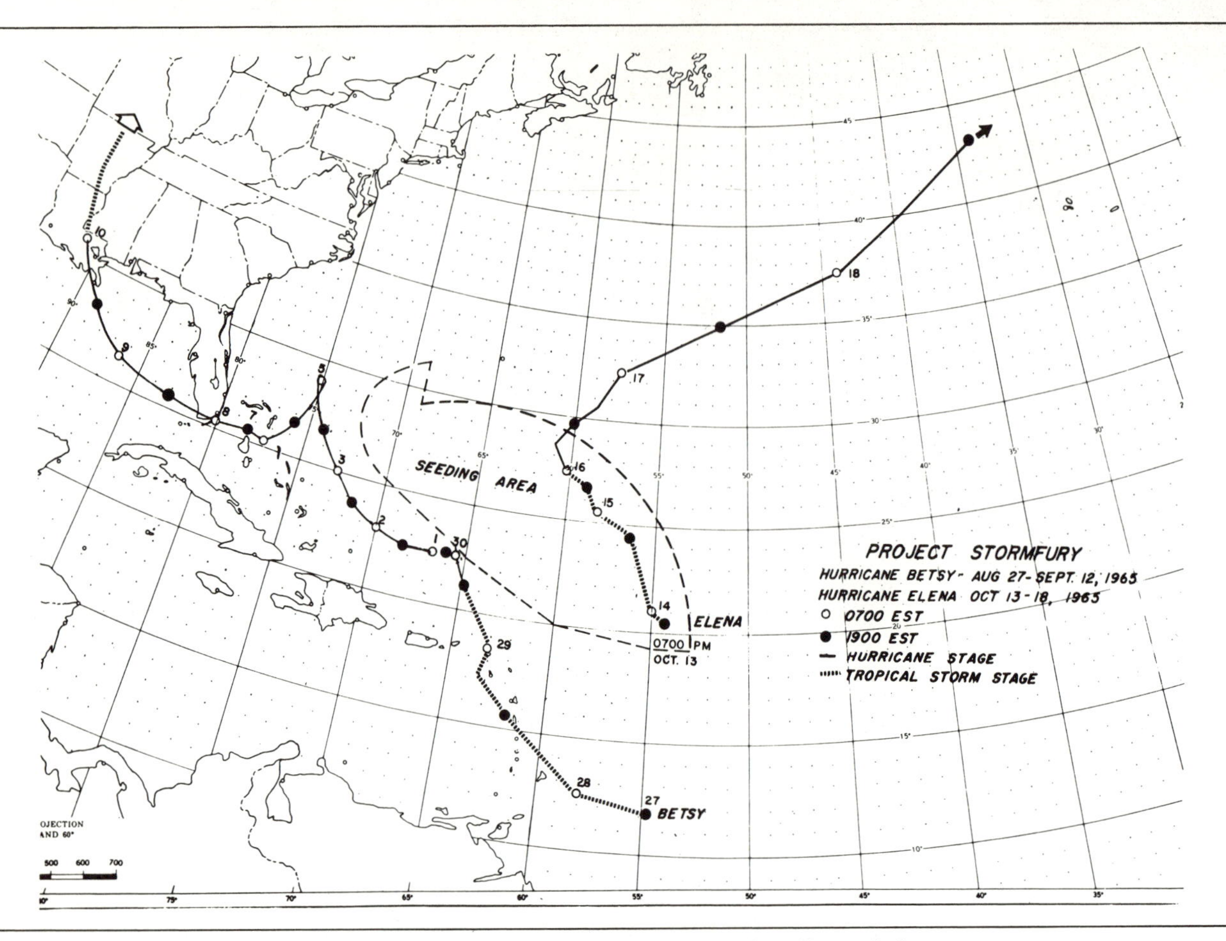

Figure 9 Region in which ground rules permit hurricane seeding by Stormfury (for period 15 August–31 August). These charts were compiled for each two-week period of the hurricane season from climatological records of storm tracks. Southern border excludes all recorded storms striking populated areas within 36 hours. Northern border is 700-mile arc from Roosevelt Roads, Puerto Rico. Tracks of hurricanes Betsy and Elena, 1965, are shown. Stormfury aircraft were deployed to Puerto Rico in anticipation of experimenting on these storms, but neither one was possible to seed.

communication) makes a miniature "hurricane" suggesting some of the key features of the real phenomenon. Differences in its structure and formation appear to be reproduced by a controllable variation in input conditions.

Another approach involves the lower boundary. There, two possible means of restricting the sources of energy required to stoke the hurricane engine involve modifications of the sea surface. Malkus and Riehl (1960) showed, from both theory and budget calculations, that the pressure gradients necessary to sustain hurricane force winds cannot develop without abnormally large fluxes of latent heat from the sea to the in-spiralling air currents of the tropical storm. Recent tritium sampling by Östlund (1966) suggests that local evaporation may provide a much larger fraction of the hurricane's water vapor fuel even than had previously been inferred. For an example, his calculations on Betsy in 1965 in an intense phase (winds exceeding 100 knots) indicate that in the eyewall cloud, the locally evaporated water substance may be in a two-to-one ratio to that imported from the storm's surroundings. This implies a continuous exchange of water substance by evaporation-precipitation in the inflowing air, with that penetrating to the inner core raining out about three times its initial precipitable water content. The validity of this conclusion and its implications for the storm's water and energy budget warrant careful consideration and testing. If the deduction regarding the water source stands up to the tests, its importance is far reaching for all phases of hurricane study, forecasting, and possible modification attempts. Further evidence of a crucial relationship between local evaporation and hurricane intensity was suggested by Östlund's tritium samples on a weaker day of hurricane Betsy (maximum winds less than 80 knots). On this occasion, no appreciable gradient in tritium count was found in approaching the storm center, suggesting the absence of the extra oceanic source. If the local source proves a necessary condition for the maintenance of an intense storm, a

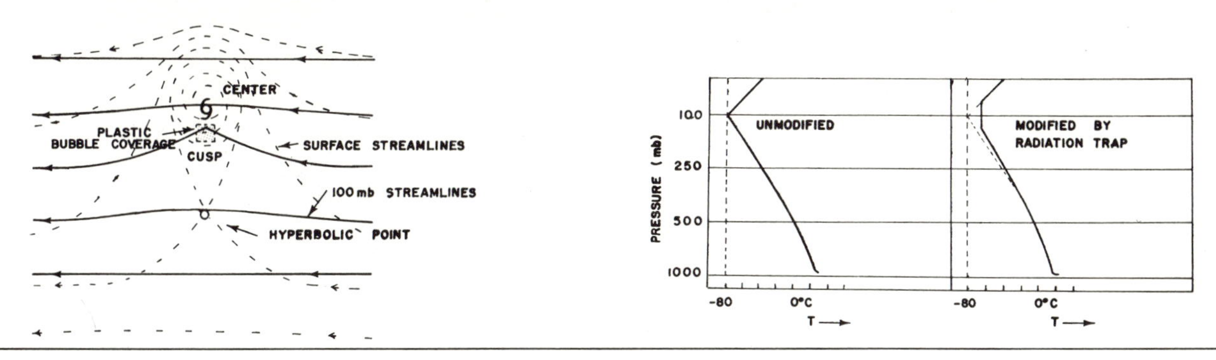

Figure 10a Surface (dashed) and 100 mb streamlines (solid) in typical hurricane showing position of cusp point relative to center and to hyperbolic point. If plastic bubbles are disseminated in rectangular region near cusp, they should presumably move with the storm.

Figure 10b Hypothetical temperature versus pressure profiles in a hurricane in the region marked "cusp" in Figure 10a before (left) and after (right) plastic bubbles have been disseminated.

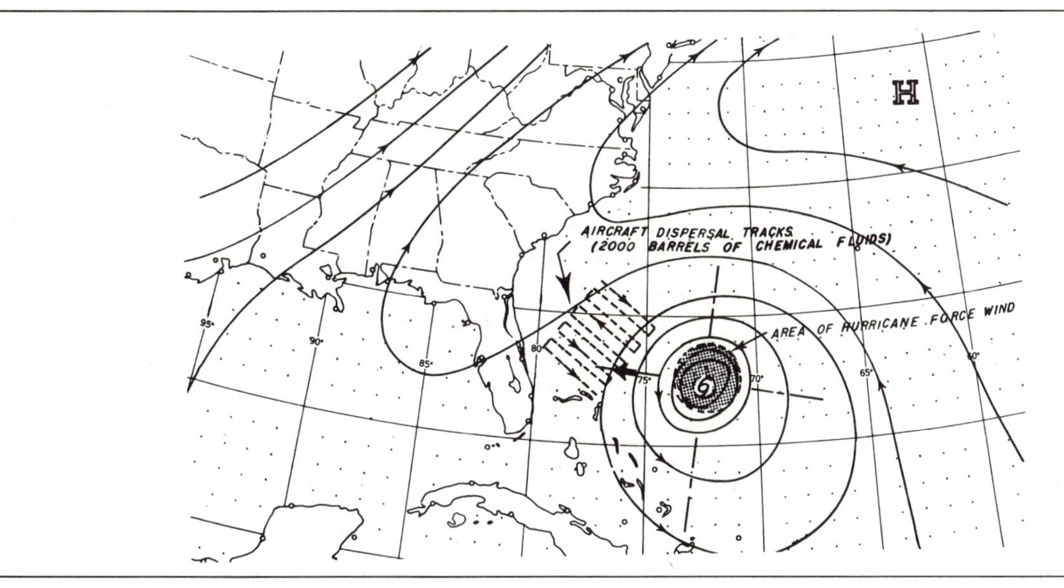

Figure 11 Diagram illustrating possible method of distributing a fluid chemical on the sea surface in advance of a hurricane approaching the U.S. coastline. The point of these very rough calculations is that a reasonable number of aircraft and amount of chemical may be involved, if such a chemical can be found to withstand wind and wave action.

promising modification method lies in the suppression of evaporation in the storm core, regardless of whether this energy transfer plays the role of the chicken or the egg in the causal cycle.

There are several immiscible fluids which, when applied to a water surface, spread out in a monomolecular layer and strongly inhibit evaporation (La Mer, 1962). Recent developments (Dressler, 1962; Grundy, 1962) indicate that some of these chemical films are able to maintain and restore themselves on large reservoirs with winds up to about 20–35 knots. This is a long way from the stormy seas of a full hurricane, but with our growing technology it is conceivable that the development can be extended. If further experimentation shows this approach feasible, large applications of such fluids to the water surface near a coastline might critically reduce the oceanic source of energy needed to sustain hurricane winds and thus reduce materially the destruction at the coast.

One further question is whether such an operation, even if theoretically effective, would be logistically and economically feasible. Preliminary estimation, illustrated in Figure 11, implies that it may be, when weighed against hurricane destruction measured in hundreds of millions of dollars. Here a fleet of thirty C-130 (Hercules) aircraft would be able to spray a total of some 1,500 to 1,800 tons of chemicals on an area 100 by 120 miles in advance of the hurricane. This region should be strategically located to encompass the initial area of hurricane force winds as the storm approaches the coastline. The numbers are based on figures by Timblin et al. (1962) of the U.S. Reclamation Bureau, for the amount of hexadeconal required to maintain a film on a 1,000 acre lake in winds up to 15 knots. This gross approximation, of course, leaves many questions of feasibility and applicability unanswered, but it does point to the need for further study along these lines.

Another approach to the modification of the sea surface has been suggested by Suomi (personal communication). He proposes that a mechanical means be used to stir up the oceanic thermocline layer so as to bring to the surface the colder water that is usually at 50–100 meter depth. This would be implemented by the buoyancy added when some of the dissolved gases in the water are released. He estimates that by dragging a long cable strung between two submarines, or even towed by two tugs, the surface temperature could be lowered by several degrees. Logistically this could pose some very difficult, perhaps insurmountable, problems. First, however, preliminary computations and limited experimentation are needed to assess the value and feasibility of this line of attack.

CONCLUSION

No conclusion may be drawn now about the usefulness of the present Stormfury approach to hurricanes. This must await operationally successful execution of the experiments planned for 1966. Analyses of these data should show whether further experiments along these lines will be valuable.

It has been said that if resources comparable to those directed toward the space effort were devoted to hurricane modification, these menaces could be eradicated. This statement contains a basic error. If engineering difficulties can be overcome, we know from scientific assurance that man can rendezvous in space and reach the moon. There remain only questions of detail as to how and when. In hurricane modification, the basic scientific questions remain unanswered; the question of "how" is not a matter of detail, since the question of "whether" still looms unanswered. It is to this problem that our major additional efforts must be devoted.

Field experimentation on the full-scale hurricane and its components must be backed and guided by laboratory and theoretical-numerical experiments and by the development of modification techniques and their delivery systems. All these must be regarded as a many-pronged scientific program, as a necessary investment if understanding of causality and eventual control are to be attained. An experimental approach to atmospheric problems is costly, but it is cheap compared either to the cost of the space program or to the destruction wrought by severe storms. The experimental approach may be the most effective way, if not the only way, to break through some of the bottlenecks obstructing progress in the science of meteorology, as has so often been the case in the science of physics.

REFERENCES

Averitt, J. M. and Ruskin, R. E., 1967. Cloud particle replication in Stormfury tropical cumulus. Project Stormfury Reports, 5–65, U.S. Navy and E.S.S.A., Washington, D.C. *J. Appl. Meteor. 6*, 88–94.

Dressler, R. G., 1962. An engineering approach to reservoir evaporation control. In *Retardation of Evaporation by Monolayers*, ed. V. La Mer. Academic Press Inc., New York. 203–211.

Grundy, F., 1962. Some problems of maintaining a monomolecular film in reservoirs affected by winds. In *Retardation of Evaporation by Monolayers*, ed. V. La Mer. Academic Press Inc., New York. 213–218.

Hadlock, R. K. Personal communication.

Ito, H., 1963. Aspects of typhoon development, as viewed from observational data in the lower troposphere. *Proc. Inter-regional Seminar on Tropical Cyclones, Tokyo, 1962.* Japan Meteorological Agency, Tokyo. 103–119.

La Mer, V. (ed.), 1962. *Retardation of Evaporation by Monolayers.* Academic Press Inc., New York.

Malkus, J. S. and Riehl, H., 1960. On the dynamics and energy transformations in steady-state hurricanes. *Tellus 12*, 1–20.

Östlund, G., 1966. *Air–Sea Water Exchange in Hurricanes from Tritium Measurements.* Marine Laboratory, University of Miami.

Riehl, H., 1963. On the origin and possible modification of hurricanes. *Science 141*, 1001–1010.

Riehl, H. and Gentry, R. C., 1958. Analysis of tropical storm Frieda 1957. Preliminary report. *National Hurricane Research Project Report 17*, U.S. Dept. Commerce, Washington, D.C.

Ruskin, R. E., 1967. Measurements of water-ice budget changes at −5° C in AgI-seeded tropical cumulus. Project Stormfury Reports, 3–65. U.S. Navy and E.S.S.A., Washington, D.C. *J. Appl. Meteor. 6*, 72–81.

Simpson, J., 1967. Photographic and radar study of the Stormfury, 5 August 1965 seeded cloud. Project Stormfury Reports, 4–65, U.S. Navy and E.S.S.A., Washington, D.C. *J. Appl. Meteor. 6*, 82–87.

Simpson, J., Simpson, R. H., Andrews, D. A., and Eaton, M. A., 1965. Experimental cumulus dynamics. *Rev. Geophys. 3*, 387–431.

Simpson, J., Simpson, R. H., Stinson, J. R., and Kidd, J. W., 1966. Stormfury cumulus experiment 1965. Preliminary summary. Project Stormfury Reports, 1–65. U.S. Navy and E.S.S.A., Washington, D.C. *J. Appl. Meteor. 5*, 521–525.

Simpson, R. H., 1963. Liquid water content in squall lines and hurricanes at air temperatures lower than −40° C. *Mon. Wea. Rev. 91*, 687–693.

Simpson, R. H., 1965. Project Stormfury: An experiment in hurricane weather modification. *Geofis. Internac. 5*, 63–70.

Simpson, R. H., Ahrens, M. A., and Decker, R. D., 1963. A cloud seeding experiment in hurricane Esther, 1961. *National Hurricane Research Project Rept. 60*, U.S. Dept. Commerce, Washington, D.C.

Simpson, R. H. and Malkus, J. S., 1963. An experiment in hurricane modification: Preliminary results. *Science 142*, 498.

Simpson, R. H. and Malkus, J. S., 1964. Experiments in hurricane modification. *Scient. Amer. 211*, 27–37.

Timblin, L. O., jr., Florey, Q. L., and Garstka, W. V., 1962. Laboratory and field reservoir evaporation reduction investigations being performed by the Bureau of Reclamation. In *Retardation of Evaporation by Monolayers*, ed. V. La Mer. Academic Press Inc., New York. 177–192.

M. Neiburger

17 Weather modification and smog

The three essential ingredients in the recipe for the Los Angeles type of smog are (i) sources emitting pollution into the air, (ii) atmospheric conditions which deter or prevent rapid transport of these pollutants in the atmosphere, and (iii) solar radiation for the photochemical reactions which transform the relatively innocuous pollutants into substances which cause irritation to the eyes and the respiratory tract and damage to plants. In the Los Angeles area the first ingredient is continually present and constantly increasing. The second ingredient, the lack of dispersal by atmospheric motions, and the third, the intense shortwave radiation, depend on weather conditions, and the conditions in the Los Angeles Basin are predominantly favorable for accumulation, rather than dispersal, of pollutants and for plentiful sunshine for photochemical reactions during much of the year. The prevalence of the subtropical inversion and the preponderance of light winds or calms make it possible for objectionable concentrations to accumulate in the basin within the course of a single day, and days with low inversion tend to be cloudless and bright.

It is frequently suggested that it might be simpler to attack the problem of eliminating the second ingredient or the third rather than the first, since the first ingredient is so intimately associated with a healthy industrial and economic development.

In the present article, the various proposals for modifying the weather in order to eliminate smog are discussed. The conclusion, unfortunately, is that elimination of smog by weather modification is even more difficult or costly than control at the sources. This applies, of course, only to those proposals which have thus far come to my attention.

Meteorological Factors Leading to High Pollution

The conditions which are responsible for the accumulation and lack of dispersion of pollutants emitted into the atmosphere in the Los Angeles area are part of a large-scale pattern associated with dynamic processes involving tremendous quantities of mass and energy. The subtropical inversion which prevails over Los Angeles is characteristic of the entire eastern portion of the Pacific Ocean as well as of the subtropical oceans and the adjoining west coasts of continents throughout the world. The inversion is produced by subsidence of air circulating from the north around a high-pressure center which is present over the eastern North Pacific Ocean most of the time during the warm months and frequently at other times during the year.

The prevalence of light winds and calms over the Los Angeles Basin is likewise associated with the basin's position in relation to the general circulation of the atmosphere. The normal wind in the Los Angeles area may be regarded as a monsoon, or seasonal wind, on which is superposed a diurnal sea-land and valley-mountain wind effect. In summer, the prevailing seasonal surface wind is south-west or west, representing the tendency of the air to flow from the high-pressure area over the ocean to the thermal low over the interior desert. The addition of the effects of diurnal heating and cooling result in moderate winds from the ocean during the day, and calm or light land breezes from the north and east at night.

To illustrate the nature and significance of the inversion, Figure 1 shows the average variation of temperature with height at Long Beach at 7 AM in September. On the average, the inversion layer begins at 475 meters above sea level. Below this level the temperature decreases with height, as is normal in most places and situations in the lower atmosphere. Above 475 meters, however, the temperature increases as one proceeds upward, until the top of the inversion layer is reached, at 1,055 meters, at which point the temperature begins to decrease with height once more. It has been

Reprinted, in edited form, by permission of the author and *Science*, from *Science*, Vol. 126, pp. 637–645, 4 October 1957.

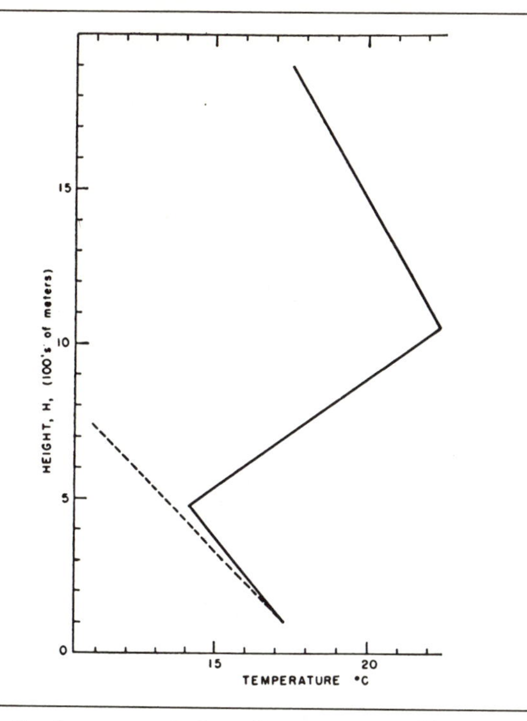

Figure 1 Average variation of temperature with height at 7 AM PST in September at Long Beach, California.

found that, as one proceeds inland from the coast to the foothills, the height of the inversion above the ground is, on the average, about constant (Neiburger, et al., 1945).

To see why the base of the inversion acts as an effective lid on the upward dispersion of pollutants, it must be remembered that, when a parcel of air rises, it cools at a rate of about 1° C per 100 meters of rise. If a parcel of air is given an upward impetus from the ground, it will arrive at any level below the inversion with a temperature only slightly lower than that of the surrounding air and, thus, will be subjected to very little downward force. Consequently, it would have a moderately long period during which it could mix with its surroundings before it sinks. On the other hand, air which rises into the inversion immediately finds itself much cooler than the air surrounding it and is quickly accelerated downward, so that it gets little opportunity to mix its contaminants with the inversion air.

The typical wind regime in the Los Angeles Basin during the warm season, which includes most days with severe smog, is illustrated in Figures 2 and 3. Figure 2 shows streamlines and isotachs (lines of equal speed) at 2:30 PM PST on 21 September 1954, when the sea breeze was near its maximum. Most of the streamlines originated near the west coast of the basin, and the wind speed was greatest there. During the afternoon and early evening the sea breeze died down and was replaced by a land breeze. Figure 3, the map for 2:30 AM on 22 September, is fairly representative of the entire period from 8:30 PM to 8:30 AM. At this time the streamlines were generally the reverse of those at 2:30 PM. The speeds were 2 miles per hour or less over most of the basin. The speeds remained low until the sea breeze replaced the land breeze at about 9:30 the following morning. The daytime hours are thus characterized by light or moderate winds from the ocean, and during the night and early morning hours gentle land breezes or calm conditions prevail. From the standpoint of air movement, the concentrations of pollutants emitted during the night and morning hours build up to high values, and then, with the onset of the sea breeze, the air containing these high concentrations is transported across the basin.

Proposals for meteorological modification are ordinarily aimed at increasing the volume into which the contaminants may spread, either by raising or eliminating the inversion or by causing the air to move more rapidly across the basin. Occasionally methods have been proposed for reducing the solar radiation below the level required for photochemical reactions. In the following sections, each of these types of proposals is discussed in turn.

Penetrating or Dissipating the Inversion

The first idea which suggests itself to one who is aware that the inversion is a prime factor responsible for the high concentration of pollutants is to consider ways of eliminating it or at least "punching a hole" in it. In considering the practicality of this idea, it is important to realize exactly what is meant by the phrase *eliminating the inversion.* It is not at all a matter of removing the air in the inversion layer. If one were to do so, of course, the result would be to have the warmer air above the inversion top immediately in contact with the cool air at the inversion base, and thus have a still more impenetrable ceiling on the upward movement of pollutants.

Elimination of the inversion can mean only one of two things: (i) cooling all of the air from the inversion base upward to a temperature below that at the inversion base, or (ii) warming all of the air below the inversion top to a temperature higher than that at the inversion top. Since there is much less air below the inversion top than there is above the inversion base, it is obvious that the simpler procedure would be to raise

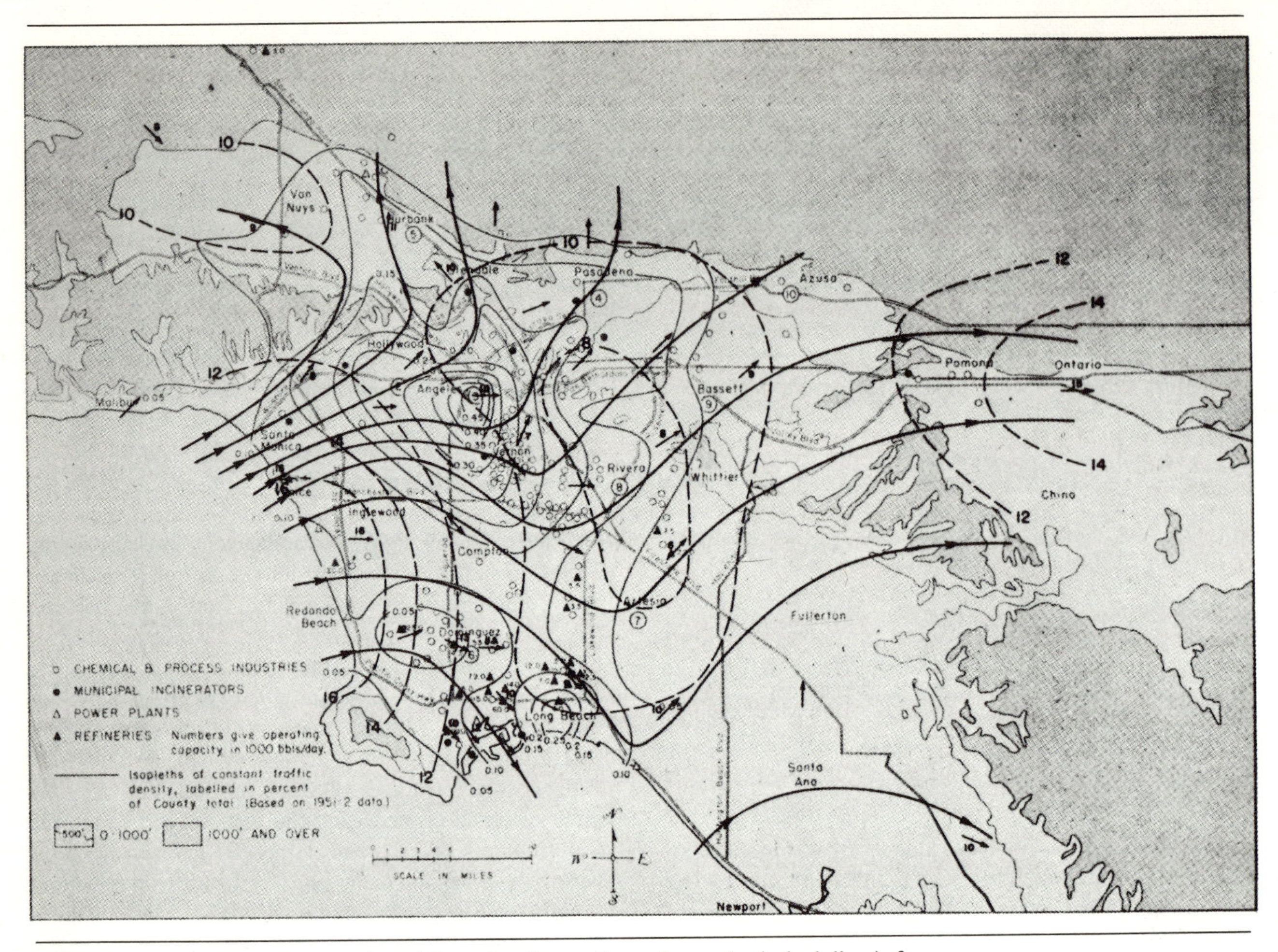

Figure 2 Streamlines (solid lines) and isotachs (lines of equal speed, dashed lines) for 2:30 PM on 21 September 1954. Speeds are given in miles per hour beside station arrows and at ends of isotachs. The principal pollution sources and traffic density are shown in the background.

the temperature of the air up to the inversion top sufficiently to eliminate the inversion. "Punching a hole" in the inversion can be interpreted intelligently only as eliminating the inversion in the same fashion over a restricted area.

It is relatively simple to compute the amount of energy required to heat the layer below the top of the inversion sufficiently to eliminate the inversion. Assuming that the energy would be injected at the ground, the process of heating would necessarily establish an adiabatic lapse rate. If we assume that the lapse rate below the base of the inversion is initially adiabatic, and that the rate of temperature increase with height in the inversion is linear, it is easily shown that the energy E required to eliminate the inversion is given by the following equation:

$$E = c_p\rho[\gamma_d(H_T - H_B) + (T_T - T_B)]\,\frac{H_T + H_B}{2} \qquad (1)$$

where c_p is the specific heat of air at constant pressure, ρ is the average density of the air below the inversion top, γ_d is the dry adiabatic rate of cooling, H_B and H_T are the heights, and T_B and T_T are the temperatures at the base and top of the inversion, respectively. For the average conditions at Long Beach at 7 AM in September, we have the following values: H_B, 475 meters; H_T, 1,055 meters; T_B, 14.1° C; T_T, 22.4° C.

Inserting these values in Eq. 1, we find that the energy required to eliminate the inversion is 330 calories per square centimeter. The area of the Los Angeles Basin is usually taken to be 4,000 square kilometers, so that the amount of energy required to eliminate the inversion over the entire basin is 1.32×10^{16} calories. It would take approximately 1.27 million tons of oil burned with 100% efficiency to produce this amount of heat. This would be equivalent to burning the amount of crude oil that is processed by all the refineries in the Los Angeles Basin during 12 days.

The preceding computation is made on the assumption that the air remains in place throughout the heating period. However, actually the process of heating itself would create horizontal motion which would bring new

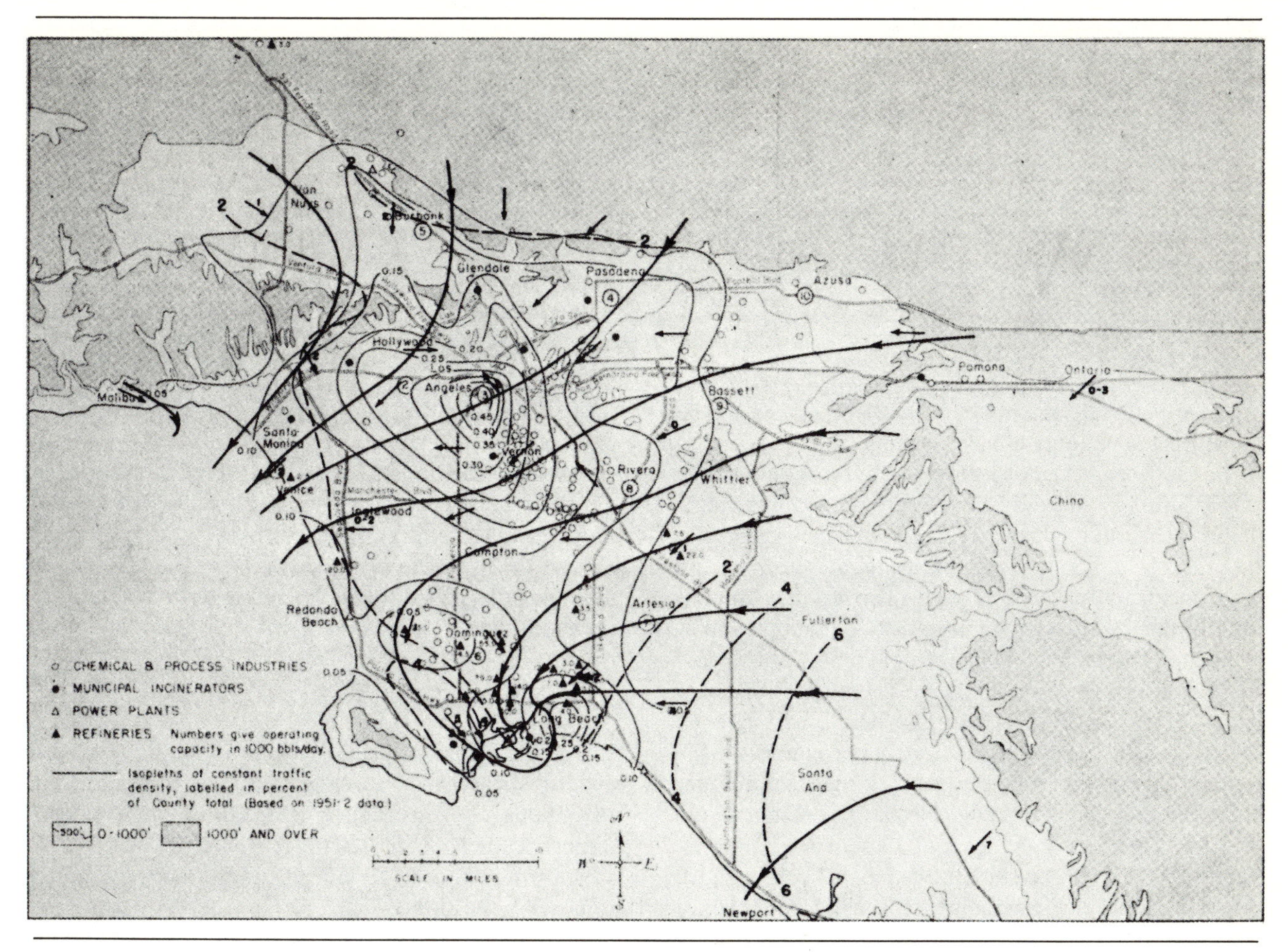

Figure 3 Streamlines and isotachs for 2:30 AM on 22 September 1954 (see Figure 2 for explanation).

cool air in over the basin and thus tend to restore the inversion or to make it necessary to use more fuel in order to eliminate it. This, indeed, is what happens every day as a result of the sun's radiation. The total energy arriving at the earth's surface from the sun ranges between 500 and 700 calories per square centimeter, per day during the smog season. While part of this energy is utilized to heat the ground, to compensate for long-wave radiation, and in evaporation and transpiration, and part of it is reflected from the ground, a considerable proportion is utilized in raising the temperature of the air.

Ordinarily, the amount of heat going into the air is not sufficient to eliminate the inversion, except in the foothills, where the layer below the inversion top is shallower than it is over the coastal plain. However, the amount of heating which it achieves is sufficient to set up the sea breeze and bring in the cooler air over much of the basin. The incoming cooler air results in a greater intensity of the inversion than would exist if the direct effect of heating were not accompanied by its indirect consequence. The divergence of the air flow in the sea breeze actually results in a lowering of the height of the inversion base from morning to evening near the shore (Neiburger, 1944).

The tremendous amount of energy required to eliminate the inversion completely leads to the suggestion that the heating be confined to a much smaller area. Since the amount of energy is proportional to the area, it is readily seen that, for instance, a 1-square-kilometer "hole" in the inversion would require 3.3×10^{12} calories, corresponding to burning about 320 tons of oil. It is often suggested that such a hole in the inversion would enable all the pollutants in the basin to pass upward. However, to cause any air to rise through the hole in the inversion, it would be necessary to heat it to the same potential temperature as that which was necessary for the elimination of the inversion. Thus, the amount of energy required would be the same, whether one attempted to get the air to pass through a small hole in the inversion or to eliminate the inversion over the area the air occupied originally.

There would be little advantage in building a stack up to the top of the inversion. Doing so might reduce the amount of radiation lost from the heated column and would surely eliminate the effect of entrainment, but the basic fact remains that all the air which one wishes to have penetrate the inversion must be heated to the potential temperature of the inversion top.

It has been suggested that if all the combustible rubbish in Los Angeles County were burned in a single large incinerator, adequate heat could be engendered to cause the pollutants from it to penetrate the inversion and, incidentally, to carry with them a large amount of entrained air containing other pollutants. It is estimated that about 14,000 tons of combustible refuse is burned in the basin. Allowing for its moisture content, we may reasonably assume a heat of combustion of about 5,000 British thermal units per pound for the refuse. Using this value, we find that the total amount of refuse collected per day is adequate to eliminate the inversion over an area of about 10 square kilometers. Obviously, this is too small an area to create much interest from the standpoint of general removal of smoggy air. However, if it were possible to carry out the combustion in such a fashion that the pollutants therefrom penetrate the inversion, this would be one excellent way of eliminating a not inconsiderable source of pollutants, particularly particulates which reduce visibility. To achieve this, it would probably be necessary to erect huge stacks to heights of 2,000 feet or more, in order to avoid entrainment and to conserve the high temperatures in the rising gases.

It has been suggested that some of the engineering difficulties in constructing stacks of this height can be avoided by having them take the form of pipes going up the slope of a mountain rather than having them be free-standing. The extra length of pipe in this case would increase the heat loss and decrease the flow rate, making it necessary to use tubes of larger diameter. The relative cost of constructing such a "smoqueduct" up the mountain, versus transporting the unburned rubbish up to incinerators at the top by truck or by conveyer belts, should be investigated if serious consideration is ever given to this scheme. Actually, it presently appears that the methods of disposing of rubbish by cut and fill or by well-designed incinerators equipped with adequate controls on the effluents represent more economical procedures.

In connection with the idea of eliminating the inversion by heating the air over the basin, it should be noted that the result would be to establish temperatures in the vicinity of 100° F over the basin for average inversion conditions, and somewhat higher values on days of low inversion which produce the most severe smog manifestations. Those who prefer these high temperatures to the smoggy conditions which are the unfortunate corollary to the otherwise more salubrious climate the inversion brings can achieve the same result with much less effort by moving to desert communities, such as Indio or Thermal.

Blowing the Smog Away

The other general proposal is to move the polluted air out of the basin. Leaving proposals to remove the mountains or dig tunnels through them for later discussion, it is of interest to compute the mass of air which would have to be moved and the energy required to produce the motion.

The average elevation of the inversion base is about 400 meters; over the 4,000-square-kilometer area below this level in the basin, there are thus about 1.6×10^{12} cubic meters of air which, under average conditions of temperature and pressure on smoggy days, weighs about 2×10^9 tons. On a day with severe smog, the inversion is lower, but one might expect that, as a minimum, one-tenth of this amount of air, or 200 million tons, is involved.

Dubridge (1955) has dramatized the problem by pointing out that this amount of air is twice the weight of all the steel produced in the United States in a year. This amount of steel, corresponding in size to a cube 1,000 feet on a side, would be easier to move than the same amount of air "because at least you could load it on freight cars and haul it away."

Moving large masses of air is indeed more difficult than hauling away an equal mass of steel, for, in addition to the problem of getting hold of it, one must displace an equal volume of air (and thus an approximately equal mass) in the place one moves it to, and one must also provide for the energy to move in the air which replaces it. Rather than attempt to solve the complete problem of initiating the motion and providing for the necessary mass exchange, we shall confine our attention to computing the rate of energy expenditure required to keep the air moving over a flat area the size of Los Angeles—that is, the rate at which energy would

Wind speed, v		Power E (kw) for various coefficients of friction C			
(m/sec)	(mi/hr)	0.01	0.05	0.10	0.50
1	2.2	5×10^4	2.5×10^5	5×10^5	2.5×10^6
2	4.5	4×10^5	2.0×10^6	4×10^6	2.0×10^7
3	6.7	1.4×10^6	6.8×10^6	1.4×10^7	6.8×10^7
4	8.9	3.2×10^6	1.6×10^7	3.2×10^7	1.6×10^8

Table 1 Power (E) required to maintain wind speed (v) against friction, for various coefficients of friction (C)

be dissipated by ground friction. This amount is far below that needed to set up and maintain the motion in actuality, for it makes no allowance for the effect of the confining mountains, the opposing pressure forces which would arise the moment the air begins to move, or the turbulence created by the fans or other propelling apparatus.

If the wind stress is S dynes per square centimeter for a wind velocity v over an area A, the rate of work done in maintaining the wind is SAv. Now, the relationship of frictional stress to wind velocity may be expressed as

$$S = C\rho v^2$$

where ρ is the air density and C is a coefficient depending on the roughness of the surface over which the wind is blowing. Sutton (1953) gives the following values of C for various surfaces, for v in centimeters per second: Ice, 0.002; close-cropped lawn, 0.005; thick grass up to 10 cm high, 0.016; thick grass up to 50 cm high, 0.032. Unfortunately no estimates are available for the value of C over the complex of surfaces comprising a city such as Los Angeles. It is reasonable to suppose that a value several times that for thick grass should apply.

The power required to keep the air speed constant is thus

$$E = C\rho Av^3$$

Taking A as 4,000 square kilometers and ρ as 1.25×10^{-3}, we get $E = 5 \times 10^{10}\ C\ v^3$ ergs per second, or $5\ C\ v^3$ kilowatts. Table 1 gives the values of E for several values of C and v. We see that if the entire Los Angeles area were as smooth as a golf green, it would require about 400,000 kilowatts to maintain a 4.5-mile-per-hour wind, and 3.2 million kilowatts to maintain a wind of 9 miles per hour against friction. For the actual character of the surface, it would seem to be reasonable to assume that the values given in the second column should be minimum, and that the actual values would probably be between those of the second and third columns. Thus the power requirements for maintaining 4.5- and 9-mile-per-hour winds over the Los Angeles Basin are probably more than 2 million and 16 million kilowatts, respectively.

For comparison, the capacity of Hoover Dam is 1.25 million kilowatts. To maintain artificially a 9-mile-per-hour wind over the Los Angeles Basin would require at least 12 Hoover Dams.

Interpreted in terms of 5,000-horsepower engines converting their energy to pure translation of the air with 100% efficiency, more than 4,000 such engines would probably be required just to overcome surface friction to maintain a 9-mile-per-hour wind. In terms of fuel consumption, this power corresponds to 1,000 tons of fuel oil per hour converted with perfect efficiency.

As stated previously, the problem of actually moving the air is far more complicated, and more requiring of energy, than simply overcoming friction. The fact that the latter process alone would require tremendous expenditures for equipment and fuel should be adequate to discourage proponents of the fan idea.

These computations likewise dispose of the proposal to build a tunnel through the mountains. If the tunnel were 100 feet in diameter, and if the air moved through it with a speed of 100 miles per hour, the amount passing through in a day would be 2.8×10^9 cubic meters, or about 0.2% of the air under the inversion when the inversion is at average height. If we consider that natural processes remove at least one-half of the air in the basin almost every day, we see that there is little likelihood that even 50 such tunnels would produce a noticeable effect on smog concentration, without taking into account the tremendous expense of digging such tunnels and the extreme difficulty of producing such velocities in tunnels several miles long.

In addition to proposals to blow away the smoggy air horizontally, there have been several proposals to achieve the same result by means of fans blowing air vertically. Usually ground-based fans blowing the smoggy air upward have been suggested, but there have also been two or three proposals to blow fresh air downward from above the inversion base by means of hovering helicopters.

In addition to the problem of the quantity of air to be moved, the difficulty with ground-based fans blowing air upward is that, except for the effect of mixing, the air blown upward would be cooled adiabatically, rapidly becoming colder than its environment and thus subjected to a downward force which would decelerate it and ultimately accelerate it downward so that it would return to the ground. Mixing would, on the one hand, rapidly reduce the velocity of the rising air so that it would not penetrate as far into the inversion, and on the other hand, raise its potential temperature so that its equilibrium position would be somewhat higher and might under some circumstances be in the inversion layer.

The basic idea of having the horizontal fans take the form of hovering helicopters blowing down warmer air from the inversion layer to mix with the polluted air near the ground has the advantage that the mixture will be warmer than the environment at the ground and will rise to some position intermediate between the level of the helicopter and the ground. For the same amount of energy there would be a larger amount of dilution by air from the inversion layer than in the case of horizontal fans blowing upward.

Again, the quantitative aspect makes these proposals unfeasible. For instance, it is estimated that a moderately large helicopter hovering would produce a jet immediately below it of velocity 20 meters per second over an area of about 250 square meters. (This change in momentum corresponds to the hovering of a helicopter weighing 12 tons.) Thus air would be brought downward at the rate of 5,000 cubic meters per second, or 1.8×10^7 cubic meters per hour. If the diluting effect of this were complete, and if it were felt over 1 square kilometer when the inversion was at a height of 360 meters, the smoggy air would be diluted at the rate of 5% per hour. If we wish to attain a dilution of 50%—that is, reduce the concentration of pollutants by half—it would require ten helicopters per square kilometer. The cost of maintaining ten, or even one, helicopter hovering over each square kilometer of the basin is obviously unreasonable, and the noise and safety hazard would be more objectionable than the smog. Ground-based fans would have to be at least as dense, similarly costly and noisy, but probably considerably safer.

Utilization of Solar Energy

Since elimination of the inversion or blowing away the smog would require prohibitive amounts of energy if the energy has to be supplied in the form of fuel or electric power, proponents of eliminating smog by weather modification have looked to the possibility of utilizing natural sources of energy. The approaches most frequently proposed have been (i) more effective conversion of solar energy to eliminate the inversion or cause greater air movements and (ii) triggering some potential instability.

The proposals to use solar energy take various forms. Some examples are painting the roofs of buildings in alternate city blocks white and black in a checkerboard pattern to promote convection; paving large areas of the basin with black asphalt for the same purpose and to eliminate the inversion; introducing black carbon dust into the air over the basin at low levels so that the air in contact with it would be heated and would rise like balloons through the inversion; and using mirrors or lenses to concentrate the sunshine.

The problem of utilizing the solar energy more effectively consists of raising the proportion of it used to heat the air, ordinarily about one-third to one-half, by reducing the amount which is reflected, lost by long-wave radiation, used to heat the ground, or used in evaporation.

The various energy-transforming processes are interdependent. For instance, if the albedo of the ground is reduced so that less solar energy is reflected back to the sky, the conversion of the energy at the ground would raise the temperature at its surface and thus increase the radiation and evaporation from the earth's surface and the conduction of heat into its interior, as well as the heating of the air by conduction and convection. Similarly, changing the thermal conductivity or heat capacity of the ground, or both, or its water content available for evaporation, would affect the surface temperature, and thus the outgoing radiation. Only a part of the energy which it is intended to tap by these modifications would be utilized for reducing the inversion or promoting convection.

As an indication of the possibilities, Table 2 shows an estimate of the average radiational and heat exchange at the ground in September, which is the month of most probable low inversion and smog in Los Angeles. Results for other months would be similar.

In column 2 is shown the average radiation received on a horizontal surface in downtown Los Angeles from sun and sky at various times of day. It will be noted that, even apart from other processes, the 330 langleys which we found previously to be required for elimination of the inversion under average conditions is not received until about 1:30 PM. Even if all the solar energy incident

Hour (PST)	Accumulated energy (langleys) from sunrise to end of hour: Solar radiation	Net long-wave radiation	Conduction to ground and evaporation	Reflected from ground	Used to heat air
7 AM	6.2	6.7	2.5	1.2	−4.2
8 AM	24.3	13.7	9.5	4.9	−3.8
9 AM	58.0	21.1	20.2	11.6	5.1
10 AM	105.9	29.1	35.0	21.2	20.6
11 AM	164.6	37.5	51.3	32.9	42.9
12 noon	233.1	46.4	69.5	46.6	60.6
1 PM	305.2	55.6	89.0	61.0	99.6
2 PM	373.7	64.7	108.2	74.7	126.1
3 PM	432.2	73.7	125.5	86.4	136.6
4 PM	474.9	82.5	139.2	95.0	158.2
5 PM	498.3	91.0	147.7	99.6	160.0
6 PM	505.1	99.2	150.5	110.0	145.4

Table 2 Estimated average radiation-heat exchange at ground in September

at the ground could be utilized to heat the air, the inversion would not be eliminated until after the time of maximum values of pollutant concentrations and smog effects over most of the basin.

In column 3 is presented the net radiational exchange between the ground and the atmosphere—that is, the (black body) radiation from the ground minus the energy radiated back to it from the water vapor and carbon dioxide in the air.

It is probable that any modification designed to increase the atmospheric heating would also, by raising the ground temperature, increase the net outgoing long-wave radiation. Thus the available solar energy would always be reduced by at least the amount in column 3. This would result in elimination of the inversion occurring, at the earliest, at about 2:30 PM, if all other energy losses were eliminated.

There are no available data on which to base a sound estimate of the average heat conducted into the ground (and buildings) or used in evaporation for the Los Angeles Basin. Homén's measurements for various surfaces in Finland suggest that, while the amount conducted into the soil and that used in evaporation separately vary greatly for rock, sand, meadows, and other surfaces, the total of the two is roughly the same for the various surfaces. This result is reasonable, for, wherever water is available for evaporation, the ground-air interface temperature is kept from rising, and thus the temperature gradient downward into the soil is smaller than it is in the absence of evaporation.

As a crude approximation of the amount of energy consumed by these two processes, the data cited by Sutton from Lettau's evaluation of measurements made by the Sven Hedin expedition were used, with slight adjustment, in column 4 of Table 2. The values are probably slightly high because of the high desert temperatures; a total of about 120 langleys for the 12-hour period might be more reasonable. However, the data obtained by Pasquill for a meadow in England were not significantly lower, and it was decided to use Lettau's data without further modification.

The average reflectivity of the complex of surfaces in the basin is likewise unknown. Measurements summarized in the Smithsonian Meteorological Tables for various kinds of surfaces occurring in the basin (grass, 14 to 33%; dry, plowed fields, 20 to 25%; sand, 18%) suggest that a value of 20% might be reasonable, and this value has been used for computing the values given in column 5. Subtracting the values in columns 3, 4, and 5 from those in column 2, we get the amount of energy available to heat the air, column 6.

The occurrence of the cessation of heating of the air between 4 and 5 PM appears to be inconsistent with the usual occurrence of the maximum temperature about 2 to 3 hours earlier. The occurrence of the sea breeze, which brings in air which has been over the ocean, may in part account for this discrepancy.

It has already been pointed out that, even if all the energy represented by columns 4 and 5 were utilized in heating the air, the inversion would not be eliminated before 2:30 PM, considerably after the smog normally is swept away from downtown Los Angeles by the sea breeze. Obviously, a complete elimination of these "losses" is impossible. Only a part of the surface of the basin is subject to being painted black to reduce the reflectivity, or with an insulating material to reduce conduction into the ground. The lawns and gardens, forming part of the attraction of the area to residents and visitors alike, could not be covered with such surfaces without objections even greater than those which the smog has aroused. If, optimistically, one-half the losses were eliminated over one-half the area the air passes over in reaching downtown, the 25% saving would be inadequate to eliminate the inversion.

Recognizing that alteration of the entire surface of the basin is impractical, some people have proposed paving selected areas in it, to eliminate the inversion locally and allow the smog to escape through the "holes" thus created. Unless the areas were very large, comprising tens or hundreds of square miles, the air ordinarily would move off them before it was heated enough to eliminate the inversion.

A variant of this proposal has as its purpose the use of solar energy to cause greater horizontal movement of the air. If the sea breeze starts earlier or is stronger, for instance, the smog will not reach as high concentrations and will move out more quickly. To accomplish this, it has been proposed that the large canyons extending upward into the mountains be cleared of brush and paved with a nonreflecting and nonconducting substance. Clearing the brush and paving would reduce the frictional resistance to flow and at the same time increase the temperature, causing the up-slope winds to begin sooner and flow faster.

Again, quantitative considerations show that this effect at best would be a minor one. For every 1 meter per second that the sea breeze is increased on the average up to the inversion base, and for a 0.5-kilometer wide canyon, the total increase in air transport would be 10^9 cubic meters per hour, or less than 1/1,000 of the air over the basin per hour. As in the case of the proposal regarding tunnels, it would require hundreds of such canyons, or, equivalently, hundreds of square miles of mountain slope to be cleared and paved to produce any considerable effect.

Triggering Potential Instability

We have seen that methods of adding energy or making better use of the natural energy input give little promise of ameliorating effects. The question arises, is there any possibility of releasing energy already in the system to produce the desired change? The answer, so far as the thermal stratification is concerned, is that the situation is extremely stable: there is no available potential energy. The large and extensive inversion is analogous to a cork raft on water, which bobs back into position no matter what the disturbance.

The wind distribution during smoggy periods, like the thermal distribution, contains no potential instability. Even at the time of maximum sea breeze, the wind shear through the inversion is very small, and during most hours the winds are light both below and above it. Thus there is no shearing instability to offset the thermal stability. With the warm layer above the cold, and with little or no kinetic energy available for redistribution, the possibilities are absolutely nil.

However, as has been pointed out by Van Ornum in proposals to the Air Pollution Foundation, there exists in the moisture distribution a possibility of altering this situation. The air above the inversion is so dry that, by evaporating water into it until it is saturated, it would be cooled to a potential temperature below that of the air below the base of the inversion. Van Ornum suggested that a string of fog nozzles be established along the mountains at, say, the 2,000-foot contour. By saturating the air at this level, a downward flow of air would be induced which at night would increase the land breeze, and in the daytime offset the sea breeze, producing a continuous flow of air from land to sea. There are many interesting aspects of Van Ornum's proposal on which to speculate, such as the wind speeds which might be achieved during the day. For instance, it might turn out that the effect would be just sufficient to invert the normal regime, producing fairly rapid flow during the night and practically stagnation during the day. However, the economics of the proposal again make unnecessary the consideration of details.

The amount of water required may be estimated in various ways. Van Ornum's own estimates have varied from 6,000 to 11,500 acre feet per day, which correspond to from 5 to 10 times the total water consumed by Los Angeles in a day. Obviously, even apart from the cost, such a large amount of water cannot be diverted to this purpose in a region of dire water scarcity. The possibility of using sea water has been mentioned. The problem of pumping this quantity of water from the sea to elevations of 2,000 feet is not insuperable, though doubtless costly, but the problem of designing fog nozzles which would operate with sea water without corroding or clogging appears to be difficult indeed, and the effect of introducing into the air about 4×10^8 kilograms of salt per day might prove as objectionable as the pollution which we are attempting to remove. The visibility would be greatly reduced, and the corrosive action of the air-borne salt would be great.

In addition to the problems involved in water supply, the cost of equipment and power would be tremendous. The height at which the fog nozzles would have to be mounted would depend on the drop size they are capable of producing. The smaller the drops, the less the height required in order to insure complete evaporation, but the more power required to break up the drops and the more nozzles required to produce the same volume of discharge. Even if towers only 100 feet high are required, the cost of constructing several per mile along the 2,000-foot contour surrounding the basin, plus the cost of the spray nozzles, the water distribution lines, the pumping stations, and the electric power installations, could surely pay for the installation of equipment for controlling all the sources of pollution in the basin.

There have been other proposals to use water spray to wash and cool the smoggy air, in order to produce a film of clean cool air at the ground, say in the lowest 20 feet where most people are. On the one hand, the difficulties of scrubbing the pollutants, particularly the gaseous ones, even from air with higher concentrations of pollutants, suggest that the cleansing action of the spray would be partial at best. On the other hand, to the extent that the drops are large enough to reach the ground, carrying pollutants with them, the water is not evaporated to cool the air. Sufficient evaporation would be required to offset the solar heating at the ground, and it would have to occur uniformly over the entire area, or else convection would mix the "cleaned" air with polluted air above. Thus the water requirement would be even greater in this case than in Van Ornum's proposal.

Reducing the Insolation

As an alternative to the reduction or elimination of the reagents which participate in the photochemical reactions to form smog, it has been proposed that attempts be made to reduce the sunlight that causes the reactions. The concentration of oxidant, and presumably of other smog effects, appears to be directly related to the intensity of the shorter-wave solar radiation received at the ground. Leighton has studied this relationship (1956). If the photochemically active portion of the sunlight were reduced sufficiently, the smog effects should be reduced below the noxious or nuisance thresholds.

A reduction of this type could be achieved by introducing an aerosol consisting of small droplets, which would scatter the sunlight, returning much of it back to the sky. The intensity of scattering depends on the drop size and the wavelength. For a given wavelength, there is a particular size of drop which maximizes the amount of scattering per mass of liquid suspended in the form of droplets. Short-wave sunlight is scattered most intensely by a given mass of oil of refractive index 1.5 if it is dispersed in drops with diameter in the range of 0.4 to 0.5 micron. Langmuir has suggested that smoke generators of the type which produce uniform drops of the required size be located along the western coast of the basin and operated for 5 or 6 hours from the time of the beginning of the sea breeze (9 AM). Assuming an average sea-breeze speed of 6 miles per hour, and about 15 miles of coast line, the area covered would be about 100 square miles, per hour. Langmuir has computed that about 10 gallons of Diol per square mile, or 1,000 gallons (4 tons) per hour would reduce the intensity of sunlight 50%. My own computation indicates that the requirement is about 10 times this amount, but in either case the consumption of oil (24 tons per day, or 240 tons per day) would not be prohibitive if it were to produce the desired effect and no others.

The question whether a 50% (say) reduction of the insolation would necessarily eliminate the undesirable smog effects is a moot one. There have been days when fog and stratus clouds limited the total radiation for the day below this fraction, and yet severe eye irritation was experienced in some parts of the basin. The peak values of oxidant on these days were generally lower than on clear days, but this is not the only instance of lack of perfect correspondence between oxidant concentration and other smog effects. To settle the question, it might be worth experimenting with smoke screens, were it clear that this solution would be otherwise feasible and acceptable.

With respect to feasibility, the location of the smoke generators along the coast presupposes that air reaching all points in the basin with intense smog has entered across the west coast the same morning. Trajectory studies (Neiburger, et al., 1956) have shown that, frequently, air arriving at sampling stations at times of peak concentrations of contaminants crossed the coast the previous afternoon and stagnated over the basin throughout the night. Furthermore, on some days in late fall, when some of the worst smog sieges occur, the winds are easterly during the day as well as at night. Thus, to handle all types of smog situations, the smoke generators would have to be dispersed widely, and the appropriate ones would have to be operated continuously for as much as 24 hours previous to times when conditions are expected to be favorable for smog.

Finally, the question of the acceptability to the community of this type of solution must be considered. Sunshine is one of the assets of the Los Angeles climate. Sacrifice of one-half of this asset, over and above the 10% reduction produced by the smog itself, will appear to most citizens as a last resort, to be taken only if it proves impossible to eliminate smog by control of the sources of the pollutants. Similarly, the severe reduction of visibility would meet with great objections. A blanket of white oil smoke would be no more attractive than the yellow-brown pall of smog to people who cherish the view of mountains and sea, and of "Catalina Island on a clear day." Langmuir states, "I have been in London during heavy fogs when the light intensity at noon was much less than it usually is at midnight, yet there was not the unpleasant irritation that exists in Los Angeles smog." But one would hope for a cure which would not so nearly resemble the ailment.

Conclusion

Thus, until proponents of abating smog by meteorological modification demonstrate the effectiveness of their schemes, it is reasonable for the agencies concerned with the solution of the problem to devote their undivided efforts to the detection and control of the sources of the pollutants responsible for the obnoxious and deleterious effects of smog.

REFERENCES

Dubridge, L. A., 1955. The air pollution problem. *Engineering and Science 19*, 18–21.

Leighton, P. A. and Perkins, W. A., 1956. *Solar Radiation, Absorption Rates, and Photochemical Primary Processes in Urban Air.* Air Pollution Foundation Rept. 14.

Neiburger, M., 1944. Temperature changes during formation and dissipation of West Coast stations. *J. Meteor. 1*, 29–41.

Neiburger, M., Beer, C. G. P., and Leopold, L. B., 1945. *The California Stratus Investigation of 1944.* U.S. Dept. Commerce, Weather Bureau, Washington, D.C.

Neiburger, M., Renzetti, N. A., and Tice, R., 1956. *Wind Trajectory Studies of the Movement of Polluted Air in the Los Angeles Basin.* Air Pollution Foundation Rept. 13.

Sutton, O. G., 1953. *Micrometeorology.* McGraw-Hill. 333 pp.

Edward A. Morris

18 Institutional adjustment to an emerging technology: legal aspects of weather modification

The law is lagging too far behind the advance of science and technology. Inevitable by-products of that advance are social and legal problems. Those problems, unless planned for now, can produce crippling legislation, unwanted litigation, and undue and unnecessary interference with scientific research.

Weather modification is one of the new fields of science which need social and legal planning now. It may require entirely new political policies, as well as laws undreamed of a generation ago. An agency of government, which incidentally would touch the everyday lives of more people than almost any other in existence today, may also be needed.

The Demand for Laws and Regulations

In the last few years there has been a sharp increase in the number of laws designed to control weather modification activities. County ordinances have even been enacted to regulate, or prevent, such activities. And, as the public becomes increasingly aware of this new science, these controls will continue to multiply. There are growing demands from various agricultural and industrial groups for laws and regulations to control the activities of private and public agencies involved in attempts to modify the weather.

On what principles should these laws and regulations be based? Only a few cases have come before the courts thus far and no general body of principles has been developed. A few general principles, however, have been suggested, such as "the greatest social utility" or "negligence of the defendant".

The rule providing "the greatest social utility" was suggested in the *Slutsky v. City of New York*, (1950) 97 NYS 2d 238 case. In that case a resort owner sought an injunction to prevent the City of New York from engaging in cloud seeding. The court, in denying the injunction, stated:

> "This court must balance the conflicting interests between a remote possibility of inconvenience to plaintiffs' resort and its guests with the problem of maintaining and supplying the inhabitants of the City of New York and surrounding areas, with a population of about 10 million inhabitants, with an adequate supply of pure and wholesome water. The relief which the plaintiffs ask is opposed to the general welfare and public good; and the dangers which plaintiffs apprehend are purely speculative. This court will not protect a possible private injury at the expense of a positive public advantage."

The New York ruling certainly represents a broad-minded and liberal attitude. Not all courts concur. Thus, in *Southwest Weather Research, Inc. v. Rounsaville*, (1958) 327 SW 2d 417, a Texas court enjoined all future cloud-seeding activities above the plaintiff's land. That case will be discussed in greater detail below.

In most lawsuits for damages the plaintiff must prove that the defendant was negligent or careless. However, some writers suggest that in the field in question, this requirement should be omitted, and that the cloud seeder should be "absolutely liable" for all damages resulting from his activities.

Lawsuits to Date

As of this writing, only eight lawsuits involving weather modification have been filed. One of these was abandoned before the court had reached a decision on the merits, and two were just recently instituted. Of the remaining five, one was decided in favor of persons desiring to enjoin weather modification activities and four were decided in favor of the defendant weather modifiers. However, three of the latter four decisions

Reprinted, with minor editorial modification, by permission of the author and editor, from *Human Dimensions of Weather Modification*, W. R. D. Sewell (editor), University of Chicago, Department of Geography Research Paper No. 105, 1966, pp. 279–288.

held that no damage was attributable to the attempted cloud seeding. Thus, we do not know what those courts would have done had it been established that the seeding did somehow modify the weather in question. These three cases are described below.

(1) *Samples v. Irving P. Krick, Inc.*, (Civil Nos. 6212, 6223, and 6224, Western District of Oklahoma, 1954). This case arose out of cloud seeding sponsored by Oklahoma City in 1953. The plaintiff, a landowner, sued for property damages incurred in a cloudburst and flood which were coincident with the cloud seeding operations. The plaintiff failed to prove to the satisfaction of the jury that the seeding could have influenced the storm, and their verdict was for defendant.

(2) *Auvil Orchard Co., Inc., et al., v. Weather Modification, Inc., et al.*, (Case No. 19268, Superior Court, Chelan County, Washington, 1956), involved cloud seeding for the prevention of hail. Flash floods had occurred on farms adjacent to the hail prevention target area. The court therefore granted a *temporary* order banning hail suppression attempts for one season. At a later date, however, after hearing expert meteorological testimony, the court refused to grant a permanent injunction. It was not convinced that cloud seeding had brought about the exceptional rainfall which caused the floods.

(3) *Adams, et al., v. The State of California, et al.*, (Docket No. 10112, Sutter County Superior Court) was decided in April 1964, after four months of trial plus twenty-six days of pretrial motions and hearings. The defendants, Pacific Gas and Electric Company and the North American Weather Consultants, operated cloud seeding generators near Lake Almanor in the headwaters of the Feather River. A damaging flood occurred in Feather River in December 1955, and owners of property damaged thereby sued for millions of dollars to recover their losses, claiming that the flood was caused, at least in part, by cloud seeding. The plaintiffs' meteorological testimony asserted that the seeding material had been blown several miles outside the target area and thus it increased the rainfall and snow pack downstream of the Lake Almanor Dam. The court found that the effects of seeding were limited to Lake Almanor, which fortunately never spilled at any time before or during the flood. Accordingly, any increase produced by cloud seeding was successfully impounded in the Lake, and damages caused by the Feather River flood could not be charged to weather modification activities.

So much for the cases in which plaintiffs failed to prove that cloud seeding caused injury to the persons suing for redress. In this connection it should be emphasized that even though proof of injury might not be made, a temporary restraining order can effectively stop an entire season's weather modification activities. This happened in the above-mentioned *Auvil Orchard* case.

(4) The one case in which cloud seeding was held to be permissible even though the plaintiff might be injured thereby was *Slutsky v. City of New York*, 97 NYS 2d 238 (1950), which was also referred to previously. The court denied an injunction on the basis of the general welfare and public good of 10 million people, as compared with a possible private injury to one plaintiff. It conceivably might have been decided differently if the resort owner had asked for monetary relief instead of seeking to bar the City of New York entirely from attempting to augment water supplies for its inhabitants.

(5) Under somewhat similar circumstances the Supreme Court of Texas did bar cloud seeding undertaken by a group of private farmers. The case is *Southwest Weather Research, Inc., v. Rounsaville*, (1958) 327 SW 2d 417. The court found that the purpose of the cloud seeding was to disperse gathering hail storms which might injure certain crops of the farmers. However, ranchers in the area desired moisture in any form, including hail, and sued to stop cloud seeding which might interfere with gathering storms. Conflicting expert and lay testimony was presented by the parties, but the trial court reached the conclusion that the cloud seeding attempts were effective and did deprive the ranchers of moisture. The Texas Supreme Court reviewed the case and found that it was proper for the trial court to issue an order in these circumstances, banning further cloud seeding until additional evidence could be produced showing that hail prevention activities would not reduce the amount of precipitation on the ranchers' lands. As far as is known, those injunctions are still in effect in Texas.

Two other cases are presently pending in Pennsylvania. One of these is criminal in nature: *Township of*

Ayr v. Fulk (No. 53 at the September 1964, Term of the Fulton County Court of Common Pleas). The Township of Ayr passed an ordinance prohibiting the operation within the Township of any cloud seeding device. Mr. Fulk set up a silver iodide generator and was convicted of violating the ordinance by a Justice of the Peace. He appealed to a higher court, where expert testimony is currently being taken. Fulk claims that he was not attempting to modify the weather in Ayr Township, but only in the nearby State of Maryland. He hopes that the expert testimony will establish that his activity could have had no effect on the weather in the Township, that the effect it had on the weather outside the Township was not harmful to anybody and that the ordinance, therefore, is arbitrary and void.

The other pending Pennsylvania case is *Pennsylvania Natural Weather Association v. Blue Ridge Weather Modification Association, et al.* It is No. 3 at the January 1965, Term of the Court of Common Pleas of Fulton County, Pennsylvania. It is not based on the ordinance, but asks that the defendants be enjoined from engaging in weather modification activities in the area within the jurisdiction of the Court. It is claimed that the weather modification activities are wrongful in at least four ways:

(1) The released chemicals are dangerous to health; (2) the seeding interferes with the rights of landowners to receive precipitation in its undisturbed character and in its natural state; (3) the seeding trespasses upon the land of the plaintiffs by invading the air space above the land; and (4) the seeding is done recklessly.

The above cases illustrate that research and guidance are needed now by the courts. Likewise, such guidance is needed by the legislators.

Legislation

Powerful lay groups of special interests with no real scientific background are exerting pressure on legislators. Neither the United States Government nor the scientific profession has made available an adequate task force of witnesses willing and ready to appear before state and local hearings to give the needed technical knowledge to the legislative deliberations. To overcome the emotional opponents of weather modification requires more than a letter from a government agency that its files contain no evidence of disasters caused by cloud seeding. There may be need for a scientific legislative advisory agency which would appear at hearings and offer technical knowledge on legislation involving science. Such an agency is needed in order to prevent unnecessary restrictions on scientific progress.

Maryland has just made it a crime, subject to three years imprisonment, to engage in any form of weather modification over that State (Maryland Senate Bill 348, 30 March 1965). The State of Pennsylvania, in November of 1965, created a law which gives to each county the option to outlaw weather modification. The Act specifically prohibits weather modification whenever the County Commissioners shall adopt a resolution stating that such action is detrimental to the welfare of the county. In Congress a bill (House Resolution 8708, 3 October 1963), was introduced, which would close the outdoor laboratory in the United States to all seeding from airplanes. Probably the proponents of such legislation believe that if research in weather modification is to be done at all, it should be relegated to barren wastes of land or some distant ocean. Unfortunately, meteorological conditions vary so much from place to place that the very experiments which fail in distant locations may be the ones which would have been successful if performed at home.

True, today most states make no attempt to completely outlaw cloud seeding. There are twenty-two states that have laws pertaining to weather modification. These laws require, in general, licensing or registration of cloud seeding operations; prior notice of operations to local governmental authorities or newspaper notices to the public in the areas to be seeded; reports by the operators to state authorities; evaluation by state authorities and universities; certain technical qualifications to operate; and financial responsibility. One state, instead of enacting proposed prohibitory legislation, by resolution of the legislature, requested Congress to investigate artificial interference with natural precipitation.

Six state laws assert sovereign rights to the moisture in the clouds or atmosphere within their respective state boundaries. The laws of four states explicitly absolve the states from liability for the weather modification activities of any private person or group. Half a

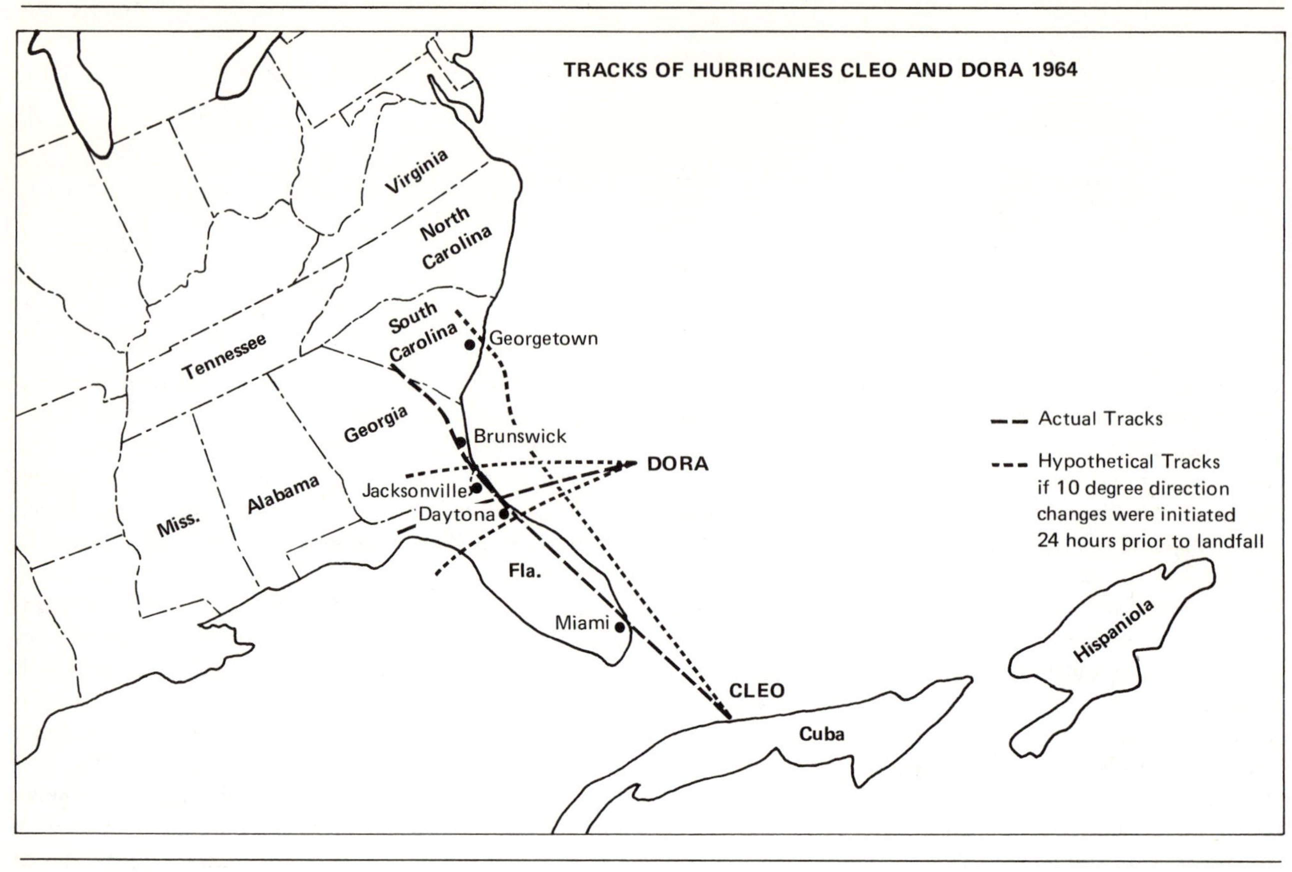

Figure 1

dozen of the States' statutes authorize research and experimentation by state agencies or universities. It appears that some of the states authorize their political subdivisions to conduct weather modification operations and to issue regulations. Nine of the states seem to regulate the methods or conditions of operation. One state regulates the manufacture, sale, lease and advertisement of weather modification equipment. Many of the state statutes express in one form or another the requirement or desirability of cooperating with the Federal Government and other states in weather modification activities. Six states expressly authorize local governmental entities or specially created districts to spend general funds or revenues raised by special taxes for weather modification operations or research.

A review of the numerous state statutes leads to the conclusion that uniform legislation is needed. It is probably too late to start considering a set of Uniform State Laws.

We now should look for the enactment of comprehensive Federal legislation in this field.

Federal Legislation

Federal legislation could be drafted, which would give reasonable protection to the public while at the same time affording encouragement to scientific progress in this field. When we consider the social and legal problems that weather modification can create, we can see innumerable reasons for federal legislation.

To enact intelligent legislation, each of the varied techniques employed in weather modification programs must be taken into consideration. The legislators must draft a bill which would cover cloud seeding from the ground, from aircraft and from rockets, hail and lightning suppression programs, fog dispersal projects, cloud electrification, hurricane modification, as well as albedo alteration of land surface, as well as others. Each project has its special requirements, which generally change with geographic location and with the characteristics of the storms being modified. Each, in turn, might create a hazard, whether real or imaginary, against which the public may seek protection.

Future Legal and Institutional Problems: Problems of Hurricane Control

A good example of the magnitude and number of problems which can arise in weather modification can be seen in the case of hurricane control. Scientists are presently attempting to find ways to steer hurricanes. To illustrate some of the problems this may create, let us assume that a means is found, perhaps by seeding one sector of a hurricane, of getting it to wobble off with a ten degree direction change. Consider then the course of Hurricane Cleo in 1964 (Figure 1).

What agency is going to be given the power to decide which way to divert a hurricane? Who will pay the damages to homes in South Carolina if the decision is made to steer it from Florida? Or, who will pay the damages to the homes in Florida if the agency decides not to steer it away from Florida? Will insurance companies continue to write coverage in South Carolina when they learn that every hurricane headed for Florida will be diverted to there? Will there be some type of agreement between the states ahead of time for a bonus payment from the State of Florida to the people of South Carolina? Will Florida refuse to pay if the scientists report that the hurricane would have hit South Carolina even if there had been no seeding? Here again it would seem that the Federal Government should handle these problems. But if the United States Government makes the payments, will the Government then undertake to pay the one and one-half billion dollars each year to everyone in the United States whose property is damaged by any type of storm?

It is difficult to suggest the ideal legal solution. One thing is certain, however: that the law should make a serious attempt to catch up with science.[1]

The Need for Original Legal Thinking

The legal profession has been taken by surprise by the rapid changes occurring in all sciences. Such advances in science are far outpacing many of the traditional rules of law. Moreover, the lawyer, by his very training, looks to past rules and precedents when he is asked to advise on entirely new and untried matters. Perhaps we are approaching in weather modification, a science in which the customs of past generations will no longer serve as adequate guides for activities of future generations.

A Congressional study is warranted. Such a study should include meetings, hearings and conferences with many interested organizations, agencies, universities and individuals engaged in various aspects of weather modification throughout the country. In addition, the problem is definitely of sufficient magnitude and importance that a committee of the legal profession should be appointed to work with similar committees of other professional societies to help enumerate all the special social and physical requirements, hazards and problems of weather modification so that, with such information, laws can be designed to strengthen, encourage and promote present and future scientific field research, while at the same time protecting the public from unqualified cloud seeders acting without regard to the best interests of the community.

Legal theories which are original and even a complete departure from prior decisions should be explored. Then the economists, the social and the political scientists can study the economic, social and political ramifications which might follow from each suggested possible legal approach.

Unfortunately, the law is slow to move. In the past, seldom has it anticipated conditions and evolved methods to remedy them. It has waited for problems to develop and then belatedly sought to make rules for solving them. We cannot afford the luxury of years of delay and deliberation in solving these problems. There is no valid reason why the law cannot modernize itself and set up machinery to prevent problems before they occur.

A New Agency to Deal with Weather Modification

Finally, the study must explore further the desirability of creating a new agency, perhaps along the lines of the Atomic Energy Commission or the Federal Communications Commission. Such an agency might cooperate with the air pollution control agencies and other agencies, and work with and coordinate the efforts of the local and corporate cloud seeding activities, as well as work with whatever international agency is created.

The agency should probably not be one which will engage in weather modification itself. The history of scientific developments shows that freedom of action is most desirable. Private as well as government research is desirable for the United States, and it should be

[1] Chief Justice Earl Warren, in his 12 February 1963 address to the Georgia Institute of Technology, stressed that the problems being created by scientific and technological advances are increasingly pressing for answers by the law, which dismally and traditionally lags behind.

encouraged. Weather modification, or any other science, should not be *operated* entirely or exclusively by the Federal Government. It is desirable to have scientists who may not agree with the policy which might be prescribed by the federal agency. They should be free to leave their Government position and either join or start a private concern where their new scientific concepts might be tested.

Conclusion

The time has come for a comprehensive study of the legal problems connected with weather modification. The scientific community must aid and guide the legal profession and the legislators in the formation of solutions to those legal problems. Without such leadership, court decisions will be made and laws drafted which will eventually result in unnecessary interference with scientific field research.

The race is on. If we do not come up with a good solution within the next few years, many of the decisions will be made by those who may not be best informed or qualified to make such decisions. We must begin the study now, and we must use the skill and money which the magnitude of the problems warrant.

SUGGESTED READING FOR PART 6

Cooper, C. F. and Jolly, W. C., 1969. *Ecological Effects of Weather Modification: A Problem Analysis.* Department of Resource Planning and Conservation, School of Natural Resources, University of Michigan, Ann Arbor. U.S. Department of the Interior Contract No. 14-06-D-6576.

Gentry, R. C., 1969. Project Stormfury. *Bull. Amer. Meteor. Soc. 50(6)*, 404–409.

Kethley, L. I., 1970. Weather modification and the hydrologic cycle. *Archiv für Meteorologie Geophysik und Bioklimatologie, Serie B, 18*, 143–154.

Riehl, H., 1963. On the origin and possible modification of hurricanes. *Science 141(3585)*, 1001–1010.

Staff, the Weather Modification Research Project of the RAND Corporation, 1969. Weather-modification progress and the need for interactive research. *Bull. Amer. Meteor. Soc. 50(4)*, 216–246.

Taubenfeld, H. J. (ed.), 1970. *Controlling the Weather: A Study of Law and Regulatory Procedures.* University Press of Cambridge and Dunellen Company, New York. 275 pp.

Part Seven

Weather Modification – Inadvertent

The topic of air pollution is introduced by Leighton's informative discussion of the subject. He refers in particular to photochemical smog in California, examining the problem from a meteorological, a topographical, and a cultural perspective. He views the introduction of control measures with some apprehension, noting how the time-lag involved in making such measures operational generally undermines the effectiveness desired.

Do large-scale weather characteristics significantly influence pollution accumulation on a local scale? Assuming they do, what parameters would you use in formulating a forecasting program for air pollution potential at a national level? With these thoughts in mind Holzworth examines pollution potential for urban areas and judges mixing heights to be of paramount importance.

Summers reinforces Holzworth's view of the significance of the mixing depth, in this case for its influence on the monthly variations of the daily cycle of pollution deposition. From his interest in the layer of air-mixing over a city he develops the concept of a ventilation coefficient as a measure of the rate of clean air intake into the city (i.e., a useful indicator of air pollution intensity).

Are people in general aware of the air pollution problem? If so, how do they react to pollution control measures? Rankin provides some surprising answers from his opinion survey of West Virginia.

That all-time favorite of the urban climatologist, the urban heat island, is given close scrutiny as is the urban precipitation controversy. Peterson's article surveys completely the literature on city climate since 1956. But what of cities in the tropics?

Philip A. Leighton

19 Geographical aspects of air pollution

> . . . this most excellent canopy, the air,
> look you, this brave o'erhanging firmament, this
> majestical roof fretted with golden fire, why, it
> appears no other thing to me than a foul and
> pestilent congregation of vapours.
>
> —HAMLET, Act II, Scene ii.

It is reasonable to suppose that man originally evolved with few if any inhibitions regarding the use of that part of his environment which he was able to capture and hold from his competitors. Only with experience, as his knowledge and numbers increased, did he come to realize that the physical requirements of life are limited and that their use must be regulated. Since earliest history he has been devising systems for the ownership, protection, and use of land and food, and, more recently, of water. Last of all to become subject to this realization and regulation is air. Here the tradition of free use is still dominant. We respect rights of ownership in land, food, and water, but except as a medium of transportation we recognize none for air.

Curiously, this divergence in attitude, or in the stage of modification of attitude, does not parallel either the urgency of man's needs or his ability to adapt his surroundings to meet those needs. He can live indefinitely away from land, he can go several weeks without food and several days without water, awake he normally eats and drinks only at intervals, and asleep he does neither, but awake or asleep his need for air is never further away than his next breath. As for ability to adapt, he can when he so wishes improve the land, he can improve and transport food and water, but except on a small scale, as in air conditioning in dwellings and other buildings or the use of wind machines in orchards, he cannot yet improve or transport air. Outdoor air in the main he only contaminates.

Reprinted, with minor editorial modification, by permission of the author and editor, from the *Geographical Review*, Vol. 56, 1966, pp. 151–174, copyright © by the American Geographical Society of New York.

Although the realization that air also is a limited resource has been slow in developing, the recognition that its contamination may easily exceed acceptable limits is not new. The first ancient who kicked a smoking ember out of his cave was taking an air-pollution control step more effective than many that are taken today, and the first air-pollution control laws on record, designed to reduce the burning of coal, were enacted in England more than six centuries ago. These attempts at control have expanded until there are now in the United States alone some 360 government agencies—local, state, and federal—partly or entirely concerned with the problem.

Despite the unremitting work of such agencies, for the most part air pollution continues to grow. Its growth has more or less paralleled man's increasing use of technology, with the result that the most technologically advanced areas of the world are also, with few exceptions, the areas of most severe air pollution. This is due, of course, to the overuse of air for waste disposal, and an excellent example, which very much involves the tradition of free use of air, is the automobile. Automobiles emit carbon monoxide, nitrogen oxides, and hydrocarbons, all of which must be diluted in air if they are not to reach adverse concentrations. The undesirable effects of nitrogen oxides begin to appear at concentrations of about 0.05 parts per million (ppm). Cruising at 60 miles per hour, the average "full sized" American automobile emits, at 25° C and 1,000 mb, about 3 liters of nitrogen oxides per minute. To dilute these below 0.05 ppm requires, for the one automobile, more than 6×10^7 liters of air per minute, a rate which is enough to supply the average breathing requirements, over the same period of time, of five to ten million people.

As a result of such prodigal uses of air for waste disposal, the employment of technology has contributed far more to the production of air pollution than to its abatement, and it is clear that the ratio must be reversed if man as a breathing organism is to retain a

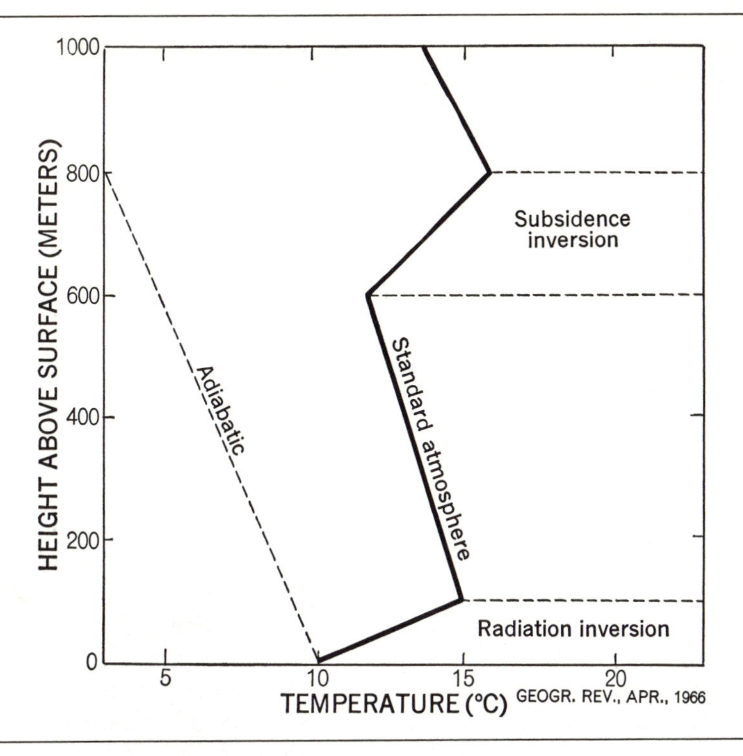

Figure 1 Temperature profile through two inversion layers.

compatible environment. But to define the extent to which the uses of air must be regulated, we must first know something about how much is available. As with man's other needs, it is a simple matter of supply and demand.

THE SUPPLY OF AIR

The height of the troposphere in the middle latitudes, 10–14 km, is about one five-hundredth of the earth's radius. This is a thin skin indeed, yet it contains about four-fifths of all the air in the atmosphere, and to man on the surface of the earth the layer of air available for waste disposal is usually only a fraction—and sometimes only a very small fraction—of the troposphere. The air supply at the surface is limited to an extent that varies with place and time, and the factors contributing to the limited surface ventilation are both meteorological and topographic.

The most common meteorological factors are inversions that limit vertical mixing of air and low winds that limit its lateral transport. An inversion is a reversal of the normal tropospheric lapse rate, or decrease in air temperature with increasing altitude above the surface, which for the United States and international standard atmospheres is 0.65° C per 100 meters (Figure 1). A parcel of air ascending in the atmosphere expands with the decreasing pressure and is thereby cooled, and when this process occurs adiabatically, the rate of cooling, or the adiabatic lapse rate, in unsaturated air is about 1° C per 100 meters. When the atmospheric lapse rate is less than this, as it is in the standard atmosphere, an ascending air parcel becomes cooler, and hence denser, than the surrounding air, and work is required to lift it against the downward force produced by the density difference. Similarly, a parcel of air being lowered in a subadiabatic temperature gradient becomes warmer and less dense than the surrounding air, producing an upward force against which work is again required. When the temperature gradient is inverted, the amount of work required to move a parcel of air across the inversion layer usually exceeds the supply available through turbulence and other atmospheric processes, and there is, in consequence, little or no mixing through the layer.

Inversions occur both at the surface and aloft. Surface inversions are most commonly produced by cooling of the ground by radiation loss, which in turn cools the surface air, and their depth, intensity, and duration are functions of the wind velocity, the nature of the surface, the transparency of the air above the surface to the emitted radiation, and the amount of insolation during the following day. The chief absorbers, in air, of the long-wave infrared emitted by a surface at ordinary temperatures are water and carbon dioxide. Hence radiative cooling is most marked when the air is dry and pure, and it increases with altitude as the amount of air overhead is reduced.

The commonest source of inversions aloft is the subsidence that normally accompanies high-pressure systems, but overhead inversions may also be

SEASON	SAN DIEGO % of days	SAN DIEGO Av. base height	SANTA MONICA % of days	SANTA MONICA Av. base height	OAKLAND % of days	OAKLAND Av. base height
Jan–Mar	38	382	47	270	22	386
Apr–June	69	465	77	323	59	327
July–Sept	90	434	92	296	85	253
Oct–Dec	55	353	61	243	45	296
Annual	63	408	69	283	53	315

Table 1 Frequency and average height of inversions below 2,500 ft along the California coast (Holzworth, Bell, and De Marrais, 1963)
(Base height in m)
Estimated from radiosonde observations taken daily at 1600 PST from June, 1957, to March, 1962

produced, both on a local scale and on an air-mass or frontal scale, by the intrusion of cold air under warm or by the overrunning of cold air by warm. In the middle latitudes subsidence inversions are most marked in the anticyclonic gradients on the easterly sides of high-pressure cells and approach closer to the surface with increasing distance from the cell center (Neiburger, Johnson, and Chien, 1961). For this reason the west coasts of the continents are subject to relatively low overhead inversions from the semipermanent marine highs, and these inversions may last for many days. Along the Southern California coast, for example, inversions below 762 m (2,500 ft), mostly due to subsidence associated with the Pacific high, exist 90% or more of the time during the summer months. The variations in average height and frequency of these inversions with season and location are summarized in Table 1.

The effect of high-pressure systems in limiting surface ventilation through subsidence inversions is enhanced by the low winds that usually accompany these systems, and also by clear skies, which promote the formation of radiation inversions. The occurrence of these conditions can be forecast, and since 1 August 1960, for the eastern United States and 1 October 1963, for the western United States the Division of Air Pollution of the United States Public Health Service has issued advisories of high air pollution potential, based on forecasts of the simultaneous occurrence, for periods of 36 hours or more over minimum areas equivalent to a 4° latitude-longitude square, of subsidence below 600 mb, surface winds below 8 knots, no winds above 25 knots up to 500 mb, and no precipitation (McCormick, personal communication). The number and regional distribution of forecast days from the initiation of the program through December 1964, are shown on Figure 2.

For the eastern United States, these forecast frequencies may be compared, if the differences in time period are kept in mind, with Korshover's (1960) estimates of the number of periods of four or more successive days of low wind resulting from stagnating anticyclones (Figure 3). Both studies agree on the absence of such conditions in the Great Plains region—perhaps to the surprise of residents of Denver—and on increasing frequency east of the Mississippi, with a maximum, though here the two charts differ, in the vicinity of eastern Tennessee. The Great Smokies, it would appear, are aptly named.

For the western United States, the forecasting program has been in operation for too short a time to permit more than tentative conclusions, but it does indicate a frequency considerably higher, in days per year, than that in the eastern states. A maximum appears in central California and perhaps another maximum in the Great Basin, extending northwest from Salt Lake City. It should be borne in mind, however, that these forecasts, like Korshover's study, are based on synoptic data and do not take local topographic effects into account; for this reason areas where local effects are important may have a much higher stagnation frequency than the charts seem to indicate.

TOPOGRAPHIC EFFECTS

Perhaps the most important effects of topography in limiting the supply of surface air are produced by drainage. Just as water drains down slopes and gullies to form rivers in valleys and lakes in basins, so the air, cooled by radiation loss, drains down those slopes at night. And like flowing water, these density or gravity flows of cold air tend to follow regular channels, which may be marked out almost as definitely as the course of a stream. The volume of air drainage, however, is much larger than that of water drainage; hence the aircourses are broader, and if the valley or basin is not too wide the flows soon collect to reach across it. The layers thus formed, further cooled by radiation loss in the valley or basin itself, become so stable that they often completely control the surface wind direction and velocity and thus control the air supply; the gradient wind is blocked out, and even the gravity flows from the surrounding slopes tend to overrun the air in the bottom (Figure 4). After sunrise thermal upslope flow soon sets in on slopes exposed to the sun, but gravity flow may continue until late morning on shady slopes, and even all day on steep northern slopes (Geiger, 1950; Defant, 1951; Leighton, 1954–55).

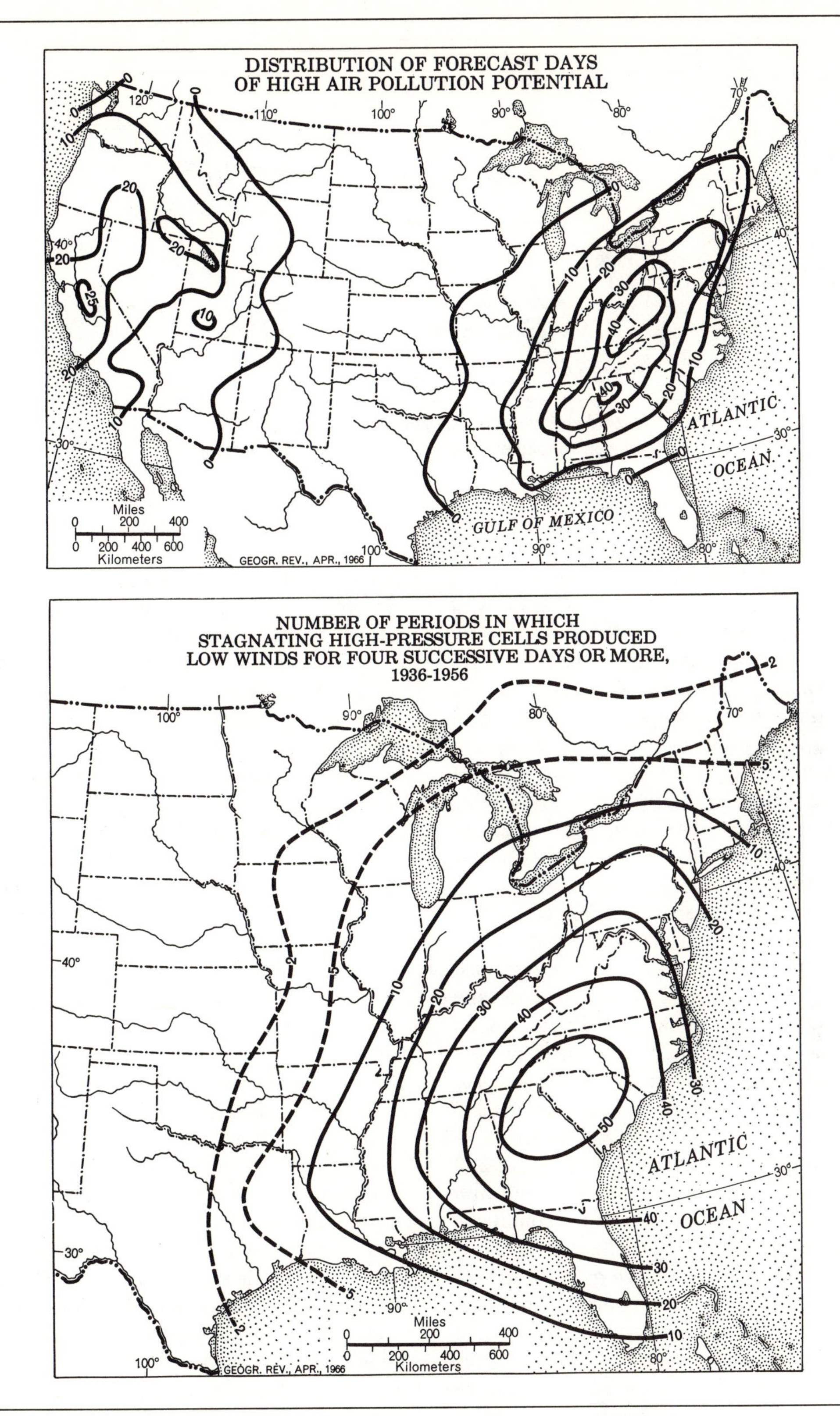

Figure 2 The air pollution potential advisory forecasts of the Division of Air Pollution, United States Public Health Service, began 1 August 1960, for the eastern United States, and 1 October 1963, for the western United States. The numbers shown on the contours indicate the number of forecast days from the initiation date in each case through December, 1964. Source: data from McCormick, personal communication.

Figure 3 Number of periods in which stagnating high-pressure cells produced low winds for four or more successive days in the eastern United States, 1936–1956. Source: Korshover (1960).

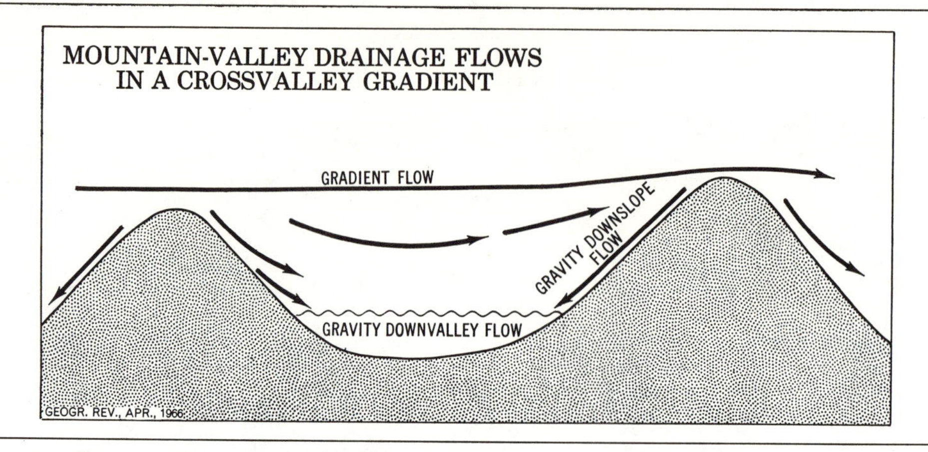

Figure 4 In the case diagramed, the gradient wind is blocked out of the valley by the bordering mountains, and the air supply on the valley floor is limited to that in the lower part of the gravity downvalley flow.

The cold layers accumulated by this process during the long nights of winter may become so deep, with inversions so intense, that they are not broken up by insolation during the short days; and when this happens, severely limited ventilation will persist until a change in weather produces gradient winds high enough, or a cold wedge strong enough, to sweep out the valley or basin. For any particular combination of topography there is usually a fairly critical gradient or synoptic wind velocity below which the local flows are dominant and above which the gradient wind is dominant. The smaller the relief, the lower is this critical velocity; for relief differences of 300–600 m it is of the order of 10–15 knots (Leighton, 1954–55).

A classic example of the consequences of unrestricted pollution in an air supply limited by both synoptic and topographic effects is found in the Copper Basin around Ducktown in the southeast corner of Tennessee. This basin, with an area of about 100 square km, lies between the Blue Ridge and the Unaka Mountains and has relief differences of as much as 600 m above its floor. It drains into the Ocoee River to the south and slopes gently upward to the northeast, and it lies in a region of maximum occurrence of synoptic conditions favoring low gradient winds (Figure 3). The local air circulation, which is dominant a large part of the time (as much as 60% in winter), consists of a low level flow that follows the drainage pattern upstream by day and downstream by night; superimposed on this is a gentle gravity flow from the periphery of the basin toward the center on clear nights, which results in pooling with strong inversions up to depths of 50–100 m (Leighton, 1954–55).

Smelting of copper ore, releasing all the sulfur and arsenic in the ore to the air as the corresponding oxides, began in the basin shortly after the Civil War and reached a maximum in 1890–1895. As a result, by the turn of the century an area of about 30 square km in the center of the basin had been completely denuded of vegetation and the remaining 70 square km had been largely denuded. Although open-hearth smelting has long since been abandoned, these areas remain bare today. Moreover, the basin has been severely eroded since it was denuded, and the bare areas are therefore still expanding (Hursh, 1948).

Another classic example, but with a happier outcome, is the international transport of polluted air by gravity flow down the Columbia Valley. The Columbia River flows from Canada into the United States in a rather narrow valley, with sides rising steeply 600–800 m above the valley floor. In 1896 a lead-zinc smelter was established in the valley at Trail, British Columbia, some 10 km north of the border, and by 1930 this smelter was emitting as much as 600–700 tons of sulfur dioxide a day. At night this sulfur dioxide was carried downstream by the gravity flow. The resultant damage to agricultural crops in the state of Washington led to international litigation, which in turn led to the formation in 1928 of an International Joint Commission and in 1935 of an Arbitral Tribunal with the dual responsibility of assessing damage and seeking a permanent solution.

The study conducted by the Arbitral Tribunal (Dean, et al., 1944; Hewson, 1945) showed that during the growing season surface concentrations of sulfur dioxide were highest rather regularly about 8:00 AM, which is about the time of day when growing plants are most sensitive. Moreover, these concentrations developed almost simultaneously at all the measuring stations, which were located from 10 to 55 km downstream from the smelter. The explanation, applicable also to somewhat similar behavior observed in the basinlike Salt Lake and Tooele valleys of Utah, is that during the preceding nights the valley or basin becomes filled with stable air, in which the gases rising from the smelter stacks soon level off to form a shallow but concentrated overhead layer. This layer is carried downstream as a long ribbon in a narrow valley or spreads out over a broader valley or basin. After sunrise, surface heating produces a superadiabatic lapse rate with strong vertical mixing, and when this turbulent layer reaches the polluted layer aloft, the pollutants are rapidly brought to the surface, producing sudden and almost simultaneous high concentrations over the areas concerned.

Both at Trail and in Utah these studies led to the adoption of methods of meteorological control, under which by continuous monitoring the hazardous periods could be anticipated and the smelter operations curtailed. In Utah the judge under whom this control method was adopted remarked on his retirement many years later that this was, to him, the most satisfactory outcome of all the cases he had tried in forty years on the bench. At Trail the need for control was reduced by recovering the sulfur dioxide and converting it to marketable products, a procedure that has since materially changed the nature of the industrial operation.

In coastal areas diurnal warming and cooling of the land, while the water temperature remains fairly constant, produce the familiar pattern of land-sea breezes, which are usually thought of as improving ventilation but which may under certain conditions restrict it. An example is found in the Los Angeles basin, where the Santa Monica Mountains to the northwest and the Sierra Madre to the north furnish shelter to the extent that local airflow is usually dominant under the subsidence inversion. This local flow consists chiefly of a gentle seaward drainage at night and a more rapid landward movement by day. But the mountains rising above the inversion layer retard the sweeping out of the basin by the landward movement, and the diurnal reversal in direction tends to move air back and forth in the basin. As a result of this entrapment, there is often some carry-over of pollutants from the day before, and pollutants emitted at night move toward or out over the sea, only to be swept back over the land the next morning. On occasion this polluted air is carried back over a neighboring area, even a fairly distant one; thus eye irritation came to Santa Barbara for the first time in January 1965, partly as the result of this process.

These effects are enhanced by a cold upwelling in the ocean along most of the California coast, which produces surface-water temperatures lower than the temperatures farther out to sea. As the surface layer of air moves over this cold water it also is cooled. One result is the familiar coastal fog of California, but a more important result, with respect to air pollution, is the additional stability the cooling imparts to the landward-moving air.

The airflow patterns in the San Francisco Bay Area illustrate another mechanism by which water may limit the air supply. During the extensive season of the semipermanent Pacific high, air cooled by the offshore ocean upwelling flows through the Golden Gate and between the hills of San Francisco to the inner bay (Figure 5). Part of this air crosses the bay and is deflected to the north and south by the east-bay hills, and part travels south and southeast over the bay itself. Meanwhile, another flow of air reaches the south-bay area by moving inland across the mountains to the west. This air, having traveled farther over land, is warmer than the air that comes down the bay, and when the two flows intersect, the warmer overrides the cooler and produces a local overhead inversion that around Palo Alto may be less than 100 m above the ground. Although the existence of this effect was demonstrated twenty years ago, its contribution to the severity of air pollution in the south-bay area remains to be determined.

Many other instances of the increase of air pollution by local topography could be cited. Winter air pollution in the Salt Lake valley is due as much to the pooling of drainage air from the Wasatch Range as it is to Utah and Wyoming coal. Air pollution at Denver, as has already been hinted, is attributed more to topographic than to synoptic limitations on the air supply. St. Louis, Pittsburgh, and Cincinnati have faced up to difficult problems created in part by local topography. In New York City the Hudson Valley and the surrounding water contribute to the problem. Mexico City suffers from pooling in the Valley of Mexico. The west coast of South America, backed by the Andes, is subject to periods of topographically limited ventilation, which increases air pollution in Santiago and Lima. In Australia the Sydney basin resembles, in a number of respects, the Los Angeles basin. The air-pollution disasters in the Meuse Valley in Belgium and at Donora, Pennsylvania, were the result of the entrapment of air

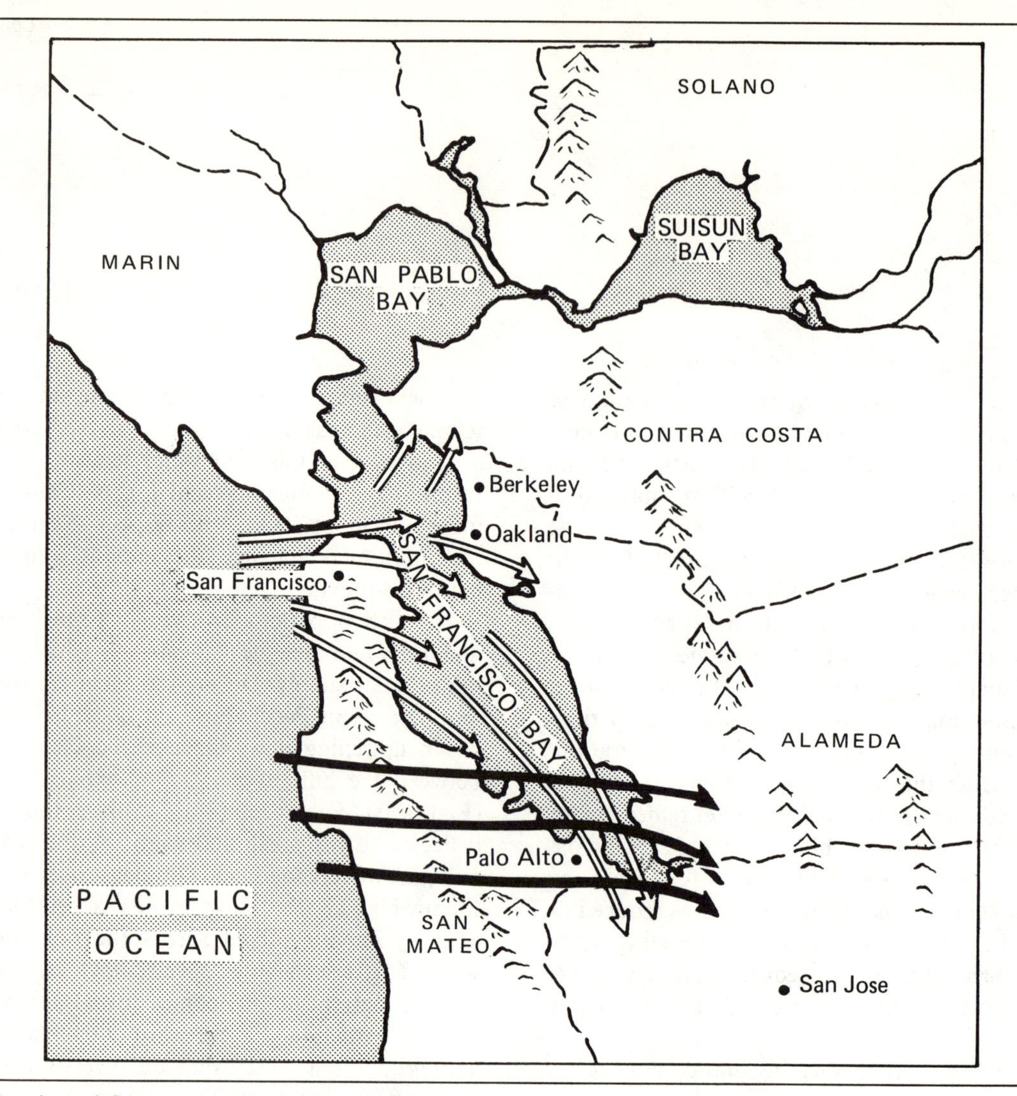

Figure 5 Daytime airflow patterns in the San Francisco Bay Area. In the southern part of the Bay Area wind coming over the mountains to the west overrides the colder air coming down the bay, producing an overhead inversion that may contribute to the severity of air pollution in the Palo Alto–San Jose area (adapted from the *Geographical Review*, Vol. 56, p. 162, 1966).

in valleys. Even the chronic problem and the repeated disasters in London may be assigned in part to topography in that the terrain offers no opportunity for drainage, and under a strong surface inversion with no gradient wind the air simply stagnates.

As urbanization and industrialization expand over the world it is interesting, and possibly beneficial, to attempt some assessment of the local air supply in areas that are still relatively empty. Although aerogeographical surveys would be required for an adequate assessment, tentative indications may be obtained merely by consulting maps and weather data. For instance, topography alone suggests that the Granby basin in Colorado would be a poor location in which to build a smelter, and both topography and weather data suggest that such places as the Santa Ynez valley in California and the Sous plain in Morocco should certainly be surveyed before any large industrial or urban development is undertaken. But one does not have to go far in this search to find that most of the unfavorable locations are already occupied. The factors that limit local ventilation are also factors conducive to habitation, and it is ironic that the areas of the world in which the air supply is on occasion most limited are often the areas in which man has chosen to build his cities.

Fortunately, poor ventilation, whether produced by general inversions and low winds or by local conditions, does not exist all the time. The sparkling clarity still enjoyed on days of good ventilation, even over large urban areas, serves to emphasize the great effect of limited air supply on the poor days, and the extent to which it increases the problems of air pollution.

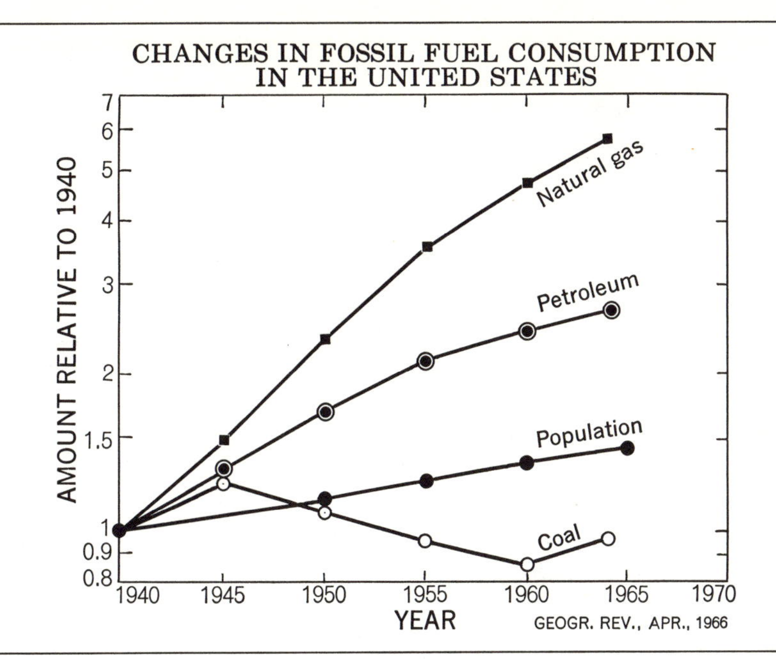

Figure 6

INCREASE OF PHOTOCHEMICAL AIR POLLUTION

The contaminants which man introduces into the surface air are of many forms; each creates its own problems, and to a large extent each problem is a case unto itself. Perhaps the least difficult of these problems are those caused by pollutants that come from only one or a few specific sources that can be pinpointed and readily controlled. Sulfur dioxide from smelters, stack dust from cement plants, fluorides from aluminum and phosphate plants, industrial smoke, and various exotic industrial gases and particulates are examples of emissions from specific sources, and control of some of them began more than half a century ago.

A more difficult group of problems, most of which remain for future solution, arise when the sources of pollution, although specific, are not fixed or for other reasons cannot be easily controlled. In this category are such things as agricultural dust, smoke from agricultural burning, airborne insecticides, and hydrogen sulfide and other obnoxious gases from sewage and organic industrial wastes.

The most difficult problems occur when the effects result from a general merging of pollutants from many diverse sources. Historically, the combustion of coal has been a major cause of general air pollution, but in the United States since World War II the overall contribution of coal to air pollution has diminished with its decreasing use, while the contributions of the hydrocarbon fuels have grown with their increasing use (Figure 6). Outstanding among the new problems created by the shift in fossil fuels is photochemical air pollution. The emissions chiefly responsible for this form of pollution are nitric oxide, together with some nitrogen dioxide, and hydrocarbons. The nitrogen oxides come from virtually every operation using fire, including internal-combustion engines, steam boilers, various industrial operations, and even home water heaters and gas stoves.

Not all the hydrocarbons emitted to the air take part in the photochemical reactions. Methane, the chief component of natural gas, is inactive. Acetylene, benzene, and the simple paraffins such as propane and butane are nearly inactive. On the other hand, all the olefins, the more complex aromatics, and the higher paraffins are reactive, though they differ widely both in rate and in products. These reactive hydrocarbons come from motor vehicles, from the production, refining, and marketing[1] of petroleum and petroleum products, and from the evaporation of solvents. Other emissions that may play some part in photochemical air pollution are aldehydes, which come chiefly from the incomplete combustion of organic materials, and sulfur dioxide. When these emissions are mixed, diluted in air, and exposed to sunlight, they undergo photochemical reactions that lead to the conversion of the nitric oxide to nitrogen dioxide, which has a brown color and may have adverse effects on plants and animals if its concentration becomes high enough. This is followed, and sometimes accompanied, by the formation of particulates that reduce visibility, of ozone and peroxyacyl nitrates (PAN) that damage plants, and

[1] Marketing emissions include such things as losses from tank trucks and service stations, evaporation losses during the filling of automobiles, and so on. In Los Angeles County alone it is estimated that these losses contributed an average of 120 tons of hydrocarbons a day to the air during the year 1963.

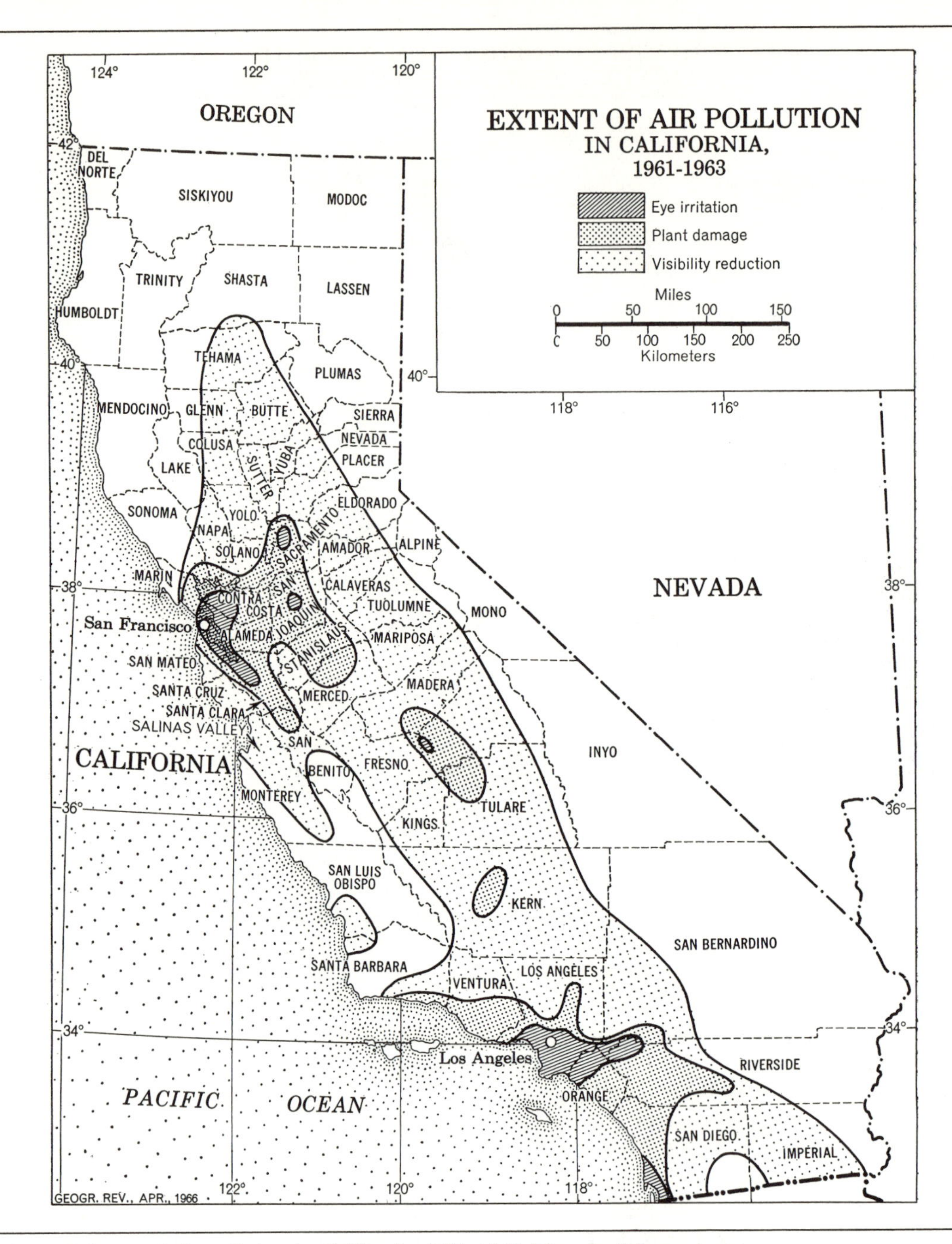

Figure 7 Extent of general air pollution in California, 1961–1963. The plant-damage areas are specific, but the eye irritation and visibility reduction may be due in part to forms of general pollution other than photochemical. Sources: for plant damage, J. T. Middleton: California against Air Pollution (California Department of Public Health, Sacramento 1961); for eye irritation and visibility reduction, local reports and personal observations up to December, 1963.

of formaldehyde and other products that, along with the peroxyacyl nitrates, cause eye irritation.

An increasing intensity of pollution is required to produce these symptoms of photochemical air pollution, lowest for visibility reduction, intermediate for plant damage, and highest for eye irritation. Accordingly, the first symptom to appear in any particular area is visibility reduction, the next is plant damage, then follows eye irritation. Similarly, the areas affected are largest for visibility reduction, intermediate for plant damage, and smallest for eye irritation. An estimate of these areas in California is shown in Figure 7. The magnitude of the problem is emphasized by the fact that the eye irritation areas comprise about 70% of the people of California, the plant damage areas 80%, the areas of general visibility reduction about 97%.

One of the most challenging aspects of photochemical air pollution is the rate at which it has grown and is

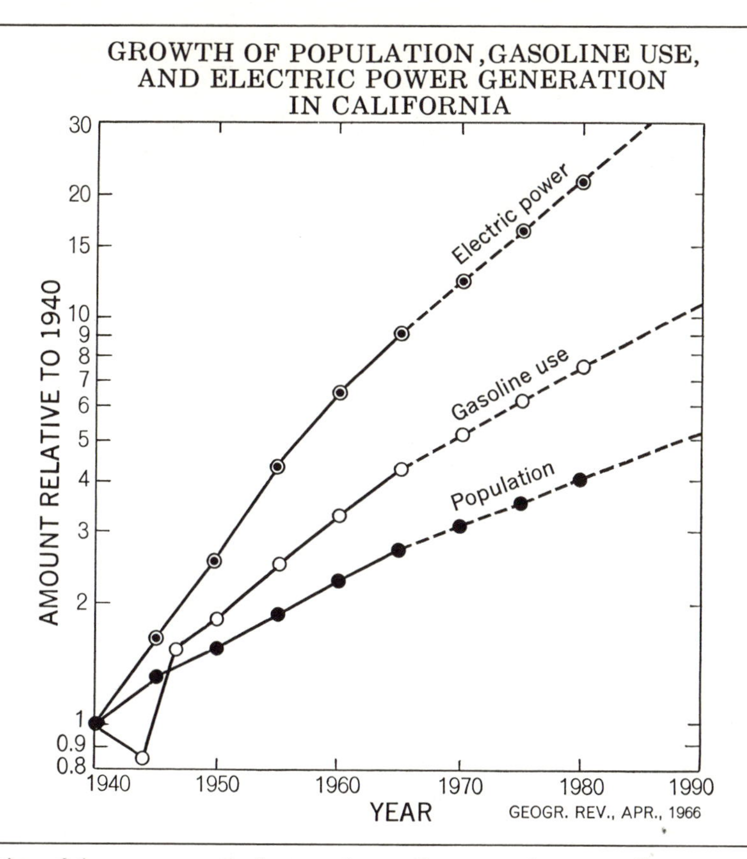

Figure 8 With the exception of the war years, the increase in gasoline use and, to a smaller extent, that of electric power generation relative to population have followed the exponential relation $A/A_{1940} = P/P_{1940}{}^n$, where A/A_{1940} is the amount of gasoline use or power production relative to 1940 and P/P_{1940} is the corresponding ratio for population. The indicated average values of n are 1.5 for gasoline use and about 2.2 for electric power, and the projections were made on this basis. Source, population projection to 1980, Financial and Population Research Section, California State Department of Finance.

growing. For example, photochemical damage to plants was first observed in an area of a few square km in Los Angeles County in 1942. In less than twenty years this area had expanded to more than 10,000 square km and new areas had appeared, bringing the total for California to nearly 30,000 square km. Photochemical pollution has now been observed in more than half the states in the United States and in an increasing number of other countries (Middleton and Haagen-Smit, 1961; Middleton, 1963).

This remarkable spread may be traced to two factors, the first of which is that nitrogen oxide and hydrocarbon emissions have increased faster than the population. The largest source of both nitrogen oxides and hydrocarbons is the automobile; in California at the present time about 60% of the nitrogen oxides and 75 to 85% of the reactive hydrocarbons, depending on how these are estimated, come from motor vehicles. Between 1940 and 1965 the population of California increased 2.7 times and gasoline use by motor vehicles in the state increased 4.3 times (Figure 8). The growth in electric-power generation, now 9.2 times what it was in 1940, has been another contributor to increasing nitrogen oxide emissions; roughly 16% of the present nitrogen oxide emissions in California come from steam-electric power plants. Hydrocarbon emissions, on the other hand, over the state as a whole have probably increased more in accordance with gasoline use.

The second factor contributing to the growth of photochemical air pollution is the relation between emission rate and the area covered by a given concentration as the pollutants are carried by the wind. This may be illustrated, for idealized conditions, by use of the box model, which assumes uniform mixing to a constant height such as an overhead inversion base, with dilution by lateral diffusion beneath that ceiling. The isopleths for a given concentration, calculated from this model (McMullen, personal communication) for various emission rates in a uniform square source (that is, an idealized city), under constant wind direction and velocity are shown in Figure 9. Starting, by definition, with the given concentration appearing at only a single point when the emission rate is unity, the areas within the isopleths are seen to increase much faster than the corresponding emission rates.

When a specific symptom of pollution has expanded

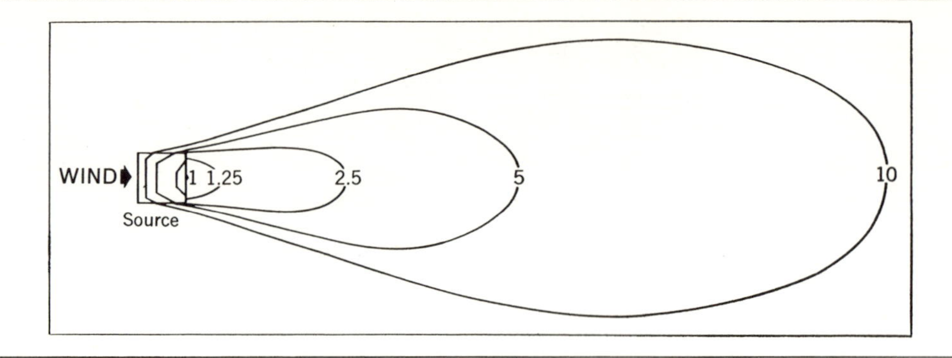

Figure 9 Area coverage by a pollutant as a function of emission rate. The figures are relative emission rates, and the curves are the corresponding isopleths of given concentration. Estimated for a uniform square source with constant wind direction and velocity under an overhead inversion at constant height.

to fill a geographical area, as is the case with plant damage in the Los Angeles basin and the San Francisco Bay Area, further increase may be expected to be in intensity rather than in extent. However, the movement of pollutants from one airshed to another is not excluded (Figure 7); the fingers of plant damage extending north and east from the Los Angeles region and southeast and east from the Bay region show that these areas are still growing, and if photochemical air pollution is not abated it may be assumed that the present visibility-reduction area is a shadow of the coming plant-damage area, and the present plant-damage area a forecast of the coming eye-irritation area.

Some of the hydrocarbons emitted to the air react much more slowly with the nitrogen oxides than others; these less reactive hydrocarbons, such as ethylene and some of the paraffins, produce ozone but little or no PAN. Accordingly, if ozone but not PAN plant damage is observed in an area, it may be taken as evidence that the pollutants have been airborne for some time and may have traveled some distance (Middleton and Haagen-Smit, 1961). Thus PAN damage is found in and around Washington, D.C., while ozone damage is observed much farther away, in areas that are in agreement with meteorological information on the trajectories of the air that has passed over Washington (Wanta and Heggestad, 1959). Similarly, ozone damage to tobacco plants in the upper Delaware Valley, with no concomitant PAN damage, suggests that the pollutants may have been transported some distance, perhaps from the Philadelphia-Trenton or New York metropolitan areas. The same situation with respect to tobacco damage in the Connecticut Valley may be due to the transport of pollutants from any of a number of centers in the Boston–New York conurbation.

THE PROSPECTS FOR CONTROL

Although photochemical air pollution is well on its way to becoming the number one form of general air pollution in the United States, a broad attack against it has thus far been mounted only in California. However, the passage by Congress on 1 October 1965, of a bill requiring the installation after September 1967, of exhaust control devices on new automobiles of domestic manufacture will expand this attack to a national scale, and in view of this prospect the California program merits examination in some detail.

In an assessment of the prospects for the abatement of photochemical air pollution by automobile controls, three factors are pertinent: the time delay or lead time; the growth in emissions over that lead time; and the degree of control likely to be achieved. To go back in time, we may now say that the visibility reduction which had become widespread in the Los Angeles basin as early as 1920 was due, in part at least, to photochemical air pollution. Reduction in the sizes of oranges and the cracking of rubber products, now known to be due to photochemical air pollution, were reported at least as early as 1930, specific plant damage was first observed in 1942, and eye irritation had appeared by 1945. The first step toward control was taken in 1948 with a California legislative act establishing air pollution control districts, and the control program in Los Angeles County was initiated shortly thereafter. Not until 1952 was the first evidence obtained that what was then known as "smog" was primarily photochemical and that the emissions chiefly responsible for it were nitrogen oxides and hydrocarbons.

The first control steps directed specifically at photochemical air pollution were applied to hydrocarbon emissions from stationary sources in the Los Angeles basin, and by 1960 these sources were about 60% controlled. In 1957–58 the elimination of home incinerators and the restriction of fuel-oil burning during the smog season achieved about a 45% control of nitrogen oxide emissions from stationary sources in the basin. The attack on hydrocarbon emissions from motor vehicles was initiated on a statewide basis in 1959. Roughly 75–80% of the reactive hydrocarbons

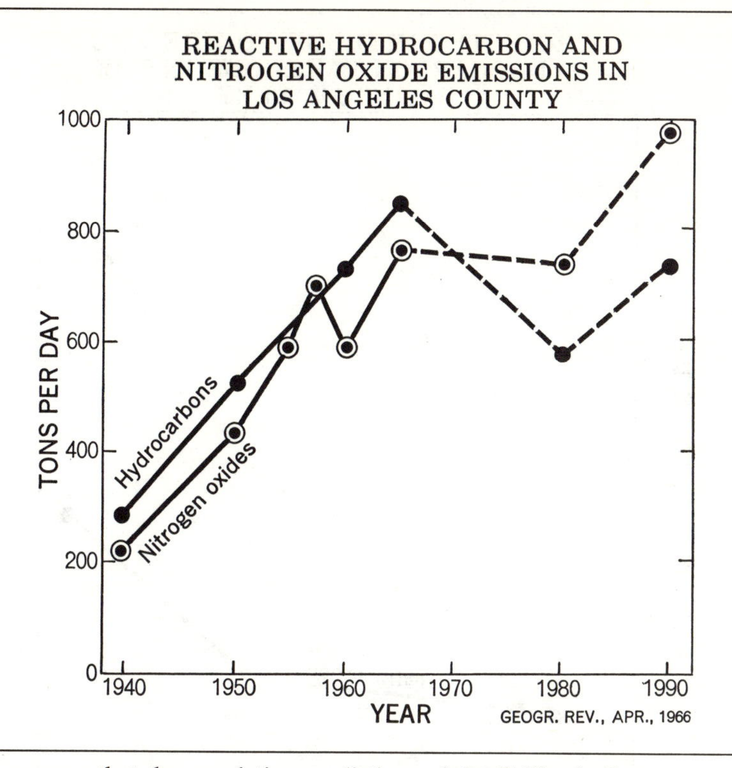

Figure 10 The projections assume that the population predictions of the California State Department of Finance will be realized; that emissions will continue to increase relative to population as they have since 1940; that motor vehicle crankcase emissions will be 80% controlled, exhaust and evaporation hydrocarbons 70% controlled, and exhaust nitrogen oxides 60% controlled by 1980; and that no other controls will be adopted. (Source for emissions to 1965, P. A. Leighton, 1964.)

emitted by automobiles come from the exhaust, 14–17% from the crankcase, and 7–8% from carburetor and fuel-tank evaporation. Installation of crankcase control devices on new cars began in 1961, but their installation on used cars has encountered complex difficulties and delays. Moreover, experience has shown that in the hands of individual owners the actual control achieved by these devices falls considerably short of the theoretical, and judgments of the degree of crankcase hydrocarbon control that will eventually be achieved range from less than 70% to about 90%.

A standard for exhaust hydrocarbons and carbon monoxide, which specifies that the hydrocarbon content under a given cycle of operation shall not exceed an average of 275 ppm, was adopted in 1960, and the installation on new automobiles of devices intended to meet this standard is beginning with the 1966 models of domestic makes. Revised standards now scheduled to take effect in 1970 will reduce the allowed exhaust hydrocarbon content to 180 ppm and will also require a reduction in evaporation losses. If the installation of devices to meet these 1970 standards is limited to new cars it will be at least 1980 before the exhaust control program as it now stands is fully effective, and judgments of the degree of exhaust hydrocarbon control that may be achieved range from 50% to 80%, the latter being the theoretical value. A standard of 350 ppm for exhaust nitrogen oxides, which is now in process of adoption, will require devices that produce a theoretical 65% control of these emissions.

What this attack has accomplished and may be expected to accomplish must be assessed in relation to the growth in sources and emissions that has occurred and may be expected over the time periods concerned (Leighton, 1964). An assessment on this basis for Los Angeles County is shown in Figure 10. Examination of the hydrocarbon curve indicates that neither the controls of emissions from stationary sources initiated after 1950 nor the crankcase controls initiated in 1961 have been sufficient to counteract the overall increase in emissions that accompanied population growth in the county. It would appear that the automobile exhaust and evaporation controls now scheduled will indeed reduce hydrocarbon emissions, even in the face of prospective growth, but if no further steps are taken, the upward climb will be resumed after the program is completed. According to the nitrogen oxide curve the controls of stationary sources initiated in 1957–58 achieved some reduction, but by about 1963 the gains had been wiped out by the process of growth. The projection indicates that the prospective control of nitrogen oxides from motor vehicles will reduce the overall

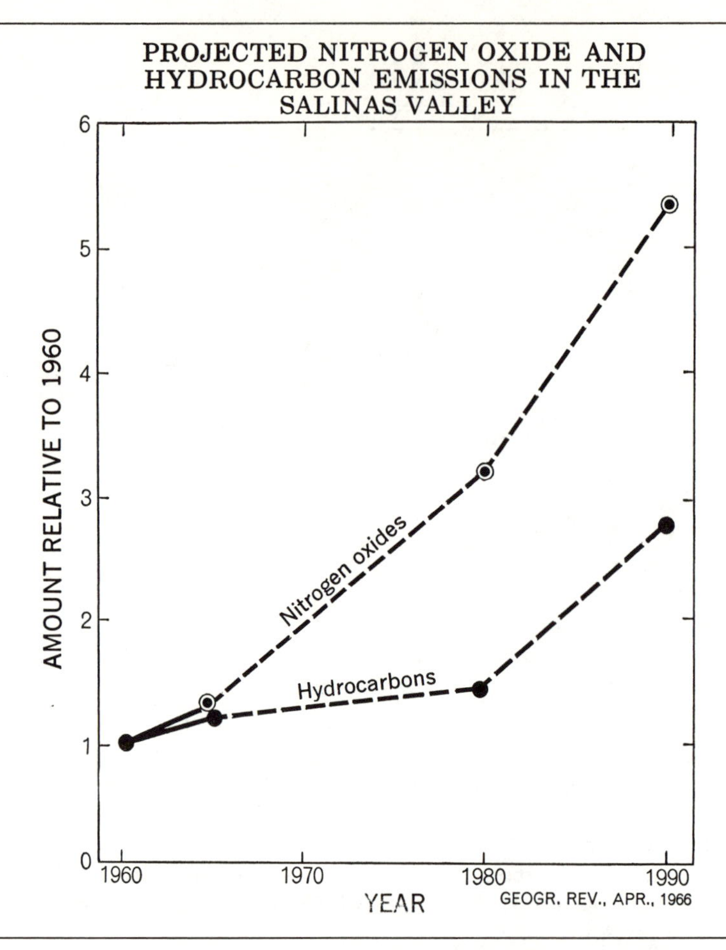

Figure 11 The projections assume that hydrocarbon emissions before controls will increase with gasoline use; that nitrogen oxide emissions from stationary sources will triple between 1965 and 1980 and will increase with population between 1980 and 1990; that vehicular nitrogen oxides will increase with gasoline use to the 1.1 power; and that the controls applied will be the same as in Figure 10.

emissions slightly between 1965 and 1980, but the growth after 1980, if no further steps are taken, will soon carry these emissions to new highs.

The level of emissions in 1940 has often been taken as the value that should be regained to eliminate photochemical plant damage and eye irritation in the Los Angeles basin. To the extent that the projections in Figure 10 are valid, it would appear that unless supplemented by other measures the California motor-vehicle control program as it now stands offers little hope of returning photochemical air pollution to its 1940 level in the basin. The program will gain some ground, but further steps will be required to hold the gain, and if such steps are not taken the situation will again deteriorate.

In the more rapidly growing areas of California the prospects of the motor-vehicle control program, taken alone, are still less optimistic. An excellent example is offered by the Salinas Valley in Monterey County. This immensely rich valley is one of the last major agricultural areas of California to be free of photochemical plant damage, but it is already suffering from visibility reduction (Figure 7), and measurements of ozone concentration indicate that the plant-damage level is being approached. The population of the valley is expected to double between 1965 and 1980, and to reach more than three times its 1965 value by 1990. Industrial expansion is being encouraged. In an industrial area at the mouth of the valley a large steam-electric power plant is now in operation, a threefold expansion of this plant has been scheduled, and the construction of an oil refinery has recently been approved. As a result of these and other industries in the valley, the contributions of stationary sources to overall pollution are high; at present, for example, about 75% of the nitrogen oxide emissions come from stationary sources, and in view of the prospective industrial growth coupled with the motor-vehicle control program this amount may be expected to increase, perhaps reaching 90% by 1980. When these factors and the predicted population increase are taken into account, the projections in Figure 11 indicate that the present California motor-vehicle control program in itself will not be sufficient to arrest the growth of photochemical pollution in the Salinas Valley.

The experience and prospects of the California control program illustrate the limitations, and the increasing challenge, that the attack on general air pollution must face in an era of growing population and increasing emissions per capita. With multiple sources and multiple types of sources in all kinds of use, controls at best are incomplete, and most of the steps regarded as practicable provide only a temporary respite from the inexorable pressure of growth. A succession of ever more severe controls is required merely to keep the situation from deteriorating, and if these are not effectively imposed the problem must eventually become one of survival.

In its broader aspects, the challenge is not limited to the air supply in specific geographical areas; it extends to the pollution of the entire atmosphere. Here the outstanding problem is the possibility of self-destruction through atmospheric radioactive contamination as the result of nuclear explosions. However, other problems also loom. There are indications, for example, that the atmospheric lead content in the Northern Hemisphere has increased with man's use of lead and its compounds until it is now about a thousand times what it probably was when our physiological responses to lead were evolved (Patterson, 1965). The carbon dioxide content of the atmosphere has increased 9% since 1890, and is reported to be currently increasing by about 0.2% a year; and it has been estimated that by the time the known reserves of fossil fuel have been burned the resultant temperature increase on earth, due to the absorption of infrared radiation by atmospheric carbon dioxide, will be sufficient to melt the polar icecaps, inundate present coastal areas, and annihilate many life forms (Conservation Foundation, 1963).

In essence these ultimate problems of general air pollution may be stated in simple terms. Whether applied to a local area or to the entire atmosphere it is a matter of maintaining the relation

$$\frac{\text{Emissions per capita} \times \text{number of persons}}{\text{Air supply}} < X,$$

where X is the maximum value to which we can accommodate. The means of maintaining this relation, however, are another matter. There is little prospect of increasing the local supply of air and none of increasing the overall supply. The per capita emissions may be reduced by controls, but, as we have seen, with increasing population the steps required become successively more severe, and the end of the process is the elimination of the sources. The accommodation coefficient X, as far as direct physiological effects are concerned, could be increased by the use of protective methods through which we breathed only purified air, but this would not help unprotected life forms or retard the other effects that must be taken into account. The remaining factor in the equation is the number of persons, and it may well be that the resource which eventually forces man to adopt population control as a requirement for survival will not be land, food, or water, but air.

REFERENCES

Conservation Foundation, 1963. *Implications of Rising Carbon Dioxide Content of the Atmosphere.* Conservation Foundation, New York.

Dean, R. S., Swain, R. E., Hewson, E. W., and Gill, G. C., 1944. Report submitted to the Trail Smelter Arbitral Tribunal. *U.S. Bur. Mines Bull. 453.*

Defant, F., 1951. Local Winds. In *Compendium of Meteorology*, ed. T. F. Malone. Amer. Meteorol. Soc., Boston. 655–672.

Geiger, R., 1950. *The Climate Near the Ground.* Translated by Stewart, M. N. et al. Cambridge, Massachusetts. 611 pp.

Hewson, E. W., 1945. The meteorological control of atmospheric pollution by heavy industry. *Quart. Journ. Royal Meteorol. Soc. 71*, 266–282.

Holzworth, G. C., Bell, G. B., and de Marrias, G. A., 1963. *Temperature Inversion Summaries of U.S. Weather Bureau Radiosonde Observations in California.* U.S. Weather Bureau, Los Angeles, and State of California, Dept. Public Health.

Hursh, C. R., 1948. Local climate in the copper basin of Tennessee as modified by the removal of vegetation. *U.S. Dept. Agri. Circ. 774.*

Korshover, J., 1960. Synoptic climatology of stagnating anticyclones east of the Rocky Mountains in the United States for the period 1936–1956. *Rept. SEC TR-A60-7*, R. A. Taft Sanitary Engineering Center, Cincinnati.

Leighton, P. A., 1954–55. Cloud travel in mountainous terrain. *Quart. Repts. 111-3 and 111-4*, Dept. Chemistry, Stanford Univ. (Defense Documentation Center AD Nos. 96571, 96486, 96487).

Leighton, P. A., 1964. Man and air in California. *Proc. Statewide Conf. Man in California, 1980's.* (Univ. California, Ext. Dept., Berkeley.) 44–77.

McCormick, R. A., *Personal Communication.* Chief, Meteorology Section, Laboratory of Engineering and Physical Sciences, R. A. Taft Sanitary Engineering Center, Cincinnati.

McMullen, R. W., *Personal Communication.* Metronics Associates, Inc., Palo Alto, Calif.

Middleton, J. T., 1963. Air conservation and the protection of our natural resources. *Proc. Nat. Conf. Air Pollution.* U.S. Dept. Health, Education, and Welfare, Washington, D.C. 166–172.

Middleton, J. T. and Haagen-Smit, A. J., 1961. The occurrence, distribution, and significance of photochemical air pollution in the United States, Canada, and Mexico. *Journ. Air Pollution Control Assn.* (*J-APCA*) *11*, 129–134.

Neiburger, M., Johnson, D. S., and Chien, C.-W., 1961. Studies of the structure of the atmosphere over the eastern Pacific Ocean in summer. *Univ. California Publications in Meteorology 1*(*1*), 1–94.

Patterson, C. C., 1965. Contaminated and natural lead environments of man. *Archives of Environmental Health 11*, 344–360.

Wanta, R. C. and Heggestad, H. E., 1959. Occurrence of high ozone concentrations in the air near metropolitan Washington. *Science 130*, 103–106.

G. C. Holzworth

20 Large-scale weather influences on community air pollution potential in the United States

In this paper "large-scale weather influences on community air pollution potential" denotes those features of dispersion over cities that may be associated with the large eddies or whorls that occur in the general circulation of the earth's atmosphere, i.e., extensive high- and low-pressure systems, anticyclones and cyclones, whose identities and life histories can be followed for at least several days. Cyclones, of course, are typified by precipitation, fast winds, and enhanced vertical motions, which produce cleansing and relatively rapid or good dispersion. Such conditions are also common along the "frontal" zones that separate large air masses with contrasting properties. Anticyclones, on the other hand, are usually characterized by fair weather, slow winds, and inhibited vertical motions, which result in relatively slow or poor dispersion. Anticyclones may be classified as cold or warm, the latter being much more potent in terms of sustained slow dispersion.

Mean Sea-Level Pressure Patterns

Some idea of the sizes of cyclones and anticyclones, and their geographical and seasonal occurrence over the Northern Hemisphere may be gained from the normal patterns of sea-level pressure (U.S. Weather Bureau, 1952) for January and July (Figures 1 and 2). Although the typical sizes of individual pressure systems are somewhat smaller than indicated by the normal patterns, they do occasionally approach such large dimensions. The low-level circulation is roughly parallel to the isobars; around cyclones it is counterclockwise, and around anticyclones it is clockwise, both typically with some cross-isobaric component in the lower levels toward lower pressure. This commonly results in mass convergence around cyclones, which produces general rising motion, and in divergence around anticyclones, which causes general sinking or subsidence of the atmosphere. Subsidence is important because it causes warming aloft, tending to stabilize the atmosphere and to limit the extent of vertical mixing. Wind speeds are approximately inversely proportional to isobaric spacing.

Figure 1 shows that on the average in January the northern parts of the oceans are dominated by huge cyclones, the Aleutian and Icelandic Lows. The subtropical oceans experience relatively high pressure and the continents are overlain by anticyclones, the very large Siberian High and comparatively weak high pressure over the United States. Thus, in winter pressure over the oceans is relatively low and over the continents it is relatively high; this pattern is reversed in summer.

In July (Figure 2) the oceans are almost completely covered by gigantic anticyclones, the North Pacific High and the Atlantic or Azores High, which extends well into the eastern United States. Eurasia is covered by a very large but weak low-pressure area with several centers. Pressure over the United States is comparatively low, especially over the southwestern desert. The latter is a Thermal Low since it results from marked heating of the atmosphere through several thousand feet. Unlike conventional lows, it is not characterized by stormy weather, it contains no fronts, and it remains nearly stationary.

Figures 1 and 2 reveal certain geographical regions and seasons in which the occurrence of these large-scale weather features is naturally preferred. The regions between the predominant mean systems are often influenced by migrating high- and low-pressure systems. Much of the United States lies in this "in-between" region. A better depiction of some weather systems that affect the United States is presented in individual weather maps.

The Cold Anticyclone

Figure 3 illustrates the situation on 16 October 1948, at 0730 EST (Air Weather Service, 1948). An extensive cyclone is centered north of the Great Lakes with a

Reprinted, with minor editorial modification, by permission of the author and editor, from the *Journal of the Air Pollution Control Association,* Vol. 19, 1969, pp. 248–254.

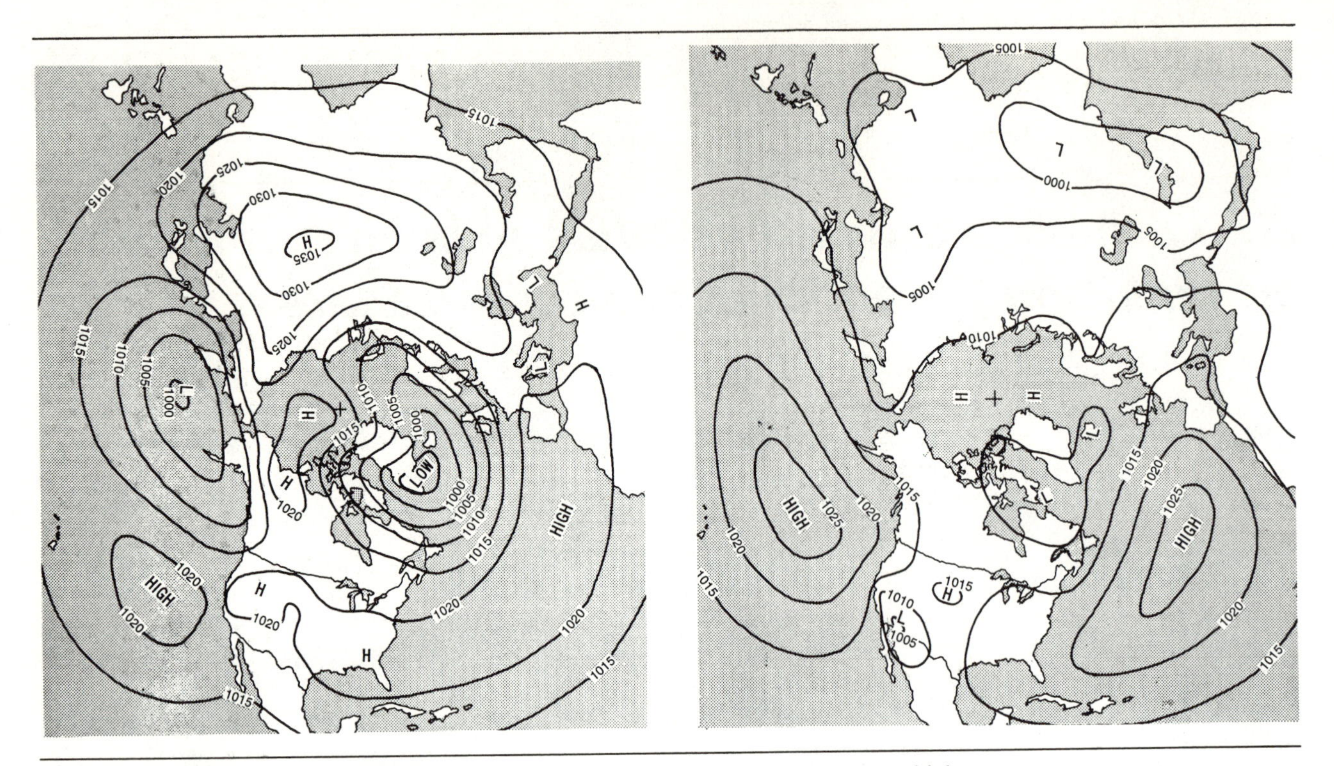

Figure 1 Isopleths of normal sea-level pressure (millibars), January. H and L denote high- and low-pressure centers.

Figure 2 Isopleths of normal sea-level pressure (millibars), July.

cold front extending south through the middle of the United States and then westward. The counter-clockwise circulation around the low merges with the clockwise circulation around the highs to the east and west; the circulation is intense through a deep layer over a very large region, e.g., where the isobars are spaced close together. The cold front is very sharp with surface temperatures immediately ahead of it in the 60's and 70's dropping into the 40's just behind it. Accordingly, the anticyclone centered in southwestern Canada is of the cold type. Especially in its forward or eastern and southern portions, i.e., behind the cold front, the atmosphere is relatively cold through a deep layer. The rapid transport of cold air aloft into this region tends to maintain a comparatively fast rate of temperature decrease with height, making the atmosphere unstable and enhancing vertical mixing. Since wind speeds normally increase with height over cold anticyclones, vertical mixing also enhances horizontal mixing by the downward transport of faster winds from aloft. Clearly, on the morning of 16 October 1948, dispersion is excellent over a very wide region from the Rockies through the Plains, into the Mississippi Valley, and through the Great Lakes. The only likely large land areas of relatively slow or limited dispersion are along the ridge of high pressure over the East Coast and over the high-pressure area in the vicinity of southwestern Canada and northwestern United States. In these areas the winds are slow, i.e., where the isobaric spacing is large, and some subsidence aloft is indicated.

In concert with the large-scale weather features and from practical considerations, the occurrence of poor dispersion may be considered significant whenever it remains over a given large area for *at least* one day. Therefore, the two areas of slow dispersion on the morning of 16 October are of no great concern since they and the other major weather features are advancing quickly. For instance, in 24 hours the high that was centered over southwestern Canada slipped 1,000 miles southeastward, the high-pressure area that was along the East Coast moved off-shore, the deep Canadian Low intensified and moved 800 miles northeastward, and the cold front advanced an average of 500 miles east and south through the middle of the country. This type of rapid movement is characteristic of large-scale systems associated with cold-type anticyclones.

The Warm Anticyclone

The other type of anticyclone, that during which episodes of undesirable air quality have been found to occur (Niemeyer, 1960; Boettger, 1961; Holzworth, 1962; Miller and Niemeyer, 1963; Schrenk, et al., 1949; Korshover, 1967), is the warm high, so named because temperatures at levels above a thousand or so feet are warm compared with those of adjacent large-scale systems. Although temperatures near the surface in warm highs *are* warm in the afternoon, at night they often are cool because of the occurrence of intense

Figure 3 Synoptic weather map, 16 October 1948, 0730 EST. Isopleths of sea-level pressure (millibars), continuous lines; weather fronts, pipped heavy lines; H and L, high- and low-pressure centers.

Figure 4 Synoptic weather map, 26 October 1948, 0730 EST. Donora, Pennsylvania, marked by large dot. Otherwise, same as Figure 3.

surface-based radiation inversions. Warm highs are also called stagnating anticyclones because they usually move very slowly or remain quasi-stationary and are characterized by slow winds through a deep layer and widespread sluggish dispersion. Figure 4 shows the weather pattern on 26 October 1948, near the beginning of large-scale stagnation conditions associated with the air pollution disaster at Donora, Pa. (Schrenk, et al., 1949). Except along the north Atlantic Coast, the eastern United States is dominated by a warm anticyclone. This warmth is not apparent at the surface at 0730 EST because of widespread surface inversions, but aloft it extends through a depth of several miles; cold air lies over the Rockies and off the Atlantic Coast. The warmth aloft in the warm high is derived largely from slow but sustained and widespread subsidence, in which the air is heated adiabatically, i.e., by compression. The wide spacing of isobars around the warm high is indicative of slow winds, which typically extend through a depth of several miles.

Historical weather maps (Air Weather Service, 1948) show that in 4 days the warm anticyclone moved little. By 30 October it had shifted eastward slightly and a cold front (in Figure 4, the one through Washington State and southwest Canada) approaching the western Mississippi Valley was causing increased dispersion, but very limited dispersion conditions persisted over the eastern one-fourth of the nation. This air pollution incident ended on 31 October as the cold front continued eastward and rain showers developed in advance of it.

Although stagnation situations like that of 26–30 October 1948, are not an everyday occurrence, neither are they completely unique. For example, Korshover (1967) found that in a 30-year period in states east of the Rocky Mountains, there were 227 stagnation episodes, about $7\frac{1}{2}$ per year, that lasted at least 4 days. It should be pointed out, however, that the stagnation areas could have been as small as roughly one-third the area of Florida. Thus protracted stagnation over very large areas occurs less frequently than indicated by Korshover. Yet, the frequency is not as rare as the occurrence of Donora-type disasters, as experience in the National Air Pollution Potential Forecasting Program has demonstrated.

The Donora disaster was not due just to a large emission of toxic pollutants and large-scale conditions of limited dispersion, but also to local details, meso- and micro-scale, of the particular source-dispersion-receptor configurations that favored high pollutant concentrations at Donora. Small-scale dispersion processes are highly dependent upon local topography, particularly during large-scale conditions of poor dispersion.

Mixing Heights and Vertically Averaged Wind Speeds

Up to this point the general characteristics of large-scale weather systems, particularly those associated with poor atmospheric dispersion, have been described qualitatively. It is appropriate now to describe quantitatively some specific dispersion parameters. The basic parameters that are being used in the National Air Pollution Potential Forecasting Program and in the preparation of a national air pollution potential climatology are: (1) the urban morning mixing height,

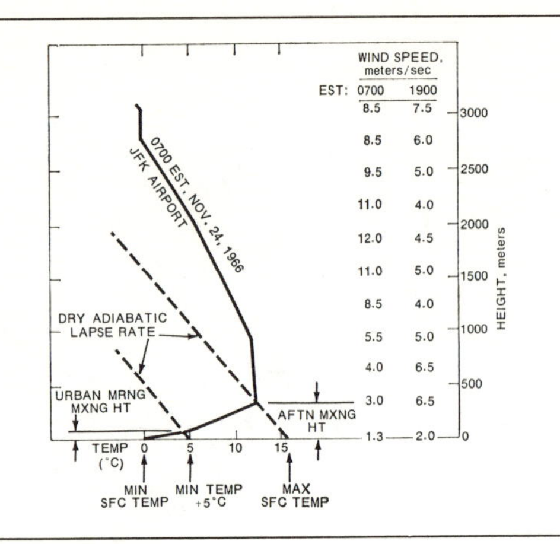

Figure 5 Temperature and wind speed data at J. F. Kennedy Airport, N.Y., 24 November 1966, 0700 EST, and method of calculating afternoon and urban morning mixing heights.

(2) the vertically averaged wind speed through the morning mixing height, (3) the afternoon mixing height, and (4) the average speed through the afternoon mixing height. Physically, a mixing height is that height through which relatively vigorous vertical mixing occurs.

Calculation Method

Mixing heights are not measured directly but can be estimated from routine Weather Bureau observations, as illustrated in Figure 5. The meteorological observations were taken at Kennedy International Airport, located on Long Island, on Thanksgiving Day 1966 during the widely publicized air pollution incident that occurred in New York and other eastern cities (Fensterstock and Fankhauser, 1968). The vertical temperature profile at 0700 EST is used to calculate both the morning and afternoon mixing heights. Notice the great stability through the lowest 900 meters, an intense surface-based inversion topped by an isothermal layer, the latter due largely to subsidence. The (urban) morning mixing height is defined as the height above the surface where the dry adiabatic extension of the surface minimum temperature plus 5° C intersects the observed temperature profile. "Dry adiabatic" refers to the temperature change caused by expansion or compression of a dry gas without an exchange of heat with its environment. In the atmosphere the expansion of a rising dry air parcel results in cooling at the rate of 1° C per 100 meters; a subsiding parcel is warmed at the same rate. In the definition of the morning mixing height the "plus 5° C" is specified arbitrarily to allow for urban-rural and downtown-suburban differences of morning surface temperature (Mitchell, 1962), since Weather Bureau soundings are made at airports in rural or suburban surroundings; it also allows for some solar heating of the surface after sunrise. The dry adiabatic lapse rate is used under the assumption that heat input at the surface creates instability, which results in vertical mixing, which in turn tends to establish an adiabatic lapse rate. In Figure 5 the morning mixing height is 100 meters.

The afternoon mixing height is defined as for the morning, except that the maximum afternoon surface temperature is used instead of the minimum plus 5° C. Differences of afternoon surface temperature at urban and rural locations typically are smaller than morning differences. In Figure 5 the afternoon mixing height is about 300 meters.

The average wind speed through a mixing height is just the average of the speeds within the mixing layer. For mornings, the wind speeds aloft observed at 0700 EST are used. Accordingly, in Figure 5 the average speed through the morning mixing layer is just the surface speed, 1.3 m/sec. For afternoons, the wind speeds aloft at 1900 EST are used since that is the observation time nearest to mid-afternoon. In Figure 5 the average speed through the afternoon mixing layer is 4.2 m/sec since the mixing height extends only through the first wind level above the surface.

Admittedly, mixing height calculations are rough estimates as revealed by detailed temperature soundings over cities (Davidson, 1967). Nevertheless, for the purpose of quantitatively describing the large-scale aspects of dispersion, the morning and afternoon mixing heights and average wind speeds are highly useful. These parameters have also been used successfully in an urban diffusion model (Miller and Holzworth, 1967).

Typical Monthly Mean Values

Compared with mean values for many locations in the United States, the mixing height and average wind speed values at New York on Thanksgiving 1966 (Figure 5) are low. For comparison the monthly mean mixing height and average wind speed values are presented in Figures 6 and 7 for Washington, D.C.; St. Cloud, Minn.; Denver, Colo.; and Oakland, Calif. Monthly mean values for 7 other locations have been given previously (Holzworth, 1967). All values are based on observations for 5 years, 1960–64, except those for Washington, which are for 1961–64. Values for mixing height and average wind speed with which precipitation occurred are excluded from the means, since the dry adiabatic assumption (no condensation) may not be applicable.

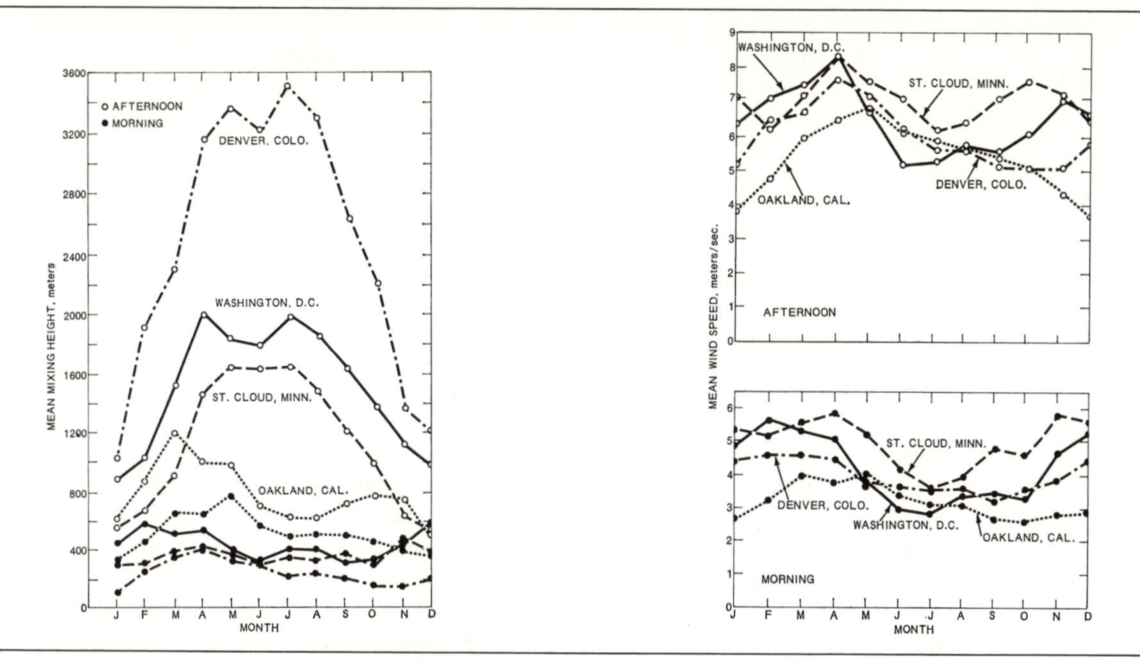

Figure 6 Monthly mean afternoon (open circles) and urban morning (solid circles) mixing heights. All data for 1960–1964, except Washington for 1961–1964.

Figure 7 Monthly mean values of vertically averaged wind speed through morning and afternoon mixing layers. Data same as Figure 6.

Figure 6 shows that for the stations considered here the monthly mean afternoon mixing heights are highest and the morning mixing heights are lowest at Denver in every month. Throughout the year at Denver the mean afternoon mixing heights vary over a very large range, from about 1,000 meters in January to 3,500 meters in July; by comparison, the morning mixing heights vary only from 100 to 400 meters. In the National Air Pollution Potential Forecasting Program the critical mixing height values that have been used are 500 meters for the morning and 1,500 meters for the afternoon (National Meteorological Center, 1967). Thus, in terms of monthly averages, vertical mixing at Denver is very limited at night throughout most of the year and in the afternoon is practically unlimited, except during the colder months. These mixing height characteristics for Denver are typical of the high plateau and mountain areas of the West.

At Washington and St. Cloud the mean afternoon mixing heights follow the same general pattern throughout the year as at Denver, although the heights are considerably lower at Washington and St. Cloud. The morning mixing heights at Washington and St. Cloud are generally around 200 meters greater than at Denver. The mixing height characteristics for St. Cloud are typical of the northern Central States.

In some respects the characteristics of the Oakland mixing heights are quite different from those at Denver, Washington, and St. Cloud. At Oakland in summer the unusually low afternoon mixing heights of around 650 meters and the small morning-afternoon variation of about 150 meters are typical of the California Coast; they are due to the intense subsidence inversion that occurs on the eastern side of the Pacific Anticyclone.

Figure 7 shows the monthly means of the vertically averaged wind speeds through the morning and through the afternoon mixing heights. Like the mixing heights, the values of afternoon mean speeds are consistently greater than morning speeds; the differences average roughly 2 m/sec (1 m/sec = 1.94 knots = 2.24 miles/hr) throughout the year at each location. The significance of these mean wind speeds may be judged from their use in the National Air Pollution Potential Forecasting Program in which the critical value is 4 m/sec for both morning and afternoon.

Of the locations shown, the only one with monthly mean afternoon speeds of 4 m/sec or less is Oakland where the criterion is barely met in December and January. At the other locations no monthly mean afternoon speed is less than 5 m/sec; at St. Cloud none is as slow as 6 m/sec.

In the morning the mean speeds are 4 m/sec or less at Oakland in every month; they are less than 3 m/sec in 5 months. At Denver and Washington the mean morning speeds are less than 4 m/sec in about half of the months with the slowest speeds around 3 m/sec. At St. Cloud the slowest mean speeds in the morning are 3.6 and 4.0 m/sec.

In summary the fastest morning and afternoon mean speeds occur at St. Cloud and tend to offset the shallow mixing heights there. The slowest morning and afternoon mean speeds occur at Oakland where mean

	Frequency of episodes (Number)	Total number episode-days (Days)	Longest episode (Days)	Longest episode (Season)
St. Cloud, Minn.	1	3	3	Spring
Washington, D.C.	18[a]	39[a]	3	Autumn
Denver, Colo.	16	40½	4½	Winter
Oakland, Cal.	54	204½	12	Winter

[a] Prorated to 5 years, based on 4 years of records

Table 1 Episodes of at least 2 days duration during which morning and afternoon mixing heights were 1,500 m or less and average wind speeds were 4.0 m/sec or less without significant precipitation, based on 5 years of records

afternoon heights are mostly very low. Although the Oakland morning mixing layers are higher in most months than at the other locations, they are generally more than offset by very slow morning winds. Thus, of the four locations considered, it is concluded that in the mean, limited dispersion occurs most often at Oakland and least at St. Cloud.

Episodal Features

The foregoing conclusion is borne out by the episodal nature of limited dispersion conditions based on individual daily values of morning and afternoon mixing heights and vertically averaged wind speeds. An episode of limited dispersion may be defined as no mixing layer higher than 1,500 meters and no average wind speed greater than 4.0 m/sec, with no precipitation for at least 2 days. Each period from morning to afternoon and afternoon to morning is considered 12 hours. Thus, for an episode of 2 days the critical conditions must have been satisfied in five successive mixing height and wind speed computations. As shown in Table 1, these conditions occurred only once in 5 years at St. Cloud. That episode lasted 3 days and occurred in the spring. Episodal data for Washington and Denver are similar, both having a total of about 40 episode-days. Oakland had 54 episodes which totalled 204½ episode-days.

Figure 8 Synoptic weather map, 1 December 1962, 1300 EST. Precipitation areas stippled. Otherwise, same as Figure 3.

Spatial Distributions during an Air Pollution Incident

As an example of the mixing height and average wind speed values during an extensive air pollution incident, consider the episode of 27 November–5 December 1962, in the eastern United States, for which pollution conditions and air pollution potential forecasts have been described (Lynn, et al., 1964). In terms of air pollution potential forecasts this episode was exceptional in that it lasted up to 7½ days over Massachusetts, southern New York State, and Pennsylvania; at times it covered all or most of the New England States, New York, New Jersey, Pennsylvania, Maryland, West Virginia, Ohio, Kentucky, Indiana, Michigan, Illinois, and Missouri.

Figure 8 shows the synoptic weather situation at 1300 EST, 1 December 1962 (Weather Bureau, 1962), near the middle of the episode. Almost all of the eastern states are overlain by the warm anticyclone centered over Lake Erie and extending southwestward into the Gulf of Mexico. As in the Donora incident (compare Figures 4 and 8), this is a classical stagnation situation with the high-pressure area nearly stationary, slow winds near the surface (note wide spacing of isobars) and through a very thick layer aloft, cloud-free skies that enhance the formation of nocturnal surface-based inversions, and subsidence that further increases atmospheric stability. Consequently, the mixing height and average wind speed values were very low.

Figure 9 shows isolines of morning mixing height and average wind speed through the mixing layer on 1 December 1962, corresponding to the synoptic situation of Figure 8. As mentioned, critical morning mixing heights have been considered to be 500 meters and average wind speeds, 4 m/sec. Mixing heights were 200 meters or less and average wind speeds 4 m/sec or less over most of the area covered by the warm high;

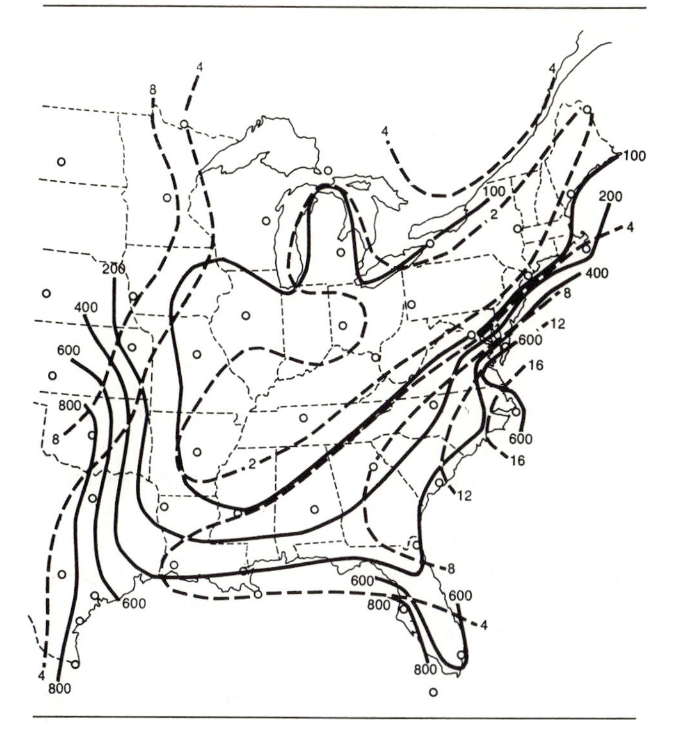

Figure 9 1 December 1962, isopleths of urban morning mixing height (meters), solid lines, and vertically averaged wind speed (meters/sec) through the mixing layer, dashed lines. Based on data (not given) for Weather Bureau upper air observing stations shown by small circles.

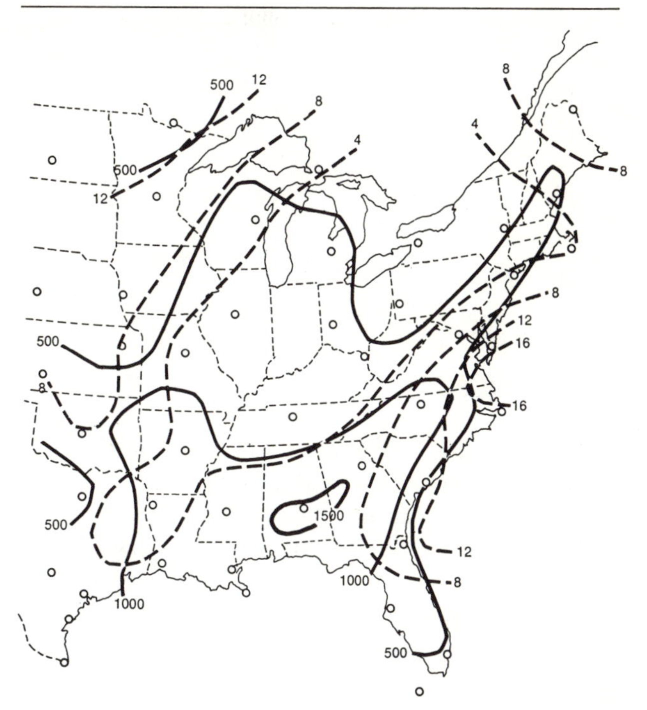

Figure 10 1 December 1962, isopleths of afternoon mixing height and vertically averaged wind speed. Otherwise, same as Figure 9.

in fact, the values were less than 100 meters and 2 m/sec over a surprisingly large area. Around the periphery of the high, relatively high mixing heights were caused variously by cloudiness and windiness, induced by stormy or unsettled weather.

By afternoon (Figure 10) the average wind speeds had increased noticeably in the vicinity of the western Great Lakes and Maine, but otherwise the area with morning speeds less than 4 m/sec changed very little. For the area with afternoon wind speeds of 4 m/sec or less, the afternoon mixing heights barely exceeded 1,000 meters only in the vicinity of Arkansas; they were less than 500 meters over Ohio, Pennsylvania, New York, and Vermont. Since the critical afternoon mixing height is 1,500 meters, it is clear that throughout 1 December 1962, dispersion was slow over a large populous area of the eastern United States. With similar dispersion conditions on other days of the long episode, it is surprising that the consequences were not disastrous.

Summary

Widespread conditions of slow atmospheric dispersion, lasting the order of days, usually occur with warm anticyclones. Such high-pressure systems are typified by slow movement, sluggish winds, subsidence, and inhibited vertical mixing. These characteristics are variously illustrated for the well-known air pollution incidents at Donora, Pennsylvania in 1948 (Schrenk, et al., 1949), in the eastern United States in 1962 (Lynn, et al., 1964), and again at Thanksgiving 1966 (Fensterstock and Fankhauser, 1968). It is shown that the large-scale dispersion features can be assessed in terms of morning and afternoon mixing heights and vertically averaged wind speeds through these layers. Mean monthly values of these variables are presented for St. Cloud, Minn.; Washington, D.C.; Denver, Colo.; and Oakland, Calif. In general the mean morning mixing heights are a few hundred meters throughout the year; the afternoon values in winter range from several hundred to about 1,000 meters, and in summer, from about 600 meters at Oakland to over 3,000 meters at Denver. The mean wind speeds are roughly 2 m/sec faster in the afternoon than in the morning. In general, the fastest speeds occur at St. Cloud and the slowest at Oakland.

Using mixing height and wind speed data for individual days, the episodal nature of limited dispersion is assessed. An episode of limited dispersion is defined as no mixing height greater than 1,500 meters, no average wind speed greater than 4.0 m/sec, and no precipitation, all for at least 2 days. In 5 years the total number of such episode-days at St. Cloud was 3; at Washington and Denver, about 40; and at Oakland, it was over 200.

REFERENCES

Air Weather Service, 1968. *Northern Hemisphere Historical Weather Maps, Oct. 1948.*

Boettger, C. M., 1961. Air pollution potential east of the Rocky Mountains: fall 1959. *Bull. Amer. Meteorol. Soc. 42(9)*, 615–620.

Davidson, B., 1967. A summary of the New York urban air pollution dynamics research program. *J. Air Pollution Control Assoc. (J-APCA) 17(3)*, 154–158.

Fensterstock, J. C. and Fankhauser, R. K., 1968. Thanksgiving 1966 air pollution episode in the eastern United States. *PHS Pub. 999-AP-45*. Cincinnati. 45 pp.

Holzworth, G. C., 1962. A study of air pollution potential for the western United States. *J. Appl. Meteor. 1(3)*, 366–382.

Holzworth, G. C., 1967. Mixing depths, wind speeds, and air pollution potential for selected locations in the United States. *J. Appl. Meteor. 6(6)*, 1039–1044.

Korshover, J., 1967. Climatology of stagnating anticyclones east of the Rocky Mountains 1936–1965. *PHS Pub. 999-AP-34*. Cincinnati. 15 pp.

Lynn, D. A., Steigerwald, B. J., and Ludwig, J. H., 1964. The November–December 1962 air pollution episode in the eastern United States. *PHS Pub. 999-AP-7*. Cincinnati. 23 pp.

Miller, M. E. and Holzworth, G. C., 1967. An atmospheric diffusion model for metropolitan areas. *J-APCA 17(1)*, 46–50.

Miller, M. E. and Niemeyer, L. E., 1963. Air pollution potential forecasts—a year's experience. *J-APCA 13(5)*, 205–210.

Mitchell, J. M., jr., 1962. The thermal climate of cities. In *PHS Symposium Air Over Cities*, SEC Tech. Rept. A62-5. Cincinnati. 131–145.

National Center for Air Pollution Control, 1968. *Forecasting Air Pollution Potential.* National Center for Air Pollution Control, Cincinnati.

National Meteorological Center, 1967. The air pollution potential forecast program. *Wea. Bur. Tech. Memo WBTM-NMC 43*, National Meteorological Center, Suitland, Maryland. 8 pp.

Niemeyer, L. E., 1960. Forecasting air pollution potential. *Mon. Wea. Rev. 88(3)*, 88–96.

Schrenk, H. H., Hieman, H., Clayton, G. D., Gafafer, W. M., and Wexler, H., 1949. Air pollution in Donora, Pa. *PHS Bull. 306*. U.S. Govt. Print. Off., Washington, D.C. 173 pp.

U.S. Weather Bureau, 1952. Normal weather charts for the northern hemisphere. *Wea. Bur. Tech. Paper 21*, October, U.S. Govt. Print. Off., Washington, D.C.

Weather Bureau, 1962. Daily weather map. Washington, D.C.

Peter W. Summers

21 The seasonal, weekly, and daily cycles of atmospheric smoke content in central Montreal

Since the end of World War II there has been a tremendous growth of the oil, mining, paper, and other industries to the point where Canada now ranks 6th in manufacturing output for the nations of the world, but still only 27th in total population. This industrial growth has produced a remarkable growth in urban populations, with some of the medium sized cities doubling their population during the last decade. The two largest industrial areas are in the St. Lawrence River Valley between Montreal and Kingston, and around the west end of Lake Ontario in the Oshawa Toronto-Hamilton-Niagara belt (Katz, 1961). Neither one of these areas has yet reached the point of forming a megalopolis (Landsberg, 1962), but they are in regions where the local topography of the valley or shoreline has important meteorological consequences affecting the diffusive capacity of the atmosphere. Along with this growth and industrialization many cities in Canada are now experiencing air pollution problems.

Local residents have long been aware that Montreal is a smoky city, but the only published data up until 1960 were dustfall figures. Comparison with other cities shows that downtown Montreal has the highest dustfall in Canada and approaches that recorded in some of the worst areas in the United States and the United Kingdom before active reduction measures were taken (Katz, 1963).

In January 1960, measurements of soiling index were commenced at the offices of Weather Engineering Corp. of Canada Ltd. in Central Montreal. It soon became obvious that the readings were very high and on a par with published data from the smokiest cities in the U.S.A. But, perhaps more important was the discovery that these high readings were often associated with meteorological conditions very different from the classical anticyclonic pattern (Denison, et al., 1960). Some mobile sampling confirmed that the high soiling index was general in the downtown area. Early in 1961, the McGill University Observatory installed a smoke sampler on the University Campus and another on top of Mount Royal. A preliminary analysis of these data was published by Summers in 1962. In the present paper, a more detailed analysis is made of the time variations of the soiling index from all three locations during the period January 1960–April 1963.

Reprinted, with minor editorial modification, by permission of the author and editor, from the *Journal of the Air Pollution Control Association,* Vol. 16, 1966, pp. 432–438.

Sampling Location and Procedures

Throughout this paper the following abbreviations will be used for the three samplers:

WEC—Weather Engineering sampler located near the intersection of Crescent and Burnside Streets in residential-commercial district of Central Montreal.

McG—McGill University sampler located on the 2nd floor of Macdonald Physics Building on the campus.

CBC—McGill University sampler located in the Canadian Broadcasting Commission transmitter building on top of Mount Royal.

The location of these stations with respect to Mount Royal is shown in Figure 1, and the height of the air intakes is indicated in feet above m.s.l. The WEC and McG intakes are 35 ft above ground, and for the CBC 10 ft. Standard AISI automatic smoke samplers were used (Hemeon, et al., 1953). Although these do have serious limitations (Stalker, et al., 1960; Katz, et al., 1958; Sullivan, 1962; Sanderson and Katz, 1963), they do offer the best index of suspended particulates available for meteorological analysis.

For the purposes of this study the months will be grouped on a seasonal basis as follows:

Winter season—December through March (ground frozen and snow covered).

Summer season—June through September (no heating of buildings).

Transition season—April, May, October, November.

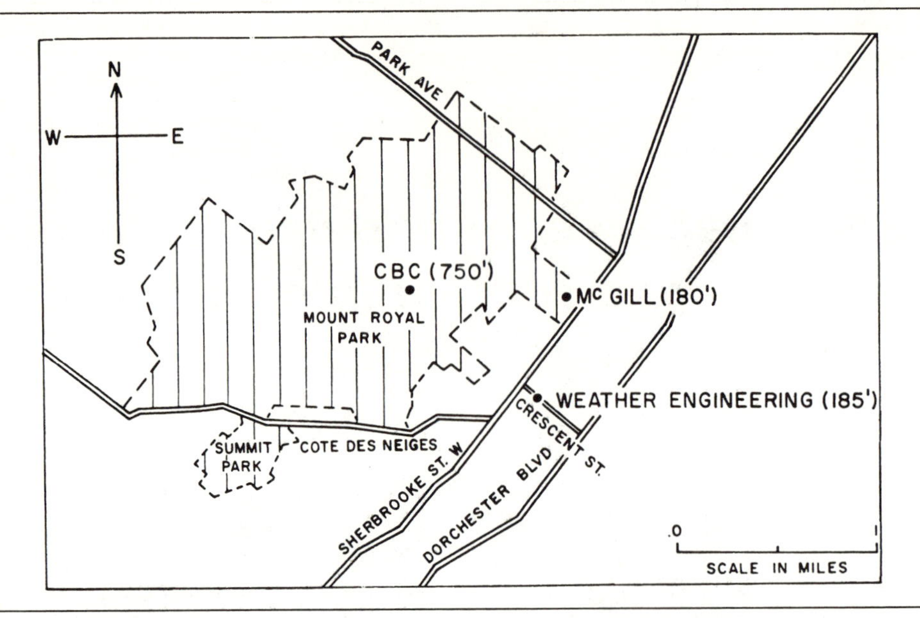

Figure 1 Map showing location of the three smoke samplers with respect to Mount Royal Park.

In order to minimize the errors in measuring the soiling index (Stalker, et al., 1960), the samplers were operated on a 2-hour sampling time in the summer season and on a 1-hour sampling time during the remainder of the year.

All the soiling index data were placed on IBM punched cards, along with meteorological data for the Montreal International Airport, thus allowing tabulations to be performed on the IBM 1410 computer at the McGill Computing Center.

Mean Annual Soiling Index

All available data were averaged to obtain the mean monthly values of soiling index. These 12 monthly means were then used to obtain the annual means shown in Table 1. The most striking feature is the large variation between the three stations. Although only half a mile away from WEC, the McG location has 46% less smoke. The prevailing wind in Montreal is west-southwest to west and Figure 1 shows that with such winds the air arriving at McG has spent the last two miles of its trajectory passing over, or skirting around Mount Royal Park, and therefore picking up little in the way of smoke (Summers, 1962).

The soiling index atop Mount Royal is only 21% of that at WEC, and 40% of that at McG. This reduction is due partly to surrounding parkland, but mainly due to the effects of elevation.

A comparison between Montreal and other Canadian cities for which data has been published is shown in Table 2. It must be remembered that data from other cities will also be a function of location, but the comparison suggests that Central Montreal is one of the smokiest parts of Canada.

Table 1 The mean annual soiling index in central Montreal

Location:	WEC	McG	CBC
Mean annual soiling index:	1.97	1.07	0.42

The Seasonal Cycle

The seasonal cycle of soiling index is shown in Figure 2. Each point is the mean monthly value and the vertical line gives the absolute range between the highest and lowest monthly average during the 3- or 4-year record.

Table 2 Comparison of mean annual soiling index in Montreal with other Canadian cities. Source: Munn and Ross, (1961); Katz (1963)

City	District	Soiling Index
Ottawa (WRB)[a]	Commercial	2.2
Montreal (WEC)	Residential-commercial	2.0
Windsor		1.8
Ottawa (SVH)[b]	Commercial	1.7
Vancouver	Commercial	1.1
Montreal (McG)	Residential-commercial	1.1
Winnipeg	Central business	0.8
Harrow, Ont.		0.6
Montreal (CBC)	Elevated park	0.4
Winnipeg	Residential	0.4

[a] WRB—War Service Records Building
[b] SVH—St. Vincent's Hospital

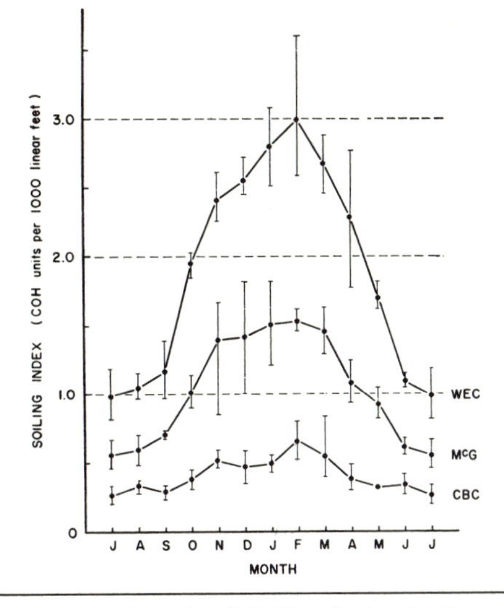

Figure 2 Seasonal cycle of Soiling Index in Central Montreal.

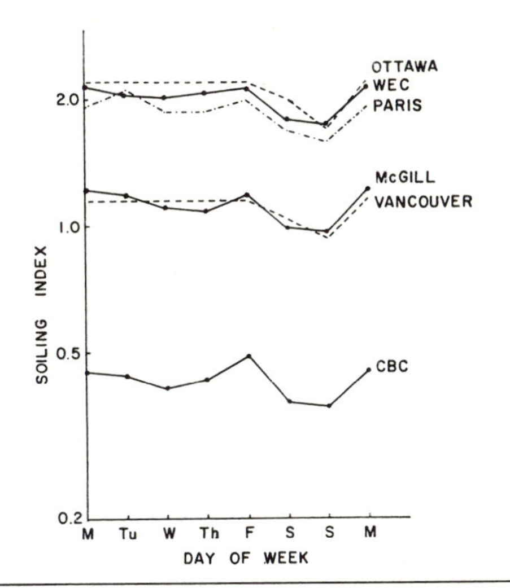

Figure 3 Weekly cycle of Soiling Index in Montreal compared to three other cities.

All three stations have a similar seasonal trend. The lowest soiling index occurs in the summer months and the highest readings in the winter months. Both the spring and the fall can be considered as periods of rapid transition between the two main regimes.

The Weekly Cycle

The weekly cycle of pollution has been studied in many other cities. In Leicester, England (Department of Scientific and Industrial Research, 1945), the ratio of both the smoke and SO_2 content of atmosphere on Sundays and bank holidays compared to other days averaged between 0.55 and 0.87, depending on the season. A year-round reduction in smoke on both Saturday and Sunday has been observed in Ottawa and Vancouver (Munn and Ross, 1961; Munn, 1961). In Paris, a significant reduction in smoke occurs on Saturday, with a further reduction on Sunday (Grisollet and Pelletier, 1957). Twenty years of measurements taken in downtown Toronto show a 2.8% increase of incoming solar radiation on Sunday compared to weekdays (Mateer, 1961).

A comparison of the three Montreal stations with Paris, Ottawa, and Vancouver on a year-round basis is shown in Figure 3. Note that the soiling index is shown on a logarithmic scale to emphasize the similar shape of the curves. There is some tendency for highest readings to occur at the beginning of the week (Monday or Tuesday) and again on Friday. All cities show a similar reduction on Sunday amounting to between 17 and 23% of the mean weekday value, and a slightly smaller reduction on Saturday. For all three Montreal stations the differences among weekdays is statistically insignificant, but the reduction on both Saturday and Sunday is statistically significant at the 97%, or greater, confidence level. Thus the shutting down of many industries and most commercial activities does lead to a significant reduction in downtown smokiness over the weekend.

The Weekday Daily Cycle

Since there is a significant reduction in the soiling index on the weekend, the daily cycle will be considered for weekdays only. Combining all weekdays, a sufficiently large sample of data can be obtained to study the daily cycle on a month-by-month basis. Saturday and Sunday data can be grouped separately on a seasonal basis, but only in the summer are there sufficient data with a small enough standard deviation to make meaningful comparisons. These data will be considered later.

Figure 4 shows the weekday diurnal variations of soiling index at each of the three stations in Montreal for each month of the year. Both WEC and McG have the same basic well defined daily cycle throughout the year but with important changes in both amplitude and times of the maxima from month to month. This basic cycle consists of:

Two main maxima—one shortly after sunrise and another near or just after sunset.

Two main minima—one during the early morning hours and another during the early afternoon.

A secondary maximum between 2100 and 0000 EST from October through May.

The late evening peak may also exist during the summer, but is not detected due to the lower time resolution of a 2-hour sampling period and because the main maximum also occurs later in the evening.

The daily cycle of soiling index at CBC is not nearly

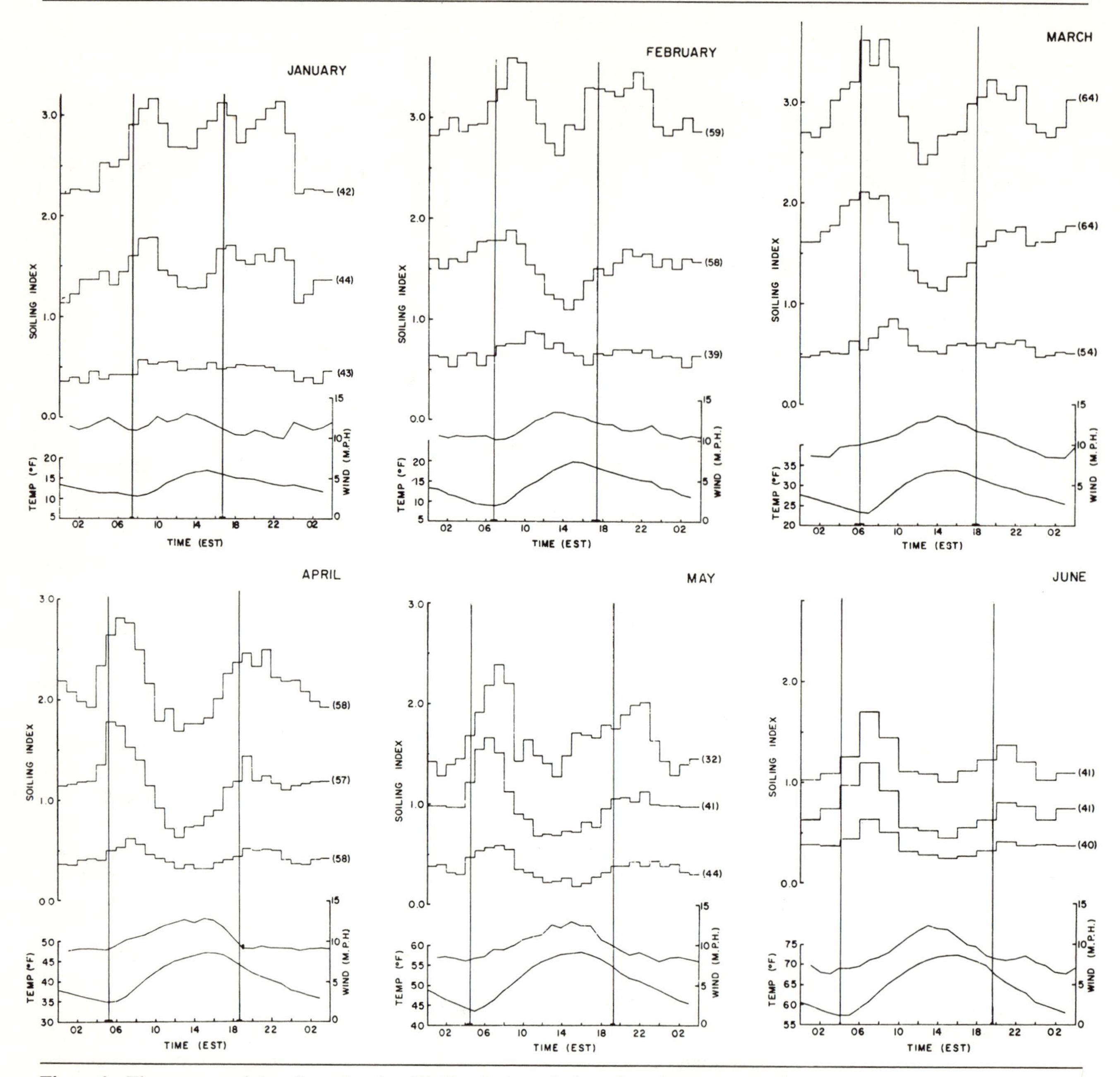

Figure 4 The mean weekday diurnal cycle of Soiling Index, windspeed, and temperature for each of the months January through December. On each of the above figures the upper histogram is for WEC, the middle one for McG, and the lower for CBC. The number in parenthesis at the right-hand end is the number of days data used. The bottom two curves show the diurnal variation of wind and temperature, each curve being attached to its appropriate scale. The range of time of sunrise and sunset for the month is indicated by the horizontal bar on the abscissa, and the vertical line gives the mean time for the month.

Table 3 Average time lag of morning soiling index maximum at CBC

Season	Behind McG, hr	Behind WEC, hr
Winter	$2\frac{1}{2}$	2
Transition	$1\frac{3}{4}$	$1\frac{1}{4}$
Summer	1	$\frac{1}{2}$

so well defined. The only significant point to show up in all months of the year is a rather poorly defined morning maximum. An interesting feature of this morning peak at CBC is that it lags behind the peak at the lower elevations by as much as 3 hours in some months. This time lag is evaluated for each of the seasons, and Table 3 shows that the lag is greater in winter than in summer.

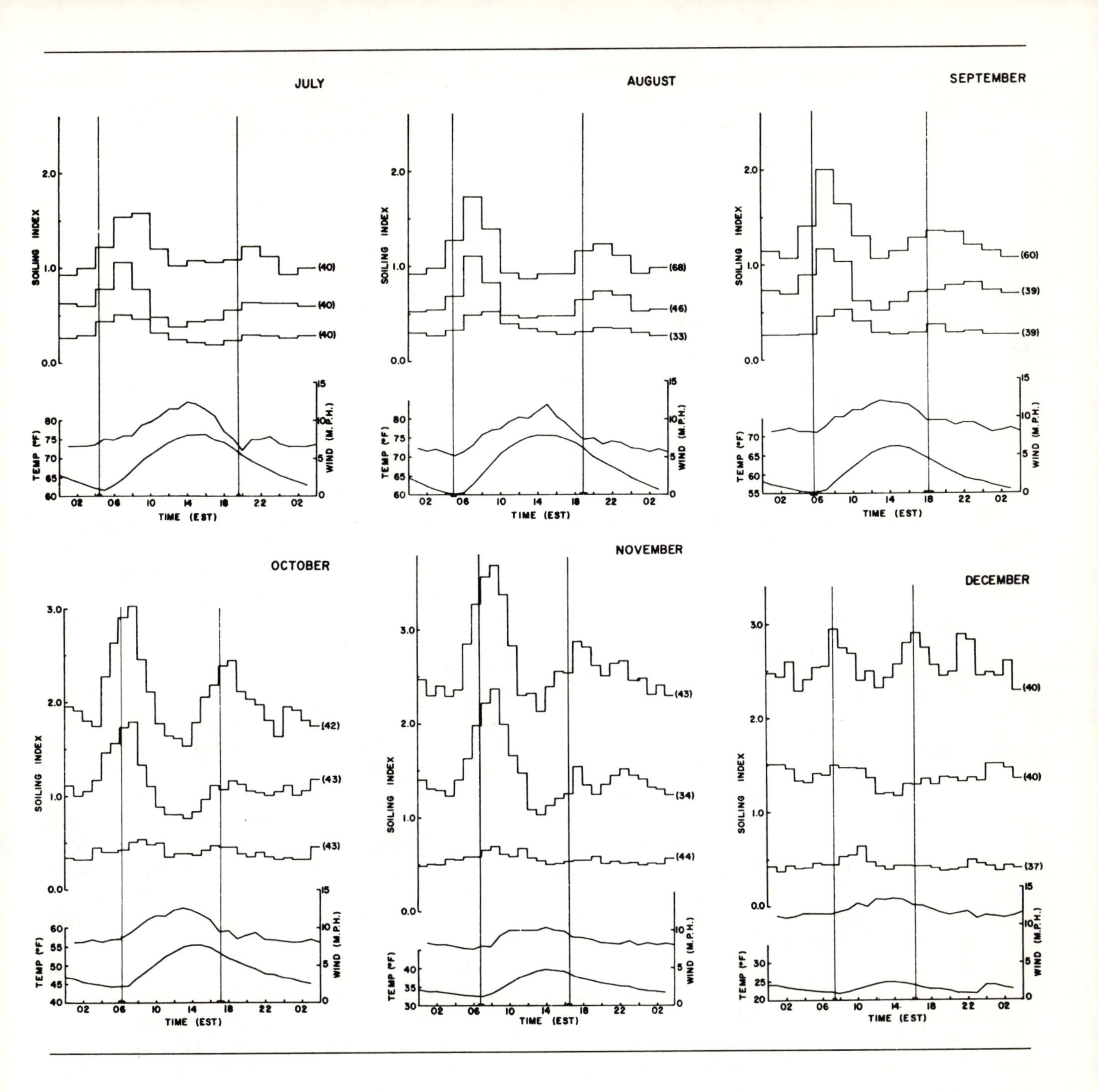

Further inspection of Figure 4 shows that the daily range at both WEC and McG has a marked yearly cycle. Also it can be seen that the morning and evening peaks are approximately of equal value in the winter months, but in the summer the morning peak is more prominent. It therefore appears that the months with a small diurnal range correspond to months with a small difference between the morning and evening peaks. This then suggests two basic types of daily smoke pollution cycle:

Type A—Small daily range (amplitude at WEC <45%, at McG <60%). Small difference between magnitude of morning and evening peaks.

Type B—Large daily range (amplitude at WEC

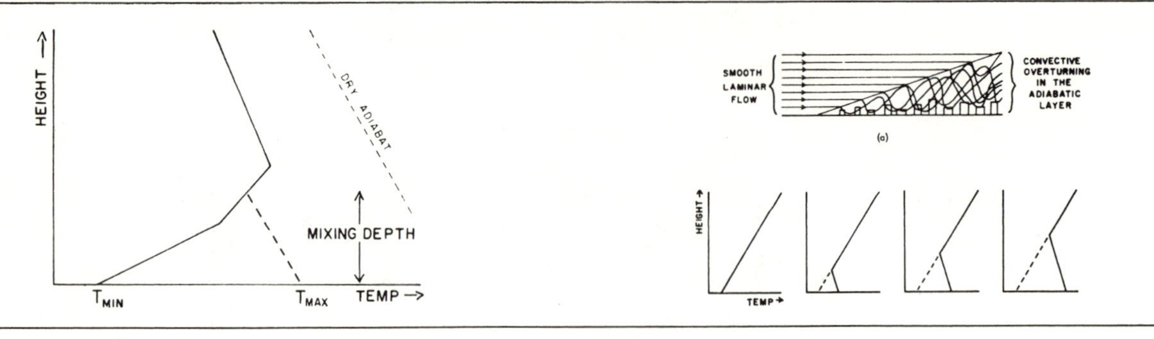

Figure 5 Calculation of maximum mixing depth.

Figure 6 Schematic representation of the build-up of the adiabatic mixing layer, and successive stages in the modification of the vertical temperature profile as the air moves in over the city from left to right.

>45%, at McG >80%). Morning peak considerably higher than evening peak (by >20% at WEC, by >25% at McG).

Type A occurs during the winter months December through March, and Type B during the remainder of the year. The change from Type B to Type A is very sudden in the late fall. In the spring the change from A back to B is more gradual.

The Weekend Daily Cycle

The diurnal variations of soiling index at McGill on weekdays compared to Saturday and Sunday during the summer months are shown in Figure 8. The shapes of these curves and the relative magnitude are almost identical with those for WEC (Summers, 1962). The mean summer weekday soiling index is 0.69, on Saturday it is 0.52 (a reduction of 25%), and on Sunday it is 0.47 (a reduction of 32%). Both of these reductions are statistically significant at the 99.5% confidence level.

The general shape of the Saturday curve is very similar to the weekday curve with the reduction spread fairly uniformly over the whole 24 hours. This is due to the reduction in many industrial and commercial activities on Saturday. Another interesting feature is the complete disappearance of the morning peak on Sunday morning when man's activities are at a minimum.

The Mixing Depth Concept

Daytime Vertical Mixing

During anticyclonic weather the lowest several thousand feet of the atmosphere are stable due to subsidence. This stability is reinforced overnight by the formation of intense surface based nocturnal inversions of temperature in the lowest few hundred feet of the atmosphere. Daytime solar heating destroys at least the lowest part of this inversion and strong vertical mixing takes place through a so-called mixing depth. The depth of the mixed layer can be estimated from the tephigram by following the dry adiabat through the surface temperature up to the point where it intersects the upper air sounding curve. The estimated maximum mixing depth is calculated as above using the maximum temperature, as illustrated schematically in Figure 5. This idea was first used in the Los Angeles area, and then during a prolonged anticyclonic spell over the Eastern United States in October 1956 (Pack and Hosler, 1958). The idea was extended to climatological studies of air pollution potential (Holzworth, 1962).

The nearest radio-sonde station to Montreal is at Maniwaki, about 130 miles to the northwest. The mean monthly maximum mixing depth was calculated from the Maniwaki data, which can be taken as typical of conditions over southwestern Quebec and compared to data from San Antonio, Texas and Los Angeles (Holzworth, 1962).

In Los Angeles, the maximum mixing depth is lowest during the late summer and fall when the subsidence inversion due to the "Pacific High" cell is most intense. At Maniwaki the daytime mixing depth is least during the winter months December through March when the ground is frozen or snow covered.

Nighttime Vertical Mixing

A large city acts as a heat source at night, and therefore stable air moving over the city is heated below, leading to the development of a shallow adiabatic mixing layer. This is illustrated in Figure 6. The modification of the vertical temperature profile is shown in Figure 6. This is similar to Figure 5, except that T_{min} is replaced by the surface temperature in the country and T_{max} by the increased surface temperature over the city. This model is developed in full elsewhere, but the major results are summarized below.

Assuming that the excess radiational loss is small compared to the heat input (H), and making the simplifying assumption that the wind (u) is constant

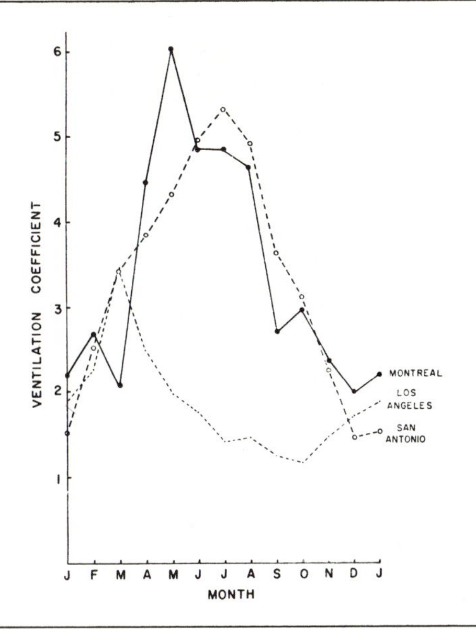

Figure 7 Monthly variation of the afternoon ventilation coefficient.

with height, then the height of the mixing depth (h) at the center of the city is given by:

$$h = \left[\frac{2HL}{u\alpha\rho c_p}\right]^{0.5} \quad (1)$$

where L is the distance from the edge to the center of the city along the wind direction, and α is the difference of the country air lapse rate from the dry adiabatic. [c_p is the specific heat of air at constant pressure and ρ is the density.]

If the smoke emitted into the atmosphere over the city is uniformly mixed in the vertical within this mixing depth, then the steady-state equilibrium concentration of smoke (C) at the center of the city is given by:

$$C = \frac{QL}{uh} \quad (2)$$

where Q is the rate of smoke production, and h is given by eq. 1. This same relation has been obtained from a simpler model (Smith, 1961), but h and L have slightly different interpretations. Combining eqs. 1 and 2:

$$C = Q\left[\frac{L\alpha\rho c_p}{2Hu}\right]^{0.5} \quad (3)$$

Thus the smoke concentration is directly proportional to the rate of production, proportional to the square root of the size of the city, proportional to the square root of the stability (as measured by α), and inversely proportional to the square root of the wind speed.

Ventilation Coefficient

For any given city h and u are the direct meteorological factors which determine the equilibrium smoke concentration. The product (uh) in the denominator of eq. 2 is a measure of the rate at which clean air is being brought in at the upwind end of the city, or a ventilation coefficient. The smoke concentration is therefore inversely proportional to the ventilation coefficient.

The product of the mean monthly maximum mixing depth and the mean wind speed gives a measure of the seasonal variation in the maximum daytime ventilation shown in Figure 7. Ventilation is at a minimum in Los Angeles during the late summer and fall, which combined with the long hours of sunshine, produces the well known photochemical smog problem. In the Montreal area, daytime ventilation is at a minimum during the winter, coinciding with the time of year when the production of most forms of pollution is at a maximum. However the stronger winds at this time of year prevent the ventilation coefficient being reduced below the lowest values occurring in Los Angeles.

DISCUSSION

The ideas of the previous sections can now be applied to describe, at least subjectively, the main features of the diurnal variations shown in Figure 4.

At the time of the early morning minimum in soiling index, which occurs between 0000 and 0400 EST, conditions appear to be in equilibrium, with the rate of production of smoke exactly balanced by the rate of ventilation.

The nighttime mixing layer is well established at this time, and since there is no abrupt change in meteorological conditions, the rise in soiling index which commences some time between 0300 and 0600 EST, and always before sunrise, must be due to an increase in smoke production. Many industries preparing for the day's activities will be turning up their furnaces, as will be the early risers. As the city gradually comes to life, more and more smoke sources become operative and smoke concentrations continue to rise until a peak is reached just after sunrise. Now the solar radiation acts as a strong heat source causing the surface temperature, and hence the height of the mixing depth, to rise rapidly. Since there is now no further increase in smoke production, the concentration will therefore begin to decrease. This decrease in concentration will continue

Season	h, ft	u mph	Ventilation coefficient $(uh) \times 10^{-4}$
Winter	344	10.8	0.37
Transition	332	9.8	0.33
Summer	250	8.5	0.21

Table 4 Mean seasonal nighttime mixing depth, wind speed, and ventilation coefficient at McG during the 3 hours preceding the morning peak in soiling index (based on period March 1962 to April 1963)

until the maximum mixing depth occurs in the afternoon (Holzworth, 1962). Then, toward sunset, vertical mixing decreases and the lowest layers in the atmosphere begin to stabilize in the country air. This will mean a limited mixing depth begins to form over the city again and the soiling index begins increasing until man's daytime activities start to close down. Smoke production continues to decrease slowly in the late evening as the nighttime mixing layer becomes well established, and so the cycle starts all over again in the early morning. These considerations explain why the evening soiling index peak is relatively much higher in winter than in summer—because in winter ventilation decreases before production, while in summer the reverse is true.

Also from Figure 7 and Table 4 it can be seen that the difference between nighttime and daytime ventilation is much greater in summer than in winter, hence the greater diurnal range in soiling index in summer.

The second evening maximum in soiling index is interesting. Since there is no apparent discontinuity in surface meteorological conditions at this time, a temporary increase in production is a possible explanation. One unique feature of Montreal is the very large number of apartment buildings in the heavily built-up central area. Each one of these has its own incinerator

Figure 8 Diurnal variations of Soiling Index at McGill on weekdays compared to Saturday and Sunday during the summer months.

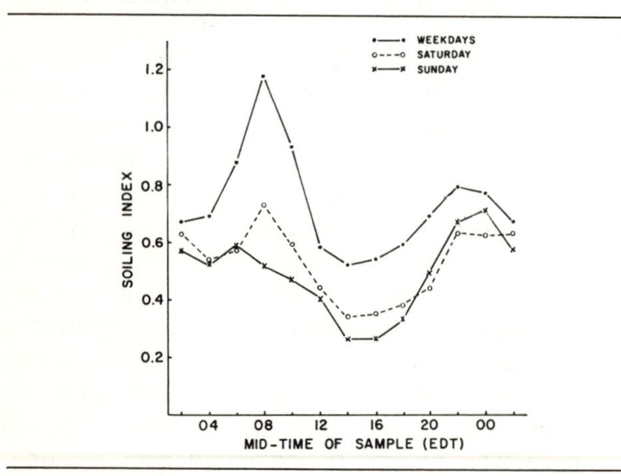

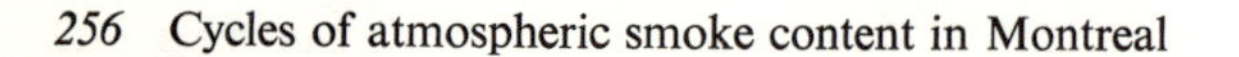

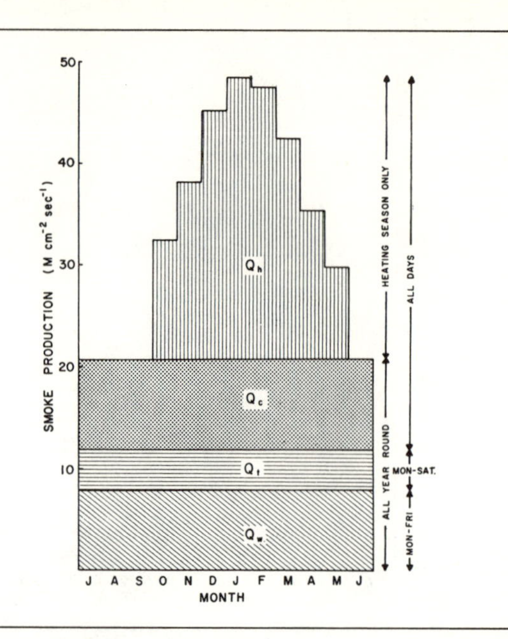

Figure 9 Diagram showing the contribution, from various smoke source categories, to the morning Soiling Index peak at McGill.

which is usually fired up during the late evening. There is also the possibility that the second evening maximum is related in some way to the development or decay of the nocturnal low-level jet stream.

If at all times the nighttime mixing depth included the top of Mt. Royal, then the morning peak at CBC would be expected to occur at about the same time as at either McG or WEC. However on many occasions the top of the smog layer is below CBC (Summers, 1965). Thus the maximum soiling index at CBC will occur as the top of the mixing layer rises through this level, which will occur more rapidly in summer than in winter, Hence the shorter average time lag between the maximum at CBC and the maximum at the lower elevations in the summer months.

Meteorologically, Sunday is no different from any other day of the week, and so the lack of a morning peak in Figure 8 is due to the absence of the increase in smoke production which occurs on other days. In fact the data contained in Figures 3 and 8 have been analyzed together with meteorological data to estimate the contribution from several smoke source categories to the total smoke content of the air at McG, utilizing the simple nighttime ventilation model described earlier. The method used is described in full elsewhere (Summers, 1965), but the final result is shown in Figure 9, where the various smoke source categories are defined as follows:

Q_h = production of smoke from fuel used for *heating* of buildings (note heating season in Montreal is October to May inclusive).

Q_c = production of smoke from *continually* operating

industrial and commercial establishments, water heating and incinerators.

Q_w = production due to *weekday* only commercial and industrial operations.

Q_t = production due to vehicular *traffic*.

To summarize: The hypothesis of a continuous nighttime "fumigation" process over a large urban area is proposed, in which case the morning peak in soiling index can be explained by an increase in smoke production as the city comes to life. The seasonal, weekly, and daily cycles of smoke pollution in Central Montreal combined with this simple urban ventilation model yield a useful estimate of the contributions from various smoke source categories. The concept of a ventilation coefficient is introduced and its seasonal variation gives some insight into the various types of diurnal variations of soiling index observed in Montreal.

REFERENCES

Denison, P. J., Power, B. A., and Summers, P. W., 1960. Analysis of air pollution levels in Montreal related to meteorological variables. *Presented Can. Branch Royal Meteorol. Soc.* (unpublished).

Dept. Scientific and Industrial Research, 1945. Atmospheric pollution in Leicester—A scientific survey. *Dept. Scientific and Industrial Research, Atmos. Poll. Res. Tech. Paper No. 1.* HMSO, London.

Grissollet, H. and Pelletier, J., 1957. La pollution atmospherique au centre de Paris et ses relations avec quelques facteurs climatologique. *La Meteorologie 393.*

Hemeon, W. C. L., Haines, G. F., jr., and Ide, H. H., 1953. Determination of haze and smoke concentrations by filter paper samplers. *J. Air Pollution Control Assoc.* (*J-APCA*) *3*, 22–28.

Holzworth, G. C., 1962. Some physical aspects of air pollution. *Seventh Conf. on Ind. Hygiene and Air Poll.* Univ. Texas, Austin.

Katz, M., 1961. Air pollution as a Canadian regional problem. *Resources for Tomorrow Conference*, Montreal.

Katz, M., 1963. Air pollution in Canada—current status report. *Amer. Journ. Pub. Health 53*, 173.

Katz, M., Sanderson, H. P., and Ferguson, M. B., 1958. Evaluation of air-borne particulates in atmospheric pollution studies. *Analyt. Chem. 30*, 1172–1180.

Landsberg, H. E., 1962. City air—better or worse. *Symposium Air over Cities, SEC Tech. Report A62-5*, 1.

Mateer, C. L., 1961. Note on the effect of the weekly cycle of air pollution on solar radiation at Toronto. *Int. Journ. Air and Water Poll. 4*, 52–54.

Munn, R. E., 1961. The interpretation of air pollution data, with examples from Vancouver. *CIR-3454, TEC-351*, Met. Branch, Dept. Transport, Canada.

Munn, R. E. and Ross, C. R., 1961. Analysis of smoke observations at Ottawa, Canada. *J-APCA 11*, 410.

Pack, D. H. and Hosler, C. R., 1958. A meteorological study of potential atmospheric contamination from multiple nuclear sites. *2nd U.N. Geneva Conf. on Peaceful Uses of Atomic Energy. A/Conf. 15P/426*, 265.

Sanderson, H. P. and Katz, M., 1963. The optical evaluation of smoke or particulate matter collected on filter paper. *J-APCA 13*, 476–482.

Smith, M. E., 1961. The use and misuse of the atmosphere. *Adv. Papers, Int. Symposium on Chem. Reactions in the Lower and Upper Atmosphere, 273.* San Francisco.

Stalker, W. W., Park, J. C., and Keagy, D. M., 1960. Developments and use of the AISI Automatic Smoke Sampler. *J-APCA 12*, 303–306.

Sullivan, J. L., 1962. The calibration of smoke density. *J-APCA 12*, 474–478.

Summers, P. W., 1962. Smoke concentrations in Montreal related to local meteorological factors. *Symposium Air over Cities, SEC Tech. Rept. A62-5*, 89.

Summers, P. W., 1965. An urban heat island model; its role in air pollution problems, with applications to Montreal. *1st Canadian Conf. on Micromet.* Toronto.

Robert E. Rankin

22 *Air pollution control and public apathy*

The importance of public involvement in programs of air pollution control has been stressed by a number of writers in the field, most recently by Sterner (1967) in his comments to the Third National Conference on Air Pollution. As he points out, public apathy appears to be a critical barrier in the development and implementation of effective control programs. Along the same line, Dixon and Lodge (1965), in a digest of the report of the AAAS Air Conservation Committee, state that, "The best program in the world can fail in the face of opposition or apathy on the part of the public." It is the purpose of this paper to take a more analytical look at this problem, particularly in the light of recent data from an opinion survey in West Virginia. If such apathy exists, effective countermeasures require that we examine it more fully and understand its sources. Collective handwringing will do little to meet the problem.

Although a relatively young field, there have now been a number of studies assessing public attitudes and opinions on the subject of air pollution. Much of this research has attempted to document the level of public awareness and concern, with some attention to the perception of causes and effects. While, as de Groot (1967) has shown, the percent of the population considering air pollution a problem varies rather directly with the level of pollution in the neighborhood, a relatively stable 65% of various metropolitan populations have reported awareness of the problem. In a recent study by the author (1968), attitudes of residents of Charleston, West Virginia, and three smaller communities in the industrialized segment of the Kanawha Valley were surveyed. Approximately 900 persons within the corporate limits of Charleston and 500 from the three outside communities were randomly selected from lists and personally interviewed. Interviews were conducted during the summer and fall of 1966.

Reprinted, with minor editorial modification, by permission of the author and editor, from *Journal of the Air Pollution Control Association*, Vol. 19, 1969, pp. 565–569.

Two kinds of questioning provided an indication of the degree of awareness and concern evidenced by the several populations. In the first of these, respondents were asked, "As you see it then, is there any air pollution in (city or town)? How about here in your neighborhood? Would you say that there is air pollution around here?" Table 1 presents the answers to both of these questions. The second source of data was a question borrowed from the St. Louis study (Schusky, 1965), and used to probe the seriousness with which Kanawha Valley residents viewed air pollution relative to other community problems. This question was worded as follows:

> I will mention a few problems which different communities are facing. How would you rate each of these for (city or town)? First, recreation areas and programs—would you rate these very serious, somewhat serious, or not serious?

The other problems brought up were unemployment, air pollution, race, juvenile delinquency, traffic and parking, and medical care for low-income families. The respondent was also given an opportunity to mention any other problem which he considered serious, although few did. The same question was then asked with regard to the respondent's own neighborhood—"right here where you live, instead of the entire city (town)." Table 2 gives the responses to this question, both for the entire community and the neighborhood.

Two points are worth noting from these data. First there is important confirmation of de Groot's observation, based on the Buffalo and St. Louis studies, that people are less likely to perceive air pollution as a problem in their neighborhood than they are in the community-at-large. This appears to be the case even where conditions of pollution are relatively homogeneous throughout the area. The consistency of this finding suggests that such differential perception may be a major, and relatively universal, means of coping with the threat of air pollution. As such it could be an

	Charleston %	So. Charleston %	Nitro %	Montgomery %
Air pollution in city				
Yes	90.9	94.4	97.8	94.0
No	5.3	4.9	0.6	4.8
Don't know	3.8	0.7	1.6	1.2
Air pollution in neighborhood				
Yes	70.8	89.6	84.7	92.8
No	25.4	9.7	15.3	6.0
Don't know	3.8	0.7		1.2
N[a]	918	203	196	102

[a] N equals size of sample. Except where indicated Ns are the same throughout.

Table 1 Is there any air pollution in (city or town)? In your neighborhood?

Problem	Charleston % City	Charleston % Neighbor.	So. Charleston % City	So. Charleston % Neighbor.	Nitro % City	Nitro % Neighbor.	Montgomery % City	Montgomery % Neighbor.
Recreation								
Very serious	46.3	39.3	22.1	26.6	11.5	10.5	53.1	51.3
Somewhat serious	32.6	25.0	48.3	28.0	43.7	40.3	30.9	33.3
Not serious	13.1	28.4	23.4	42.7	38.8	45.9	12.3	11.5
Don't know	7.4	7.3	6.2	2.8	6.0	3.3	3.7	3.9
Unemployment								
Very serious	19.7	10.7	4.8	4.2	2.2	1.6	21.0	19.5
Somewhat serious	26.5	13.0	20.0	10.4	25.5	23.8	33.3	28.6
Not serious	38.8	63.4	59.3	75.5	63.1	70.7	34.6	42.8
Don't know	15.0	12.9	15.9	9.8	9.2	3.9	11.1	9.1
Air Pollution								
Very serious	66.7	35.2	84.8	61.0	78.7	55.8	82.7	79.5
Somewhat serious	20.1	29.8	12.4	30.5	19.1	37.0	11.1	12.8
Not serious	9.7	32.5	2.1	8.5	2.2	7.2	5.0	6.4
Don't know	3.4	2.5	0.7	0.0	0.0	0.0	1.2	1.3
Race Problems								
Very serious	6.0	2.8	0.7	0.0	0.0	0.0	3.7	1.3
Somewhat serious	20.9	7.0	12.5	7.2	0.0	2.8	9.9	10.2
Not serious	66.8	85.7	81.9	92.8	100.0	97.2	86.4	88.5
Don't know	6.3	4.5	4.9	0.0	0.0	0.0	0.0	0.0
Juvenile Delinq.								
Very serious	14.7	8.4	2.8	2.1	0.0	0.0	2.5	0.0
Somewhat serious	38.8	20.3	38.6	16.8	38.6	30.4	23.4	24.4
Not serious	36.5	63.9	46.2	74.8	58.1	69.1	64.2	70.5
Don't know	10.1	7.3	12.4	6.3	3.3	0.5	9.9	5.1
Traffic and Parking								
Very serious	64.2	28.9	25.5	16.8	4.3	2.8	76.5	70.5
Somewhat serious	22.4	26.2	42.1	28.0	52.2	41.4	14.8	16.7
Not serious	10.7	42.7	30.3	53.8	43.5	55.8	7.4	11.5
Don't know	2.7	2.2	2.1	1.4	0.0	0.0	1.2	1.3
Medical Care–Low Income Families								
Very serious	14.2	8.2	8.3	2.1	0.0	0.0	9.9	5.1
Somewhat serious	20.0	10.6	20.7	10.5	14.7	8.3	16.0	14.1
Not serious	34.0	54.2	31.0	62.9	79.2	81.8	39.5	44.9
Don't know	31.9	27.0	40.0	24.5	6.0	9.9	34.6	35.9

Table 2 How would you rate problems in (city or town) and here in the neighborhood?

expression of a common pattern observed in World War II bombing studies and disaster research—the development of a strong sense of personal invulnerability to major threat. Second, and more directly relevant here, these data show that air pollution is both perceived and considered a "serious" or "somewhat serious" problem by the great majority of the several populations studied. Taken as a whole, around 90% or more appear both aware and concerned. Clearly such results, together with those of earlier studies, indicate that whatever the nature of public apathy, it is not a lack of awareness of the problem.

A related issue concerns public conceptions of pollution control. It appeared to the writer significant that while considerable attention has been focused on public awareness of air pollution there has been relatively little work in the area of control attitudes. As a consequence, a major emphasis in the Kanawha Valley study concerned this issue. Granted that a significant proportion of urban populations is aware of air pollution, and at least moderately knowledgeable about its consequences, the critical question becomes—what can be done about it?

Several types of questioning were utilized to explore attitudes relative to the issue of pollution control. Questioning began with the following:

> Do you think it's possible for air pollution around here to be greatly reduced, or would you say it probably cannot be cut down very much below present levels?

Table 3 gives the replies and shows that a sizable majority in the area feels that pollution can be greatly reduced. Charleston gave the lowest figure (64.5%), while Nitro had the highest (92.9%). A rather significant number in Charleston indicated that they didn't know whether it could be reduced or not.

	Charleston %	So. Charleston %	Nitro %	Montgomery %
Can be greatly reduced	64.5	75.4	92.9	88.2
Cannot be reduced below present	14.5	10.8	4.1	2.0
Don't know	21.0	13.8	3.0	9.8

Table 3 Can air pollution around here be greatly reduced or not?

In order to explore the relationship between the perceived seriousness of air pollution and beliefs in the possibility of control, the data from Tables 2 and 3 were cross-tabulated for the Charleston sample. Table 4 shows the rather interesting result. Of those considering pollution "very serious," 72.4% believe that significant reductions can be made in pollution levels whereas only 39% of those viewing community pollution as "not serious" believe this. The chi square indicates a highly significant relationship—the more serious the rating the more likely the respondent is to feel that control is possible. Whether knowledge or information follows concern or vice versa cannot be

	Seriousness rating, %		
	Very serious	Somewhat serious	Not serious
Can be greatly reduced	72.4	58.3	39.0
Cannot be reduced below present	12.0	17.1	26.8
Don't know	15.6	24.6	34.2
N	635.0	193.0	90.0

$X^2 = 47.29$; df $= 4$; p < 001

Table 4 Possibility of control of air pollution by rating of seriousness in the community (Charleston only)

	Charles-ton %	So. Charleston %	Nitro %	Mont-gomery %
Do something about pollution	93.6	94.6	99.0	93.1
Things all right as they are	6.4	5.4	1.0	6.9

Table 5 Do you think it's a good thing to do something about air pollution in the Kanawha Valley?

	Charles-ton %	So. Charleston %	Nitro %	Mont-gomery %
Yes	10.2	11.8	11.7	7.8
No	77.0	82.3	82.1	82.3
Don't know	12.8	5.9	6.2	9.9

Table 6 Do you think that doing something about air pollution would have any bad effects?

	Charles-ton %	So. Charleston %	Nitro %	Mont-gomery %
Cost of one dollar				
Yes	82.2	91.6	91.3	93.1
No	13.5	7.9	2.0	4.9
Don't know	4.3	0.5	6.7	2.0
Cost of five dollars				
Yes	64.5	83.7	No	80.4
No	27.0	16.3	Data	16.7
Don't know	8.5			2.9

Table 7 Would you be in favor of doing something about air pollution if it cost (one dollar) (five dollars)?

determined from these data, however it is of some interest that those most concerned do not perceive the situation as hopeless.

Granted that pollution can be reduced, the appropriate follow-up appeared to be—should it be reduced? Respondents in all communities were asked,

> Do you think it's a good thing to do something about air pollution in the Kanawha Valley or do you feel that things are all right the way they are and it's better to leave them alone?

By presenting clear alternatives it could be determined whether a significant number would prefer the status quo as opposed to change with whatever risks that might entail. The results rather emphatically supported change, as shown in Table 5. Since it was possible that the rather loaded wording may have contributed to the strong endorsement on the last question, the issue was approached again—this time calling attention to the possible negative consequences of control. The question read, "Do you think that doing something about air pollution would have any bad effects here in the Valley?" Table 6 shows that the more negative phrasing produced greater variability in the responses. Worth mentioning is the fact that when those who thought "doing something" would have bad effects were asked to indicate what these might be, over 90% mentioned economic issues, i.e., loss of jobs, plants leaving, etc. Perhaps more significant however was the fact that in all areas studied over three-fourths of the respondents could think of no bad effects—and this in an Appalachian region where unemployment and economic health have been salient issues in recent years.

Three additional questions were used to assess the incentive value of cleaner air. Two of these were similar to questions asked in other surveys of this type. They were:

> Would you be in favor of doing something about air pollution if it cost you a dollar a year, either in a small tax or higher prices for things you buy? Would you be in favor of doing something if it cost five dollars a year, either in a small tax or higher prices for things you buy?

The replies to both these questions are found in Table 7. While a clear majority would accept either cost, there is less agreement at the five dollar level, especially in Charleston. The third question in this series attempted to strike more directly at the broader economic issue commonly raised in connection with air pollution control. The respondent was asked: "Would you be in favor of doing something about air pollution if it meant that a few people would have to find other work?" While a majority would still favor controls, Table 8 shows that this condition produced the greatest resistance. The percent of "No" and "Don't Know" responses both increase over those given to the two questions concerned with dollar costs.

It is apparent that most respondents in these populations tended to think that something could be done about air pollution and, further, that something should

	Charles-ton %	So. Charleston %	Nitro %	Mont-gomery %
Yes	58.3	73.4	90.3	76.5
No	28.6	20.2	7.6	16.7
Don't know	13.1	6.4	2.1	6.8

Table 8 Would you be in favor of doing something about air pollution if it meant that a few people would have to find other work?

	Charles-ton %	So. Charleston %	Nitro %	Mont-gomery %
Yes	54.1	60.6	70.9	55.9
No	23.9	29.6	25.0	26.5
Don't know	22.0	9.8	4.1	17.6

Table 9 Do you think anything will be done about air pollution here in the Kanawha Valley?

	Charles-ton %	So. Charleston %	Nitro %	Mont-gomery %
Noticeably improved	41.8	44.3	19.9	46.1
About same as now	40.9	40.9	62.2	35.3
Worse than now	14.7	12.3	17.9	15.3
Don't know	2.6	2.5		3.3

Table 10 In the next five years would you judge that the condition of the air in the Valley will be:

be done. What of the public's feeling about the actualities? The issue here concerns their view of the chances that anything will be done about the problem. Two questions focused on this—the first asked, "Do you think anything will be done about air pollution here in the Kanawha Valley?" The second sought to reduce the ambiguity somewhat by specifying a time period and providing three alternatives. This was worded as follows: "In the next five years would you judge that the condition of the air in the Valley will be noticeably improved, about the same as now, or worse than now?" These data are reported in Tables 9 and 10. A significant proportion of those interviewed appeared less than optimistic about the future control of air pollution. Approximately one-fourth felt that nothing will be done while the projections for the next five years are even more gloomy. In most of the communities studied, well over half guess that either things will stay about the same or else become worse. The fact that such an attitude is so widespread may have greater consequences for control programs than the finding that such programs have the endorsement of a very large sector of the population. One might read in this a kind of apathy, perhaps in part the result of past unfulfilled expectations, which could be particularly difficult to overcome. As control programs take shape and become effective it may be difficult, particularly during the early stages, for a segment of the population to accept the fact that things are getting better. Awareness of improvement may become a more important issue than awareness of pollution.

One further line of questioning in the Kanawha Valley study dealt with the respondent's actual knowledge of existing attempts to control air pollution. Four questions dealing with factual topics relative to air pollution control were included. The questions are listed below:

> Would you happen to know whether the State (West Virginia) has ever done anything about cleaning up the air and keeping it clean? Could you say what that was?
>
> How about the federal government—has the federal government ever done anything about air pollution? Do you recall anything about it?
>
> Has anything happened locally—say in the past 10 or 12 months or so—with regard to dealing with air pollution? What was that?
>
> Would you happen to know whether auto manufacturers have done anything about air pollution in the past few years? What have they done?

In all cases the appropriate answer to the question was, of course, in the affirmative. Activity relative to each question had been treated repeatedly in the local media, and in the case of the second and fourth questions, at the national level. Table 11 gives the "Yes," "No," and "Don't Know" responses to each of the above questions as well as the percentages of "Yes" responders and of the total sample correctly identifying the control activity.

It is apparent that the general level of public information was low in all communities studied. All areas considered, from two-thirds to over 90% of the respondents were apparently completely unaware of existing control activities, while in most cases only very small numbers were able to recall any factual information about control. Thus in Charleston, only 7.6% knew anything about federal action while only 13.8% were able to mention devices on autos.

In broad outline then, while most people in the Kanawha Valley felt that air pollution can and should be controlled, they were not especially optimistic about the effectiveness of such control. They showed little awareness or knowledge of current efforts in the area of control. These findings suggest that an area of public apathy does indeed exist, but that it appears related more to questions of what can and will be done about air pollution, than to simple awareness of the existence of the problem. Further evidence supporting this view

Item	Charleston %	So. Charleston %	Nitro %	Montgomery %
State done anything				
Yes	26.9	26.6	14.3	26.5
No	34.3	39.4	70.4	59.8
Don't know	38.8	34.0	15.3	13.7
% of "yes" responses correctly identifying (% of total given in parentheses)	37.3 (10.0)	25.9 (6.9)	21.4 (3.1)	37.0 (9.8)
Federal government done anything				
Yes	22.3	21.2	11.7	33.3
No	32.6	37.4	71.4	51.0
Don't know	45.1	41.4	16.8	15.7
% of "yes" responses correctly identifying (% of total given in parentheses)	34.0 (7.6)	58.1 (12.3)	47.8 (5.6)	47.0 (15.7)
Local action				
Yes	21.5	29.6	7.1	20.6
No	44.1	45.8	78.6	68.6
Don't know	34.3	24.6	14.3	10.8
% of "yes" responses correctly identifying (% of total given in parentheses)	32.6 (7.0)	65.0 (19.2)	28.6 (2.0)	52.4 (10.8)
Auto manufacturers				
Yes	19.3	30.0	3.6	16.7
No	34.4	42.9	85.7	62.7
Don't know	46.3	27.1	10.7	20.6
% of "yes" responses correctly identifying (% of total given in parentheses)	71.7 (13.8)	72.1 (21.7)	71.4 (2.6)	76.5 (12.7)

Table 11 Responses to four information items concerning air pollution control activity.

may be found in the data on complaints. As in prior studies only a small percentage of respondents report having actually complained (about 5% in Charleston). A much greater number however, indicate that they had felt like complaining (in Charleston about 24%). Of greatest interest however were the reasons given for not complaining (even though feeling like it). These are shown in Table 12. Close to half of this group said that it doesn't do any good to complain, while another sizable number didn't know where to complain. In other words, the average citizen, while recognizing the problem, was unfamiliar with what could be done, or what has been done, and appeared apathetic or pessimistic regarding his own role and the likelihood of control.

Reason given	Charleston %	So. Charleston %	Nitro %	Montgomery %
Doesn't do any good to complain	40.7	52.8	66.7	35.3
Didn't know where (or to whom) to complain	26.5	17.1	14.3	47.0
No opportunity to complain	7.3	8.6		5.9
Opposed to complaining	4.5	4.3	4.8	11.8
Problem is being worked on	4.5	2.8		
Didn't bother respondent personally	8.5			
Don't know and other	8.0	14.4	14.2	
N	177	70	21	34

Table 12 Reasons given for not complaining (by those who felt like complaining but didn't)

These findings are of some concern, but hardly surprising. For the typical city-dweller, air pollution exists as a kind of shapeless and impersonal force—at best a nuisance and at worst a serious threat to health and property. Apart from moving out of the area there appears little that he can do but endure it, and moving is often not a meaningful alternative. Unlike many epidemics and other personal threats, there are few actions that he can take as an individual to ward off or lessen the danger. Recent public information programs have focused largely on attempts to increase awareness of the problem, however, data from the Kanawha Valley study, as well as those reviewed recently by de Groot, strongly suggest that such an approach may have reached a point of rapidly diminishing returns.

Since information campaigns have generally emphasized the undesirable consequences of air pollution, e.g., health hazards, property deterioration, etc., research on the role of threat appeals and fear in changing attitudes and inducing public action would seem an appropriate note on which to close. In general such research has shown that the arousal of anxiety or concern over a particular condition or state of affairs is (1) likely to be effective only when followed rather closely by recommended actions which serve to reduce the fear or anxiety, (2) likely to lead to defensive avoidance of the topic when no effective means are available to reduce the threat and (3) more likely to result in following recommended actions when the level of fear elicited by the communication is low or moderate rather than maximal. Such findings carry significant implications for information programs in the air pollution field. While the public shows considerable awareness of pollution, such programs have not, for the most part, been effective in communicating actions which might help reduce the threat. Without such appropriate action-oriented suggestions, continued emphasis on the nature and scope of the threat may have little effect, and indeed, could be self-defeating in the long run. The indications are that when people have no effective response to meet threat, they tend either to avoid the issue, or protect themselves by withdrawing into a shell of personal invulnerability. The apparently widespread tendency, to see pollution in the community but not in one's immediate neighborhood, suggests that such a process may have already begun.

REFERENCES

DeGroot, I., 1967. Trends in public attitudes towards air pollution. *J. Air Pollution Control Assoc. (J-APCA) 17(10)*, 679–681.

Dixon, J. P. and Lodge, J. P., 1965. Air conservation report reflects national concern. *Science 148*, 1060–1066.

Rankin, R. E., 1968. *Air Pollution and the Community Image.* Terminal Rept. to U.S. Public Health Service, Grant AP-00460-01.

Schusky, J., 1965. *Public Awareness and Concern with Air Pollution in the St. Louis Metropolitan Area.* Prepared for the Division of Air Pollution, U.S. Public Health Service. Contract PH86-63-131.

Sterner, J. H., 1967. Effective involvement of the general public in the control of air pollution. *Proc. 3rd National Conf. on Air Pollution.* U.S. Public Health Service Publ. No. 1649. 439–441.

James T. Peterson

23 *The climate of cities: a survey of recent literature*

INTRODUCTION

As metropolitan areas expand, they exert a growing influence on their climate. An increasing amount of scientific literature is being devoted to analyses of data on urban climates, often comparing urban data with data from nearby rural areas to show the differences between "natural" conditions and those influenced by man. These studies also contribute to such areas as the effects of urban environment on health, the influence of meteorological parameters on urban diffusion, and the possible global climatological consequences of increased atmospheric pollution.

The standard review of urban climate consists of the paper by Landsberg (1956) and his supplementary articles in 1960 and 1962. The purpose of this report is to review the recent literature in this field, primarily that in English, to note the areas of agreement and difference with Landsberg's earlier summaries (Table 1) and to point out the aspects of urban climatology for which more detailed information is now available. This survey concentrates on the most frequently discussed aspects of city-country climatic differences: temperature, humidity, visibility, radiation, wind, and precipitation. Also included is a discussion on urban particulate concentrations. A review of urban concentrations of various gaseous pollutants has recently been published by Tebbens (1968) and thus will not be discussed in this report.

Previous summaries of literature on urban climates include an extensive article by Kratzer in 1937 (revised in 1956 and translated into English), containing references to 533 works, and a bibliography by Brooks (1952) listing 249 references. Books by Geiger (1965) on microclimatology and Chandler (1965) on the climate of London, from which several examples are included herein, also refer to a number of other urban studies. Chandler is currently compiling a bibliography on urban climate under the auspices of the World Meteorological Organization (WMO); this bibliography will reference well over 1,000 articles.

TEMPERATURE

Of all the urban-rural meteorological differences, those of air temperature are probably the most documented. That the center of a city is warmer than its environs, forming a "heat island," has been known for more than a hundred years and continues to receive considerable attention in the literature. Many aspects of a heat island have been studied, such as possible reasons for its occurrence; diurnal, weekly, and seasonal variations; relation to city size; and dependence on topography. This section reviews some of these characteristics of urban-rural air temperature relationships.

Nighttime Differences

The fact that a city is warmer than its environs is seen most readily in a comparison of daily minimum temperatures. As Landsberg (1956) pointed out, such comparisons often show temperature differences of 10° F and occasionally differences as great as 20° F. However, since nocturnal temperatures are dependent on topography, a fraction of these differences, sometimes a large fraction, can often be ascribed to terrain features.

Numerous measurements of urban heat islands have been made, frequently by use of automobiles to obtain many observations within a short time period. An example of a London temperature survey associated with clear skies, light winds, and anticyclonic conditions is presented in Figure 1 (Chandler, 1965). This figure shows certain features common to most heat islands.

Reprinted, with minor editorial modification, by permission of the author and the Environmental Protection Agency, from *The Climate of Cities: A Survey of Recent Literature, National Air Pollution Control Administration Publication No. AP-59*, 1969, 48 pp.

Element	Comparison with rural environs
Temperature	
Annual mean	1.0 to 1.5° F higher
Winter minima	2.0 to 3.0° F higher
Relative humidity	
Annual mean	6% lower
Winter	2% lower
Summer	8% lower
Dust particles	10 times more
Cloudiness	
Clouds	5 to 10% more
Fog, winter	100% more
Fog, summer	30% more
Radiation	
Total on horizontal surface	15 to 20% less
Ultraviolet, winter	30% less
Ultraviolet, summer	5% less
Wind speed	
Annual mean	20 to 30% lower
Extreme gusts	10 to 20% lower
Calms	5 to 20% more
Precipitation	
Amounts	5 to 10% more
Days with < 0.2 inch	10% more

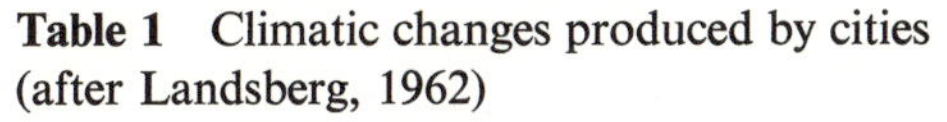

Table 1 Climatic changes produced by cities (after Landsberg, 1962)

The temperature anomalies are generally related to urban morphology. The highest temperatures are associated with the densely built-up area near the city center; moreover, the degree of warming diminishes slowly, outward from the city's heart, through the suburbs and then decreases markedly at the city periphery. The effect of topography is also evident in this example. Urban warming is reduced along the Thames River, in the smaller nonurbanized valleys, and near the city's higher elevations.

Steep temperature gradients at a city's edge have been measured during clear, calm conditions at Hamilton, Ontario (Oke and Hannell, 1968), and Montreal, Quebec (Oke, 1968). These investigators found temperature changes of 3.8 and 4.0° C km^{-1}, respectively, which they regarded as typical values for moderate to large cities.

Some recent studies have indicated that the mean annual minimum temperature of a large city may be as much as 4° F higher than that of surrounding rural areas. Chandler (1963, 1966) reported on two studies of London, which showed differences of 3.4 and 4.0° F in mean minimum temperatures at urban and rural sites. The first study was based on data from 1921 to 1950 for several stations in and around London; the second compared 1959 data for one downtown and one rural location and applied a correction for the difference in elevation at the stations. In another study, Woollum (1964) and Woollum and Canfield (1968) presented data for several stations in the vicinity of Washington, D.C., for a 20-year period; mean minimum temperatures for each season were approximately 4° F higher in downtown areas than in outlying regions.

Although the city heat island as indicated by minimum temperatures can be readily detected year-around, the investigations in London by Chandler (1963, 1966) and in Reading, England, by Parry (1966) indicated that the greatest temperature differences occur in summer or early autumn. Woollum (1964) also found that the mean differences between the warmest and coldest stations of his network were greatest in fall and summer, but that the greatest extreme differences between these stations occurred in winter (see also Landsberg, 1956).

Daytime Differences

The heat island of a city can be detected during the day, but much less readily than during the evening. The slight daytime temperature differences observed are often difficult to distinguish from those due to the effects of topography. In some instances daytime city temperatures may even be lower than those of the suburbs. For example, Landsberg (1956) presented 1 year of data from city and airport observations in Lincoln, Nebraska, a location essentially free from complicating terrain factors. Daily maxima in the cold season showed little difference between the two sites. During the warm season, however, the airport was more frequently warmer than the downtown site. Such results are not the rule, however, and Landsberg also points out cases of daily maxima that are higher in the city. Similar examples have been given by Chandler (1963, 1966); his data showed that the annual average maximum temperatures of London were 0.6 and 1.1° C higher than those in the outlying areas. Munn, et al., (1969) also readily detected a daytime heat island at Toronto, Canada, using daily maximum temperatures.

A recent report by the Stanford Research Institute (Ludwig, 1967; Ludwig and Kealoha, 1968) presents perhaps the most comprehensive documentation of urban daytime temperatures to date. The authors made about twelve auto traverses each at San Jose, California; Albuquerque, New Mexico; and New Orleans, Louisiana, during daytimes in the summer of 1966. Although

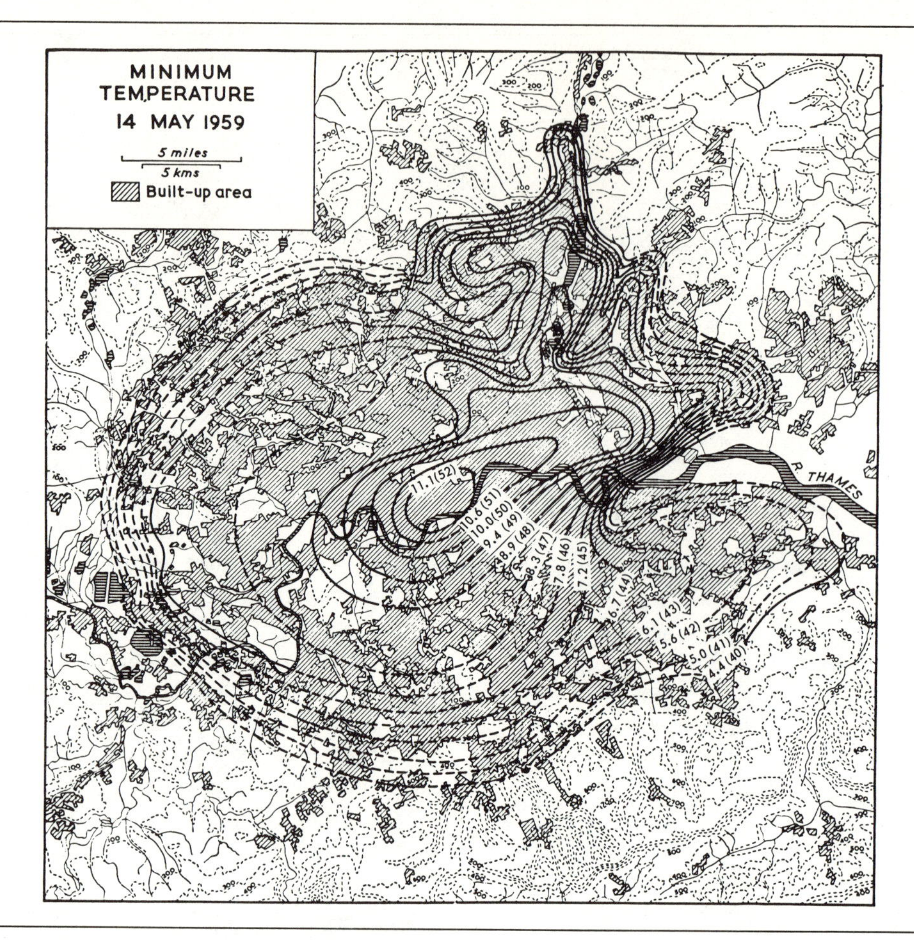

Figure 1 Minimum temperature distribution in London, 14 May 1959, in °C and °F in brackets (from Chandler, 1965).

the temperature anomalies resulting from topographical influences in these cities were greater than those from the heat island, the downtown areas were approximately 0.5° C warmer than the suburbs despite the effects of topography.

In the summer of 1967 SRI extended the study to Dallas, Ft. Worth, and Denton, Texas (population 35,000), where they made 20, 4, and 2 surveys, respectively. When the 1966 and 1967 data were combined with data from three surveys each at Minneapolis, Minnesota, and Winnipeg, Canada (Stanford U. Aerosol Laboratory, 1953a, 1953d, 1953e), and one at London, the investigators found that for these 67 cases the city's warmest part near the downtown area averaged 1.2° C above the typical areas of its environs, with a standard deviation of about 1.0° C. This average value is higher than that observed by other studies, and Ludwig and Kealoha pointed out two possible reasons why investigations of the daytime heat island may underestimate its magnitude. First, they noted that at ground level the highest temperatures of a city do not occur in the central area of tall buildings but rather near that part of the downtown area with "densely packed three- to five-story buildings and parking lots." Second, they noted that temperatures observed at suburban airports were higher than those of a true rural environment, either grass-covered or forested.

The diurnal and seasonal heat island variations discussed above are illustrated in Figure 2, taken from Mitchell (1962). This figure shows hour-by-hour monthly averages of temperature at an urban and a suburban site in Vienna, Austria. The temperature of the city is higher at night during both February and July, but this urban-rural difference is greater in July. In the daytime, however, the city-rural temperature differences are small during July and consistently small and positive during winter.

Annual Differences

The average annual temperatures of a city and its environs, calculated from the daily maxima and minima, also reflect the presence of the urban heat island. Table 2 (Landsberg, 1960) lists the average annual urban-rural temperature differences for several large cities. To this can be added the average value for London of 1.3° C based on the two studies by Chandler (1963, 1966). Although Woollum and Canfield (1968) do not state a specific number for the mean annual urban-rural temperature differences of Washington,

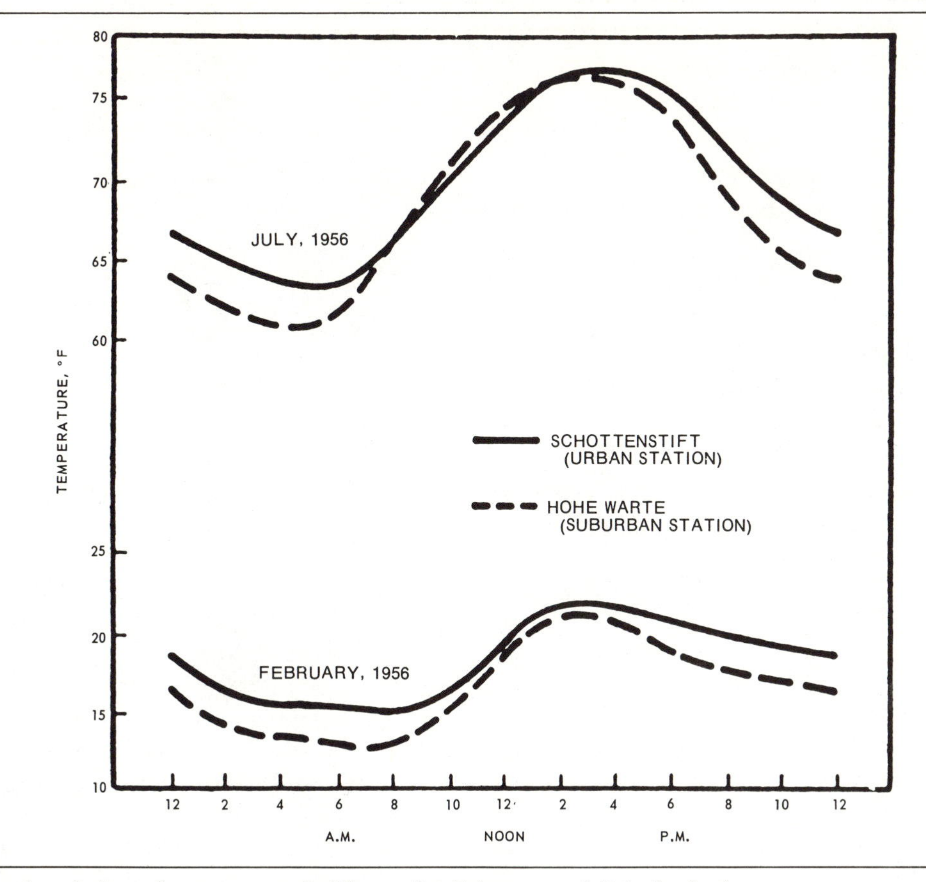

Figure 2 Diurnal variation of temperature in Vienna for February and July for both an urban and suburban station (from Mitchell, 1962).

D.C., their recent data for that city indicate that it should be at least 1.0° C, after consideration of the elevation changes of more than 200 feet within the area.

Effects of City Size

Several urban studies have considered the effect of city size on the magnitude of the heat island. For example, Mitchell (1961, 1962) showed that during this century most major U.S. metropolitan areas have been both expanding and warming. While the temperature increase may be partly due to global climatic conditions, the amount of warming is well correlated with city growth rate. Dronia (1967) compared temperature trends from 67 paired locations around the world, each pair representing an urban and a rural site, usually separated by several hundred kilometers. He found that in the first five decades of this century the urban areas warmed by 0.24° C more than the rural locations. Similarly, Lawrence (1968) noted that from the late 1940's to early 1960's mean daily minimum temperatures at the Manchester, England airport increased by about 2.0° F relative to nearby rural stations as the urban area expanded beyond the airport. Landsberg (1960) gave 30 years of data for Los Angeles and San Diego which showed that as the difference in population between those cities increased so did the difference between their mean temperatures.

The relation between city size and urban-rural temperature difference is not linear, however; sizeable nocturnal temperature contrasts have been measured even in relatively small cities. For example, in more than 20 surveys of Palo Alto, California (population 33,000), Duckworth and Sandberg (1954) found that the maximum temperature difference of the survey area was 4 to 6° F. Hutcheon, et al. (1967) measured the temperature distribution in Corvallis, Oregon (population 21,000), on two occasions and noted a definite heat island, with maximum temperature differences of 13 and 10° F. Sekiguti (1964) observed a heat island in Ina, Japan (population 12,000). Finally, a heat island effect resulting even from a small, isolated building

Table 2 Annual mean urban-rural temperature differences of cities, ° C

Chicago	0.6	Moscow	0.7
Washington	0.6	Philadelphia	0.8
Los Angeles	0.7	Berlin	1.0
Paris	0.7	New York	1.1

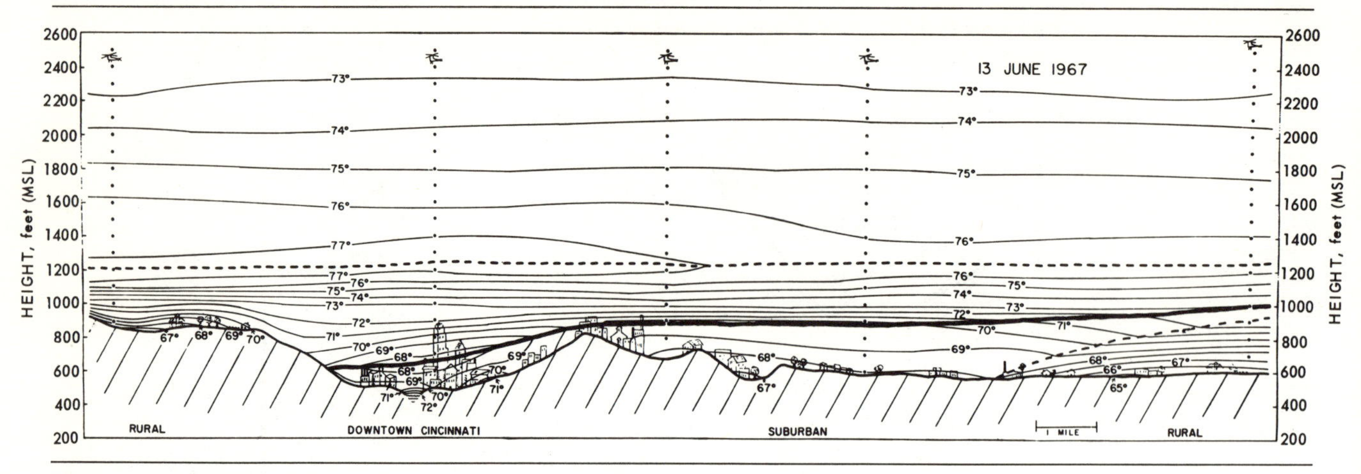

Figure 3 Cross section of temperature (°F) over metropolitan Cincinnati about one hour before sunrise on 13 June 1967. The heavy solid line indicates the top of the urban boundary layer and the dashed lines indicate a temperature discontinuity with less stable air above. Wind flow was from left to right (from Clarke, 1969).

complex after sunset has been detected (Landsberg, 1968). In contrast, during the two daytime surveys at Denton, Texas, Ludwig and Kealoha (1968) reported no appreciable difference in the maximum temperatures at the center of town and at its outskirts.

Although general relationships have been developed between heat island magnitude and some parameter representing city size, be it area, population, or building density, Chandler (1964, 1966, 1967b) has emphasized that the heat island magnitude at a given location often depends strongly on the local microclimatic conditions. He noted (1968) that data from several English towns showed that the strength of the local heat island was strongly dependent upon the density of urban development very near the observation point, sometimes within a circle as small as 500 meters radius. During nights with strong heat islands, the correlation between the heat island and building density was usually greater than 0.9.

Vertical Temperature Profile Differences

Investigators need detailed knowledge about the vertical distribution of temperature near urban areas to accurately determine the dispersion of pollutants. In two recent investigations, one over New York City (Davidson, 1967; Bornstein, 1968) and one over Cincinnati (Clarke, 1969) helicopters have been used to measure the three-dimensional, nocturnal temperature patterns over a city. Besides recording multiple elevated inversions over New York City, Davidson and Bornstein observed that a ground-based inversion was present over the outlying areas, while over the city temperatures were generally higher than those over the countryside from the surface up to about 300 meters. At heights around 400 meters temperatures over the city were generally lower than those of the surrounding area. This finding is similar to those of Duckworth and Sandberg (1954), who noted a "crossover" of the urban and rural temperatures on about half of their wiresonde data.

Clarke has studied vertical temperature distributions in Cincinnati both upwind and downwind of the city. Figure 3 shows an example from one of his surveys. On clear evenings, with light consistent surface winds, he found a strong surface-based inversion upwind of the urban area. Over the build-up region, lapse conditions occurred in the lowest 200 feet, while downwind of the urban area a strong inversion was again observed at the surface. Above this inversion, weak lapse conditions prevailed, which Clarke interpreted as the downwind effect of the city. This "urban heat plume" was detectable by vertical temperature measurements for several miles in the lee of the city.

Other authors have investigated the vertical distribution of nighttime urban temperatures with tower-mounted instruments. DeMarrais (1961) compared lapse rates in Louisville between 60 and 524 feet with typical rural profiles. During the warm half of the year, while surface inversions were regularly encountered in the country, nearly 60% of the urban observations showed a weak lapse rate and another 15% showed weak lapse conditions above a super-adiabatic lapse rate. Munn and Stewart (1967) instrumented towers at 20 and 200 feet in central Montreal, suburban Ottawa, and a rural location near Sarnia, Canada. They noted that inversions occurred more frequently and were stronger over the country than over the city. Both DeMarrais, and Munn and Stewart, however, found little difference between rural and urban daytime temperature profiles.

In another study of the vertical distribution of temperature, Hosler (1961) compiled statistics on the frequency of inversions based below 500 feet above station elevation for selected United States localities and discussed the dependence of these data on season, cloud cover, wind speed, time of day, and geographic location. Since he used radiosonde data, which are usually taken at an airport on the outskirts of a city, his results are generally representative of suburban locations and thus underestimate the inversion frequencies of rural sites and overestimate those of cities.

Importance of Meteorological Factors

In addition to city size the magnitude of the urban heat island has been shown to depend upon various meteorological parameters. An early study of this type was made by Sundborg (1950), who investigated the relation between the temperature difference between Uppsala, Sweden, and its rural surroundings and meteorological variables measured at the edge of the city. He derived two equations by regression analysis for daytime and nighttime conditions at Uppsala, based on more than 200 sets of data. Correlation coefficients between observed and calculated temperatures were 0.49 and 0.66 for day and night, respectively. At night, wind speed and cloud cover were the most important variables for determining the heat island magnitude.

Chandler (1965) made a similar study based on temperature differences of daily maxima and minima at city and country sites for London and meteorological data observed at London airport. The multiple correlation coefficients for the four equations are 0.608, 0.563, 0.286, and 0.114 for nighttime (summer and winter) and daytime (summer and winter), respectively, an indication that the equations are much better estimators for nocturnal than for daytime conditions. The magnitude of the heat island was shown to depend on wind speed and cloud amount at night, whereas no meteorological variables were found to be particularly significant during the day.

Ludwig and Kealoha (1968) estimated the magnitude of a city's heat island by using the near-surface temperature lapse rate, which was usually measured in the environs of the city or at a radiosonde facility of a nearby city. They found this single variable to be highly correlated with the heat island magnitude and thus provided a simple, accurate method for predicting urban-rural temperature differences. These investigators compiled data from 78 nocturnal heat island surveys from a dozen cities and estimated the heat island magnitude by subtracting a typical rural temperature from the highest temperature in the city's center. Examples of their results, stratified by city population, are as follows:

$$\Delta T = 1.3 - 6.78\gamma; \text{ population } < 500{,}000 \quad (1)$$

$$\Delta T = 1.7 - 7.24\gamma \quad \text{,,} \quad 500{,}000 \text{ to } 2 \text{ million} \quad (2)$$

$$\Delta T = 2.6 - 14.8\gamma \quad \text{,,} \quad > 2 \text{ million}, \quad (3)$$

where the lapse rate, γ, is the temperature change with pressure (°C mb^{-1}), i.e., a surface-based inversion is represented by negative γ. Correlation coefficients between ΔT and γ for the three cases are -0.95, -0.80, and -0.87, and the root mean square errors are ± 0.66, ± 1.0, and $\pm 0.96°$ C, respectively. Thus, the resulting equations of this example will usually predict ΔT to within $\pm 2.0°$ C.

Another meteorological parameter that influences heat island development is wind speed. When the regional wind speed is above a critical value, a heat island cannot be detected. Table 3, taken from Oke and Hannell (1968), summarizes several reports on the relation between city population (P) and this critical wind speed (U). Log P and U are highly correlated (0.97), and their relationship is described by the regression equation:

$$U = -11.6 + 3.4 \log P \quad (4)$$

The authors used this equation to estimate the smallest-sized city that would form a heat island. When U = O, equation (4) yields a population of about 2,500. Although they had no data to test this estimate, they recognized that the scatter of their data points increased at the smaller populations and that sometimes even small building complexes produced measurable heat island effects.

The location of the highest city temperatures also depends on local meteorology. Munn, et al. (1969), noted that at Toronto, Canada, the position of the daytime heat island was strongly influenced by the regional and lake breeze windflow patterns of the area and was often displaced downwind of the city center.

City	Author	Year of survey	Population	Critical wind speed $m \cdot s^{-1}$
London, England	Chandler (1962a)	1959–61	8,500,000	12
Montreal, Canada	Oke, et al.[a]	1967–68	2,000,000	11
Bremen, Germany	Mey	1933	400,000	8
Hamilton, Canada	Oke, et al.[a]	1965–66	300,000	6–8
Reading, England	Parry (1956)	1951–52	120,000	4–7
Kumagaya, Japan	Kawamura	1956–57	50,000	5
Palo Alto, California	Duckworth, et al.	1951–52	33,000	3–5

[a] Unpublished.

Table 3 Critical wind speeds for elimination of the heat island effect in various cities

Causes of the Heat Island

It is generally accepted that two primary processes are involved in the formation of an urban heat island, both of which are seasonally dependent (see, for example, Mitchell, 1962). First, in summer the tall buildings, pavement, and concrete of the inner city absorb and store larger amounts of solar radiation (because of their geometry and high thermal admittance) than do the vegetation and soil typical of rural areas. In addition, much less of this energy is used for evaporation in the city than in the country because of the large amount of run-off of precipitation from streets and buildings. At night, while both the city and countryside cool by radiative losses, the urban man-made construction material gradually gives off the additional heat accumulated during the day, keeping urban air warmer than that of the outlying areas.

In winter a different process dominates. Since the sun angle at mid-latitudes is low and lesser amounts of solar radiation reach the earth, man-made energy becomes a significant addition to the solar energy naturally received. Artificial heat results from: combustion for home heating, power generation, industry, transportation, and human and animal metabolism. This energy reaches and warms the urban atmosphere directly or indirectly, by passing through imperfectly insulated homes and buildings. This process is most effective when light winds and poor dispersion prevail.

Many authors have investigated the magnitude of man-made energy in metropolitan areas. In two often-cited older studies for Berlin and Vienna (see Kratzer, 1956), the annual heat produced artificially in the built-up area equalled 1/3 (for Berlin) and 1/6 to 1/4 (for Vienna) of that received from solar radiation. More recently, Garnett and Bach (1965) estimated the annual average man-made heat from Sheffield, England (population 500,000), to be approximately 1/3 of the net all-wave radiation available at the ground. Bornstein (1968) reported results from a similar study of densely built-up Manhattan, New York City. During the winter the amount of heat produced from combustion alone was 2-1/2 times greater than that of the solar energy reaching the ground, but during the summer this factor dropped to 1/6.

In addition to the two seasonal primary causes of heat islands, other factors are important year-around. The "blanket" of pollutants over a city, including particulates, water vapor, and carbon dioxide, absorbs part of the upward-directed thermal radiation emitted at the surface. Part of this radiation is re-emitted downward and retained by the ground; another part warms the ambient air, a process that tends to increase the low-level stability over the city, enhancing the probability of higher pollutant concentrations. Thus, airborne pollutants not only cause a more intense heat island but alter the vertical temperature structure in a way that hinders their dispersion.

Reduced wind speed within an urban area, a result of the surface roughness of the city, also affects the heat island. The lower wind speeds decrease the city's ventilation, inhibiting both the movement of cooler outside air over the city center and the evaporation processes within the city.

Tag (1968) used a numerical model of the energy balance at the atmosphere-ground interface to investigate the relative importance of albedo, soil moisture content, soil diffusivity, and soil heat capacity on urban-rural temperature differences. The author found that during the day the lower city values of albedo and soil moisture caused higher urban temperatures, whereas the higher diffusivity and heat capacity of the city surface counteracted this tendency. At night, however, the relative warmth of the city was primarily the result of the higher urban values of soil diffusivity and heat capacity.

HUMIDITY

Even though little research on humidity has been done, the consensus of urban climatologists is that the average relative humidity in towns is several percent lower than that of nearby rural areas whereas the average absolute humidity is only slightly lower in built-up regions. The main reason to expect differences in the humidity of urban and rural areas is that the evaporation rate in a city is lower than that in the country because of the markedly different surfaces. The countryside is covered with vegetation, which retains rainfall, whereas the floor of a city is coated with concrete, asphalt, and other impervious materials that cause rapid run-off of precipitation. Although the city's low evaporation rates result from the shortage of available water and the lack of vegetation for evapotranspiration, some moisture is added to urban atmospheres by the many combustion sources.

Variations of relative humidity within metropolitan areas resemble those of temperature, since the spatial temperature changes of a city are significantly greater than those of vapor pressure. Thus, because of the heat island, relative humidities in a city are lower than in the suburbs and outlying districts. The humidity differences are greatest at night and in summer, corresponding to the time of greatest heat island intensity (Chandler, 1967a). Other studies have yielded similar findings. Sasakura (1965) reporting on 1 year of data from Tokyo showed that the mean relative humidity in the city center was 5% lower than that in the suburbs, a value that concurs with Landsberg's average figure. Chandler (1965) also found a 5% difference in the relative humidity of a downtown and a rural site near London. In other work (1962b, 1967a) he has presented nocturnal relative humidity profiles across London and maps of spatial distribution for Leicester, England. These show the dependence of relative humidity variations on the form of the city's heat island, which in turn depends upon the density of the built-up complex. Typical humidities of 90 to 100% were noted in rural areas during conditions favorable to heat island formation, whereas in the heart of the city humidity values were approximately 70 to 80%. Because of the temperature dependence, when the magnitude of the heat island was small, the urban-rural humidity differences were also small.

Although Chandler (1965) found that the mean annual vapor pressure in London was slightly lower (0.2 millibars) than that at a nearby rural location, he (1962b, 1967a) frequently observed that at night the urban absolute humidity was higher than in the outlying regions. Furthermore, variations of humidity within the city often directly corresponded to building density, especially when the meteorological conditions were conducive to heat island formation. Typically under these conditions urban vapor pressures were about 1.5 to 2.0 millibars higher than those in the country. The corresponding relative humidities were about 80 and 90% for the city and country, respectively. Chandler attributed these higher urban absolute humidities to the low rate of diffusion of air near ground level between tall city buildings during nights with light winds. This air with its high daytime moisture was trapped in the city canyons and remained there into the evening, keeping the absolute humidity high.

The summer surveys at Dallas by Ludwig and Kealoha (1968) recorded slightly lower values of absolute humidity within Dallas than outside it, with the greatest differences measured in the afternoon. During the morning hours the observed patterns were not well developed and could not be related to the urban complex.

VISIBILITY

The atmosphere of metropolitan areas is usually characterized by increased concentrations of pollutants, which cause a difference between the visibilities of urban and rural regions. In this section, the effect of cities on atmospheric particulates is discussed first, followed by examples of visibility contrasts between city and country, with emphasis on the effects of fog. Readers interested in further information on these subjects should consult Robinson (1968), who presents a comprehensive review of practical and theoretical aspects of visibility, or the summary by Holzworth (1962), which also discusses visibility trends and their analysis.

Atmospheric Particulates

A consequence of metropolitan areas is increased concentrations of atmospheric particles, such as smoke and combustion products. Landsberg (1960) stated that on the average the number of particulates present over urban areas is 10 times greater than that over rural environs. A more recent study by Horvath (1967), who measured the number and size distribution of submicron particles over Vienna and a nearby mountainous region, concurred with this figure. Although the particulate concentrations in Vienna varied considerably, often because of differing meteorological conditions, the general shapes of the size distribution curves of both the urban and rural samples were similar.

Summers (1966) also showed that the number of particles in an urban atmosphere depends on industrial activity. In central Montreal the soiling index (Hemeon, et al., 1953) was approximately 20% lower on Saturday and Sunday than on weekdays. Furthermore, in midwinter, the time of maximum heating requirements, this index was 2 to 3 times that of midsummer. Similar weekly and annual variations were also found by Weisman, et al. (1969) for Hamilton, Ontario. In addition to industrial sources, urban particulate concentrations depend on meteorological conditions, particularly ventilation. In a summary of the relation between smoke density over Montreal and various weather factors, Summers (1962) showed that winter snow cover was particularly important in that it produced low-level atmospheric stability and frequent temperature inversions.

As a result of air pollution and the associated high aerosol concentrations, visibilities are lower and occurrences of fog are higher in a city than outside the metropolitan area. Fog is more frequent within urban regions because many atmospheric particulates are hygroscopic (Byers, 1965). Thus water vapor readily condenses on them and forms small water droplets, the ingredients of fog. An analogous example was described by Buma (1960), who analyzed visibility data at Leeuwarden on the Netherland coast. For similar relative humidities the visibility was much lower when the wind was from the continent (with high concentrations of condensation nuclei) than when the wind was from the sea (with low concentrations of nuclei). An example of the relationship between air pollution and visibility has been given by Georgii and Hoffman (1966), who showed that for two German cities low visibilities and high concentrations of SO_2 were highly correlated when low wind speeds and low-level inversions prevailed. McNulty (1968) pointed out that between 1949 and 1960 the occurrence of haze as an obstruction to visibility at New York City increased markedly as a result of increased air pollution.

Urban-Rural Visibility Contrasts

As part of his general summary Landsberg (1956) presented visibility data from Detroit Municipal Airport (6 miles from downtown) and the Wayne County Airport (17 miles from town). During conditions conducive to the formation of city smogs (winds of 5 miles per hour or less) visibilities less than 1 mile were observed an average of 149 hours per year at the Municipal Airport but only 89 hours at the rural site. The cause of these low visibilities was listed as smoke in 49 of the observations at the Municipal Airport but in only 5 at the County Airport; most of the occurrences of low visibility were during the late fall and winter. Landsberg (1960) summarized urban-rural fog differences by noting that metropolitan areas had 100% more fog in winter and 30% more in summer. In another study, Smith (1961) compiled the number of occurrences of visibilities less than 6 1/4 miles for locations throughout England in the afternoon, the time of least likelihood of fog (Figure 4). He found that industrial areas reported low visibilities on two to three times more days than did the rural areas.

Although fog generally occurs more frequently in metropolitan areas, this is not true for very dense fog. Chandler (1965) attributed the high frequencies of fog within a city to atmospheric pollution and relatively low wind speeds, but the extra warmth of a city often prevents the thickest nocturnal fogs from reaching the densities reported in the outlying districts. Table 4, presented by Chandler from data of Shellard (1959), shows the estimated hours per year of various density fogs in the vicinity of London, based on four observations per day. The high frequency of fog and the low frequency of very dense fog in the city center are evident. These same general relationships were also detected by Brazell (1964) using similar London data.

Effects of Air Pollution Control

A few recent reports indicate that the visibility in many locations has improved during the last two decades. The better visibilities of major U.S. cities have been associated with local efforts at air pollution abatement and substitution of oil and gas for soft coal in production of heat (Holzworth, 1962; Beebe, 1967).

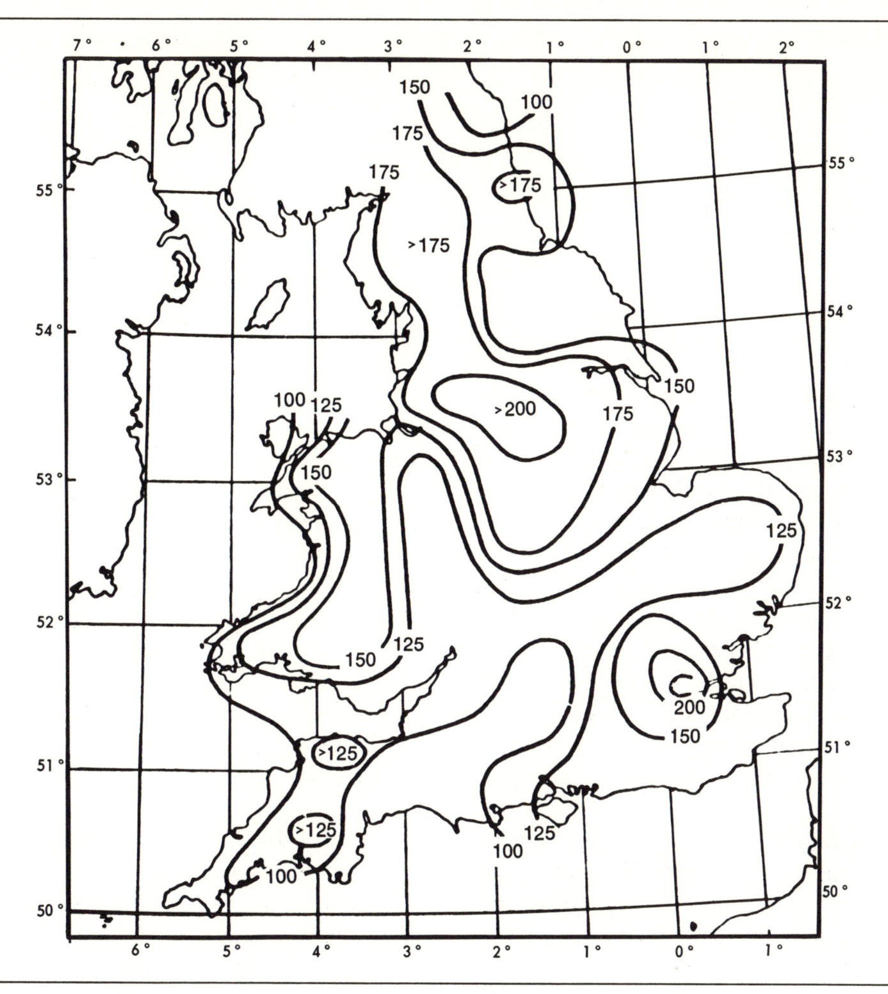

Figure 4 Average number of days per year with afternoon visibilities less than 6¼ miles in England and Wales (from Smith, 1961, *Meteorological Magazine*, Vol. 90, with permission of the Controller of Her Britannic Majesty's Stationery Office).

	Hours per year with visibility less than			
	40 m	200 m	400 m	1,000 m
Kingsway (central)	19	126	230	940
Kew (inner suburbs)	79	213	365	633
London Airport (outer suburbs)	46	209	304	562
Southeast England (mean of 7 stations)	20	177	261	494

Table 4 Fog frequencies in London

Brazell (1964), Wiggett (1964), and Freeman (1968) have suggested that London's improved visibility may be due to enforcement of the air pollution ordinances of 1954 and 1956. Similarly, Atkins (1968) and Corfield and Newton (1968) found that visibility has also improved near other English cities as a result of air pollution legislation. In another study of London, Commins and Waller (1967) compiled data showing that the particulate content of that city's atmosphere has decreased. Measurements from downtown London showed that the average smoke concentration from 1959 to 1964 was 32% lower than from 1954 to 1959.

Visibility is not improving at all United States cities. A study by Green and Battan (1967) has shown that from 1949 to 1965 the frequency of occurrence of poor visibility at Tucson, Arizona, definitely increased and was significantly correlated with that city's population.

RADIATION

The blanket of particulates over most large cities causes the solar energy that reaches an urban complex to be significantly less than that observed in rural areas. The particles are most effective as attenuators of radiation when the sun angle is low, since the path length of the radiation passing through the particulate material is dependent on sun elevation. Thus, for a given amount of particulates, solar radiation will be reduced by the largest fraction at high-latitude cities and during winter. Landsberg (1960) summarized the average annual effect of cities on the solar radiation they received as follows: the average annual total (direct plus diffuse) solar radiation received on a horizontal surface is decreased by 15 to 20%, and the ultraviolet (short wavelength) radiation is decreased by 30% in winter and by 5% in summer.

De Boer (1966) based a recent study on this topic on 2 years of global solar radiation measurements at six stations in and around Rotterdam, Netherlands. The study showed that the city center received 3 to 6% less radiation than the urban fringe and 13 to 17% less than the country. Chandler (1965), also reporting on the solar energy values in the heart of smoky urban areas, observed that from November to March solar radiation at several British cities was 25 to 55% less than in nearby rural areas. In addition, the central part of London annually received about 270 hours less of bright sunshine than did the surrounding countryside because of the high concentration of atmospheric particulates.

Further emphasizing the dependence of the transfer of solar radiation on the air's smoke content, the study of Mateer (1961) showed that the average annual energy received in Toronto, Canada, was 2.8% greater on Sunday than during the remainder of the week. Moreover, the Sunday increase during the heating season, October through April, was 6.0% but was only 0.8% in all other months.

The investigations of atmospheric turbidity by McCormick and Baulch (1962) and McCormick and Kurfis (1966), which were based on aircraft measurements of the intensity of solar radiation, provided data on the variation of solar energy with height over Cincinnati. These authors observed that pollutants over the city, which often had a layered structure and were dependent on the vertical temperature profile, significantly reduced the amount of solar energy that reached the city surface. In addition, they discussed changes of the vertical variation of turbidity (or solar radiation) from morning to afternoon, from day to day, and from clean air to polluted air. Roach (1961) and Sheppard (1958) have also studied attenuation of solar radiation by atmospheric dust particles. They concurred that most radiation scattered by these particles is directed forward and thus attenuation of total solar radiation is primarily due to absorption. Roach estimated that over "heavily polluted areas" absorption of solar energy by the particles was of sufficient magnitude to cause atmospheric heating rates in excess of 10° C per day. A discussion of optical properties of smoke particles is presented in a report by Conner and Hodkinson (1967).

The introduction of smoke controls in London during the mid-1950s has afforded an opportunity to check the radiation-smoke relation. Monteith (1966) summarized data on particulate concentration and solar energy at Kingsway (central London) for the years 1957 to 1963. During this time smoke density decreased by $10 \mu g \cdot m^{-3}$ while total solar radiation increased by about 1%. The average smoke concentration of $80 \mu g \cdot m^{-3}$ at Kew (inner suburbs) represents an energy decrease of about 8%, and in the center of town, where smoke concentrations average 200 to $300 \mu g \cdot m^{-3}$, the income of solar radiation is about 20 to 30% less than that in nearby rural areas. Similarly, Jenkins (1969) reported that the frequency of bright sunshine in London also increased in recent years after implementation of the air pollution laws. During the period 1958 to 1967, the average number of hours of bright sunshine from November through January was 50% greater than that observed from 1931 to 1960.

Measurements of ultraviolet radiation in downtown Los Angeles and on Mt. Wilson (Nader, 1967) showed its dependence on the cleanliness of the atmosphere. Attenuation of ultraviolet radiation by the lowest 5,350 feet of the atmosphere averaged 14% on no-smog days; when smog was present, attenuation increased to a maximum of 58%. Reduced values of ultraviolet radiation in Los Angeles were also measured by Stair (1966). He presented an example in which the effect of smog was to decrease the amount of ultraviolet radiation received at the ground by 50%, and he also noted that on "extremely smoggy days" the decrease may be 90% or more.

	0100 GMT		1300 GMT	
	Mean speed	Excess speed	Mean speed	Excess speed
December–February	2.5	−0.4	3.1	0.4
March–May	2.2	−0.1	3.1	1.2
June–August	2.0	−0.6	2.7	0.7
September–November	2.1	−0.2	2.6	0.6
Year	2.2	−0.3	2.9	0.7

Table 5 Average wind speeds at London airport and differences from those at Kingsway, $m \cdot s^{-1}$

WIND

The flow of wind over an urban area differs in several aspects from that over the surrounding countryside. Two features that represent deviations from the regional wind flow patterns are the differences in wind speeds in city and country and the convergence of low-level wind over a city. These differences occur because the surface of a built-up city is much rougher than that of rural terrain—exerting increased frictional drag on air flowing over it—and because the heat island of a city causes horizontal thermal gradients, especially near the city periphery. The excess heat and friction also produce more turbulence over the urban area. These general ideas were discussed by Landsberg (1956, 1960); he stated that the annual mean surface wind speed over a city was 20 to 30% lower than that over the nearby countryside, that the speed of extreme gusts was 10 to 20% lower, and that calms were 5 to 20% more frequent. Since then several investigations have refined and expanded the studies summarized by Landsberg. Readers interested in a comprehensive and detailed review of wind flow over a city are directed to a recent paper by Munn (1968).

A difficulty in estimating urban-rural wind differences is selecting representative sites within the city from which to take measurements. Most observations of urban wind flow have been taken either from the roofs of downtown buildings, usually several stories high, or from parks or open spaces, whereas very few data have been obtained at street level in the city center, that place where most human activity occurs. However, these conventional measurements are generally representative of the gross wind flow patterns over a city, and as long as their limitations are recognized they can be useful for urban-rural comparisons.

Wind Speed

Although recent reports have concurred that the average wind speed within a city is lower than that over nearby rural areas (Frederick, 1964; Munn and Stewart, 1967; Graham, 1968), the study reported by Chandler (1965) for London shows significant variation from this general rule. Although his analysis is based on only 2 years of data, it indicates that differences in urban and rural wind speeds depend on time of day, season, and wind speed magnitude. Some of these relations are brought out in Table 5 (from Chandler, 1965), which summarizes the mean wind speeds at London Airport (on the fringe of the city) and indicates their excess over the values recorded at Kingsway (in central London). The data show that when the regional wind speeds are light (typically at night) the speeds in downtown London are higher than those at the airport, whereas when wind speeds are relatively high, higher speeds are recorded at the airport. This is evident in comparison of the daytime and nighttime wind speeds given in Table 5.

Chandler (1965) attributes this diurnal variation of urban influence to the diurnal differences of regional wind speed and atmospheric stability. At night when surface winds are relatively calm, the stability is much greater in the country, where inversions are common, than in the metropolitan area, where lapse conditions may prevail. This relative instability in the city, combined with the greater surface roughness, enhances turbulence and allows the faster moving winds above the urban area to reach the surface more frequently; thus at night the average city wind speeds tend to be greater than those of the country. During the day, however, with faster regional winds, the frictional effect of the rough city surface dominates the turbulence effect, and lower wind speeds are observed within the built-up area.

The critical value of wind speed that determines whether the urban winds will be faster or slower than those of the country is highest during summer nights and during both day and night in winter (about 5.0 to 5.5 $m \cdot s^{-1}$). These are times of relatively high atmospheric stability. The lowest values of critical speed (about 3.5 $m \cdot s^{-1}$) occur during summer and fall days. Moreover, the magnitude of the decrease in urban wind speeds is greatest during spring days and least during spring nights.

Chandler (1965) summarized the wind speed statistics for London by noting that the annual average urban and suburban speeds were lower than those of the outlying regions but only by about 5%. However, the annual differences between wind speeds in the center of London and in the outlying areas were somewhat greater. For these locations, when speeds were more than

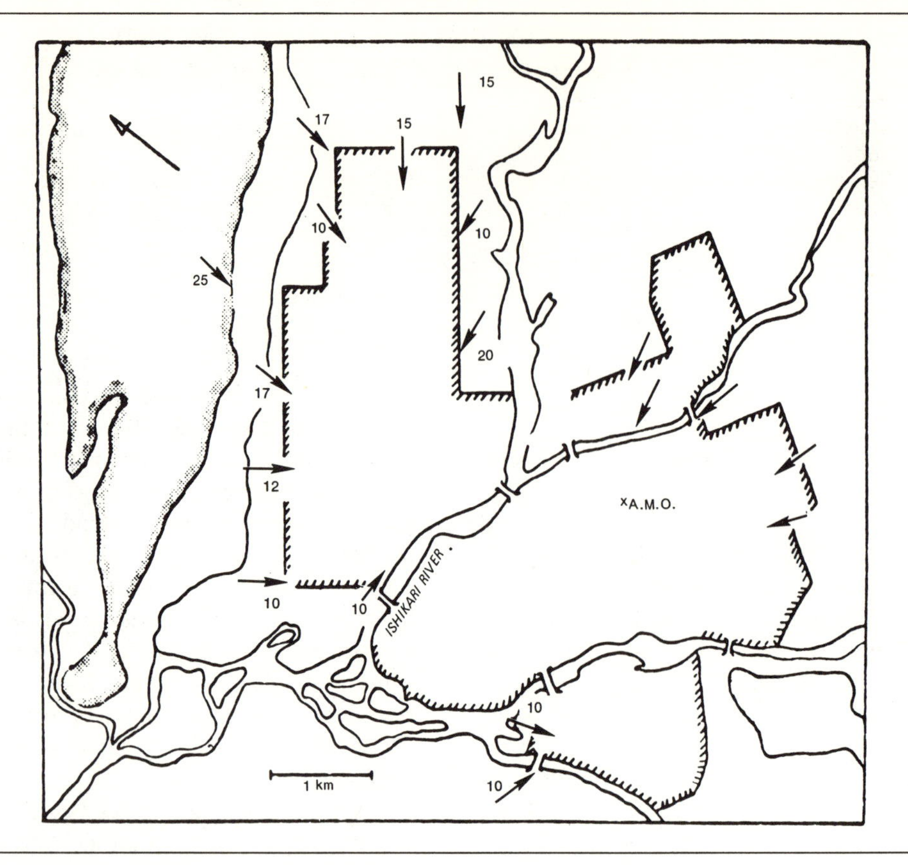

Figure 5 The direction of wind flow around Asahikawa, Japan, 26 February 1956, as deduced from formation of rime ice on tree branches (from Okita, 1960).

$1.5\ m \cdot s^{-1}$ the average difference was about 13%, whereas for speeds greater than $7.9\ m \cdot s^{-1}$ the mean difference was only 7%. In summer, little difference between average city and rural wind speeds was evident. Light regional winds ($1.5\ m \cdot s^{-1}$ or less), which increase in speed over the city, occur more frequently in summer and compensate for the occurrences of strong rural winds, which decrease over the city. Finally, in contrast to the earlier figures reported by Landsberg, the London data showed that fewer calms and light winds occurred in the city center than in rural regions.

Wind Direction

Past research on the direction of wind flow over urban areas has been primarily concerned with detecting and measuring a surface flow in toward the urban complex. It has been surmised for some time that if a city is warmer than its environs the warm city air should rise and be displaced by cooler rural air. However, this inflow is weak and occurs only in conjunction with well-developed heat islands, which in turn are dependent on certain meteorological conditions. Since direct measurements of the inflow require a coordinated set of accurate observations, few such investigations have been made.

Pooler (1963) analyzed wind records from the Louisville local air pollution study and determined that there was indeed a surface inflow of air toward the city. This inflow was the dominant feature of the surface wind flow pattern when the regional winds were weak, a situation that could result, for example, from a weak large-scale pressure gradient. Georgii (1968) reported on urban wind measurements in Frankfort/Main, Germany. During clear, calm nights an inflow toward the city center was detected, with a convergent wind velocity of up to 2 to $4\ m \cdot s^{-1}$. He also noted that when the large-scale surface geostrophic wind speed reached 3 to $4\ m \cdot s^{-1}$ a local city circulation was prevented from becoming established, although the increased roughness of the city was still affecting the wind regime.

Measurements of the wind flow around an oil refinery in the Netherlands have been reported by Schmidt (1963) and by Schmidt and Boer (1963). Although the area involved ($4\ km^2$) was much smaller than that of a moderate-sized city and the heat produced by the refinery per unit area was considerably greater than that of a city, the general relationships are of interest. A cyclonic circulation occurred around the area, with convergence toward the center. Furthermore, ascending air was detected over the center of the area, with descending currents over the surroundings. The greatest vertical velocities measured in the vicinity of the maximum heat production were about $15\ cm \cdot s^{-1}$.

Vertical wind velocities were obtained from low-level

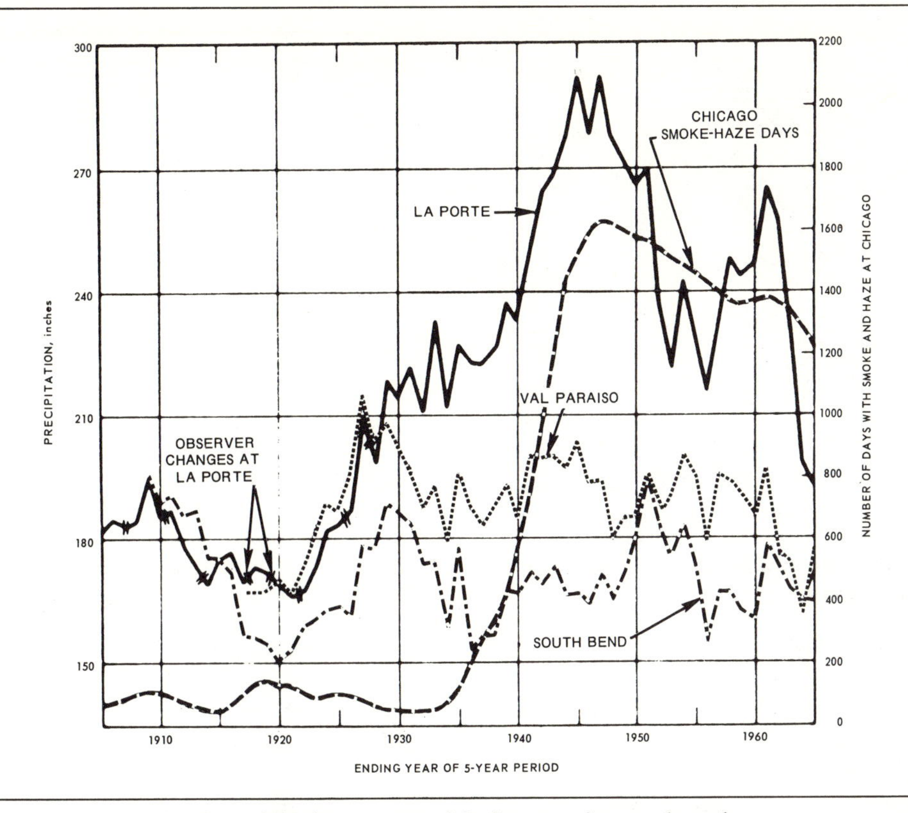

Figure 6 Five-year moving totals of precipitation at several Indiana stations and smoke-haze days at Chicago (from Changnon, 1968a).

tetroon flights over New York City by Hass, et al. (1967), and Angell, et al. (1968). They observed an upward flow over densely built-up Manhattan Island and downward motion over the adjacent Hudson and East Rivers. They ascribed this flow pattern to the urban heat island, to the barrier effect of the tall buildings, and to the relatively cool river water. Low-level convergence over a city has also been detected by Okita (1960, 1965), who applied two novel techniques. For one study he utilized the fact that rime ice formed on the windward side of tree trunks and the thickness of the ice was proportional to wind speed. For the other (Figure 5), he observed the smoke plumes from household chimneys. By observing these features around the periphery of Asahikawa, Japan, he deduced the local wind flow patterns and determined that when the large-scale wind flow was weak there was a convergence over the city.

As a final consideration of urban wind flow, Chandler's observations (1960, 1961) on the periphery of London and Leicester are of interest. Chandler noted that when a well-developed nocturnal heat island formed with a strong temperature gradient near the edge of the built-up area, winds flowed inward toward the city center, but the flow was not steady. Rather, movement of cold air from the country toward the city occurred as pulsations, the strongest winds occurring when the temperature gradient was strongest.

PRECIPITATION

A city also influences the occurrence and amount of precipitation in its vicinity. For several reasons an urban complex might be expected to increase precipitation. Combustion sources add to the amount of water vapor in the atmosphere, higher temperatures intensify thermal convection, greater surface roughness increases mechanical turbulence, and the urban atmosphere contains greater concentrations of condensation and ice nuclei. Since no continuous, quantitative measurements of these various parameters have been made in conjunction with the few urban precipitation studies, the relative significance of these factors is not easy to establish. Landsberg (1956, 1958) gave several European examples of urban-rural comparisons and concluded that the amount of precipitation over a city is about 10% greater than that over nearby country areas. More recent studies have shown that his conclusion may be an oversimplification and that the greatest positive anomalies occur downwind of the city center.

The Effects of Cities

The effects of cities on precipitation are difficult to determine for several reasons. First, very few rural areas remain undisturbed from their natural state. Second, there is a lack of rain gauges in metropolitan

	Chicago	La Porte	St. Louis	Tulsa	Champaign-Urbana
Annual precipitation	5	31	7	8	5
Warmer half-year precipitation	4	30	a	5	4
Colder half-year precipitation	6	33	a	11	8
Rain days ≥ 0.01 or 0.1 inch					
Annual	6	0	a	a	7
Warmer half-year	8	0	a	a	3
Colder half-year	4	0	a	a	10
Rain days ≥ 0.25 or 0.5 inch					
Annual	5	34	a	a	5
Warmer half-year	7	54	a	a	9
Colder half-year	0	5	a	a	0
Annual number of thunderstorm days	6	38	11	a	7
Summer number of thunderstorm days	13	63	21	a	17

[a] Data not sufficient for comparison.

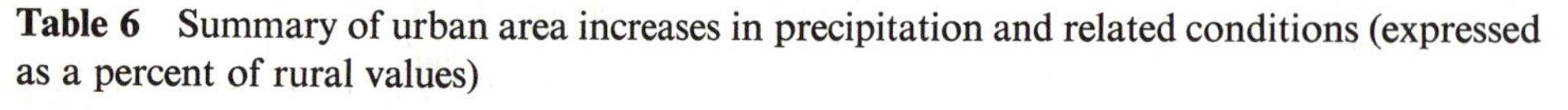

Table 6 Summary of urban area increases in precipitation and related conditions (expressed as a percent of rural values)

areas, especially instruments with long-term records and uniform exposure throughout the record. Third, many cities are associated with bodies of water or hilly terrain, and these also affect the patterns of precipitation. Finally, the natural variability of rainfall, particularly the summer showers of mid-America, further complicates the analysis of urban-rural precipitation differences. An example of the difficulty of rainfall analysis is the study of Spar and Ronberg (1968). They observed that the record from Central Park, New York City, showed a significant decreasing trend of precipitation of 0.3 inch per year from 1927 to 1965. This trend was not substantiated by data from other nearby sites; at Battery Place, the nearest station, a small rainfall increase was observed during the period.

A very striking example of the effect of the Chicago urban region on local precipitation has been documented by Changnon (1968a). He showed that at La Porte, Indiana, some 30 miles downwind of a large industrial complex between Chicago and Gary, Indiana, the amount of precipitation and the number of days with thunderstorms and hail have increased markedly since 1925. Furthermore, the year-to-year variation of precipitation at La Porte agrees generally with data on the production of steel and number of smoke-haze days at Chicago (Figure 6). During the period from 1951 to 1965, the positive anomalies at La Porte were 31% for precipitation, 38% for thunderstorm days, 246% for hail days, and 34% for days with precipitation ≥0.25 inch. These results are summarized in Table 6. Changnon concluded that these observed differences were a real effect of the industrial area and that they represent the general size of precipitation increase that is possible as a result of man's activities. However, because of the effect of Lake Michigan and its associated lake breeze in channeling the pollutants around the south end of the lake toward the La Porte area and inhibiting dispersion, the large differences detected in this study probably are not representative of American cities in general.

Changnon (1961, 1962, 1968b, 1969) has summarized precipitation data for several other midwestern cities and detected positive increases, but not nearly as pronounced as those at La Porte. In St. Louis, Chicago, Champaign-Urbana, Illinois, and Tulsa, Oklahoma, the city precipitation was 5 to 8% higher than the average of nearby rural stations. At St. Louis and Champaign-Urbana the rainfall maxima were downwind of the urban center. The increase of 5% at Chicago proper represents a relative maximum over the city and is distinct from the higher precipitation rates downwind of Chicago at La Porte. Spatial data were not available for Tulsa. Table 6 (from Changnon, 1968b) shows that the percentage precipitation increase is greater during the cold season and that increases in the warm season result from more days of moderate rain and thunderstorms.

Authors do not totally agree on the distribution of precipitation over cities. The rainfall minimum over the Missouri half of metropolitan St. Louis detected by Changnon (1968b) was also noted by Feig (1968), who observed that a map of isohyets of annual precipitation over the eastern United States showed that minimum precipitation areas occur around most cities having no obvious geographic influences. The annual precipitation pattern at Washington, D.C. (Woollum and Canfield, 1968), also shows low values downtown along the Potomac River, with the greatest amounts of precipitation on the north to northwest side of town and with relative maxima in both the eastern and western suburbs (Figure 7). On the other hand, Dettwiller (1968) has indicated that the effect of a city is to increase

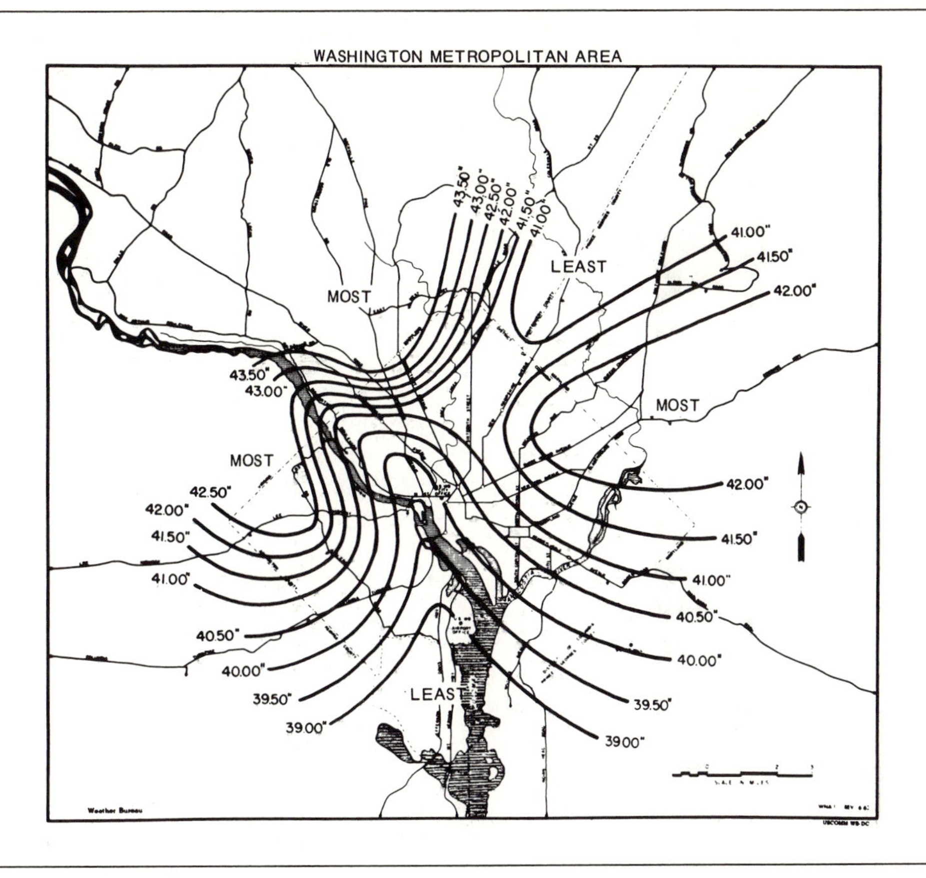

Figure 7 Mean annual precipitation (inches) over the Washington, D.C., metropolitan area (from Woollum and Canfield, 1968).

precipitation. He showed that from 1953 to 1967 the average rainfall in Paris was 31% greater on weekdays than on Saturdays and Sundays.

Another reason for believing that more precipitation falls over metropolitan districts is that heavy thundershowers sometimes occur over an area that roughly coincides with the urban complex (Staff, 1964; Chandler, 1965; Atkinson, 1968). However, it is always difficult to determine whether a thundershower was a natural event or whether the city actually influenced it in any way.

The Important Urban Factors

Although it is hard to determine the relative importance of the several urban factors that affect precipitation, the extensive studies by Changnon (1968b) led him to the following conclusions about the city-precipitation relationship. Two factors were probably most effective in enhancing precipitation at La Porte—high concentrations of ice nuclei from nearby steel mills and added heat from local industrial sources. Because of the high frequency of nocturnal thunderstorms and hailstorms at La Porte, he concluded that the thermal and frictional effects were probably the most significant. In addition, the precipitation maximum associated with Champaign-Urbana, an area with little industry and a minimum of nuclei sources, also indicated that the thermal and frictional factors could produce significant differences.

Many recent studies have concentrated on measuring condensation and ice nuclei and determining their possible influence on cloud development and precipitation. Measurements showing that cities are an important source of nuclei were reported by Mee (1968). Typical concentrations of cloud droplets and condensation nuclei at the convective cloud base near Puerto Rico were 50 cm^{-3} in the clean air over the ocean, about 200 cm^{-3} over the unpolluted countryside, and from 1,000 to 1,500 cm^{-3} immediately downwind of San Juan. Abnormally high concentrations were detected for at least 100 miles downwind of the city. In another study, Squires (1966) pointed out that measurements of condensation nuclei over Denver showed that the concentrations of these nuclei produced by human activities were similar to the natural concentration there. Moreover, Telford (1960), Langer (1968), and Langer, et al. (1967), observed that industrial areas, and in particular steel mills, are good sources of ice nuclei. Finally, Schaefer (1966, 1968a) and others (Morgan, 1967; Morgan and Allee, 1968; Hogan, 1967) have shown that lead particles in ordinary auto exhaust form effective ice nuclei when combined with iodine vapor.

Even though cities are generally recognized as good sources of nuclei, the net effect of such nuclei cannot be definitely determined. For example, a few ice nuclei added to supercooled clouds may enhance rainfall, a principle used by commercial cloud seeders. On the other hand, rain that falls from warm clouds is dependent on a number of large drops within a cloud so that coalescence may be effective. If large numbers of condensation nuclei are introduced into a warm cloud, many small drops will form and thus rainfall may be inhibited (Gunn and Phillips, 1957; Squires and Twomey, 1960).

Many examples in the literature show that rainfall can be artificially increased, whereas only a few show decreases. Fleuck (1968) statistically analyzed results of the Missouri seeding experiments by the University of Chicago group and concluded that seeding of clouds in that area suppressed precipitation. In addition, reports of an investigation of the trend of rainfall in eastern Australia (Warner and Twomey, 1967; Warner, 1968) concluded that as the amount of smoke from burning sugar cane in the area increased during the past 50 years, rainfall correspondingly decreased by 25%. Therefore, an accurate determination of the effect of a city on precipitation in its neighborhood requires knowledge not only of the number and type of nuclei being introduced by the city but also of such factors as the concentration of natural nuclei and vertical temperature profiles over the city.

Recent findings have further documented Landsberg's statement that it is becoming increasingly difficult to find undisturbed rural areas with which to compare cities to determine urban-rural meteorological differences. Both direct (Schaefer, 1968b) and indirect (Gunn, 1964; McCormick and Ludwig, 1967; Peterson and Bryson, 1968; Volz, 1968) measurements have shown that the particulate content of the atmosphere, even in remote areas, is increasing as a result of greater human activity.

Other Precipitation Elements

Two other precipitation elements of interest are hail and snowfall. Changnon's study (1968b) of the La Porte anomaly found that from 1951 to 1965 the number of hail days was 246% greater than that of surrounding stations. Results from similar investigations at other midwestern cities were not definitive. Landsberg (1956) cited a few instances in which snowfall over an urban area was lighter than that at nearby rural locations, presumably because of higher temperatures over the cities. Potter (1961) found similar results for Toronto and Montreal, Canada.

SUGGESTED FUTURE RESEARCH

Although this review indicates that a large volume of research has been conducted in urban climatology, several areas warrant emphasis in the future. Among the following suggestions for further study, many were brought out at the recent international conference on the climate of cities at Brussels.

Few studies have been made of the vertical structure of the urban atmosphere, which includes such features as the variation with height of temperature, wind flow, pollutants, and radiation. More such information is needed for a better understanding of the transport and diffusion of pollutants over metropolitan areas. Likewise, additional urban wind measurements are needed at a variety of sites with different exposures, to delineate the fine structure of wind flow throughout a city.

Several aspects of urban-rural radiation differences should be further investigated. In several studies of the effects of a polluted atmosphere on ultraviolet radiation in Los Angeles, the effects were found to be significant. Observations from other cities are now needed to show the amount of attenuation of ultraviolet radiation at different locations. Similarly, studies should be conducted to determine the net effect of a polluted city atmosphere on the total radiation balance of a city, including visible and thermal infrared wavelengths, since net radiation comprises a major fraction of a city's energy budget. In particular, urban-rural differences of infrared radiation should be measured to determine whether reduced values of solar radiation at the city surface resulting from atmospheric pollution are compensated for by an increase in infrared energy.

A cause-and-effect relationship between a city and precipitation has not yet been found. Although several factors are believed to be important, the role of anthropogenically produced dust particles is of special interest. The influence of these particles on the physics of precipitation and the optimum concentration for modifying precipitation have not yet been determined. Also, investigations should be undertaken to ascertain whether city-induced precipitation anomalies of a size similar to that at La Porte, Indiana, are widespread or whether this example is primarily the result of local influences, such as Lake Michigan.

Further research should be directed toward determing exactly upon which parameters urban-rural temperature differences depend. For example, the relative importance of the type and density of local buildings and gross city size on the temperature at a given city location has yet to be definitely established. Similarly, the local cooling produced by parks and greenbelts and the extent of this cooling into nearby neighborhoods should be measured. Such information would be useful in city planning and land use studies, for example.

Finally, these two questions should be studied: How far downwind does a city influence climate, and to what extent does a city in the tropics modify its climate? The former question has implications in larger-scale studies of inadvertent weather modification; the latter is important since nearly all research in urban climatology has been done at cities in temperate climates.

REFERENCES

Angell, J. K., Pack, D. H., Hass, W. A., and Hoecker, W. H., 1968. Tetroon flights over New York City. *Weather 23*, 184–191.

Atkins, J. E., 1968. Changes in the visibility characteristics at Manchester/Ringway Airport. *Meteorol. Mag. 97*, 172–174.

Atkinson, B. W., 1968. The reality of the urban effect on precipitation—a case study approach. *Presented at W.M.O. Symp. on Urban Climates and Building Climatology*. Brussels, Belgium. October 1968.

Beebe, R. G., 1967. Changes in visibility restrictions over a 20 year period. *Bull. Amer. Meteorol. Soc. 48*, 348.

Bornstein, R. D., 1968. Observations of the urban heat island effect in New York City. *J. Appl. Meteor. 7*, 575–582.

Brazell, J. H., 1964. Frequency of dense and thick fog in central London as compared with frequency in outer London. *Meteorol. Mag. 93*, 129–135.

Brooks, C. E. P., 1952. Selective annotated bibliography on urban climate. *Meteor. Abstracts Bibliog. 3*, 734–773.

Buma, T. J., 1960. A statistical study of the relationship between visibility and relative humidity of Leeuwarden. *Bull. Amer. Meteorol. Soc. 41*, 357–360.

Byers, H. R., 1965. *Elements of Cloud Physics*. University of Chicago Press. 191 pp.

Chandler, T. J., 1960. Wind as a factor of urban temperatures—a survey in north-east London. *Weather 15*, 204–213.

Chandler, T. J., 1961. Surface effects of Leicester's heat-island. *E. Midland Geographer 15*, 32–38.

Chandler, T. J., 1962a. London's urban climate. *Geographical Journal 128*, 279–302.

Chandler, T. J., 1962b. Temperature and humidity traverses across London. *Weather 17*, 235–242.

Chandler, T. J., 1963. London climatological survey. *Int. Journ. Air Water Poll. 7*, 959–961.

Chandler, T. J., 1964. City growth and urban climates. *Weather 19*, 170–171.

Chandler, T. J., 1965. *The Climate of London*. London, Hutchinson University Library, publishers. 292 pp.

Chandler, T. J., 1966. London's heat island. In *Biometeorology II*, Proc. of Third Int. Biometeor. Congr., Pau, France, September 1963. London, Pergamon Press. 589–597.

Chandler, T. J., 1967a. Absolute and relative humidity of towns. *Bull. Amer. Meteorol. Soc. 48*, 394–399.

Chandler, T. J., 1967b. Night-time temperatures in relation to Leicester's urban form. *Meteorol. Mag. 96*, 244–250.

Chandler, T. J., 1968. Urban climates: inventory and prospect. *Presented at W.M.O. Symp. on Urban Climates and Building Climatology*. Brussels, Belgium. October 1968.

Changnon, S. A., 1961. Precipitation contrasts between the Chicago urban area and an offshore station in southern Lake Michigan. *Bull. Amer. Meteorol. Soc. 42*, 1–10.

Changnon, S. A., 1962. A climatological evaluation of precipitation patterns over an urban area. In *Symposium: Air over Cities*. U.S. Public Health Service, Taft Sanitary Eng. Center, Cincinnati, Ohio, Tech. Rept. A62-5. 37–67.

Changnon, S. A., 1968a. The LaPorte weather anomaly—fact or fiction? *Bull. Amer. Meteorol. Soc. 49*, 4–11.

Changnon, S. A., 1968b. Recent studies of urban effects on precipitation in the United States. *Presented at W.M.O. Symp. on Urban Climates and Building Climatology*. Brussels, Belgium. October 1968.

Changnon, S. A., 1969. Recent studies of urban effects on precipitation in the United States. *Bull. Amer. Meteorol. Soc. 50*, 411–421.

Clarke, J. F., 1969. Nocturnal urban boundary layer over Cincinnati, Ohio. *Mon. Wea. Rev. 97*, 582–589.

Commins, B. T. and Waller, R. E., 1967. Observations from a ten-year study of pollution at a site in the city of London. *Atm. Env. 1*, 49–68.

Conner, W. D. and Hodkinson, J. R., 1967. Optical properties and visual effects of smoke-stack plumes. *Publ. 999-AP-30, Public Health Service, U.S. Dept. Health, Education, and Welfare*, Cincinnati, Ohio.

Corfield, G. A. and Newton, W. G., 1968. A recent change in visibility characteristics at Finningley. *Meteorol. Mag. 97*, 204–209.

Davidson, B., 1967. A summary of the New York urban air pollution dynamics research program. *J. Air Pollution Control Assoc. (J-APCA) 17*, 154–158.

DeBoer, H. J., 1966. Attenuation of solar radiation due to air pollution in Rotterdam and its surroundings. *Koninklijk Nederlands Meteorologisch Institute, Wetenschappelijk Rapport W.R. 66-1*, de Bilt, Netherlands. 36 pp.

DeMarrais, G. A., 1961. Vertical temperature difference observed over an urban area. *Bull. Amer. Meteorol. Soc. 42*, 548–554.

Dettwiller, I., 1968. Incidence possible de l'activite industrielle sur les precipitations a Paris. *Presented at W.M.O. Symp. on Urban Climates and Building Climatology*. Brussels, Belgium. October 1968.

Dronia, H., 1967. Der Stadteinfluss auf den Weltweiter Temperaturtrend. *Meteorologische Abhandlugen 74*, 1.

Duckworth, F. S. and Sandberg, J. S., 1954. The effect of cities upon horizontal and vertical temperature gradients. *Bull. Amer. Meteorol. Soc. 35*, 198–207.

Feig, A. M., 1968. An evaluation of precipitation patterns over the metropolitan St. Louis area. In *Proc. First Nat. Conf. on Wea. Mod.*, Albany, N.Y., Amer. Meteorol. Soc. 210–219.

Flueck, J. A., 1968. A statistical analysis of Project Whitetop's precipitation data. In *Proc. First Nat. Conf. on Wea. Mod.*, Albany, N.Y., Amer. Meteorol. Soc. 26–35.

Frederick, R. H., 1964. On the representativeness of surface wind observations using data from Nashville, Tennessee. *Int. Journ. Air Water Poll. 8*, 11–19.

Freeman, M. H., 1968. Visibility statistics for London/Heathrow Airport. *Meteorol. Mag. 97*, 214.

Garnett, A. and Bach, W., 1965. An estimation of the ratio of artificial heat generation to natural radiation heat in Sheffield. *Mon. Wea. Rev. 93*, 383–385.

Geiger, R., 1965. *The Climate near the Ground.* Cambridge, Massachusetts. 611 pp.

Georgii, H. W. and Hoffman, L., 1966. Assessing SO_2 enrichment as dependent on meteorological factors. *Staub, Reinhaltung der Luft* (in English) *26*(*12*), 511.

Georgii, H. W., 1968. The effects of air pollution on urban climates. *Presented at W.M.O. Symp. on Urban Climates and Building Climatology*. Brussels, Belgium. October 1968.

Graham, I. R., 1968. An analysis of turbulence statistics at Fort Wayne, Indiana. *J. Appl. Meteor. 7*, 90–93.

Green, C. R. and Battan, L. J., 1967. A study of visibility versus population growth in Arizona. *J. Ariz. Acad. Sci. 4*, 226.

Gunn, R., 1964. The secular increase of the world-wide fine particle pollution. *J. Atm. Sci. 21*, 168–181.

Gunn, R. and Phillips, B. B., 1957. An experimental investigation of the effect of air pollution on the initiation of rain. *J. Meteor. 14*, 272–280.

Haas, W. A., Hoecker, W. H., Pack, D. H., and Angell, J. K., 1967. Analysis of low-level constant volume balloon (tetroon) flights over New York City. *Quart. Journ. Royal Meteorol. Soc. 93*, 483–493.

Hemeon, W. C. L., Haines, G. F., jr., and Ide, H. H., 1953. Determination of haze and smoke concentrations by filter paper samplers. *J-APCA 3*, 22–28.

Hogan, A. W., 1967. Ice nuclei from direct reaction of iodine vapor with vapors from leaded gasoline. *Science 158*, 800.

Holzworth, G. C., 1962. Some effects of air pollution on visibility in and near cities. In *Symposium: Air Over Cities.* U.S. Public Health Service, Taft Sanitary Eng. Center, Cincinnati, Ohio, Tech. Rept. A62-5. 69–88.

Horvath, H., 1967. A comparison of natural and urban aerosol distribution measured with the aerosol spectrometer. *Env. Sci. Technol. 1*, 651–655.

Hosler, C. R., 1961. Low-level inversion frequency in the contiguous United States. *Mon. Wea. Rev. 89*, 319–339.

Hutcheon, R. J., Johnson, R. H., Lowry, W. P., Black, C. H., and Hadley, D., 1967. Observations of the urban heat island in a small city. *Bull. Amer. Meteorol. Soc. 48*, 7–9.

Jenkins, I., 1969. Increase in averages of sunshine in central London. *Weather 24*, 52–54.

Kawamura, T., 1964. Analysis of the temperature distribution in Kumagaya city—a typical example of the urban climate of a small city. *Geog. Rev. Japan 27*, 243.

Kratzer, P., 1956. *Das Stadtklima*. Braunschweig, Friedrich Vieweg and Sohn. (English trans. available through ASTIA, AD 284776.) 221 pp.

Landsberg, H. E., 1956. The climate of towns. In *Man's Role in Changing the Face of the Earth*, ed. W. L. Thomas. Chicago, Ill., University of Chicago Press. 584–606.

Landsberg, H. E., 1960. *Physical Climatology*. 2nd Revised Ed. DuBois, Penn., Gray Printing Co. 446 pp.

Landsberg, H. E., 1962. City air—better or worse. In *Symposium: Air Over Cities*. U.S. Public Health Service, Taft Sanitary Eng. Center, Cincinnati, Ohio, Tech. Rept. A62-5, 1–22.

Landsberg, H. E., 1968. Micrometeorological temperature differentiation through urbanization. *Presented at W.M.O. Symp. on Urban Climate and Building Climatology*. Brussels, Belgium. October 1968.

Langer, G., 1968. Ice nuclei generated by steel mill activity. In *Proc. First Nat. Conf. on Wea. Mod.* Albany, N.Y., Amer. Meteorol. Soc. 220–227.

Langer, G., Rosinski, J., and Edwards, C. P., 1967. A continuous ice nucleus counter and its application to tracking in the troposphere. *J. Appl. Meteor. 6*, 114–125.

Lawrence, E. N., 1968. Changes in air temperature at Manchester Airport. *Meteorol. Mag. 97*, 43–51.

Ludwig, F. L., 1967. Urban climatological studies. *Interim Rept. No. 1, Contr. OCD-PS-64-201*. Stanford Res. Inst., Menlo Park, Calif. (AD 657248).

Ludwig, F. L. and Kealoha, J. H. S., 1968. Urban climatological studies. *Final Rept., Contr. OCD-DAHC-20-67-C-0136*. Stanford Res. Inst., Menlo Park, Calif.

Mateer, C. L., 1961. Note on the effect of the weekly cycle of air pollution on solar radiation at Toronto. *Int. Journ. Air Water Poll. 4*, 52–54.

McCormick, R. A. and Baulch, D. M., 1962. The variation with height of the dust loading over a city as determined from the atmospheric turbidity. *J-APCA 12*, 492–496.

McCormick, R. A. and Kurfis, K. R., 1966. Vertical diffusion of aerosols over a city. *Quart. Journ. Royal Meteorol. Soc. 92*, 392–396.

McCormick, R. A. and Ludwig, J. H., 1967. Climate modification by atmospheric aerosols. *Science 156*, 1358–1359.

McNulty, R. P., 1968. The effect of air pollutants on visibility in fog and haze at New York City. *Atm. Env. 2*, 625–628.

Mee, T. R., 1968. Microphysical aspects of warm cloud. *Presented at ESSA Atm. Phys. Chem. Lab. Symp. on Wea. Mod.* Sept. 1968. Boulder, Colorado.

Mey, A., 1933. Die stadteinfluss auf den temperaturgang. *Das Wetter 50*, 293.

Mitchell, J. M., jr., 1961. The temperature of cities. *Weatherwise 14*, 224–229.

Mitchell, J. M., jr., 1962. The thermal climate of cities. In *Symposium: Air Over Cities*, U.S. Public Health Service, Taft Sanitary Eng. Center, Cincinnati, Ohio, Tech. Rept. A62-5. 131–145.

Monteith, J. L., 1966. Local differences in the attenuation of solar radiation over Britain. *Quart. Journ. Royal Meteorol. Soc. 92*, 254–262.

Morgan, G. M., jr., 1967. Technique for detecting lead particles in air. *Nature 213*, 58–59.

Morgan, G. M., jr. and Allee, P. A., 1968. The production of potential ice nuclei by gasoline engines. *J. Appl. Meteor. 7*, 241–246.

Munn, R. E., 1968. Airflow in urban areas. *Presented at W.M.O. Symp. on Urban Climates and Building Climatology*. Brussels, Belgium. October 1968.

Munn, R. E. and Stewart, I. M., 1967. The use of meteorological towers in urban air pollution programs. *J-APCA 17*, 98–101.

Munn, R. E., Hirt, M. S., and Findlay, B. F., 1969. A climatological study of the urban temperature anomaly in the lakeshore environment at Toronto. *J. Appl. Meteor. 8*, 411–422.

Nader, J. S., 1967. Pilot study of ultraviolet radiation in Los Angeles, October 1965. *Public Health Serv. Publ. 999-AP-38*. U.S. Dept. HEW, Nat. Cent. for Air Poll. Contr., Cincinnati, Ohio.

Oke, T. R., 1968. Some results of a pilot study of the urban climate of Montreal. *Climat. Bull. (McGill Univ.) 3*, 36–41.

Oke, T. R. and Hannell, F. G., 1968. The form of the urban heat island in Hamilton, Canada. *Presented at W.M.O. Symp. on Urban Climates and Building Climatology*. Brussels, Belgium. October 1968.

Okita, T., 1960. Estimation of direction of air flow from observation of rime ice. *J. Meteorol. Soc. Japan 38*, 207–209.

Okita, T., 1965. Some chemical and meteorological measurements of air pollution in Asahikawa. *Int. Journ. Air Water Poll. 9*, 323–332.

Parry, M., 1956. Local temperature variations in the Reading area. *Quart. Journ. Royal Meteorol. Soc. 82*, 45–57.

Parry, M., 1966. The urban heat island. In *Biometeorology II*, Proc. of Third Int. Biometeor. Congr., Pau, France, Sept. 1963. London, Pergamon Press. 616–624.

Peterson, J. T. and Bryson, R. A., 1968. Atmospheric aerosols: Increased concentrations during the last decade. *Science 162*, 120–121.

Pooler, F., jr., 1963. Air flow over a city in terrain of moderate relief. *J. Appl. Meteor. 2*, 446–456.

Potter, J. G., 1961. Changes in seasonal snowfall in cities. *Canadian Geographer 5*, 37–42.

Roach, W. T., 1961. Some aircraft observations of fluxes of solar radiation in the atmosphere. *Quart. Journ. Royal Meteorol. Soc. 87*, 346–363.

Robinson, E., 1968. Effect on the physical properties of the atmosphere. In *Air Pollution*, ed. A. C. Stern. Vol. 1, 2nd Ed. New York, Academic Press. 694 pp.

Sasakura, K., 1965. On the distribution of relative humidity in Tokyo and its secular change in the heart of Tokyo. *Tokyo J. of Climat. 2*, 45.

Schaefer, V. J., 1966. Ice nuclei from automobile exhaust and iodine vapor. *Science 154*, 1555–1557.

Schaefer, V. J., 1968a. Ice nuclei from auto exhaust and organic vapors. *J. Appl. Meteor. 7*, 148–149.

Schaefer, V. J., 1968b. New field evidence of inadvertant modification of the atmosphere. In *Proc. First. Nat. Conf. on Wea. Mod.*, Albany, N.Y., Amer. Meteorol. Soc. 163–172.

Schmidt, F. H., 1963. Local circulation around an industrial area. *Int. Journ. Air Water Poll. 7*, 925–926.

Schmidt, F. H. and Boer, J. H., 1963. Local circulation around an industrial area. *Berichte des Deutschen Wetterdienstes 91*, 28.

Sekiguti, T., 1964. City climate in and around the small city of Ina in central Japan. *Tokyo Geog. Papers 8*, 93.

Shellard, H. C., 1959. The frequency of fog in the London area compared with that in rural area of east Anglia and south-east England. *Meteorol. Mag. 88*, 321–323.

Sheppard, P. A., 1958. The effect of pollution on radiation in the atmosphere. *Int. Journ. Air Poll. 1*, 31–43.

Smith, L. P., 1961. Frequencies of poor afternoon visibilities in England and Wales. *Meteorol. Mag. 90*, 355–359.

Spar, J. and Ronberg, P., 1968. Note on an apparent trend in annual precipitation at New York City. *Mon. Wea. Rev. 96*, 169–171.

Squires, P., 1966. An estimate of the anthropogenic production of cloud nuclei. *J. Rech. Atmos. 2*, 299–308.

Squires, P. and Twomey, S., 1960. The relation between cloud drop spectra and the spectrum of cloud nuclei. In *Physics of Precipitation*, Amer. Geophys. Union Monog. 5, 211–219.

Staff, River Forecast Center (Tulsa) and Dist. Meteor. Office (Kansas City), 1964. Cloudburst at Tulsa, Oklahoma, July 27, 1963. *Mon. Wea. Rev. 92*, 345.

Stair, R., 1966. The measurement of solar radiation, with principal emphasis on the ultraviolet component. *Int. Journ. Air Water Poll. 10*, 665–688.

Stanford University Aerosol Laboratory and The Ralph M. Parsons Co., 1952. Behavior of aerosol clouds within cities. *Jt. Quart. Rept. No. 2, Oct.–Dec. 1952.* 100 pp. (AD 7261).

Stanford University Aerosol Laboratory and The Ralph M. Parsons Co., 1953a. Behavior of aerosol clouds within cities. *Jt. Quart. Rept. No. 3, Jan.–Mar. 1953.* 218 pp. (AD 31509).

Stanford University Aerosol Laboratory and The Ralph M. Parsons Co., 1953b. Behavior of aerosol clouds within cities. *Jt. Quart. Rept. No. 4, Apr.–June 1953.* 196 pp. (AD 31508).

Stanford University Aerosol Laboratory and The Ralph M. Parsons Co., 1953c. Behavior of aerosol clouds within cities. *Jt. Quart. Rept. No. 5, July–Sept. 1953.* 238 pp. (AD 31507).

Stanford University Aerosol Laboratory and The Ralph M. Parsons Co., 1953d. Behavior of aerosol clouds within cities. *Jt. Quart. Rept. No. 6, Vol. I, Oct.–Dec. 1953.* 246 pp. (AD 31510).

Stanford University Aerosol Laboratory and The Ralph M. Parsons Co., 1953e. Behavior of aerosol clouds within cities. *Jt. Quart. Rept. No. 6, Vol. II, Oct.–Dec. 1953.* 187 pp. (AD 31711).

Summers, P. W., 1962. Smoke concentrations in Montreal related to local meteorological factors. In *Symposium: Air over Cities*, U.S. Public Health Service, Taft Sanitary Eng. Center, Cincinnati, Ohio, Tech. Rept. A62-5. 89–113.

Summers, P. W., 1966. The seasonal, weekly, and daily cycles of atmospheric smoke content in central Montreal. *J-APCA 16*, 432–438.

Sundborg, A., 1950. Local climatological studies of the temperature conditions in an urban area. *Tellus 2*, 222–232.

Tag, P. M., 1968. *Surface temperature in an urban environment*. M.S. Thesis, Dept. of Meteor., The Pennsylvania State Univ., University Park, Penn. 69 pp.

Tebbens, B. D., 1968. Gaseous pollutants in the air. In *Air Pollution*, ed. A. C. Stern. Vol. 1, 2nd Ed. New York, Academic Press. 694 pp.

Telford, J. W., 1960. Freezing nuclei from industrial processes. *J. Meteor. 17*, 676–681.

Volz, F. E., 1968. Turbidity at Uppsala from 1909 to 1922 from Sjostrom's solar radiation measurements. *Meddelanden, Ser. B, 28*. Sver. Meteor. Hydrolog. Inst., Stockholm.

Warner, J., 1968. A reduction in rainfall associated with smoke from sugar-cane fires—an inadvertent weather modification? *J. Appl. Meteor. 7*, 247–251.

Warner, J. and Twomey, S., 1967. The production of cloud nuclei by cane fires and the effect on cloud droplet concentration. *J. Atmos. Sci. 24*, 704–706.

Weisman, B., Matheson, D. H., and Hirt, M., 1969. Air pollution survey for Hamilton, Ontario. *Atm. Env. 3*, 11–23.

Wiggett, P. J., 1964. The year-to-year variation of the frequency of fog at London (Heathrow) Airport. *Meteorol. Mag. 93*, 305.

Woollum, C. A., 1964. Notes from a study of the microclimatology of the Washington, D.C., area for the winter and spring seasons. *Weatherwise 17*, 263–271.

Woollum, C. A. and Canfield, N. L., 1968. Washington metropolitan area precipitation and temperature patterns. *ESSA Tech. Memo. WBTM-ER-28*. Garden City, N.Y. 32 pp.

SUGGESTED READING FOR PART 7

Bach, W., 1970. An urban circulation model. *Archiv für Meteorologie Geophysik und Bioklimatologie, Serie B, 18*, 155–168.

Bach, W., 1972. *Atmospheric Pollution.* McGraw-Hill, New York. 144 pp.

Davies, J. C., III, 1970. *The Politics of Pollution.* Western Publishing Company. 231 pp.

Landsberg, H. E., 1970. Man-made climatic changes. *Science 170(3964)*, 1265–1274.

McCormick, R. A., 1969. Meteorology and urban air pollution. *World Meteor. Org. Bull. 18*, 155–165.

Oke, T. R. and East, C., 1971. The urban boundary layer in Montreal. *Boundary-Layer Meteor. 1*, 411–437.

Terjung, W. H. and collaborators, 1970. The energy balance climatology of the city–man system. *Annals,* Association of American Geographers *60(3)*, 466–492.

Part Eight

Climate and Man

Of what interest is weather to the construction industry? Is there an economic relationship? Russo's article provides the answers for the U.S.A.

Is our attitude to the urban snow hazard somewhat antiquated? Could we be more efficient? Rooney's assessment of the impact of snow in seven urban areas of the United States uncovers, largely through content analyses, a hierarchy of disruptions and relates these to critical physical-weather properties.

To what extent do the paintings of a region reflect the actual climatic environment? Are artists, famed for their emotional sensitivity, sensitive also to this aspect of life? The last article gives the result of a content analysis of a rather unusual but nevertheless valid nature; Neuberger shows the close relationship between regional climate and its depiction in art.

John A. Russo, Jr.

24 The economic impact of weather on the construction industry of the United States

Introduction

The U.S. Weather Bureau (E.S.S.A.) has long provided services to the general public and to users with special interests (e.g., in aviation, marine, and agricultural activities). In a continuing effort to broaden its services to those segments of the economy that are particularly vulnerable to weather, the Weather Bureau recently sponsored a study to determine the nature and magnitude of losses due to weather in the construction industry of the United States, and the potential capability of present and future weather forecast accuracy and meteorological services to reduce these losses.

In this study, efforts were directed toward:

determining those operations of the construction industry influenced by weather,

determining the type of weather information needed by the construction industry for specific operations,

assessing the availability of this information to the industry,

estimating the nationwide construction-industry dollar loss due to weather,

estimating the reduction in weather-associated dollar loss likely to be derived by the industry from the appropriate use of *existing and improved* weather information, and

determining the most effective methods for educating the construction industry regarding construction-oriented weather information.

Construction Industry Operations Influenced by Weather

The effects of weather are felt by the construction industry through lost or inefficient work days, idle equipment, ruined material, etc. To determine the magnitude and extent of the weather's influence on construction operations, many days were spent in the field with construction personnel observing the numerous operations of the industry. The information gathered, when combined with that obtained through surveys and an extensive literature search (ASHVE, 1955; Bristow, 1955; Court, 1958; Hendrick, 1959; Landsberg, 1960; Merritt, 1944; Potter, 1952; Thom, 1956; Waters, 1957), resulted in the conclusion that 43 major construction operations are weather sensitive to some extent. Table 1 lists these weather sensitive operations, along with an estimate of the point at which each operation becomes critically affected by 13 weather elements or conditions. The construction operations shown in Table 1 represent a composite taken from the four major categories of construction (residential, general building, highway, and heavy and specialized) and are approximately arranged according to actual time sequence. It is clearly shown in the table that weather sensitivity is evident through almost the entire spectrum of construction operations, that is, from planning operations such as surveying, through such assembly operations as concrete paving, to finishing operations such as landscaping and painting.

Weather Information Needs of the Construction Industry

While weather-initiated decisions in the construction industry may be made at any hour of the day or night, the study revealed that weather information requirements reach a maximum at about 2–3 PM and 6–7 AM. Planning and scheduling of the following day's activities, including labor, equipment, and material usage, are often based on the weather forecast available in the afternoon. Such a schedule is tentative in many cases, and the final decision, made early the next morning, is heavily influenced by the observed weather conditions at the work site(s) during the 7–8 AM period.

It was also found that the weather information required by the construction industry for these short

Reprinted, with minor editorial modification, by permission of the author and the American Meteorological Society, from the *Bulletin of the American Meteorological Society*, Vol. 47, 1966, pp. 967–972.

Item	Operation	Rain	Snow and sleet	Freezing rain	Low temperatures (° F)	Low temperature and high wind (chill factor)	High temperature	High temperature and high humidity (temperature—humidity index)	High wind (mph)	Dense fog	Ground freeze	Drying conditions	Temperature inversion	Flooding and abnormal tides
1	Surveying	L[a]	L	L	0– –10	Low temperature/high wind equivalent to "Chill Factor" = 1000–1200 ("Very Cold"–"Bitter Cold")	Temperatures above 90° F	High temperature/high humidity equivalent to temperature—humidity index (THI) = 77	25	x[b]	—	—	—	—
2	Demolition and clearing	M	M	L	0– –10				15–35	x	x	—	x	—
3	Temporary site work	M	M	L	0– –10				20	x	x	—	—	—
4	Delivery of materials	M	M	L	0– –10				25	x	—	—	—	—
5	Material stockpiling	L	L	L	0– –10				15	x	—	—	—	—
6	Site grading	M	M	L	20– 32				15–25	x	x	x	—	—
7	Excavation	M	M	L	20– 32				35	x	x	x	x	—
8	Pile driving	M	M	L	0– –10				20	x	x	—	—	x
9	Dredging	M	M	L	0– –10				20	x	x[c]	—	—	x
10	Erection of coffer dams	M	L	L	32				25	x	x	x	—	x
11	Forming	M	M	L	0– –10				25	—	x	—	—	—
12	Emplacing reinforcing steel	M	M	L	0– –10				20	—	x	—	—	—
13	Quarrying	M	M	L	32				25–35	x	x	x	x	—
14	Delivery of pre-mixed concrete	M	L	L	32				35	x	x	—	—	—
15	Pouring concrete	M	L	L	32				35	—	x	x	—	—
16	Stripping and curing concrete	M	M	L	32				25	—	x	x	—	—
17	Installing underground plumbing	M	M	L	32				25	—	x	x	—	—
18	Waterproofing	M	M	L	32				25	—	x	—	—	—
19	Backfilling	M	M	L	20– 32				35	x	x	x	—	—
20	Erecting structural steel	L	L	L	10				10–15	x	—	—	—	—
21	Exterior carpentry	L	L	L	0– –10				15	—	—	—	—	—
22	Exterior masonry	L	L	L	32				20	—	x	x	—	—
23	External cladding	L	L	L	0– –10				15	—	—	—	—	—
24	Installing metal siding	L	L	L	0– –10				15	—	—	—	—	—
25	Fireproofing	L	L	L	0– –10				35	—	—	—	—	—
27	Roofing	L	L	L	45				10–20	—	—	x	—	—
28	Cutting concrete pavement	M	M	L	0– –10				35	—	x	—	—	—
29	Trenching, installing pipe	M	M	L	20– 32				25	—	x	x	—	—
30	Bituminous concrete pouring	L	L	L	45				35	x	x	x	—	—
31	Installing windows and doors, glazing	L	L	L	0– –10				10–20	—	—	—	—	—
38	Exterior painting	L	L	L	45– 50				15	x	—	x	—	—
39	Installation of culverts and incidental drainage	M	L	L	32				25	—	x	x	—	x
40	Landscaping	M	L	L	20– 32				15	x	x	x	—	—
41	Traffic protections	M	M	L	0– –10				15–20	x	x	—	—	—
42	Paving	L	L	L	32– 45				35	x	x	x	—	—
43	Fencing, installing lights, signs, etc.	M	M	L	0– –10				20	x	x	—	—	—

[a] L indicates light; M indicates moderate.
[b] x indicates operation is affected by this condition but critical limit is undeterminable.
[c] x water freeze, critical limit undeterminable.

Table 1 Critical limits of weather elements having significant influence on construction operations

(0–24 hour) time periods must be *much more detailed* than that made available through public dissemination media.

For weather information to be most useful to any industry, it should be geared specifically to the needs of all components of the industry. The weather *service* requirements of the construction industry were found to be analogous to that of the aviation industry, where the *timing* of critical weather events is emphasized and a constant vigilance with regard to the critical limits is maintained through timely updatings and revisions. While the service needs of these two industries are quite similar, the pertinent weather conditions and their critical limits are quite different.

A detailed analysis was made of the weather elements of significance, their threshold limits, important decision points, and types of decisions required for each of the 43 construction operations listed in Table 1. Condensation of the operation-by-operation list resulted in the determination of a "weather product package," which encompasses the requirements of the entire industry. This weather product package comprises *two primary forecasts per day* (6–7 AM and 2–3 PM) of the *timing and magnitude or intensity* of the critical weather events (in probabilistic form, where applicable), a *constant watch on the weather* with construction industry problems in mind, and *timely revisions* whenever required. The weather product package, with reference to the weather conditions listed in Table 1, is classified into six time periods of interest:

local observations of current weather over an entire state or multi-state area;

warnings, revisions, and modifications, when necessary, at least *two* hours in advance;

zero to twenty-four hour forecasts of the timing and intensity (or magnitude) of weather events critical to individual operations;

five-day forecasts of general characteristics of the weather for planning of indoor-outdoor work, equipment maintenance, and perishable materials operations;

thirty-day outlooks; historical *climatological information* — for longer-range planning and bidding

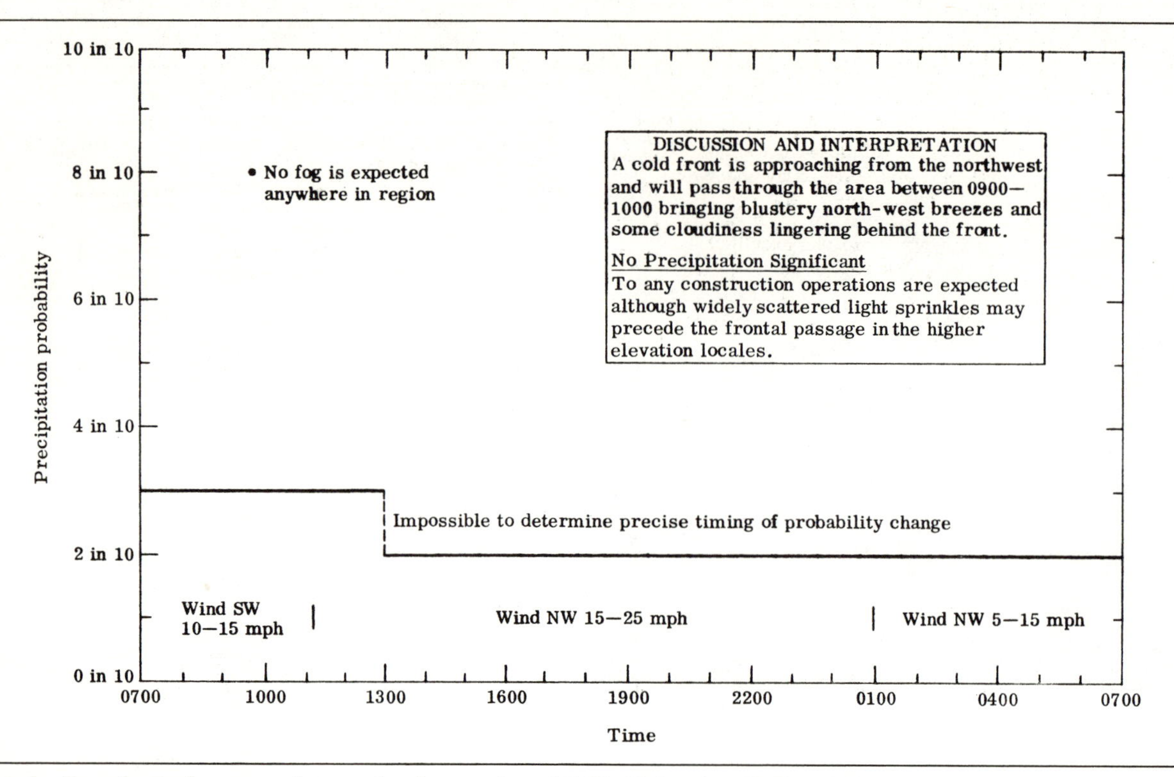

Figure 1 Experimental construction weather forecast issued 0700, 22 October 1965: 0–24-hr forecast of precipitation, wind, fog.

To substantiate the findings, an experimental forecast format was formulated and forecasts were issued on a real-time basis to local construction firms. The combination of consulting with, and issuing forecasts to, a variety of construction personnel (e.g., concrete supplier, industrial builder) resulted in the development of an experimental construction-weather service. Through this experimental service and personal contact with contractors, important information was compiled concerning the deficiencies in the present Weather Bureau forecast system *from the construction industry's point of view*, the type of information desired, the time it is most useful, and the reason it is needed. During the experimental period it was found that the most convenient way to present weather information is via time-oriented graphical products. Examples of the zero to 24 hour and five-day forecast components of the weather product package are given in Figures 1 and 2 and in Table 2. Note in Figures 1 and 2 that the timing of weather events is emphasized, and that the uncertainties in the forecasts are reflected by either probabilities (precipitation) or error limits (temperature).

Availability of Weather Products

The availability status of each of the weather products required by the construction industry was assessed according to current Weather Bureau availability. It was concluded that some of the products deemed essential to the construction industry *are now made available to the industry* by the Weather Bureau through mass dissemination media and teletypewriter circuits. Other products are *considered by forecasters* in their daily routines, *but are not disseminated*, and still others, *due to their specialized nature, would require additional effort to generate as well as to disseminate.*

It appears that a *mixture of government and private weather services* is needed to completely and satisfactorily fulfill the weather information needs of the construction industry. While some of the critical weather condition forecasts could be issued by the Weather Bureau, the constant weather-watch aspect of the

Table 2 Experimental construction weather forecast issued 0600 22 October 1965—summary, 5-day forecast

Date	Sky condition	Precipitation probability	Minimum temperature (° F)	Maximum temperature (° F)
22 Oct.	Contained in 24-hr forecast			
23 Oct.	Clear	1/10	32 ± 3	50 ± 3
24 Oct.	Increasing cloudiness	1/10 AM 3/10 PM	30 ± 4	55 ± 4
25 Oct.	Cloudy	6/10 rain	34 ± 5	50 ± 5
26 Oct.	Partly cloudy	3/10 showers	33 ± 6	40 ± 6

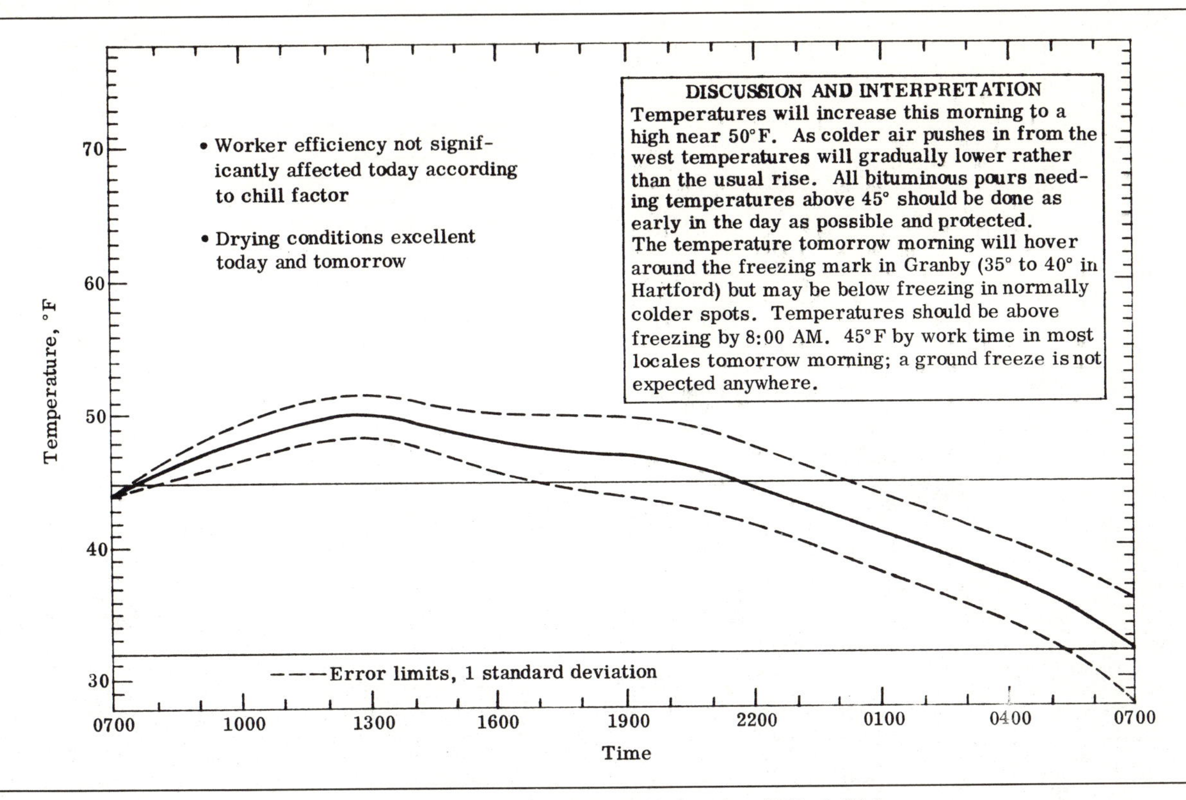

Figure 2 Experimental construction weather forecast issued 0700, 22 October 1965: 0–24-hr forecast of temperature, chill factor, drying conditions, ground freeze.

Construction category	Annual volume	Potentially weather sensitive				Total sensitive (per cent of annual volume)
		Perishable material	On-site wages	Equipment	Overhead and profits	
Residential	17.2	0.960	1.624	0.073	2.141	4.8 (27.9)
General building	29.7	1.928	4.079	0.222	2.670	8.9 (30.0)
Highways	6.6	1.666	1.633	0.773	0.727	4.8 (72.7)
Heavy and specialized	12.5	1.875	3.125	2.500	2.500	10.0 (80.0)
Repair and maintenance	22.0	2.674	3.996	1.386	3.143	11.2 (50.9)
Total (rounded)	88.0	9.1	14.4	5.0	11.2	39.7 (45.1)

Table 3 Distribution of total annual construction volume and the proportion considered *potentially* weather sensitive ($ billions)

service, *with warnings and revisions and rapid dissemination to the user*, is the responsibility of private weather forecasting services.

Nationwide Construction Industry Dollar Losses Due to Weather

United States construction expenditures totaled about $88 billion in 1964 (Dooley, 1964; *Hartford Times*, 1965) —more than 10% of the gross national product. Of this total volume of construction, it is estimated that about 45% (*Construction Review*, 1962, 1964a, 1964b, 1965; Sumichrast, 1964; USDL, 1963a, 1963b; USGC, 1958), or $39.7 billion, is spent in potentially weather sensitive areas of construction (i.e., outdoors, or work requiring perishable materials, etc.). Table 3 gives the distribution of the total annual construction volume among the four major categories of construction, plus repair and maintenance as a fifth category, and the corresponding proportion of the volume considered *potentially* weather sensitive.

Detailed analysis of pertinent weather conditions over the United States for a five-year period showed that the weather effects on the construction industry may be classed into two separate, but overlapping, categories. These are the seasonal effects brought about by weather typical of the winter season, and the intermittent day-to-day weather changes in all seasons.

The economic impact of seasonal weather effects was assessed by analyzing available data on seasonal unemployment in the construction industry, and by estimating the economic effects of the various components contributing to a decrease in the annual construction dollar volume due to adverse winter weather (*Construction Review*, 1965; ILO, 1964; USDL, 1963b). The seasonal variation in United States construction expenditures which peaks in August and reaches a minimum in February is caused by both the planned seasonal cycle of the construction industry brought about by the inherent seasonal demand of the consumer and adverse winter weather. Analysis of the annual construction cycle of construction dollar volume in the southwestern United States (where adverse weather effects may be considered negligible) resulted in an evaluation of the extent of the seasonal demand effects in that area. Projection of this calculation to the United States as a whole, on a proportional basis, resulted in an estimate of the decrease in annual construction dollar volume due to adverse weather alone.

To assess the economic impact of intermittent day-to-day weather, the annual potentially weather-sensitive construction volume ($39.7 billion) was distributed on a daily basis. The weather over the entire United States was represented by five years (1959–1963) of hourly observations of the important weather conditions shown in Table 1 for six United States cities. The six cities: New York, Chicago, Portland, Oreg., Los Angeles, Dallas, and Atlanta, each located in a different climatic regime, represented near-average temperature and precipitation conditions for their respective regions, as well as 10–15 % of the total annual U.S. construction expenditure. For each day of the five year period at each of the six cities, and according to pre-defined rules simulating construction industry decision processes, the consequences of working or not working were tabulated in terms of lost time, idle equipment, inefficiency, etc. Using the daily distribution of the potentially weather-sensitive dollar volume, these values were converted from a frequency basis to a dollar basis.

The total United States dollar loss due to weather was quantitatively evaluated at a *minimum of $3 billion* annually. The maximum possible dollar loss was estimated to be *as high as $10 billion*. This wide range results from a highly speculative estimate of the decreased construction volume due to seasonal weather effects.

Potential Savings with Appropriate Use of Weather Information

Under the assumption that the recommended *weather-service information is made available* to the construction industry and is appropriately used, it is estimated that, with the present forecast accuracy, a potential annual *savings of $0.5 to $1.0 billion* is possible. This represents approximately 10 to 17% of the estimated weather-caused loss. The maximum savings achievable if the forecast products for the shorter (0–24 hour) time periods were 100% accurate are estimated to be $0.8 to $1.3 billion, or $300 million more than is obtainable with present forecast accuracy.

Weather Education and the Construction Industry

Weather products and services can be of value only if decision makers within the construction industry appreciate their economic potential and know how to use them effectively. There is a definite need for additional education of industry members regarding weather and its relation to their operations. Recommended outlets are weather-use courses in college curricula, and self-teaching manuals and kits for those in the field.

The ability to interpret and to use effectively weather forecasts and services is essential if the construction industry is to work with the Weather Bureau and with private weather consultants to benefit fully from the specialized service it needs.

Summary and Conclusions

Of the total United States construction expenditure of $88 billion (in 1964), it is estimated that about 45%, or $39.7 billion, is spent in potentially weather sensitive areas of construction.

Considering only the potentially weather sensitive expenditures, the total construction industry dollar loss due to weather, throughout the entire United States, is quantitatively evaluated at a minimum of $3 billion annually. The maximum possible dollar loss is estimated to be as high as $10 billion.

Under the assumption that weather-service information such as that shown in Figures 1 and 2 and in Table 2 is made available to the construction industry through the Weather Bureau and private weather forecast services, and is appropriately used, it is estimated that with the present forecast accuracy a potential annual savings of $0.5 to 1.0 billion is possible. The maximum savings achievable by the industry, assuming 100% accuracy of all forecast products for the shorter (0–24 hour) time periods, are estimated at $0.8 to 1.3 billion, or $300 million more than is obtainable with the present forecast accuracy.

REFERENCES

ASHVE, 1955. Heating, ventilating, air conditioning guide. *Amer. Soc. Heating and Ventilating Engineers 33*, 122–128.

Bristow, G. C., 1955. How cold is it? *Weekly Weather and Crop Bull. Natl. Summary*. 2 pp.

Construction Review, 1962. Materials used in federal office building construction. *Construction Review 8(10)*.

Construction Review, 1964a. Material and equipment for civil works construction. *Construction Review 10(6)*.

Construction Review, 1964b. Material used in private and public housing construction. *Construction Review 10(8)*.

Construction Review, 1965. Various industry-wide statistics. *Construction Review 11(1 and 2)*.

Court, A., 1958. Wind chill. *Bull. Amer. Meteorol. Soc. 34*, 487–493.

Dooley, W. G., 1964. $88 billion year in prospect for 1964. *The Constructor*, Jan. 16–18.

Hartford Times, 1965. 1964 construction sets record. *Hartford Times*, Feb. 20.

Hendrick, R. L., 1959. An outdoor weather-comfort index for the summer season in Hartford, Conn. *Bull. Amer. Meteorol. Soc. 40*, 620–623.

ILO, 1964. Practical measures for regularization of employment in the construction industry. *International Labor Organization, Rept. 3, Building, Civil Engineering and Public Works Committee*. 87 pp.

Landsberg, H. E., 1960. Bioclimatic work in the Weather Bureau. *Bull. Amer. Meteorol. Soc. 41*, 184–187.

Merrit, A., 1944. Construction methods in sub-arctic. *Pacific Builders and Engineers 50*, 33–34.

Potter, A. R., 1952. Sub-zero weather complicates well-servicing operations. *World Petroleum 23*, 88–89.

Russo, J. A. Jr., Trovern-Trend, K., Ellis, R. H., Koch, R. C., Howe, G. M., Milly, G. H., and Enger, I., 1965. The operation and economic impact of weather on the construction industry of the United States. Final Rept. 7665-163 Contract Cwb-10948, The Travelers Research Center, Inc., Hartford, Conn., 102 pp plus 12 appendices.

Sumichrast, M., 1964. Rising sand costs are boosting sale price of homes. *NAHB Journ. Homebuilding 18*, 30–32.

Thom, E. C., 1956. Measuring the need for air conditioning. *Air Conditioning, Heating, and Ventilating 53*, 65–70.

USDL, 1963a. Labor and materials requirements: Highway construction, 1958 and 1961. *Mon. Labor Rev.*, U.S. Dept. Labor. August.

USDL, 1963b. *Mon. Labor Rev.*, U.S. Dept. Labor. Jan.–Dec.

USGC, 1958. *Blueprint for Profit*. U.S. Gypsum Co. 113 pp.

Waters, J. W., 1957. Weather limitations to the construction of industrial establishments. *Meteorol. Monographs, Amer. Meteorol. Soc. (Boston) 2(9)*, 37–52.

John F. Rooney, Jr.

25 *The urban snow hazard in the United States: an appraisal of disruption*

Traditionally the mid-latitude American city has coped with snow and ice in an organized but inefficient manner. Even in this age of galloping technological advance, most of our northern cities annually experience the crippling impact of at least one severe snowstorm. The brunt of the disruption, ironically, occurs in areas that are "well prepared" to handle any snow emergency. The continuing trend toward urban sprawl has been a major factor in accentuating the difficulties that stem from an occasional snowstorm. Distances separating urban dwellers from their everyday affairs and transactions have lengthened, and dependence on both private and mass transportation facilities has increased. By introducing snow or ice into an urban setting with hypersensitive movement patterns, any form of chaos may be precipitated.[1]

The present study is designed to assess the impact of snow in urban areas, using several lines of investigation. Snow's disruptive effects on man are analyzed, with an emphasis on the identification of the critical physical-environmental variables (amount and kind of snow, wind, temperature, terrain, and so on). Then the role of community adjustment and adaptation is examined, and, finally, attitudes concerning the snow hazard are probed, by means of interviews, to gain an understanding of the adaptations and adjustments that are characteristically made. The study focuses on seven selected cities, but the findings presumably have much wider application.

THE SNOW HAZARD

The snow hazard may be defined as comprising all the perils that snow and ice present, both in themselves and in association with other weather conditions. The hazard is also influenced by the nature of the terrain and the kinds of road-surfacing materials used in the area in question.

Most of man's activities are to some degree sensitive to weather. Certain of them are highly sensitive to cold (highway construction, outdoor painting), some to wind (structural-steel erection, roofing installation, recreation), and many others to precipitation, especially snow (Russo, et al., 1965). Transportation is the activity most critically affected by snow, but a host of others also suffer, including construction, merchandising, manufacturing, agriculture, power supply, communications, recreation, and public-health and safety services. The catalogue of problems caused by weather provides ample evidence of the disruptive impact of snow. It is noteworthy that snow cover is cited as a hindrance to transportation more often than any other meteorological phenomenon, and that surfacing icing and glaze are considered to be only slightly less severe (Rapp and Huschke, 1964).

The snow hazard has a number of important implications for urban geography. Perhaps the most notable is the impediment it poses to spatial interaction, both within the city and between the city and its tributary area. Normal movement—for example, the journey to work, shopping trips, and travel to participate in recreational activities—is often changed or curtailed. As many individuals rearrange their plans and patterns of action, the density and direction of traffic flow are also altered. In the snow hazard's most extreme form it may sever tributary connections for extended periods, cutting off supply and distribution lines and thereby resulting in emergency situations.

From an economic standpoint snow may induce heavy financial losses. The cost of combating snow and ice, fixed capital and overhead costs for schools, factories, and stores, and damage to property are some

Reprinted, with minor editorial modification, by permission of the author and editor, from the *Geographical Review*, Vol. 57, 1967, pp. 538–559, copyrighted by the American Geographical Society of New York.

[1] Quite apart from snow and ice, consider the effect of an accident on an expressway in Chicago, New York, or Los Angeles, especially during rush-hour periods.

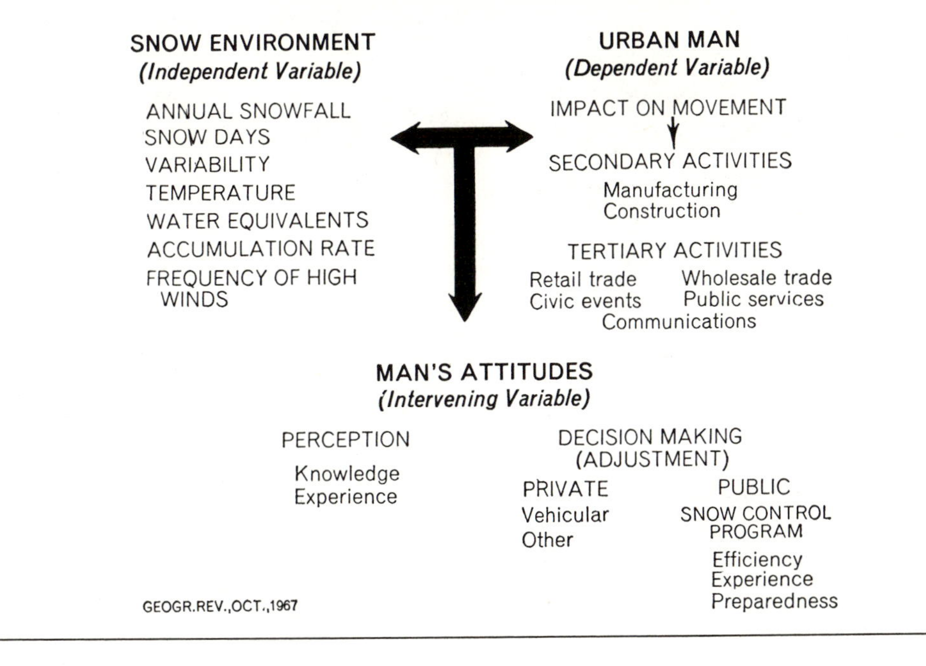

Figure 1

of the direct losses. A diversion of expenditures is also a common economic response; money that might have been spent on clothing or recreation may be reallocated to snow-control equipment for homes and cars, for instance.

Some of the myriad relationships that exist between "urban man" and his snow environment are depicted in Figure 1. This diagram attempts to portray the typical urban dweller confronted by his snow environment, which includes not only snow but the associated conditions (wind, air temperature, and other forms of precipitation) that often affect the degree of difficulty he experiences. Within this framework, man's accumulated contacts with snow hazards will influence the action he takes. His decisions will be affected by the kind of adjustments he has made to counteract the hazard, both independently and as a part of the total urban population. These adjustments are influenced in turn by his attitudes concerning snow and ice, which have been shaped by his perception of these phenomena.

ASSESSMENT OF DISRUPTION

The investigation of the disruptions caused by snow covers a period of ten years, 1953–1963. This period is long enough to provide an adequate sample of conditions and recent enough so that a reasonable amount of information was available.

Seven cities served as the basic laboratory in which the effects of snow were examined. Cheyenne and Casper, Wyoming, and Rapid City, South Dakota, were the pilot sites. After testing for relationships between disruption and the snow environment in these cities, four more sites—Green Bay and Milwaukee, Wisconsin; Muskegon, Michigan; and Winona, Minnesota—were selected for study (Figure 2).[2] In addition to being in another section of the country, they represent different types of snow environments and public adjustments. Man's attitudes concerning the snow hazard and his subsequent adjustments to it were investigated in each city, but extensive interviewing was confined to Cheyenne, Casper, Rapid City, and Winona.

Six of the cities studied are of medium size, ranging in population from 25,000 to 65,000. One larger place, Milwaukee (*ca.* 750,000), was included so that the measures developed could be tested at a locale of authentic urban stature. Medium-size cities were selected for several reasons. It was felt that cities of this size would provide a less complicated setting than larger cities, and, in fact, they proved to be excellent laboratories, though their problems do not approach those of huge metropolitan areas such as New York, Chicago, and Boston. The seven sites chosen for study were large enough to possess a nearly complete span of urban functions, yet small enough to permit a careful analysis of snow-caused disturbances. At the same time, distances between sections of the cities were long enough to augment disruption when snow and ice were present.

[2] Figure 2 also shows the location of ten other sites that were investigated more generally to test the significance of the moisture content of snow.

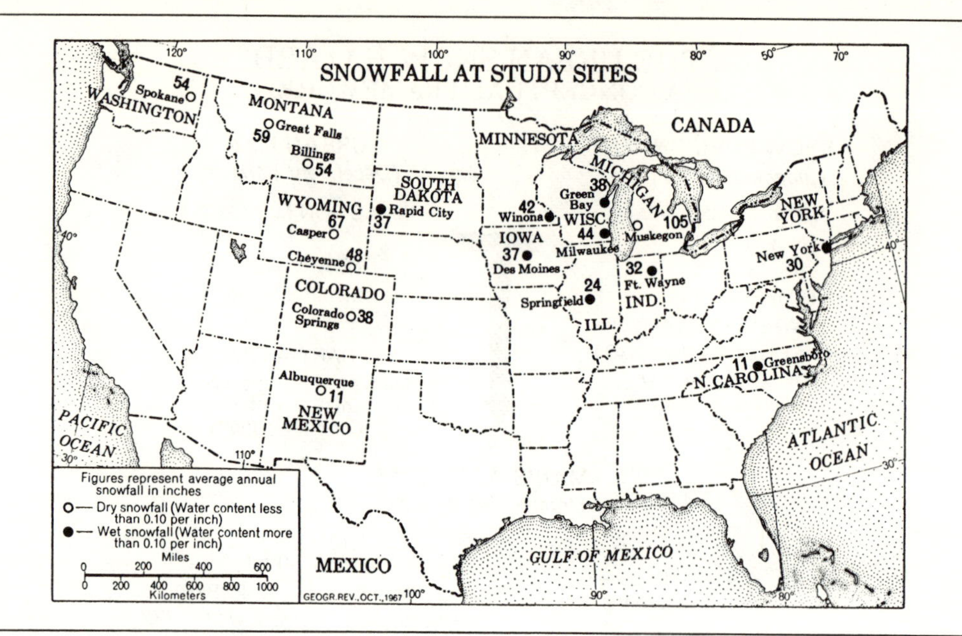

Figure 2

Methods of Investigation

Information concerning disruption was taken chiefly from daily newspapers and public records.[3] From this material it was possible to classify the impact of all snow days[4] under consideration. Each snow day (event) was rated as to severity of impact within a hierarchy or scale of disruption.

Some measure of data reliability was thought desirable. The use of newspapers to identify trends and to form a basis for generalization falls under recognized research techniques of "content analysis"—that is, the categories of analysis used to classify the content are clearly and explicitly defined so that other individuals can apply them to the same content to verify the conclusions; the analyst is not free to select and report merely what strikes him as interesting, but must methodically classify all the relevant material in his sample. By using this technique to analyze communications media, much of the ambiguity is eliminated.

Interviews with private individuals and with proprietors of commercial establishments were used to probe attitudes and adjustments concerning the snow hazard (Rooney, 1966). The questions were designed to reveal the respondents' attitudes toward snow and ice and to determine their ability to cope with the hazard personally and as members of a group. Merchants were asked to comment about the effect of snow on their business, and to discuss the municipal snow-control program. Interviewing was done on an area-sampling basis. A total of 255 personal interviews were conducted, and 61 firms made up the commercial sample.

Patterns of Snowfall

Before analyzing the disruption attributable to snow, it is necessary to examine the snow patterns of the seven cities. In terms of mean annual accumulation (ten-year averages),[5] Muskegon leads with 105 inches; Casper has 67; Cheyenne, 48; Milwaukee, 44; Winona, 42; Green Bay, 38; and Rapid City, 37. The Midwestern cities are characterized by a concentration of snowfall from December through March, with January the peak month. In the Rocky Mountain area and adjacent plains, most of the fall comes later, and March and April are usually the snowiest months. In general, snow is wetter in the Midwest and in the East than it is in the

[3] The major newspapers used were the *Wyoming Eagle* (Cheyenne), the *Casper Tribune-Herald*, the *Rapid City Daily Journal*, the *Winona Daily News*, the *Green Bay Press Gazette*, the *Milwaukee Journal*, and the *Muskegon Chronicle*. The public records that proved most helpful were police accident reports, parking-meter revenue summaries, and street maintenance reports. Telephone company files and information compiled by public utility firms were among the private sources consulted. The most dependable sources were newspapers, police and city records, and the data on power failures kept by public utility companies. School attendance records were too general for more than token use, and since the telephone company records were available only on out-of-city calls, they proved to be of limited value.

[4] A "snow day" is one on which one inch or more of snow is recorded. In the case of a storm that lasts more than one day, the entire snow period is considered as one snow day.

[5] The ten-year averages were within two inches of the long-term averages at all sites except Muskegon, which was 33 inches in excess of the long-term average of 72 inches.

Rocky Mountain region. The water equivalent of ten inches of snow at the Wisconsin, Minnesota, and South Dakota sites is approximately one inch, while at Casper and Cheyenne fifteen to sixteen inches of snow are required to produce one inch of water.

General Impact

Most of the difficulties caused by snow stem from disruption of transportation facilities. Even small accumulations may effectively curtail movement and contribute to accidents. Disruptions of retail trade, industrial production, school attendance, construction, civic events, and numerous other activities can be traced largely to the impairment of movement. In the disruption model developed here, highway transportation constituted the major link between the cities and their tributary areas. Although rail transportation is also vital, its functions are more important on a long-term basis; curtailment of rail facilities for a day or two does not generally produce the repercussions associated with disruption of highway transportation. The airlines that serve the study sites currently play only a minor role in the total transportation pattern.

To assess accurately the troublesome effects of snow in urban areas, some kind of categorization is desirable. The hierarchy of disruption presented in Table 1 rests on the assumption that snow unleashes its most damaging effects against transportation. The orders of disruption are ranked along a scale ranging from first order, the most severe, to fifth order, the minimal, which designates assumed but unvalidated inconvenience.

Urban snow disruption is of two kinds: internal, when interchange within the city itself is hampered; and external, when conditions affect the relationships between a city and its tributary area. First-order disruptions may occur in either or both of these situations. So far as internal activity is concerned, the complete restriction of mobility is normally the most serious problem that can be attributed to snowfall, since most functions characteristic of urban areas require movement from one section of the city to another (journey to work, shopping, appointments, and so on).

An inspection of the effects of snow situations on other forms of urban and tributary activity serves to confirm transportation curtailments. For example, if schools were dismissed and a number of business establishments closed, we have verification that traffic flow throughout the city was greatly impeded. In the same vein, the postponement of athletic contests between teams from the city schools and those from other institutions in or near the tributary area provides additional evidence. Another inconvenience associated with snow concerns electric power and communications. On rare occasions a wet clinging snow may result in widespread breakage of cables and wires. In some cases electricity and telephone service may have been disrupted for extended periods, in which event first-order categorization seems warranted.

The hierarchy of disruption is arranged largely on an economic basis. First-order disruptions generally, but not always, cost more than second-order ones, second-order more than third, and so on. However, cost is not always an ideal measure. Thirty accidents do not necessarily result in a greater monetary loss than twenty, and the infrequent fatality associated with a lower-order disruption may produce a more serious loss than might occur in some first-order situations, depending on the value placed on human life.

A combined internal-external first-order or "paralysis" disruption generally finds a community in a state that resembles suspended animation. Vehicular and pedestrian movement is at a standstill; most stores, schools, and offices are either empty or closed. Air, highway, and rail transport are severely hampered, and on occasion the city is completely cut off from its surrounding area.

Perhaps an example will best serve to illustrate both the conditions that may obtain in first-order disruptions and the operation of the classification system. In Cheyenne, snow began to accumulate at 7:00 PM on 8 April 1959, and continued throughout the following day, reaching a total of 8.4 inches. Curtailment of internal and external movement was severe enough to merit a combined first-order disruption rating. The wet snow, accompanied by winds of 15 to 25 miles an hour, halted city bus service and brought private transportation to a standstill. Abandoned automobiles blocked streets throughout the city. Traffic in and out of Cheyenne was confined mainly to emergency vehicles, attempting to get aid to the more than three hundred motorists stranded on United States Highway 30 to the west of

ACTIVITY	1st ORDER (Paralyzing)	2nd ORDER (Crippling)	3rd ORDER (Inconvenience)	4th ORDER (Nuisance)	5th ORDER (Minimal)
Internal					
Transportation	Few vehicles moving on city streets City agencies on emergency alert, Police and Fire Departments available for transportation of emergency cases	Accidents at least 200% above average Decline in number of vehicles in CBD Stalled vehicles	Accidents at least 100% above average Traffic movement slowed	Any mention Traffic movement slowed	No press coverage
Retail trade	Extensive closure of retail establishments	Major drop in number of shoppers in CBD Mention of decreased sales	Minor impact		No press coverage
Postponements	Civic events, cultural and athletic	Major and minor events Outdoor activities forced inside	Minor events	Occasional	No press coverage
Manufacturing	Factory shutdowns Major cutbacks in production	Moderate worker absenteeism	Any absenteeism attributable to snowfall		No press coverage
Construction	Major impact on indoor and outdoor operations	Major impact on outdoor activity Moderate indoor cutbacks	Minor effect on outdoor activity	Any mention	No press coverage
Communication	Wire breakage	Overloads	Overloads	Any mention	No press coverage
Power facilities	Widespread failure	Moderate difficulties	Minor difficulties	Any mention	No press coverage
Schools	Official closure of city schools Closure of rural schools	Closure of rural schools Major attendance drops in city schools	Attendance drops in city schools		No press coverage
External[a]					
Highway	Roads officially closed Vehicles stalled	Extreme-driving-condition warning from Highway Patrol Accidents attributed to snow and ice conditions	Hazardous-driving-condition warning from Highway Patrol Accidents attributed to snow and ice conditions	Any mention, for example, "slippery in spots" warning	No press coverage
Rail	Cancellation or postponement of runs for 12 hours or more Stalled trains	Trains running 4 hours or more behind schedule	Trains behind schedule but less than 4 hours	Any mention	No press coverage
Air	Airport closure	Commerical cancellations	Light plane cancellations Aircraft behind schedule owing to snow and ice conditions	Any mention	No press coverage

[a] Warnings are the key to this classification. They provide excellent indicators because they are widely publicized.

Table 1 Hierarchy of disruptions: internal and external criteria

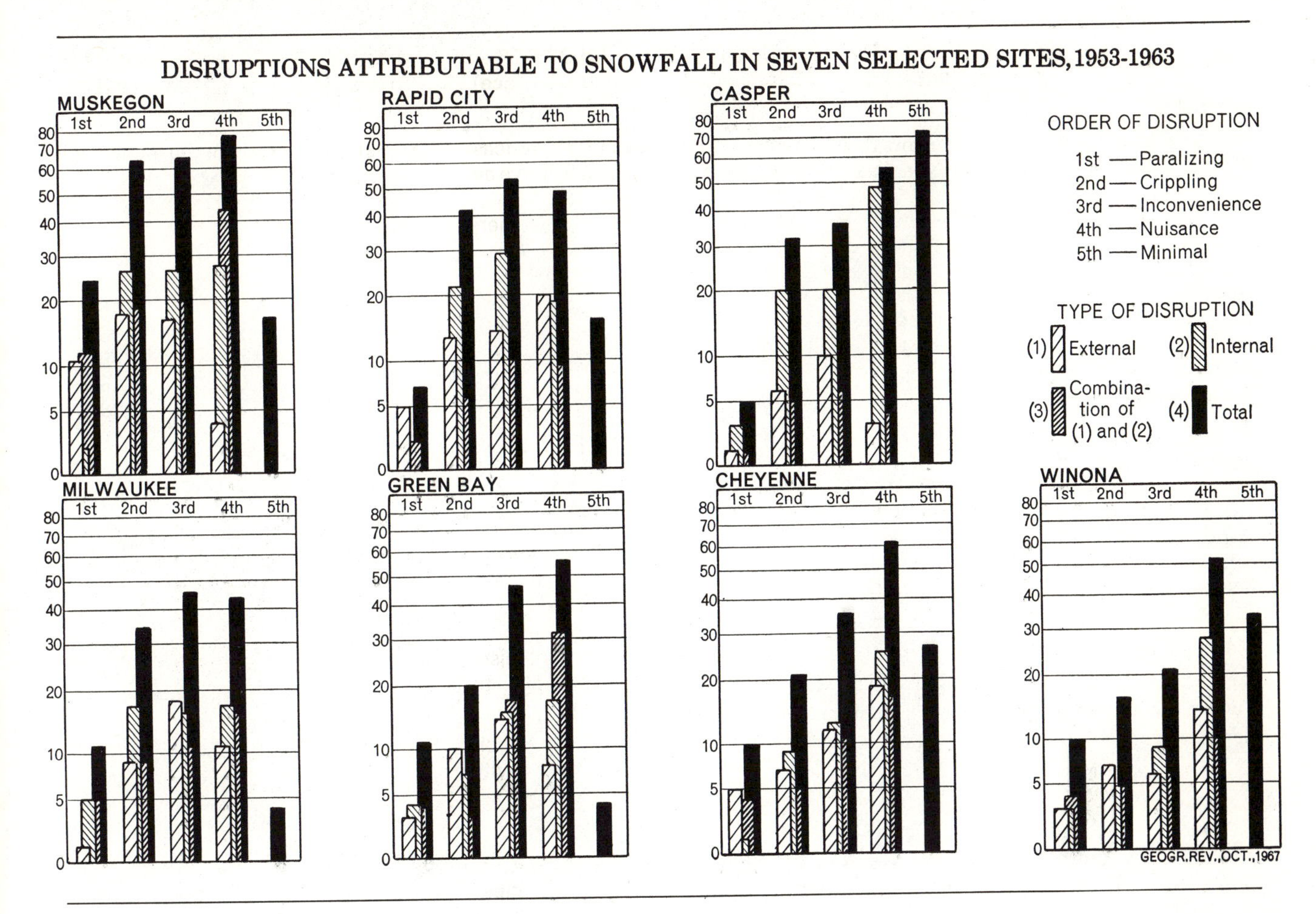

Figure 3

the city. All outbound air traffic was grounded, and no planes were able to land at the municipal airport. City and rural schools closed their doors, and retail trade was heavily curtailed. The ineptness of the Cheyenne snow-control program was all too evident on this occasion. Crews began the removal job the evening of the ninth, too late to combat effectively the impact of the storm. It was like beginning sandbag operations after the stream has crested.

Classification as second- or third-order disruption is based mainly on city motor-vehicle-accident data and on reports on road conditions in the tributary area filed by the State Highway Patrol. Other criteria include the condition of rail and air transportation, activity in the Central Business District, traffic jams, postponements, and school-attendance patterns. Disruption of the fourth order occurs when the impact of snow is not sufficient to cause a breakdown of the activities affected. A fifth-order classification designates an inconvenience so trivial as not to merit mention in the press.

Analysis of Disruption

Disruption in the seven cities is summarized graphically in Figure 3. In general, lower-order disturbances ("inconvenience," "nuisance," and "minimal") are the rule, and paralyzing disruptions occur only at intervals. However, certain differences among the cities are conspicuous. The Midwestern communities tend to have a higher percentage of paralyzing and crippling situations than Casper and Cheyenne do. Both Wyoming sites and Winona (for a different reason) demonstrate a greater clustering of disruptions at the lower end of the scale. The profiles of the two largest cities, Milwaukee and Muskegon, are quite similar, and show greater concentrations of first- and second-order situations. Rapid City is also characterized by a higher proportion of crippling storms, though not of the paralysis variety.

Basically, frequency of disruption increases with annual accumulation of snow (Table 2). The relationship is not perfect, but it is nevertheless significant. Muskegon, which has recorded an annual average of 105 inches of snow, has experienced 206 higher-order disruptions, or an average of 20.6 a year over the past ten years. Among the Midwestern sites with snowfall averages ranging from 37 to 44 inches, the pattern of snow-caused inconvenience is similar, with Winona the only exception.

The relationship between accumulation and disruption is less apparent when the Western sites are

SITES	SNOW ENVIRONMENT: Mean annual snowfall (in inches)	Snow days per year	DISRUPTIONS IN 10-YEAR PERIOD: No. of 1st, 2nd, and 3rd order	Intensity per 10 inches of snow	No. of 1st and 2nd order	Intensity per 10 inches of snow	No. of 1st order	Intensity per 10 inches of snow
	10-year averages							
Casper	67	18.5	84	1.25	36	0.54	6	0.09
Cheyenne	48	12.3	87	1.82	40	0.83	14	0.29
Spokane	54	13.5	129	2.39	70	1.30	25	0.46
Great Falls	59	14.0	106	1.81	77	1.31	2	0.03
Billings	54	13.2	99	1.83	55	1.02	15	0.28
Albuquerque	11	3.3	34	3.27	19	1.83	5	0.48
Colorado Springs	39	11.1	76	1.95	37	0.95	11	0.28
Muskegon	105	20.2	206	1.96	120	1.14	37	0.35
Rapid City	37	11.2	119	3.21	56	1.51	9	0.24
Milwaukee	44	10.2	118	2.68	55	1.25	16	0.36
Green Bay	39	10.8	101	2.59	38	0.97	15	0.38
Winona	42	13.1	62	1.48	35	0.83	14	0.33
Springfield	24	9.1	76	3.18	44	1.83	16	0.68
Des Moines	37	10.5	105	2.84	57	1.54	15	0.41
Fort Wayne	32	9.4	119	3.72	72	2.25	16	0.50
Greensboro	11	3.4	54	5.04	48	4.48	21	1.96
New York	30	7.2	96	3.20	71	2.36	42	1.39

Table 2 Relationship between snow environment and disruptions of the first, second, and third order

considered. Cheyenne (48 inches) averages 8.7 higher-order disruptions in the normal year and Casper (67 inches) experiences only 8.4. To search out the causes of this discrepancy, ten additional cities were investigated, on a ten-year sampling basis similar to that used for the original seven sites. Western sites in the sample were Spokane, Great Falls, Billings, Colorado Springs, and Albuquerque. To supplement the Midwestern sites, New York City, Fort Wayne, Des Moines, Springfield (Illinois), and Greensboro (North Carolina) were selected.

The graphic comparison of disruption and the snow environment suggests that basic differences exist between the Western disruption patterns and those characteristic of the areas east of the High Plains (Figure 4). Most of the Western cities experience fewer snow problems. This generalization applies regardless of whether magnitude or intensity measurements are used.

Why do these differences exist? It appears that the most important factor is the lower water content associated with the majority of snowfalls in the West. The drier snows generally present a less formidable barrier to movement and hence on the average tend to cause less disruption.[6] However, this is not to say that in a given situation (for example, abundant snow buffeted by blizzard-force winds) dry snow is incapable of producing as much difficulty as the wetter variety.

[6] "Dry snow" refers here to snow with a water content of less than one inch per ten inches of snow.

A second factor that reduces disruption at the Western sites is the way in which the hazard is perceived. The view that regards snow as an element that must be coped with by the individual has produced a much more comprehensive range of personal adjustments than are commonly found farther east.[7] Such adjustments are particularly effective at reducing disruption of the second and third orders. "Western perception," on the other hand, has contributed to the pathetic ineptitude of public adjustment in the region. This inability of the public sector to react has created a greater vulnerability to severe snow conditions than exists in the East. As a result, the Western cities are no better than their Eastern counterparts in reducing first-order disruption, and under most circumstances their recovery period is considerably longer.

The graphs also suggest that disruption, though it increases with annual snowfall, does so at a diminishing rate. Inversely, the intensity of disruption decreases with increasing annual accumulations. The matched pairs of Albuquerque-Greensboro and Cheyenne-Milwaukee are clearly demonstrative of such a pattern. A larger sample of north-south cities would probably further validate this relationship.

[7] A considerable percentage of those interviewed in the West carried emergency provisions such as canned food and blankets in their automobiles. Also, many more Westerners than Easterners equipped their cars with snow tires or chains, and made greater use of professional advice provided by the United States Weather Bureau and the State Highway Patrol.

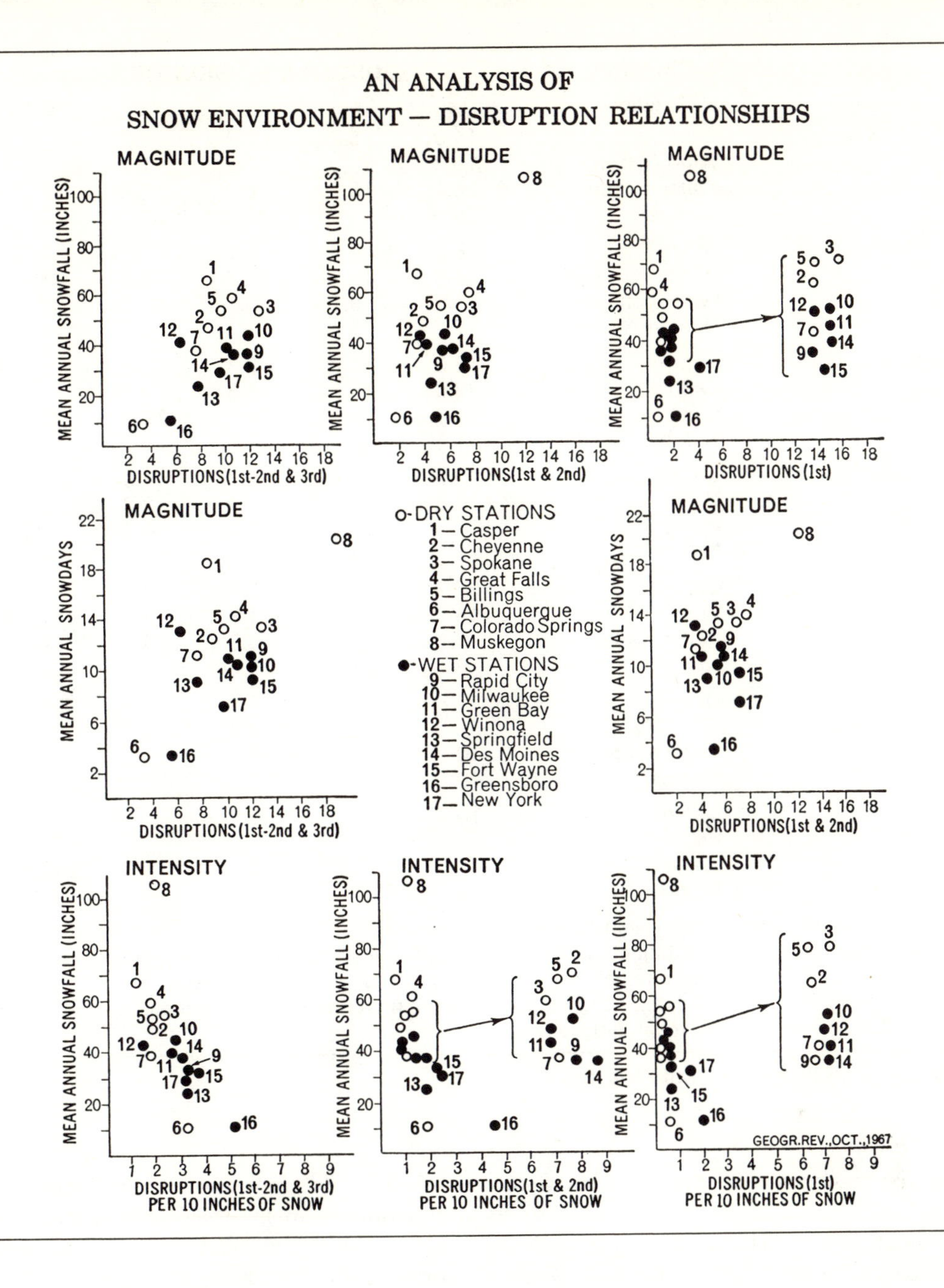

Figure 4

Specific Causes of Disruption

The relationship between disruption and snow depth, water content, and wind in individual storms at the seven original sites is presented in Table 3. The data illustrate a strong correlation between depth of snow and curtailment of human activity. Storms that resulted in first-order situations almost invariably registered accumulations of five inches or more. Only on rare occasions—for instance, a sleet storm, drifting of snow on the ground from a previous fall, extremely high winds—did lesser accumulations produce widespread havoc.

Most of the snowfalls that caused first-order disruption accumulated during twenty-four-hour periods. Such rapid falls generally resulted in conditions that were extremely difficult to deal with. Only in Muskegon did a significant number of paralysis situations develop over a period of two or more days; on several occasions it snowed continuously for as long as nine days, creating conditions of severity unknown at any of the other sites. However, even during some of these prolonged falls, crews were able to prevent complete curtailment by maintaining operations around the clock.

Although deep snow and severe disruption are strongly correlated, deep snow by itself often tends to produce only moderate, and sometimes even minimal, disruption. Most accumulations exceeding five inches were accompanied by winds of more than fifteen miles per hour. However, on many occasions when winds

SITES	CRITERIA (Averages)	ORDER OF DISRUPTION 1st	2nd	3rd	4th	5th
Muskegon	Depth (in)	14.0	5.7	3.7	2.0	1.3
	Wind (mph)	15.2	12.1	11.2	11.7	13.8
	Water content[a]	16.1	16.2	16.2	12.1	15.2
	Number[b]	25	53	47	57	17
Rapid City	Depth	8.8	2.3	2.1	1.5	1.4
	Wind	24.8	15.5	11.7	13.3	15.8
	Water content	10	9.1	10.1	9.6	10.1
	Number	7	33	35	22	11
Milwaukee	Depth	12.0	4.2	2.7	1.6	1.4
	Wind	23.5	16.6	13.3	14.3	7.5
	Water content	11.3	10.9	11.7	12.1	12.2
	Number	11	26	34	29	4
Green Bay	Depth	6.9	2.9	2.7	1.8	1.4
	Wind	16.7	15.1	11.4	9.8	8.3
	Water content	10.5	11.7	10.6	10.3	10.2
	Number	11	14	33	42	4
Casper	Depth	7.7	4.2	3.1	3.1	1.7
	Wind	15.9	12.1	11.8	12.9	12.2
	Water content	13.7	14.9	16.8	13.6	14.8
	Number	5	27	29	57	69
Cheyenne	Depth	8.8	5.0	3.3	2.0	1.5
	Wind	25.6	15.6	14.3	13.7	13.6
	Water content	10.9	12.7	16.2	13.9	12.3
	Number	10	16	33	36	27
Winona	Depth	10.7	5.8	3.6	2.8	1.7
	Wind					
	Water content	11.6	11.2	10.2	12.9	12.6
	Number	10	11	19	44	34

[a] Amount of snow equivalent to one inch of water.
[b] Represents only the highest-order disruption for any snow day. For example, if a snow day is rated 1st order internal and 3rd order external, only the former is included in the sample. The fact that combination disruptions are not counted twice in these computations also decreases the number in the sample.

Table 3 Disruption in relation to physical variables

were light and the snow was dry, the problems were substantially reduced.

Wind. A significant statistical difference exists between the impact of snow in association with winds of fifteen miles per hour or more, and that accompanied by winds of lesser velocity.[8] Nearly all first-order occurrences developed with wind velocities in excess of fifteen miles per hour (Table 4). On the other hand, lower-order disruptions characteristically were associated with lower wind velocities.[9]

A reexamination of Table 3 reveals average wind velocities to be substantially higher in first-order situations than in lower-order disruptions. Winds of more than fifteen miles per hour are common in first-order disruptions at all sites, and exceed twenty-three miles per hour at Milwaukee, Rapid City, and Cheyenne. In addition, a precipitous drop in wind velocities is associated with second- and third-order cases at all sites, and particularly in these three cities.

Further evidence in support of the impact of high winds was obtained through personal interviews at four of the seven sites. Respondents were asked to select from a list of six types of snow conditions the two they felt presented the most serious hazard in their area. The responses are summarized in Table 5. Snow in association with high winds was ranked as the most severe type of hazard in Cheyenne, Rapid City, and Casper, while deep snow in itself was given low priority. The respect afforded ground blizzards in those locations also demonstrates the role of wind. Such evidence, though not conclusive, does substantiate the data procured from newspapers and public records. It

[8] For the purposes of this study a statistically significant association is one (measured by chi-square, representing the sum of the differences of the observed distribution and one that might be expected if there were no association between the given variables) that had only one chance in twenty or less of arising purely by sampling variability or by chance (0.05 level of significance).

[9] Chi-square tests were significant at the 0.01 level of probability (that is, there is less than one chance in a hundred that the association is attributable to chance) at all sites except Casper and Muskegon.

		ORDER OF DISRUPTION			
SITES	WIND VELOCITY	1st	2nd and 3rd	4th and 5th	Total
Muskegon	Data incomplete				
Rapid City	≥ 15	9	48	20	77
	< 15	0	62	54	116
Milwaukee	≥ 15	13	58	21	92
	< 15	3	44	38	85
Green Bay	≥ 15	14	32	15	61
	< 15	1	54	75	130
Casper	≥ 15	4	26	43	73
	< 15	2	52	87	141
Cheyenne	≥ 15	13	35	39	87
	< 15	1	38	67	106
Winona	Data unavailable				

Table 4 Effect of wind in promoting disruption

Sites	Number of persons interviewed	Number making estimate	Snow and high winds	Snow and sleet	Snow over ice	Ground blizzard	Extremely deep snow	Snow and extreme cold
Cheyenne	43	43	1	3	2	5	4	6
Casper	45	45	1	6	3	2	4	5
Rapid City	62	55	1	3	2	4	5	6
Winona	21	20	2	1	4	6	3	5

Table 5 Ranking of severe hazards by persons interviewed

also indicates an awareness on the part of the respondents of what proved to be the most significant cause of disruption.

Water Content. Snow made heavier and stickier owing to relatively greater moisture content often tends to present a more formidable hazard, as we have seen. In Cheyenne it was found that wetter snow resulted in a considerably greater amount of disruption. Although a statistically significant difference between the impact of "wet" and "dry" snow was not characteristic of the other sites, it appears, again, that moisture content of snow is responsible for areal differences in disruption. The Midwestern and Eastern cities, where higher-order disruptions are considerably more frequent, experience snow-fall which on the average is about 50% wetter than that recorded at the Western sites.

Temperature. The role of air temperature is difficult to measure.[10] An analysis of high, low, and mean temperatures provides little evidence concerning the function of temperature in maximizing or minimizing disruption. Mean temperatures associated with all orders of disruption did not vary more than four degrees at any of the sites.

Temporal Variation. Another element that tends to complicate and obscure the relationship between disruption and the snow environment is the time of occurrence. Temporal variation as used here refers to the rate of fall and the time of day when it occurs. The rate of fall governs the ability of snow-control operations to keep pace; the time of fall often means the difference between a fourth-order and a second-order disruption. For example, three inches immediately preceding the morning rush hour can cause great difficulty, whereas the same amount accumulating during the late evening or early morning hours might result in minimal inconvenience. Weekend snows produce less internal disruption, but generally more external difficulties, and the reverse is true of snows that occur during the week.

The Factors Combined

To summarize the impact of snow and associated weather conditions on human activity, probability matrixes have been constructed for the Western and Midwestern sites (Table 6). The data demonstrate the relative contribution to disruption of snow depth, wind, water content, and temporal variation.

Fundamentally, the matrix for the Midwestern sites illustrates that the probability of severe disruption increases with depth. High winds are shown to be an important catalyst in promoting the hazard at all snow depths. For example, the odds against a three-inch snow causing a crippling disruption are five to one; add wind and they drop to two and a half to one. Wet snows are not consistent in promoting additional havoc in the

[10] At temperatures below 15° F salt has little effect on ice, and when the temperature approaches 0° F even calcium chloride is ineffective. A temperature hovering around the freezing mark is critical, for variation in either direction can make the difference between very slippery or merely wet pavements.

DEPTH IN INCHES	NO. OF SNOW DAYS	DEPTH ALONE 1st	2nd	3rd	4th	5th	WINDS OF ≥ 15 MPH 1st	2nd	3rd	4th	5th	WATER EQUIVALENTS OF ≥ 0.10/INCH 1st	2nd	3rd	4th	5th	WEEKEND SNOWS (TIME FACTOR) 1st	2nd	3rd	4th	5th
Midwestern Sites																					
1–1.99	341	.00	.10	.25	.48	.17	.00	.20	.28	.44	.08	.00	.11	.30	.54	.05	.00	.08	.22	.48	.22
2–2.99	159	.02	.20	.31	.38	.09	.04	.32	.36	.28	.00	.02	.22	.39	.30	.07	.00	.21	.34	.40	.05
3–3.99	86	.04	.19	.44	.31	.02	.08	.41	.44	.07	.00	.06	.23	.39	.30	.02	.02	.16	.46	.34	.02
4–4.99	51	.04	.26	.39	.31	.00	.17	.58	.17	.08	.00	.08	.25	.37	.30	.00	.04	.29	.35	.32	.00
5–5.99	45	.07	.44	.44	.05	.00	.10	.70	.20	.00	.00	.12	.55	.30	.03	.00	.04	.45	.42	.09	.00
6–6.99	30	.20	.43	.23	.14	.00	.30	.40	.30	.00	.00	.27	.43	.20	.10	.00	.14	.47	.19	.20	.00
7–7.99	30	.27	.40	.30	.03	.00	.44	.38	.18	.00	.00	.38	.38	.21	.03	.00	.20	.42	.35	.03	.00
8–8.99	8	.13	.50	.37	.00	.00	.33	.67	.00	.00	.00	.25	.50	.25	.00	.00	—	—	—	—	—
9–9.99	12	.50	.33	.17	.00	.00	1.00	.00	.00	.00	.00	.60	.20	.20	.00	.00	.50	.25	.25	.00	.00
10 and over	65	.52	.42	.06	.00	.00	.64	.36	.00	.00	.00	.54	.42	.04	.00	.00	.46	.48	.06	.00	.00
Western Sites																					
1–1.99	151	.00	.06	.11	.33	.50	.00	.08	.06	.44	.42	.00	.07	.21	.33	.39	.00	.05	.10	.40	.45
2–2.99	73	.03	.15	.22	.38	.22	.07	.11	.29	.32	.21	.05	.10	.15	.45	.25	.02	.08	.24	.34	.32
3–3.99	45	.02	.09	.39	.35	.15	.06	.18	.35	.35	.06	.00	.20	.20	.30	.30	.00	.15	.35	.40	.10
4–4.99	26	.04	.27	.19	.50	.00	.08	.42	.08	.42	.00	.00	.25	.50	.25	.00	.04	.18	.24	.54	.00
5–5.99	15	.20	.33	.27	.20	.00	.00	.38	.38	.24	.00	.25	.00	.50	.25	.00	.17	.33	.25	.25	.00
6–6.99	13	.00	.38	.38	.24	.00	.00	.20	.60	.20	.00	.00	.50	.00	.50	.00	.00	.20	.40	.40	.00
7–7.99	17	.18	.30	.30	.22	.00	.50	.33	.17	.00	.00	.38	.38	.24	.00	.00	.25	.38	.12	.25	.00
8–8.99	1	1.00	.00	.00	.00	.00	1.00	.00	.00	.00	.00	1.00	.00	.00	.00	.00	—	—	—	—	—
9–9.99	1	.00	1.00	.00	.00	.00	.00	1.00	.00	.00	.00	.00	1.00	.00	.00	.00	—	—	—	—	—
10 and over	13	.39	.39	.15	.07	.00	.67	.33	.00	.00	.00	.67	.33	.00	.00	.00	.25	.75	.00	.00	.00

Table 6 Probability of experiencing disruption from snow and selected associated weather conditions

Midwest, but this is because most snows there are fairly wet. Expectably, weekend snows are generally less disruptive than their weekday counterparts. In the West the role of depth and wind are also apparent from the matrix. Furthermore, wet snows appear to cause significantly more difficulty than dry ones do.

THE ROLE OF ADJUSTMENT AND ATTITUDE

Most cities that lie within the "snow belt" (the area north of 35° N, excluding the Pacific states) have some form of snow-control program. Expenditures for this service range from the $22 million spent by New York City in 1963–1964 to the few hundreds of dollars appropriated annually by many of the smaller communities in the southern part of the belt. The total expenditures for snow control in the United States can only be estimated. The American Public Works Association has been using a figure of $100 million a year, but considers this to be much too low, especially since salt costs alone run to $44 million. The following statement by Lockwood of the American Public Works Administration (1965) indicates the difficulties encountered in compiling figures for snow removal costs: "One of the main things we uncovered in our study was the general lack of basic operating and especially cost data. Surveys we made were generally inconclusive because data received were not comparable. We obviously were comparing apples with pears. The fact that data on snow operations are limited is understandable. Snow is treated as an emergency and record keeping has a secondary priority." But costs are only one measure of man's attempt to counteract the snow hazard. Other important facets of control comprise organization, communication with the public, coordination among city agencies, and among state, city, and suburban agencies.

It is impossible within the scope of the present investigation to assess precisely the positive effects of public adjustment to the snow hazard. However, by analyzing and comparing the snow-control programs and the disruption patterns of the cities studied, a number of insights can be gained.

Snow- and ice-control operations at the Midwestern sites run the gamut from very good to marginal.[11] By present technological standards the programs of Milwaukee, Muskegon, and Winona rank high. The Green Bay program is better than average; that of Rapid City is definitely marginal. By Midwestern or Eastern standards, neither Casper nor Cheyenne is unduly concerned with snow and ice control.

Since the snow environments at Green Bay and Rapid City are basically similar, it is possible to gauge the effect that adjustment has on the impact of the hazard. Green Bay spends more money per capita, has an efficient alert system, and has a well-organized course of action that calls for operations to begin *during* the storm when accumulations reach two inches or more. On the other hand, Rapid City has no formal plan, and generally postpones action until after the snowfall has ceased. Moreover, the city's police department is apparently unwilling to enforce snow-emergency regulations.

The more efficient snow-control operation at Green Bay has resulted in holding down higher-order disruptions to 2.59 per ten inches of snow, as compared with the figure of 3.21 for Rapid City (Table 2). In

[11] The quality of a snow-control program can be evaluated on the basis of expenditures, alert systems, deployment of men and equipment, organization, public relations, and the degree of cooperation that exists among the various agencies necessary to its success.

percentage terms, Green Bay experiences less than 80% as much disruption. Whether or not this reduction can be attributed solely to snow-control differences is a question that can be answered only by the investigation of snow problems at many additional sites, or perhaps by a concentrated economic analysis of one or two cities. It is interesting that nearly half of those interviewed in Rapid City considered the city program to be woefully ineffective. The majority were in favor of immediate improvement, another indication of the general dissatisfaction with the current quality of snow-control service.

Newspaper coverage of storms in Milwaukee and Muskegon revealed that their alert and well-organized public works departments minimized the impact of snow on numerous occasions. Many first-order situations failed to materialize in Muskegon owing to the diligence of the city and county officials.

The Winona program presents an opportunity to measure community attitudes toward the disruptive impact of snow. Snow control in that city was viewed as "excellent" to "very good" by more than 80% of those queried. In addition, fewer than 10% felt improvements in the program were warranted. If these views can be accepted as factual, and not simply as statements of community pride, it would seem that Winonans are satisfied with the present level of snow-caused inconvenience which their city experiences.

The Winona case suggests that the "satisficer" notion is of considerable value in accounting for the curious relationship that exists between man and the snow hazard (Simon, 1957). Perhaps the esthetic values of snow make us somehow reluctant to wage all-out war against it. If the "optimizer" approach were applied to the snow hazard, it would emphasize strategy designed to eliminate completely the negative impact of snow. Further evidence to support the applicability of the satisficer concept comes from the field of snow control. Innovations designed to improve snow control have lagged behind, and those that have emerged are being adopted very slowly. Essentially we are coping with snow in much the same way we did thirty years ago. Our attitudes are simply reflected in our actions. If demand existed for real innovations in snow control, they would be forthcoming. Most of the improvements developed thus far (radiant heat, snow melters, street flushing devices) have been rejected owing to their high cost and to the lack of knowledge concerning the losses attributable to snow.

Individuals spend considerably more to protect themselves and their property from the snow hazard than they spend as members of the public sector. Snow tires alone cost the citizens of Rapid City, Casper, and Cheyenne combined more than $600,000 annually.[12] Added to that are expenditures for tire chains, shovels, snow brooms and scrapers, sand, salt, and numerous ice-melting compounds. Since people are willing to funnel these amounts into personal snow control, it seems reasonable to suppose that they would favor additional public expenditure. In fact, the majority of those queried did support the idea of program improvement. However, although an occasional clamor is heard after an unusually severe winter, memories tend to be short when the time for increased appropriations is at hand. This pattern is analogous to the attitudes that often prevail with respect to other hazards, particularly floods (Hart, 1957).

Attitudes can promote or reduce disruption, largely through their effects on adjustment. Most of the persons queried in the seven cities tended to underestimate the hazard potential of snow, considering it to be more of a nuisance than a serious problem[13]—as, indeed, it is most of the time. In the case of minor storms, these attitudes probably lessen disruption; that is, people exhibit little concern for two, three, or four inches of snow and go about their business normally. On the other hand, they are apt to be grossly unprepared for a severe storm, and thus to experience substantially more disruption.

THE ROLE OF PERCEPTION

Man's attitude toward the snow hazard can be partly explained in terms of his perception of the phenomenon,

[12] This amount is derived from the number of car registrations and the percentage of vehicle owners who say they install snow tires. The estimate allows for a tire life of three years.

[13] This statement does not apply to Muskegon, Milwaukee, and Green Bay, where extensive interviewing was not conducted. There is reason to believe that people hold the hazard in higher esteem in the Midwest, as evidenced by the existence of more sophisticated snow-control programs in that area.

but this perception is difficult to measure. Awareness of any element varies not only among individuals and groups, but with the same individuals at different points in time and space. "In any society, individuals of similar cultural background, who speak the same language, still perceive and understand the world differently" (Lowenthal, 1961, p. 255). "The specialized literature is replete with examples of difference in hazard perception" among the experts themselves (Burton and Kates, 1964, p. 424). Even with these drawbacks, the study of perception can be extremely valuable in identifying the geographical implications of the milieu, particularly that part of it which contains an uncertain hazard element.

In talking with people in the various cities a tendency was detected, especially in the West, to minimize the potential danger associated with snow. Statements such as these were common: "The legendary blizzard of 1949 was some sort of oddity that will probably never happen again" (Calef, 1950). "Snow doesn't bother or interfere with us or our business." "I view snow and ice as a challenge, something to break the monotony of the everyday routine." Perhaps the slogan emblazoned on the façade of the Engineering Building at the University of Wyoming is symbolic of prevailing attitudes. It reads: "Strive On, the Control of Nature Is Won, Not Given."

Perception of any hazard is based largely on experience. The slower pace of life in the Western cities may account in part for the rather low priority granted to the snow hazard there. This does not explain the views that prevail with respect to the "blizzard of 1949," or to the other storms since that have resulted in loss of life and widespread disruption. As many as three or four first-order disruptions may occur during the course of any winter season, yet the population remains largely apathetic and ill prepared, at least as a public body.[14] An examination of additional sites, particularly those with only small and highly variable amounts of snow, should provide greater insight concerning snow-hazard perception. "On the spot" interviews should also be useful.

FUTURE NEEDS

The snow hazard demands more attention. Additional research is needed, both in the physical parameters—accumulations, moisture content, wind velocities, and so on—and in the evaluation of local adjustments at the private and public levels.[15]

An even more pressing need is for improved public service in the field of snow control. A single storm can cost any of the cities investigated considerably more than they spend on snow removal each year. Organization, coordination, and public relations are integral parts of more effective snow-control programs. Disruption could be substantially reduced if more funds were available, and if present funds were more efficiently allocated. An accurate benefit-cost analysis might produce guidelines for intelligent decisions on expenditures.

We are confronted by an interesting challenge. As a technically advanced urban society we have at our disposal the organizational and inventive abilities to deal successfully with snow, a menace that often severely hampers activity in our major centers. Why do we not use them?

[14] The political ideology in Wyoming and western South Dakota that stresses the role of the individual may be reflected in the inefficiency of public snow-removal programs. However, speculation along this line is difficult to substantiate.

[15] The writer is currently engaged in further research on this matter at sites in the south–central and southeastern United States. An analysis of disruption patterns in these areas should provide answers concerning the benefits to be expected from the maintenance of various levels of snow-control programs.

REFERENCES

Burton, I. and Kates, R. W., 1964. The perception of natural hazards in resource management. *Natural Resources Journ. 3*, 412–441.

Calef, W., 1950. The winter of 1948–49 in the great plains. *Annals*, Association of American Geographers *40*, 267–292.

Hart, H. C., 1957. Crisis, community, and consent in water politics. *Law and Contemporary Problems 22*, 510–573.

Lockwood, R. K., 1965. Personal communication, Assistant Director of Technical Services, American Public Works Association, May 4.

Lowenthal, D., 1961. Geography, experience, and imagination: towards a geographical epistemology. *Annals*, Association of American Geographers *51*, 241–260.

Rapp, R. R. and Huschke, R. E., 1964. Weather information: its uses, actual and potential. *The Rand Corp.*, Santa Monica, California.

Rooney, J. F., jr., 1966. *The Urban Snow Hazard: An Analysis of the Disruptive Impact of Snowfall in Ten Central and Western United States Cities.* Ph.D. thesis, Clark Univ., Worcester, Massachusetts, available from Univ. Microfilms, Ann Arbor, Michigan.

Russo, J. A., jr., Trouern-Trend, K., Ellis, R. H., Koch, R. C., Howe, G. M., Milly, G. H., and Enger, I., 1965. The operational and economic impact of weather on the construction industry of the United States. *Final Rept. 7665-163, Contract Cwb-10948, The Travelers Res. Center, Inc.*, Hartford, Connecticut. 102 pp.

Simon, H. A., 1957. *Models of Man* (2nd ed.). New York.

Hans Neuberger

26 Climate in art

Even the casual visitor to art galleries cannot fail to notice that every century has produced many artists who have chronicled the fauna and flora of their environment, the foods and fashions of their times, the furniture and architecture, the implements in daily use and the musical instruments played, as well as the sports and games that served as entertainment. In view of the powerful influence of the atmospheric environment on life on this planet in general, and on man in particular, it would be rather amazing if the artists could have remained oblivious to the weather surrounding them and merely invented whatever weather best suited their artistic purposes. Individual artists, especially modern objective painters, undoubtedly may have consciously avoided depicting the kind of weather with which they were most familiar. Nevertheless, it is the author's contention that a statistically adequate sample of paintings executed by many painters living during a given period in a given region should reveal meteorological features significantly different from those of a similar sample of paintings produced during the same epoch in a climatically different region.

To test this hypothesis, more than 12,000 paintings in 41 art museums in the United States and eight European countries were surveyed in 1967 with respect to various meteorological and other weather-related features. These paintings covered the period from 1400 to 1967, and were categorized according to the major art "schools" to which paintings are conventionally assigned in art galleries; these schools were then assumed to represent different climatic regions. The Flemish and Dutch schools were combined into the "Low-Countries" school because of the great similarity of the climate in what is now Belgium and the Netherlands; furthermore, the climate of the Atlantic seaboard of North America from Virginia northward was taken as the region equivalent of the American school because 90% of the artists from this school came from this area. The German school included Swiss, Austrian, and Bohemian artists with a handful of Polish and Jugoslavian artists.

REGIONAL CLIMATES

In each region a more or less centrally located station was selected to represent the climate of the pertinent school: New York for the American school, London for the British school, Paris for the French school, the averages of De Bilt in Holland and Uccle in Belgium for the Low-Countries school, Goerlitz for the German school, Rome for the Italian school, and Madrid for the Spanish school. The climographs for the seasonal combinations of temperature and precipitation (period 1951–60) shown in Figures 1a and 1b confirm the existence of climatic differences between regions; the hot dry summers in the Mediterranean area, the warm wet summer in Goerlitz with its cold dry winter, and the uniform wetness and large seasonal temperature amplitude in New York (Figure 1a) are in substantial contrast to one another as well as to the more moderate climates of Paris, London, and Uccle/De Bilt (Figure 1b). Although these climates underwent changes in the past five-and-a-half centuries, it is fairly safe to assume that some differences persisted throughout the period considered by the survey. Nonfulfillment of the assumption, together with the fact that many artists migrated from their place of origin (van Gogh, for example, is assigned to the Dutch school, although he painted most of his works in southern France), would tend to minimize any existing differences in meteorological features of the paintings. Therefore, whatever differences are found would be much greater if a more adequate regional assignment for the paintings had been possible.

Reprinted, with minor editorial modification, by permission of the author and editor, from *Weather*, Royal Meteorological Society, Vol. 25, 1970, pp. 46-56.

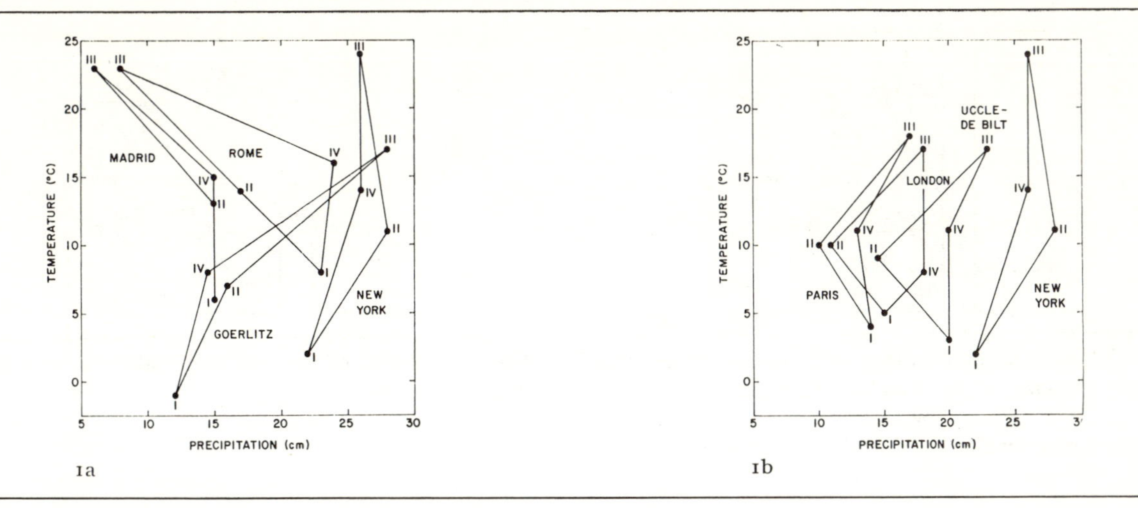

Figure 1 Seasonal climographs for various stations (1951–1960 averages).

THE OBSERVATIONS

Regarding the pictorial information gathered in this survey, the artists' names, the art school and decade of the painting, the pictorial contents, the brightness or darkness of the picture were recorded, as well as the following meteorological observations: The blueness of the painted sky was judged according to a three-step scale, denoting "1" for pale blue, "2" for medium blue, and "3" for deep blue. This simple scale, rather than a more elaborate one (Neuberger and Neuberger, 1943), was used because the blueness of painted skies may now be quite different from what it was originally, as a result of yellowing varnishes, chemical changes of the paints used, restorations undertaken, as well as the illumination at the time of viewing. The average blueness of many paintings could then be converted into percent values by taking a step average of 1.00 as zero percent and a step average of 3.00 as 100%. A similar scale was used for estimating the degree of visibility which was judged by the haziness or clarity of the landscape background in a painting.

Cloudiness was estimated according to the U.S. airways code in four categories: *clear* (less than 10% of the visible sky area covered by clouds), *scattered* (10 to 50% clouds), *broken* (60 to 90% clouds), and *overcast* (more than 90% clouds). In addition, the types of clouds were observed according to four families: high, middle, low, and convective clouds.

As a climate-related feature, the presence of a habitable edifice or a nonhabitable one (exclusive of churches) in the paintings was noted. All those painted structures were considered nonhabitable that had no walls, or no roof, or were otherwise incomplete or ruined so as to be incapable of providing suitable human shelter. It was thought that artists living in inclement climates would be less inclined to paint such structures, than those living in warm and sunny areas. Of course, it was realized that art fashions play a significant role in determining the frequency with which various features occur in the paintings of different periods; for example, depicting Roman ruins was fashionable among the European artists of the 18th century when the interest in antiquity had been reawakened. This fact, too, would tend to diminish the magnitude of regional differences.

As a first result, the analysis revealed that more than 53% of all the paintings contained meteorological information; the other paintings were mostly portraits or indoor scenes that did not show any sky, although in an average of 14% of such scenes some sky is visible through windows or doors, and many portraits were placed in an outdoor setting. By the same token, an average of 8% of all outdoor scenes showed no sky, because the horizon was placed too high or was obscured by mountains, trees, houses, etc.

REGIONAL DIFFERENCES

Regarding the relative frequency of nonhabitable structures painted, Figure 2 shows, indeed, that there are regional differences; Mediterranean artists painted most of such structures, American and British artists only very few. This statistical result derived from a total of 4,000 paintings with buildings is undoubtedly influenced by regional preferences for different pictorial contents; for example, 59% of the Italian and 55% of the Spanish paintings represent religious scenes, which are very rare in American, British (1% each), and French (9%) works of art. And most of the nativity scenes painted show nonhabitable shelters, presumably for dramatic emphasis on Christ's lowly birth. The regional correlation coefficient between religious pictures and those with nonhabitable buildings is 0.98 ± 0.01; nevertheless, one cannot entirely reject the possibility of both parameters being a function of the artists' climatic environments.

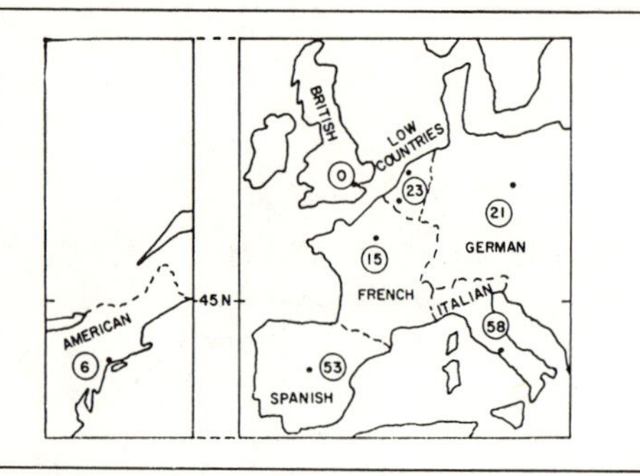

Figure 2 Number of paintings with nonhabitable structures in percent of the number of paintings with any structures. (The dots show the location of the stations considered representative of the various regions.)

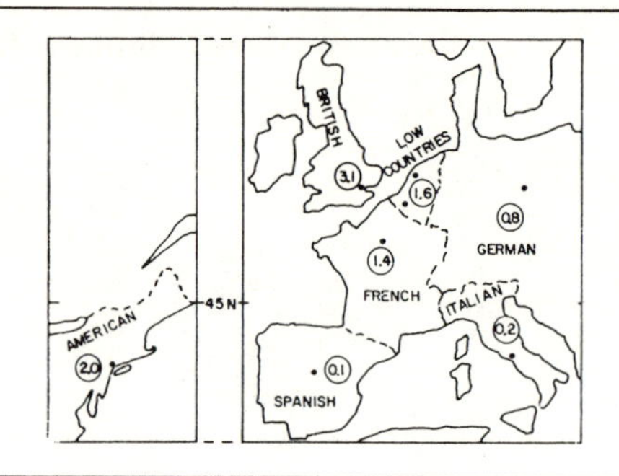

Figure 3 Ratios of frequencies of pale-blue skies to those of deep-blue skies.

In spite of the anticipated difficulties associated with the appraisal of the color of painted skies, significant differences became evident from the analysis of almost 4,500 paintings with blue skies. The ratios of the frequencies of pale-blue to those of deep-blue skies in Figure 3 show the British school to have by far the largest number of pale-blue and the fewest deep-blue skies, whereas these ratios are less than unity for the central and southern European schools which produced more deep-blue than pale-blue skies.

A very similar result was obtained from 5,600 paintings with respect to the regional distribution of the average visibilities expressed in percent (Figure 4); the clearest atmospheres were painted by the Mediterranean schools, while the haziest air is found in British paintings.

In each of the 5,800 paintings containing clouds, only the dominant cloud family was recorded. The results in Table 1 show the high clouds to be the least favored by all schools, the middle and convective clouds being the most favored families. Regional differences become apparent only with respect to the middle and low clouds between the British and the Italian schools, although any differences between these schools and the others are also statistically significant at the 1% level. Since the families of low and convective clouds contain the clouds of "bad" weather, such as nimbostratus and cumulonimbus, the combination of these cloud families is probably most likely to show regional differences in storminess (Figure 5).

More than 6,500 paintings were available for analysis of the regional distribution of cloudiness. Table 2 shows that there was not a single British painting with clear sky, whereas overcast skies were much more frequent in paintings of British than of other artists. The German, Italian, and Spanish schools painted significantly more clear and fewer overcast skies than did the other schools. As a result, these schools show a lower average cloudiness in percent of the sky area painted than the other schools (Figure 6).

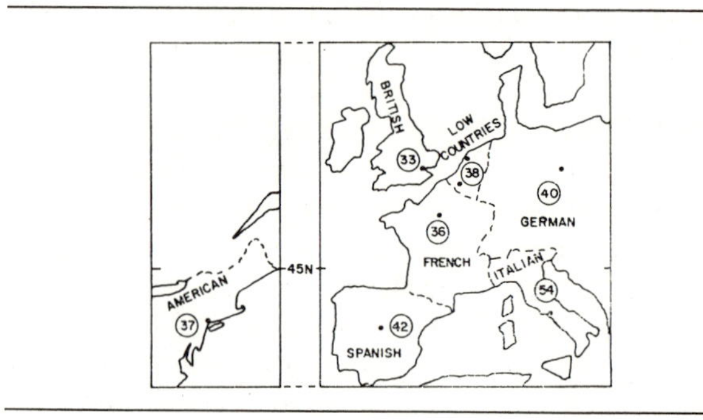

Figure 4 Average percent visibility.

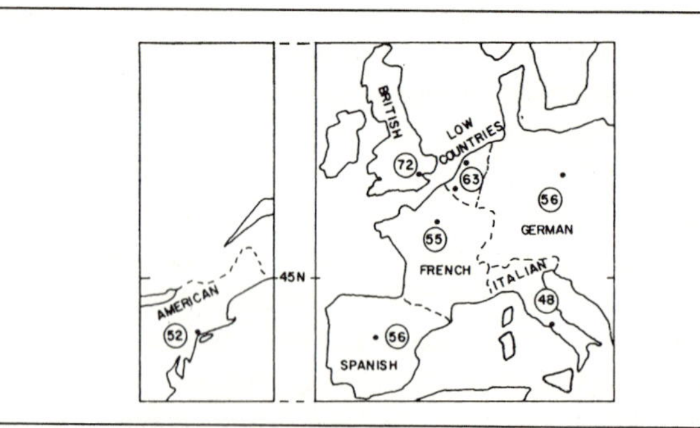

Figure 5 Relative frequencies of low and convective clouds.

Table 1 Relative frequencies (%) of cloud families by schools

	CLOUD FAMILIES			
Schools	High	Middle	Low	Convective
American	5	43	30	22
British	5	23	40	32
Low Countries	5	32	27	36
French	9	36	27	28
German	11	33	27	29
Italian	9	43	14	34
Spanish	6	38	28	28
Average	8	35	25	32

Schools	Clear	Scattered	Broken	Overcast
American	5	16	47	32
British	0	4	48	48
Low Countries	7	11	44	38
French	9	10	49	32
German	16	16	42	26
Italian	12	24	42	22
Spanish	13	13	38	36
Average	10	15	43	32

Table 2 Relative frequencies (%) of cloudiness categories

Figure 6 Average cloudiness in percent of the sky area painted.

RELATIONSHIP BETWEEN PAINTED AND OBSERVED VISIBILITY AND CLOUDINESS

To test the extent to which the regional differences in the artists' responses to their climatological environment correspond to actual climatic differences, one can determine the relationship between elements painted by artists in a recent period for which actual observations of the same elements are also available. Here, however, a difficulty arises in that a strictly synchronous comparison is not possible, because in order to obtain reasonably stable regional averages for painted features, one must consider a period of at least one hundred years. On the other hand, actual observations of visibility and cloudiness were made only in relatively recent times at the representative stations mentioned above. Nevertheless, a correlation between painted and observed meteorological features was attempted.

Because observed visibilities are reported in different categories and, moreover, are not strictly comparable to painted visibilities estimated in three steps, the observed visibilities were converted into three categories: 1. visibilities of less than $2\frac{1}{2}$ miles; 2. those between $2\frac{1}{2}$ and 11 miles; 3. those larger than 11 miles. These categories were largely dictated by the different codes used in reporting and were judged to represent the best equivalent to the scale steps of the visibility estimates in the paintings. The frequencies of the observed and painted visibility categories were then converted into percentage visibilities.

Although the observed visibilities are roughly one-third larger on the overall average than the painted visibilities, the correlation coefficient is 0.93 with a probable error of ± 0.04; this is significant at better than the 1% level. Figure 7 shows this relationship and the regression equations. Although the actual visibilities in the various regions can be computed from the painted values within the accuracy of observations, there is some uncertainty about the relationship because of the absence of any values in the middle of the range. Comparing the data of the period 1850–1967 with those for the entire period from 1400 on (Figure 4), we see that the visibilities painted by the Italian and Spanish schools are much larger for the more recent period. The reason for this change will become evident below.

The averages of the actually observed cloudiness (converted into percent of the sky area covered by clouds) and the average cloudiness painted since 1850 by the various schools yield a correlation coefficient of 0.95 with a probable error of ± 0.03 which is significant at the 0.1% level. Figure 8 shows that the regional trend of the painted cloudiness (*PC*) has a larger amplitude than that of the observed cloudiness (*OC*). When the regression equation $OC = 0.5PC + 26$ is used to compute the observed from the painted values, the computed curve is practically identical with the actually observed curve. A comparison of the cloudiness in the Mediterranean region shown in Figure 6 with that in Figure 8 shows a considerable diminishment of the cloudiness during the recent period.

In general, these results leave little doubt that the artists' collective experience of cloudiness and transparency of the air in a given region during a given period can be deducted from their paintings.

Figure 7 Relationship between average regional visibility as observed (*OV*) and as painted (*PV*) from 1850 to 1967. (A—American School; B—British; L—Low Countries; F—French; G—German; I—Italian; S—Spanish.)

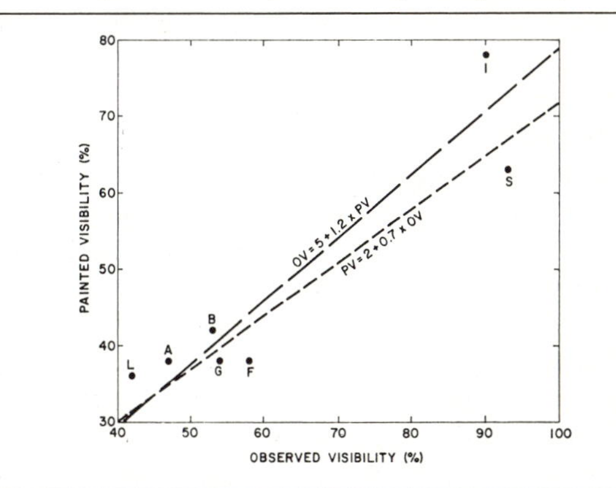

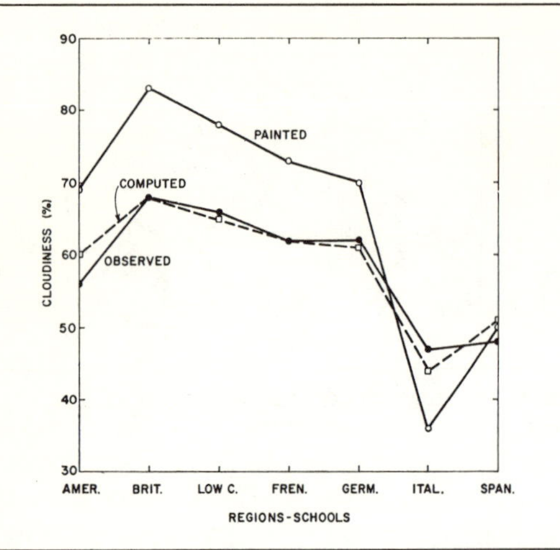

Figure 8 Average painted cloudiness (*PC*) (1850–1967) and observed cloudiness (*OC*) by regions. Dashed line represents values computed with the regression equation $OC = 0.5\ PC + 26$.

EPOCH	C	F
1400–49	29	61
1450–99	37	54
1500–49	50	85
1550–99	77	124
1600–49	84	141
1650–99	79	150
1700–49	77	130
1750–99	74	—
1800–49	65	—
1850–99	70	—
1900–67	74	—

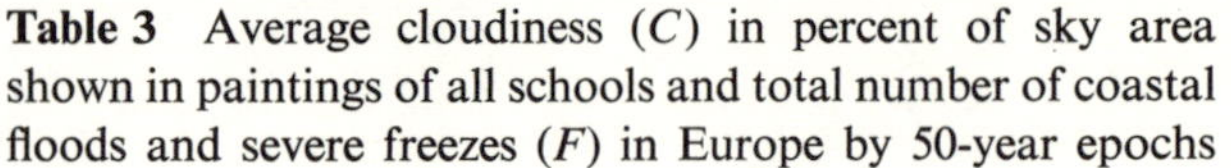

Table 3 Average cloudiness (*C*) in percent of sky area shown in paintings of all schools and total number of coastal floods and severe freezes (*F*) in Europe by 50-year epochs

THE LITTLE ICE AGE IN ART

During the period covered by this survey, the "Little Ice Age" occurred, an epoch of frequent cold, wet years, of advancing glaciers, and of increased frequency of severe freezes and storminess accompanied by a general southward displacement of storm tracks (Lamb, 1967). Chronicles of the 17th and 18th centuries (Weikinn, 1958–63) report unusual climatic events unheard of in our time, such as ice blocking the harbor of Marseilles for many weeks. The canals in Venice were frozen several times so that they could accommodate pedestrian and vehicular traffic; the Tiber River in Italy and the Ebro in Spain were frozen in several of the severe winters, and the Swedish armies crossed the frozen Baltic Sea during the 30-years' war (1618–48). In the same epoch, frequent disastrous storm surges flooded the British, Dutch, German, and Danish coastal areas around the North Sea and killed thousands of people, wiping out entire villages, and reshaping the coastline. In the beginning of the 17th century some Swiss glaciers advanced so rapidly that some villages, settled since the beginning of recorded history, were overrun by ice. Starvation was rampant in many regions of Europe and, in conjunction with war and pestilence, decimated the population.

There is strong evidence of the "Little Ice Age" having affected the entire northern hemisphere (Brooks, 1949; Lamb, 1967; Landsberg, 1958; Ludlum, 1966, 1968). Whereas the year 1850 is generally agreed upon as the approximate end of the "Little Ice Age" with a subsequent warming trend lasting well into the middle of the 20th century, its subtle emergence is more difficult to pin down. Lamb (1967) suggested that the "Little Ice Age" started around 1430 with a culmination period between 1550 and 1700; Dorf (1960), and Field (1955) set the beginning around 1600, whereas Brooks (1951) considers the start to have coincided with the advance of glaciers in Europe around 1550.

Lamb (1967), quite independently, had conceived the idea that paintings of various periods should reveal something about the climate and found from the analysis of a few hundred pictures that the cloudiness in paintings increased substantially from around the middle of the 16th century. With the present statistical material from more than 6,500 paintings with identifiable sky cover, Lamb's conclusion is completely verified as shown in Table 3. In this Table the total number of coastal floods and severe freezes in Europe as extracted from Weikinn is also given for the period from 1400 to 1749 when the publication of this series was discontinued.

According to the data, cloudiness increased slowly between the beginning of the 15th and the middle of the 16th centuries, apparently in response to the gradual development of the "Little Ice Age". Then, in the next fifty years, the greatest jump occurred in both phenomena listed; cloudiness continued to increase until the middle of the 17th century, whereas floods and severe freezes reached their maximum during the second half of that century. With the change in climatic regime around 1850, cloudiness shows another upward trend.

Considering these temporal changes, it seems reasonable to divide the total time period covered by the survey into three epochs: *1400–1549*, the preculmination period of the "Little Ice Age"; *1550–1849*, the culmination period which contains the "years without a summer" in the early part of the 19th century; and *1850–1967*, the post-culmination period in which a definite retreat of glaciers and a substantial atmospheric warming occurred. The various pictorial features were averaged over these epochs, combining all the art

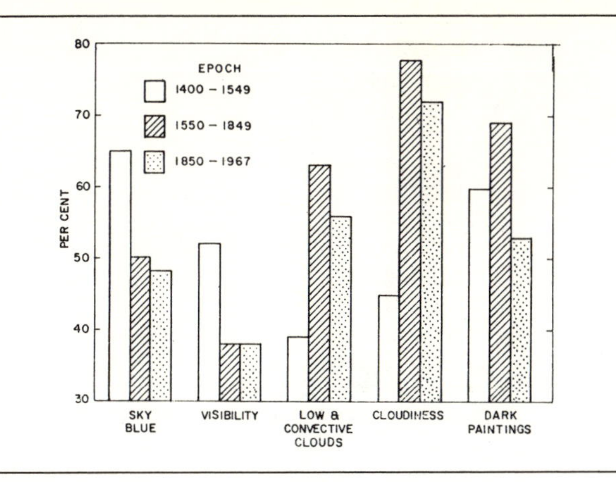

Figure 9 Epochal changes of various painting features.

schools, in order to maintain a large statistical population in each of the epochs (Figure 9).

The great drop in blueness of sky and visibility from the first to the second epoch is probably caused not only by the deterioration of the climate, but also by changes in artistic taste, which, however, may not be entirely independent of the climatic change. The failure of sky blueness and visibility to improve in the last epoch could be due to the following. This period contained various art movements such as impressionism which created "hazy" atmospheres by the uncertain contours resulting from the short brush strokes reaching their extreme in pointillism. This was also accompanied by a tendency toward paler colors, so that at least part of the diminished blueness and visibility is thereby explained. However, the subsequent expressionism and surrealism with their essentially sharp contours and bold colors should have counteracted this trend. As a climatological explanation, one could consider progressive industrialization with its attendant air pollution, which undoubtedly has diminished the blueness of the sky and the transparency of the air.

The frequency of low and convective clouds also shows a sharp change from the first to the second epoch reflecting the deterioration of the weather throughout Europe. The reversal of this trend in the third epoch may have been retarded also by increased air pollution. The same effect is evident in the amount of cloudiness (in percent). Interestingly, the frequency of dark canvases shows the same general response; lower clouds and greater cloudiness during the culmination period of the "Little Ice Age" could well have lowered the general illumination level and may thereby have induced the artist to select darker colors from his palette. The higher frequency of dark paintings in the first epoch may, in part, be a result of the aging of the paints from the earlier centuries, the yellowing of the varnishes, and the deposit of dirt on the canvases. The fresher paints and the brighter colors used by modern artists undoubtedly contribute to the relatively great drop in frequency of dark painting during the last epoch.

CONCLUSION

The results from this investigation strongly support the contention that the artist, as a conscious or subconscious chronicler of his environment, and the climate, as an all-pervasive agent in human activities and expressions, combine to reveal the artist's actual climatic experience that can be expressed as averages of climatic elements derived from his paintings.

REFERENCES

Brooks, C. E. P., 1949. *Climate Through the Ages.* McGraw-Hill, New York. 395 pp.

Brooks, C. E. P., 1951. Geological and historical aspects of climate change. In *Compendium of Meteorology*, ed. T. F. Malone. Amer. Meteorol. Soc. 1004–1018.

Dorf, E., 1960. Climatic changes of the past and present. *Amer. Scient. 48(3)*, 341–364.

Field, W. O., 1955. Glaciers. *Scient. Amer. 193(3)*, 84–92.

Lamb, H. H., 1967. Britain's changing climate. *Geographical Journal 133(4)*, 445–466.

Landsberg, H., 1958. *Physical Climatology* (2nd ed.). Gray Printing Co., DuBois, Pennsylvania. 446 pp.

Ludlum, D. L., 1966–68. *Early American Winters 1604–1820*, and *Early American Winters 1821–1870*. Amer. Meteorol. Soc., Boston, Massachusetts.

Neuberger, H. and Neuberger, M., 1943. Color memory in cyanometry. *Bull. Amer. Meteorol. Soc. 24(2)*, 47–53.

Weikinn, C., 1958–63. Quellentexte zur Witterungsgeschichte Europas von der Zeitwende bis zum Jahre 1850. *Hydrographie, Tiel 1, 1958; Teil 2, 1960; Teil 3, 1961; Teil 4, 1963.* Akademie Verlag, Berlin.

SUGGESTED READING FOR PART 8

Beckwith, W. B., 1971. The effect of weather on the operations and economics of air transportation today. *Bull. Amer. Meteor. Soc. 52*(9), 863–868.

Landsberg, H. E., 1969. *Weather and Health—An Introduction to Biometeorology*. Doubleday and Company, Garden City. 148 pp.

Maunder, W. J., 1970. *The Value of the Weather*. Methuen and Company. 388 pp.

Sewell, W. R. D., Kates, R. W., and Phillips, L. E., 1968. Human response to weather and climate. *Geographical Review 68*, 262–280.